Advances in Science, Technology & Innovation

IEREK Interdisciplinary Series for Sustainable Development

Advances in Science, Technology & Innovation (ASTI) is a series of peer-reviewed books based on important emerging research that redefines the current disciplinary boundaries in science, technology and innovation (STI) in order to develop integrated concepts for sustainable development. It not only discusses the progress made towards securing more resources, allocating smarter solutions, and rebalancing the relationship between nature and people, but also provides in-depth insights from comprehensive research that addresses the **17 sustainable development goals (SDGs)** as set out by the UN for 2030.

The series draws on the best research papers from various IEREK and other international conferences to promote the creation and development of viable solutions for a **sustainable future and a positive societal** transformation with the help of integrated and innovative science-based approaches. Including interdisciplinary contributions, it presents innovative approaches and highlights how they can best support both economic and sustainable development, through better use of data, more effective institutions, and global, local and individual action, for the welfare of all societies.

The series particularly features conceptual and empirical contributions from various interrelated fields of science, technology and innovation, with an emphasis on digital transformation, that focus on providing practical solutions to **ensure food, water and energy security to achieve the SDGs.** It also presents new case studies offering concrete examples of how to resolve sustainable urbanization and environmental issues in different regions of the world.

The series is intended for professionals in research and teaching, consultancies and industry, and government and international organizations. Published in collaboration with IEREK, the Springer ASTI series will acquaint readers with essential new studies in STI for sustainable development.

ASTI series has now been accepted for Scopus (September 2020). All content published in this series will start appearing on the Scopus site in early 2021.

Mohd Fairuz Shahidan ·
Gasim Hayder Ahmed Salih · Alessio Cardaci ·
Israa Hanafi Mahmoud
Editors

Urban Narratives: Exploring Identity, Heritage, and Sustainable Development in Cities

Springer

Editors
Mohd Fairuz Shahidan
Department of Landscape Architecture
Faculty of Design and Architecture
Universiti Putra Malaysia
Serdang, Malaysia

Alessio Cardaci
School of Engineering
Università degli Studi di Bergamo
Bergamo, Italy

Gasim Hayder Ahmed Salih
Department of Civil Engineering
Universiti Tenaga Nasional (UNITEN)
Kajang, Selangor, Malaysia

Israa Hanafi Mahmoud
Laboratorio di Simulazione Urbana
Department of Architecture and Urban Studies (DAStU)
Politecnico di Milano
Milan, Italy

ISSN 2522-8714 ISSN 2522-8722 (electronic)
Advances in Science, Technology & Innovation
IEREK Interdisciplinary Series for Sustainable Development
ISBN 978-3-031-48516-9 ISBN 978-3-031-48517-6 (eBook)
https://doi.org/10.1007/978-3-031-48517-6

This Springer imprint is published by the registered company Springer Nature Switzerland AG
The registered company address is: Gewerbestrasse 11, 6330 Cham, Switzerland

Scientific Committee

The Editors warmly thank all the Reviewers who have contributed their authority to the double-blind review process, to ensure the quality of this publication.

Preface

In an age of rapid urbanization and constant societal changes, cities have emerged as dynamic and complex entities, brimming with life, culture, and diversity. As the world's population continues to gravitate toward urban centers, the challenges faced by cities have grown increasingly daunting. Urban spaces must now confront a multitude of issues, from climate change and resource scarcity to social inequality and economic instability. Regardless of these challenges, the concept of resilience has taken center stage, shaping the way cities adapt and thrive amid adversity. Resilience encompasses cities' ability to bounce back, transform, and maintain their essence while fostering sustainable development and improved quality of life for their inhabitants.

Intended to be a guide for academics, scholars, and interested leaders, this book was designed to critically assess studies on the crucial role that architecture and the arts play in cultivating resilient urban environments. The book explores the interplay between the built environment, artistic expression, and the social fabric of cities. It also discussed unraveling the intricate relationship between architecture, art, and urban identity, showcasing how these elements converge to shape the resilience and character of our urban landscapes.

This book is a compilation of meticulously selected innovative research that was originally presented at the Cities' Identity Through Architecture and Arts (CITAA) conference, which brought together scholars from around the world to discuss cultural, historical, and economic aspects of cities. As a result, the book provides a comprehensive and global solution. The research included in this book was conducted by IEREK participants from multiple continents. The contributors to this book, a diverse group of experts spanning the fields of architecture, urban planning, arts, and social sciences, have brought their collective wisdom and experiences to create a rich tapestry of knowledge. Their observations, case studies, and reflections serve as the foundation for this investigation into the complex relationship between urban resilience and cities' artistic and architectural fabric. It provided these authors with an invaluable opportunity to exchange ideas with leading scholars, professionals, and postgraduates in the fields of architecture, arts, planning, and conservation.

The research and materials in this book are meant for people who are actively involved in making decisions as well as a diverse audience with an interest in critically analyzing all the most recent literature within the field.

A special mention should be made of the editors of this book, as well as all the authors and co-authors of the chapters, who have collectively provided the academic community with unique and increasingly valuable literature.

Mohd Fairuz Shahidan
Associate Professor, Department of Landscape Architecture
Faculty of Design and Architecture
Universiti Putra Malaysia
Serdang, Malaysia

Acknowledgments We would like to express our gratitude to the writers of the manuscripts that were chosen for inclusion in this book. We would also want to thank the reviewers who gave their experience and helpful input in order to ensure that the work is of the highest possible quality. A particular appreciation to the editors of this book for their determinations in organizing this collection and their diligence in editing it professionally. Finally, we would like to thank the ASTI and IEREK teams for their assistance in publishing the finest research papers submitted to the conference.

Introduction

Contemporary cities nowadays are a mix of technological development, art, and architecture at the crossroads. In this edited book, a group of contributions address different topics related to cities identities, cities shapes and information, heritage, arts, sustainability, vocabulary, and public regulations. While a city is an orchestration of architecture, art, and urban planning; its identity is, however, linked to human elements, geography, cultural, social, and economic integrations.

Arts and architecture are among the factors that shape the history of civilizations and allow ancient and modern cities to flourish. These factors do not stop only at buildings and constructions but also include traditions, cultural beliefs, and communal arts. By cherishing traditions and civilizations that left behind various architectural miracles, the way for future generations to develop and preserve their identity could be paved.

This edited book is the result of the work of authors and editors collaborating on developing a general understanding of the themes related to the city and its identity through art and architecture. Several sections are focusing on the shaping of the city based on its morphological formation and the cultural identities restructured in European, Middle Eastern, and international contexts.

This volume brings together thirty contributions from the latest edition that was held at the University of Pisa on December 5 and 6, 2022. The conference treated the themes of cities shapes and development of vocabularies as well as topics related to visual arts and visualization and its impact on human design of cities. The book is divided into four main parts: Part One—Identifying Cities and Formation; Part Two—Shaping Cities and Formation; Part Three—Heritage Arts: Sustainability, Identities and Public Regulations; and Part Four—Development of Cities' Vocabularies.

This book explores critical viewpoints on current urban developments and challenges while examining the historical assemblages of cities, their postcolonial identities, and architectural forms. Based on studies from all over the world, Italy, Turkey, Korea, and other countries are mentioned in this book as case studies, whereas evidence from the interplay of art, architecture, and urban planning is illustrated. The book also brings together experiences from urban design with uncertainty and formulation of environmental and social sustainability in different contexts. Meanwhile, it sheds light on the gentrification and preservation of neighborhoods in old cities and towns. Nonetheless, some contributions pose the question of arts as a way to revitalize historical contexts in some areas.

The interdisciplinarity of this volume allows the reader to grasp several debates on architectural forms while preserving its industrial or cultural heritage, on one hand. It also tackles legislative aspects of historic centers management with relationship to its challenges and preservation on the other hand. The chapters presented in this book are intended for researchers, professionals, and policymakers worldwide.

Israa Hanafi Mahmoud
Assistant Professor
Laboratorio di Simulazione Urbana
Department of Architecture and Urban Studies (DAStU)
Politecnico di Milano
Milan, Italy

Alessio Cardaci
School of Engineering
Università degli Studi di Bergamo
Bergamo, Italy

Contents

Identifying Cities and Formation

Identity as Imposed by and upon Urban Architecture

Melissa Anna Murphy

Abstract

Kevin Lynch (The image of the city (Vol. 11). MIT Press, 1960) described the image of the city through built environment elements and patterns that can be identified together at different scales. Maintaining his focus on perception but zooming further in can show that many individual identities of diverse inhabitants are embedded within any city- or district-identity. Following a relational, socio-material approach, these concepts can be linked through how urban design and architecture supports, limits, and can be changed by variable expressions of inhabitants' identities over time. This paper asks, *how can the inhabitants impose their identities upon the city?* by examining how the built environment hinders, hides, encourages, or showcases individuals' presentations of self and tactics to support their everyday lives. The study documents identity-laden spatial tactics enabled by architecture in a series of urban housing environments in three Norwegian cities. The findings are synthesized from ethnographic observation studies carried out during three research projects between 2013 and 2021. Recent trends towards large area redevelopment in Norway have resulted in a largely anonymized identity *imposed by* homogenous architecture. However, contrasting these with longer inhabited neighbourhoods from different architectural periods offer examples of how identities and the needs of daily lives can also be *imposed upon* architecture by inhabitants through spatial tactics. The analysis provides a preliminary categorization of architectural elements and characteristics that offer material potentials for individual identity-expression or identity-hindrance. The aim of the paper is to understand how architecture's stabilizing capacity can be redirected towards supporting dynamic, heterogeneous expressions of city inhabitants. This added perceptible layer in the city can redefine the image of the city so that it may also represent the ever-changing identities of its inhabitants.

Keywords

Socio-materiality · Spatial tactics · Perception

1 Introduction

In 1960, Kevin Lynch's ground-breaking work offered categories from which people can perceive and identify an "image of the city," based largely upon types of patterns that people can recognize in the urban structure. Cosmopolitan thought of the city (Sandercock, 1998) and work on the pluralism of urban societies tells us that, zooming into individual buildings, any city is composed of different individuals with multiple identities who exist together. This study builds upon the work of Lynch, but at a microscale in order to argue that architecture plays a role in the extent individual identities are represented as part of the image of the city. The study aims to open an academic discourse around how the built environment hinders, hides, encourages, and showcases individuals' presentations of self and spatial tactics. These topics are particularly pertinent in the study's context of Norway, which has seen several major waves of immigration since the 1970s, accounting for a significant amount of the country's urban growth. This chapter compiles findings from eight neighbourhoods in Oslo, Stavanger, and Drammen. Each of these cities rank among the Norway's most multicultural, with 20–38% of the populations being first- or second-generation immigrants.

M. A. Murphy (✉)
Department of Culture, Religion, and Social Studies, University of South-Eastern Norway, Drammen, Norway
e-mail: melissa.murphy@usn.no

M. F. Shahidan et al. (eds.), *Urban Narratives: Exploring Identity, Heritage, and Sustainable Development in Cities*, Advances in Science, Technology & Innovation, https://doi.org/10.1007/978-3-031-48517-6_1

1.1 Background

Today's cities are products of globalization, with increasingly diverse resident populations. Following Sorkin (1992), globalization has also led to increasingly homogeneous urban architecture around the world. The diversity of those who inhabit cities is not always a perceivable part of a city's image. At the same time, research from sociology, environmental psychology, and housing studies all show the importance of being able to present one's self identity, ambitions, and needs in the built environment (Abu-Ghazzeh, 2000; Cooper-Marcus, 1995; Twigger-Ross & Uzzell, 1996). Spatial tactics, or everyday activities that transform space, impact the city whether or not they are formalized in planning and design processes (de Certeau, 1984; Kärrholm, 2007). While many tactics may remain temporal and even invisible, others mark space and become legible in the built environment. Goffman (1959) explained how many of these markings, or *traces*, take over a symbolic role in social interaction: people seek to present themselves, their identities, and their ambitions both through how they behave, and through the material traces they display to others.

While attention to diversity in the city has been on urban agendas at least since the 1980s, the concept is perhaps only recently taking a material turn. Movements within urban planning and architecture acknowledge diversity and identity needs with increased calls for participation and user-input to formal processes that can extend so far as so-called DIY, or "do-it-yourself," architecture (Luck, 2018). Human geography has re-explained individual and group identity as developed through interactions between people and the physical spaces they inhabit (Massey, 2004). Participatory design processes are lifted for their potential to imbue value-pluralism upon the built environment (Luck, 2018). Observing material traces of use and changes to the built environment has been linked to how urban residents learn about the diversity around them (Murphy, 2017). By reframing the identity and image of the city materially, we can see that any perceivable image is shaped both by urban architecture and by the interactions the material environment sparks into being. This approach can allow us to dissect the role of the material in interactions with the built environment, much in the same manner Goffman (1959) dissected images with weight upon the social and psychological aspects. This paper argues that urban architecture's materiality largely mediates the extent to which traces of individual identities are legible in the city's image.

1.2 Architecture as an Actor

Lynch drew upon the perception of material architecture to define what contributes to the image of the city. His elements of city image are paths, edges, districts, nodes, and landmarks. These elements offer bases for people to perceive and understand patterns in a city (Lynch, 1960). Lynch's semiotic focus acknowledges that architecture affects what humans perceive. However, moving beyond semiotics, architecture can be framed as an actor that can help or hinder humans in altering it, further developing and re-developing the image of the city. Residents do not only endow meaning upon what they perceive (Lynch, 1960, p. 6), but they can also act in manners that contribute to the image by imposing aspects of their heterogeneous identities upon the city. Relational, socio-material approaches to architecture show that buildings carry with them the dynamics of the design and construction processes that bring them into existence. Buildings "work" continuously through their lives to allow, encourage, inhibit, and restrict different human behaviours and interactions (Latour & Yaneva, 2017). Tietjen (2011, p. 55) scales this approach up to urban design and "the experiential potential of urban space," explaining that when users interact with urban architecture to "perform" difference, it results in heterogeneity in the built environment. Heterogeneity in the built environment is thereby reliant on different interactions with and upon urban architecture.

Tying together the ontological understanding of architecture as an actor with the presentation of diverse identities in the city, we can frame the image of the city as being impacted by both. We know that the built environment affects how people can move throughout a city—preventing access with walls, for example. The presentation of identities in the city requires interaction with and within the built environment. Urban architecture's materiality can then be framed as mediating the interactions that occur between urban residents and the extent that they are able to interact and alter it. People can add, remove, redesign, repurpose, or transform materials to alter urban architecture. Urban architecture herein is understood as temporal, existing, and changing over time with resident actions as well as those by designers and spatial managers. The materiality of urban architecture can be such that it is easily marked and altered, allowing resident actions to leave traces. It can dually push back in resistance to some actions, inhibiting or prohibiting visible changes and traces (Murphy, 2017).

2 Methods

This study synthesizes findings from ethnographic observation studies carried out in Norwegian cities by the author in the period 2013–2021, spanning three research projects focused on urban space and housing (described in Appendix). Urban residential neighbourhoods were selected since they should be the most likely places to find individual identity-expression, due to being spaces of the everyday where people feel like they belong. Each research project that the study draws upon included a module led by this study's author aiming to understand traces left by residents in particular neighbourhoods. The neighbourhood cases studied are marked by different architectural periods and styles. An overview of the case neighbourhoods studied in each research project is in Table 1.1, while Figs. 1.1 and 1.2 place the studies on maps.

Traces were documented through photography, film, and mapping over extended periods of fieldwork. Resident and spatial management interviews further supplemented an understanding of how different traces are perceived and what they mean to local residents. Interviews were conducted by a walk-and-talk method, where an interview guide was supplemented by actively observing specific situations in the built environment. Where walk-and-talk was not possible, images from the author's observation study were shown to spark conversation.

In this study, the data across the projects has been reanalysed together using a grounded theory approach to iteratively document, define, and describe roles that architecture plays in the existence and perceivability of traces, as they contribute to the image of the city. Drawing upon a relational understanding of socio-material interactions, the visible traces observed in the environment were analysed by deconstructing them into (1) an action that produces the trace, and (2) a materiality that is necessarily affected in some manner to produce a trace. This framing provides a manner to isolate the role played by urban architecture's materiality.

3 Findings

This study's findings are separated into a short presentation of types of identity-laden traces that can be observed in the city, their perception, and analysis of categorical roles played by urban architecture. Together they demonstrate that for identity-laden traces to impose identity on the city's image, traces must be able to be created, and they must be perceivable to others.

3.1 Identity Laden Traces

Project A.1 (see Appendix) documented physical traces of residents in the built environment across different architectural styles and resulted in a series of categories of actions that leave traces. Those that impart aspects of identity are listed below, illustrated with examples in Fig. 1.3:

- Decorating and functional airing (for example of rugs or drying clothes)
- Tagging graffiti.
- Parking bicycles/scooters.
- Installing elements for personal initiatives.

Figure 1.3 demonstrates different images that these categories of actions contribute to in different neighbourhoods. Additional categories of traces found in the project but not selected for use in this study can be summarized as illegal actions which detract from a neighbourhood's image without demonstrating a particular local identity: posting commercial flyers, vandalizing, and breaking elements, littering. Further categories of bird feeding and dog walking are removed from this study as they were not found to have a consistent relationship with architectural elements beyond requiring open space.

Projects A.2 and A.3 (see Appendix) included the study of architectural cases with expressed intentions of representing spatial users. These offer data within the categories above, as well as the additional category of "Changing architectural forms and functions." Actions here can include direct actions from residents, such as shaping elements in the built environment (example in Fig. 1.4) or giving input by proxy to inform changes made by designers, municipal planners, or property owners (such as on behalf of a resident population, group, or enterprise).

Further findings from the "changing" category can be divided through what groups physically do themselves and those achieved by proxy in formal participation processes. In the former, findings from project A.3 include: resident groups renovating a common room on the first floor for social meetings with their own home furnishings, building a kitchen garden, and installing play equipment on the unused space of a common lawn. In the latter, findings include: participatory input that resulted in specific sports fields and play equipment, resident lobbying with the municipality that resulted in the redesign of a school and sports hall façade to offer more colour in the built environment (see Fig. 1.5). From project A.2, a park designed by architecture firm Helen and Hardt shows the potentials of design in collaboration with local children, where they decided the placement of reused elements from the oil industry (Fig. 1.4). Beyond the design process, youth were invited to leave traces through graffiti tagging of a long wall element in the park.

3.2 Perceiving Traces

Speaking to residents offered a picture of how these traces impart aspects of identity through perception. A selection of quotes follows below by trace category. The project, case, and anonymised informant number are included in parentheses.

Table 1.1 Overview of case neighbourhoods studied in the three research projects

Research project[a]	Norwegian City	Case neighbourhood	Architectural notes	Demographic notes
A.1	Oslo	A1-1. Tøyen public housing	1970s modernist blocks with shared yards	Rental and public housing units, highly multicultural neighbourhood with significant mix in socio-economic levels; many residents with descent from Africa, Middle East, and Indian subcontinent
A.1	Oslo	A1-2. Grunnerløkka apartment quarter	Late 1800s urban quarter with shared courtyard	Owner-occupied and rental units, multicultural neighbourhood with mix in socio-economic levels; many residents with descent from Middle East, and Eastern Europe
A.1	Oslo	A1-3. Tjuvholmen waterfront condominiums	2010s prestigious waterfront development	Owner-occupied units, high income area with low socio-economic diversity; residents predominantly Norwegian and Western European
A.2	Oslo	A2-1. Sørenga waterfront condominiums	2010s semi-prestigious waterfront development	Owner-occupied units, high income area with low socio-economic diversity; residents predominantly Norwegian and Western European
A.2	Stavanger	A2-2. Geoparken	2008 urban park in renewal area	Urban centre with high mix of socio-economic levels; draws residents, particularly youth, from the entire city
A.2	Stavanger	A2-3. Verven promenade apartments	1980s waterfront development	Owner-occupied units; middle to high income area with low socio-economic diversity; residents predominantly Norwegian and Western European
A.2	Stavanger	A2-4. Lervigbrygga condominiums	2010s prestigious waterfront development	Owner-occupied units; high income area with low socio-economic diversity; residents predominantly Norwegian and Western European
A.3	Drammen	A3-1. Fjell housing cooperatives	1970s modernist blocks with shared yards	Owner-occupied and rental units; middle to low-income area; significant mix in socio-economic levels; many residents with descent from Middle East, Indian subcontinent, and Eastern Europe

[a]See Appendix for further information about each project

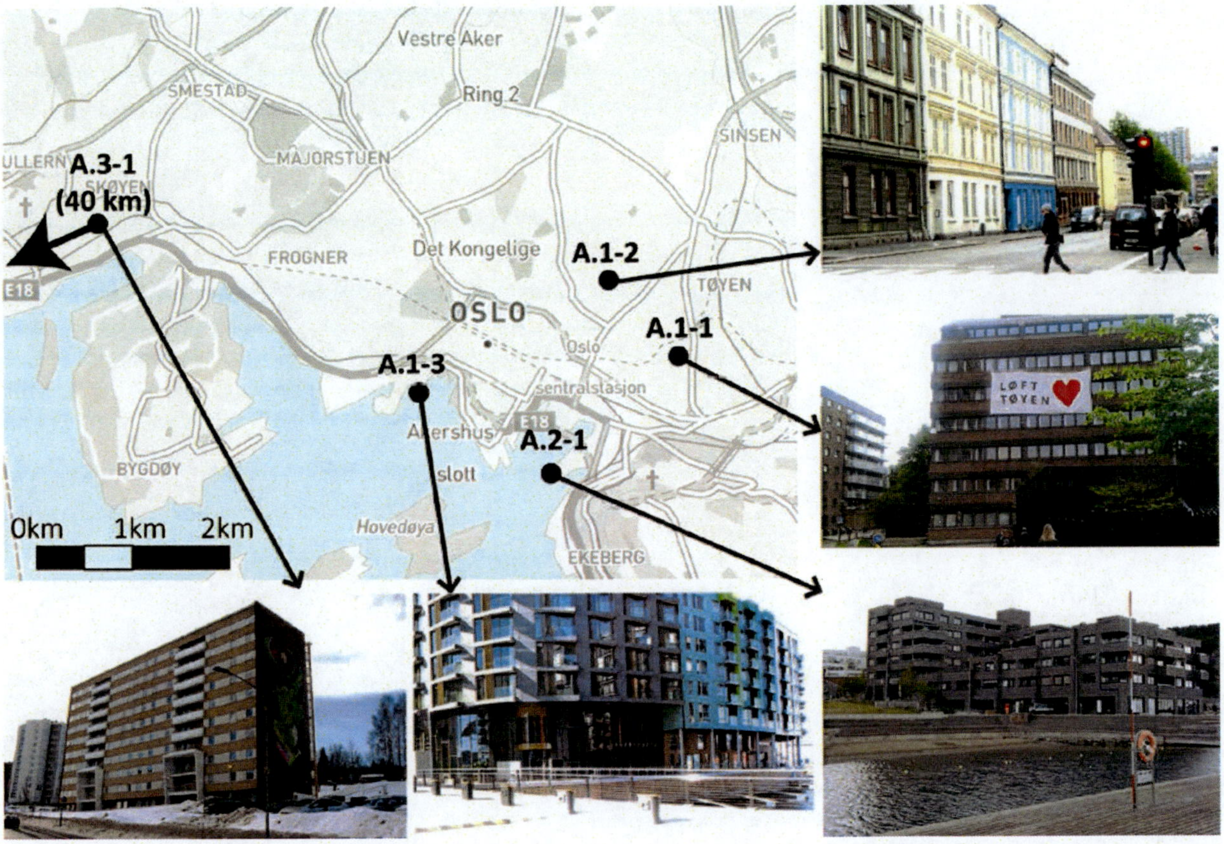

Fig. 1.1 Map and image overview of cases in Oslo and Drammen, images by author

- Decorating and functional airing: "If people dry clothes or hang a rug or whatever over their railings, I don't think it's any problem. That is just fine, it just shows that people live there, so I think it's nice—there must be people who live there too" (A.1-1:2).
- Graffiti tagging: "I'm for creativity, so… Maybe there could be like more legal painting—that the kids could also paint legally, have a board that someone paints over. I prefer street art, tagging is not so nice, but I see where it's coming from" (A.1-1:4).
- Bicycle parking: "They finally put up bicycle racks and I see people use them, so there are less parking in other places [making it look messy]," (A.1-3:1).
- Installing elements for personal initiatives: "The kitchen garden there is a private initiative…But I think it's good—it isn't taking away any area that is being used, so I think that is positive, even if it is not necessary the most pleasing setup aesthetically," (A.1-2:4).
- Changing architectural forms and functions: "The architects invited local kids to workshops to place elements, and then put up a wall where they could spray graffiti… Well the graffiti got a bit out of hand, it covers everything now and the municipality is considering redeveloping the park because some people complain," (A.2-2:3).

Themes throughout these examples reflect patterns seen in the data collected. Resident informants could talk about others, even unknown residents through the traces they had left behind. These were often connected to some form of evaluation of the trace, its quality, and what kind of people or intentions were behind it. Thus, identities are being shared through how they are represented in the built environment. A clear need for the traces to be connected to some aspect of care seemed important for the perception of others to be described as something positive. While the role of the built environment is not the object of focus in these conversations, it still comes up often—the informants could not describe the traces without alluding to other aspects of the built environment that were affected. This shows a tacit understanding among the informant that the railings, walls, bicycle racks, open unused spaces, and graffiti surfaces here are involved with the traces left upon them.

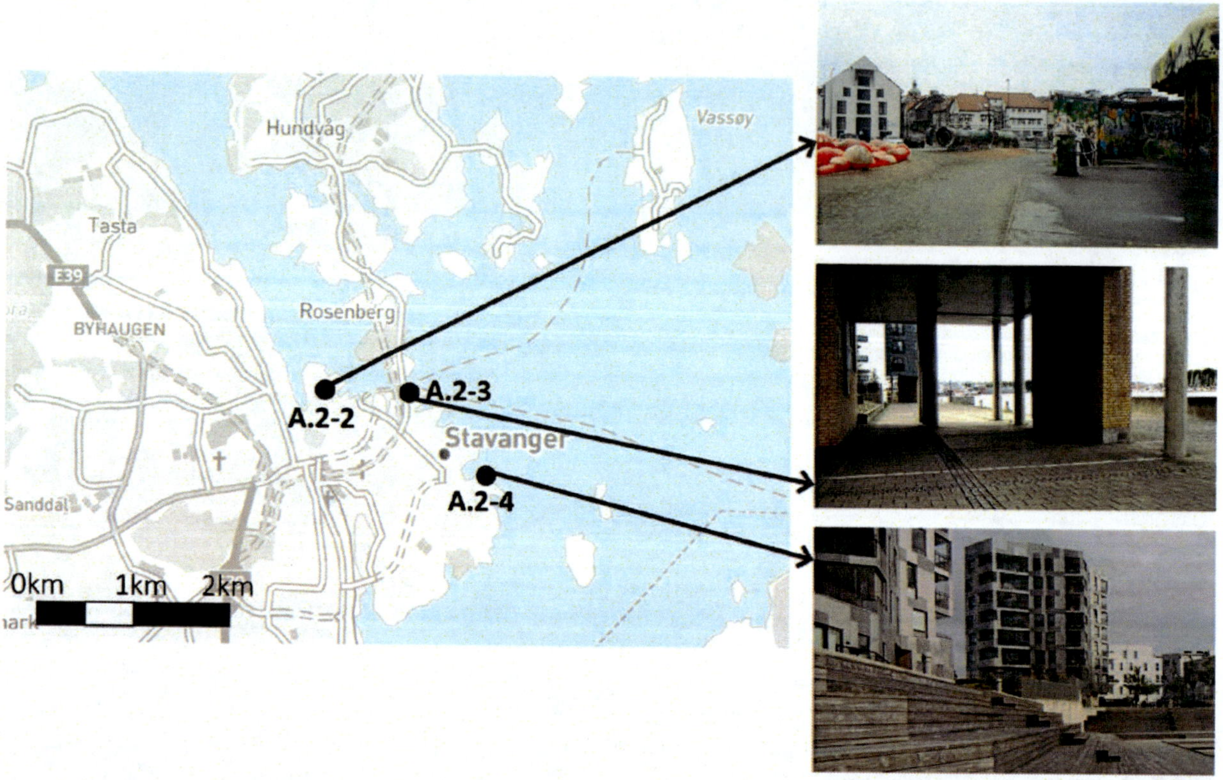

Fig. 1.2 Map and image overview of cases in Stavanger, images by author

Fig. 1.3 Traces alter the image of the built environment in three Oslo neighbourhoods (A1.1, A1.3, A1.2), photos by author

3.3 Urban Architecture and Identity

Each incidence of perceivable trace can be deconstructed into the action that was done to produce the trace and the manner it either changes elements of urban architecture or is resisted by elements of urban architecture. Deconstructing the categories of traces offers an analysis of the roles urban architecture can play in interaction with residents (see Table 1.2).

Diagramming the trace findings over a generic urban section demonstrates that these traces typically occur in quite limited zones within the built environment (see Fig. 1.6).

Summarizing the architectural elements that are changed or utilized in producing the traces found, we can arrive at three categories of architectural elements that offer potentials for enabling individual identity-expression:

- *Enabler accoutrements*: useable or appropriable amenities in urban space and on building facades. These can include rails, fences, hooks, pipes, posts, bicycle stands, and urban furniture—all typically falling within 0–1.5 m above ground in order to be reachable and useable.
- *Enabler surfaces*: reachable surfaces that can be marked or altered, serving as a canvas or background to traces

Fig. 1.4 Case A.2-1's participatory design process and invitation to spray graffiti produces a heterogenous park in Stavanger, photo by author

Fig. 1.5 Municipal representatives pushed design changes through to reflect resident input in Drammen's case A.3-1, images by author

like graffiti tagging. To enable trace actions, these must be accessible, visible, and of a material property that can be marked or altered. Accessible wall surfaces tend to be below 2 m above ground, but openable windows and balconies can extend this possibility zone.

- *Enabler spaces*: indoor or outdoor spaces with the flexibility to be changed, either by individuals, groups, enterprises, or through participatory planning and design processes. In order to enable the perception of traces, these spaces must be accessible and/or visible to others in the city.

In kind, we can group architectural elements that hinder identity traces into the lack of accoutrements in the 0–1.5 m zone, lack of accessible or markable surfaces, and lack of visible, accessible spaces that can be altered (see Fig. 1.7). In these cases, urban architecture either prevents traces from being visible or resists the actions that produce traces. Presence of the elements that hinder trace production were found to be necessary for traces to be perceivable in the cases. This reinforces the necessity of enabler materials for residents' potential to leave traces.

Further study is needed, but the findings here seem to point to different architectural styles supporting different extents of these enablers. From the six cases surveyed in these projects, pre-war architecture cases were connected to the most accoutrements. The two modernist post-war architecture cases had a reasonable amount of accoutrements together with the most enabler surfaces and the most open, accessible spaces for repurposing. The most contemporary

Table 1.2 Trace-leaving actions with the architectural elements that enable or hinder visible traces

Action (trace category)	Examples	Architectural elements changed or utilized	Architectural elements resisting
Decorating and functional airing (for example of rugs or drying clothes)	Hanging flags, flowerpots, balloons	Balconies; openable windows; railings; hooks; window boxes or ledges	Flat facades; fixed windows; absence of balconies, railings, flag holders; sunshades and screening elements; balconies blocked by overhangs or lack of distance
Parking bicycles/scooters	Designated parking, make-shift parking, undesignated parking	Bicycle racks; posts; fences and railings; trees and tree posts; window rails, pipes, downspouts, hooks on buildings	Lack of racks, posts, rails, or other appropriable elements under a height of 1 m; ground area filled with plants or water elements, indoor bicycle garages
Tagging graffiti	Graffiti tags; wall murals	Blank walls or other surfaces up to 2 m height from ground, particularly of untextured and light-coloured materials and lacking overview and permeability	Textured or colourful facades; graffiti resistance materials; windows; clear views for surveillance
Installing elements for personal initiatives	Added outdoor furniture, bicycle ramps, planters, play equipment, artwork	Flat open areas; encouraged by partial screening and intimate area divisions with hedges or similar	Heavily planted areas; water features; large, anonymous open areas
Changing architectural forms and functions	Participatory building redesign; repainting or renovating of facades; personalizing entrances, signage, and storefronts	Building plots; changeable facades; indoor or outdoor spaces that can be developed or redesigned	Solid monolith structures and facades that are not easily changed; lack of storefronts

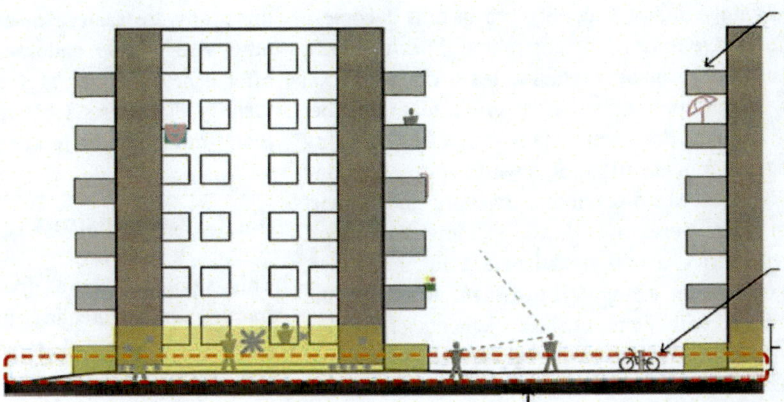

Fig. 1.6 Pre- and early post-war architecture often allows space and accoutrements that enable residents to leave traces, image by author (case A.1-1)

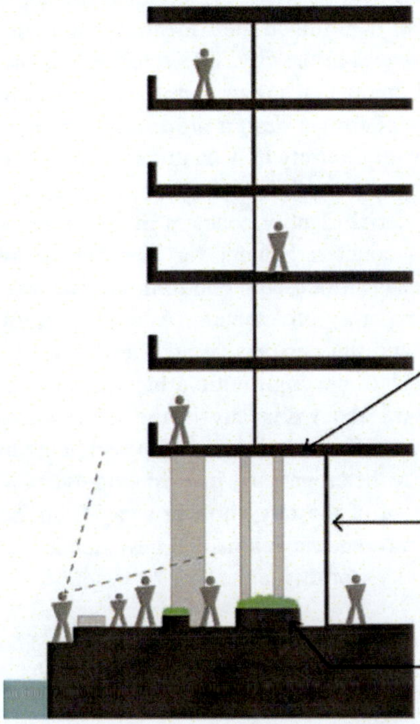

Fig. 1.7 Contemporary architecture can design away chances for residents to leave visible traces, image by author (case A1-3)

cases were built within the past 20 years, which show very few accoutrements and less open spaces. Blank surfaces were quite typical to the contemporary architecture cases, but often their enabling was mitigated with elements that prevented their access or marking. It is worth noting that the lack of enablers in the architecture of these contemporary cases aligned with the strictest maintenance and neighbourhood regulation paradigms of the study, which further inhibited trace leaving through other means than the material.

4 Discussion

While some of the categories proposed here may be more traditionally connected to identity (decorating, graffiti tagging) than others (bicycle parking), all of them were found to impart some kind of knowledge and perception of who inhabits particular neighbourhoods. Traces derived from illegal or environmentally detrimental actions, like the categories removed from this study, have been shown to contribute to misconstruing identities and breeding negative perceptions of difference (Murphy, 2017). All traces that are perceivable have the capacity of changing the image of the city, as they provide variety in the microscale between different urban neighbourhoods.

Urban architecture is both a backdrop and actor in allowing or hindering residents in leaving perceivable traces. Urban architecture can be a means of resisting trace leaving actions where certain traces are undesirable or seen to be a maintenance burden. It is largely the materiality of urban architecture that can, e.g. support surveillance, resist vandalism, discourage undesirable spatial uses (Németh & Schmidt, 2011), and make spaces defensible (Newman, 1995). Beyond the materiality, spatial management, maintenance, and social norms affect the extent people choose to leave traces in shared environments (Murphy, 2016). The cases here with the most relaxed or fragmented spatial management and the most cultural and socio-economic diversity were those where the most traces were found. Understanding the enabling potentials within urban architecture, as study opens for, we can see a that urban architecture offers a baseline material potential, a necessary precedent for being able to impose plural identities onto the image of the city. When urban architecture resists or mitigates changes and the leaving of perceivable traces, we can interpret it as architecture imposing an identity onto the city. A strongly imposed identity can in turn become a regulating normative force in the population (Sorkin, 1992), dissuading residents from attempting to leave visible marks, even looking down upon individual or group identities that stand out, as has been documented in research from housing associations (Fraser et al., 2016). It is worth noting here that the cases with least socio-economic diversity and highest income levels were those most devoid of individual traces. These also had strict, united spatial management regimes, and a lack of enabler elements by design. However, the contrasting cases with more enabler elements show that architecture can be more than the oppressive, regulating, and normalizing force described by Foucault (1977). Urban architecture can be designed to materially encourage residents to impose their identity upon it through leaving traces, this study's enabler elements become material possibilities and invitations for residents to interact.

Observing traces and the imposition of different identities upon architecture highlights the importance of flexibility and adaptability in urban architecture. While participatory design has been experimented with towards a variety of architectural goals (Luck, 2018), this study scales the concept up to how design processes and their material products can make the image of the city more malleable and connected to resident diversity also in the everyday. When resident traces can alter urban architecture, identities can be imposed upon the built environment, adding heterogeneity to the image of the city. These impositions result in details that are not only place-specific, but also user-identifiable and dynamic, changing alongside the population. It is clear that some neighbourhoods materially support impositions of identity more than others depending on materiality embedded in architectural styles, as well as spatial management and other neighbourhood regulations. Cities like Amsterdam and Berlin have recognized the need to protect and promote such identity-expressions of sub-cultures in areas where they are valued (Shaw, 2005). This study supplements urban planning and policy studies to show the role urban architecture elements can play in maintaining places where people can express themselves. It further reflects upon how different epochs in architectural style may be more or less conducive for enabling the expression of identities. Despite modernism's homogeneous hegemony in post-war architectural design, the highly regular elements here pose reasonable canvases for identity marking, in opposition to newer neighbourhoods with less accessible spaces and surfaces.

Lynch's (1960) image of the city sought legibility and over-all patterns. Bringing in the complexity of resident identities adds a layer to this that celebrates textures and variations that disturb patterns. The pluralism and dynamic of resident identities introduce the unexpected. This approach to construing the image of the city does not conflict with Lynch's, as the traces and disturbances raised in this study are easiest to see in contract and inscribed upon otherwise legible patterns. The elements pointed to here offer a new, additional scale for understanding image and identity that raises local identities into visibility and integrates them into the image of the city.

5 Conclusions

This study has built upon relational and socio-material framings of architecture, understanding architectural materials to be actors that can allow, encourage, inhibit, or prevent resident actions and interactions. Considering identity and the image of the city then turns to also considering interactions with and within urban architecture. The analysis of traces offered here show which elements of the built environment tend to be accessible for residents to leave traces, who can then imbue their identities on buildings, surfaces, and spaces in the city. This study shows that a remarkably small amount of urban architecture tends to be adaptable outside of formal design processes. However, residents can work with a variety of accoutrements and wall surfaces, when they are accessible and alterable. Raising a consciousness of material identity enablers in urban architecture may offer a counterargument for the often more oppressive, maintenance-focused controlling elements observable in contemporary architecture. Additional study could better understand the perspective and needs of residents for presenting and viewing multiple identities in cities, and how we learn about diversity in the urban every day. The material turn offered here may help architects to promote more identity, heterogeneity, interest, and dynamic texture at the eye-level of the city, moving away from the blank, homogenous facades and spaces that have dominated many images of the city worldwide the past 20 years.

Acknowledgements This project is an analysis across work done in multiple research projects, two of which were financed through the Norwegian Research Council.

Appendix

The three research projects whose data contributed to this paper are described in the following sections.

A.1 *Marking space: Negotiating room for user efficacy in residential urban spaces*

This doctoral thesis project by the author ran from 2012 to 2017, comparing traces of residents in three Oslo neighbourhood cases of different architecture epochs and styles (one pre-war tenement, one post-war modernist estate, one contemporary development). The thesis framed its findings around the extent of efficacy residents can find, accounting for property management styles, architecture, local, and cultural norms (link to pdf available online: https://nmbu.brage.unit.no/nmbu-xmlui/bitstream/handle/11250/2460695/2017-08_Melissa%20Murphy.pdf).

A.2 *DEMOSSPACE: Governing private provision of public space*

This project was funded by the Norwegian research council in the period 2016–2019 with an aim to understand the role of urban space in democracy. This author led a module that considered the role of urban design processes and spatial management decisions in contributing to residents' democratic interactions and awareness of urban diversity. Observation and interview findings from recently developed, mixed use waterfront neighbourhoods in two of the project's three Norwegian city cases (Oslo and Stavanger) are included in this paper. Project website https://uni.oslomet.no/demosspace/about/.

A.3 *PUSH: Public space in European social housing*

This European research project was funded by HERA from 2019 to 2022, looking at how the built environment interacts with people to encourage and inhibit social interactions, a relational take on the concept of publicness. This author led the Norwegian team in an investigation of a recently renewed post-war housing estate in the city of Drammen, Norway. The findings from this project which contribute to this paper include resident initiatives in the modernist housing estate and the contemporary school, sports hall, and parks contemporarily designed during the area renewal. Additional inspiration for this paper came from knowledge exchange and short field visits at project cases in Naples, Italy, Bern, Switzerland, and Copenhagen, Denmark. Project website: https://www.pushousing.eu/about.

References

Abu-Ghazzeh, T. M. (2000). Environmental messages in multiple-family housing: Territory and personalization. *Landscape Research, 25*(1), 97–115.

Cooper-Marcus, C. (1995). *House as a mirror of self: Exploring the deeper meaning of home*. Conari.

de Certeau, M. (1984). *The practice of everyday life*. University of California Press.

Foucault, M. (1977). *Discipline and punish: The birth of the prison*. Vintage.

Fraser, J., Bazuin, J. T., & Hornberger, G. (2016). The privatization of neighborhood governance and the production of urban space. *Environment and Planning A*, 844–870. https://doi.org/10.1177/0308518X15621656

Goffman, E. (1959). *The presentation of self in everyday life*. Anchor.

Kärrholm, M. (2007). The materiality of territorial production: A conceptual discussion of territoriality, materiality, and the everyday life of public space. *Space and Culture, 10*(4), 437–453.

Latour, B., & Yaneva, A. (2017). "Give me a gun and I will make all buildings move": An ANT's view of architecture. *Architectural Design Theory, 1*, 103–111. http://journals.openedition.org/ardeth/991

Luck, R. (2018). Participatory design in architectural practice: Changing practices in future making in uncertain times. *Design Studies, 59*, 139–157.

Lynch, K. (1960). *The image of the city* (Vol. 11). MIT Press.

Massey, D. (2004). Geographies of responsibility. *Geografiska Annaler. Series B. Human Geography*, 5–18.

Murphy, M. A. (2016). Framing built environment change through materials, agency and influence. *Urban Forestry & Urban Greening, 17*, 1–10.

Murphy, M. A. (2017). Dwelling together: Observable traces and controls in residential urban spaces. *Space and Culture, 20*(1), 4–23. https://doi.org/10.1177/1206331216643782

Németh, J., & Schmidt, S. (2011). The privatization of public space: Modeling and measuring publicness. *Environment and Planning-Part B, 38*(1), 5.

Newman, O. (1995). Defensible space: A new physical planning tool for urban revitalization. *Journal of the American Planning Association, 61*(2), 149–155.

Sandercock, L. (1998). *Towards cosmopolis: Planning for multicultural cities*. John Wiley & Sons.

Shaw, K. (2005). The place of alternative culture and the politics of its protection in Berlin, Amsterdam and Melbourne. *Planning Theory & Practice, 6*(2), 149–169.

Sorkin, M. (1992). *Variations on a theme park: The new American city and the end of public space*. Macmillan.

Tietjen, A. (2011). *Towards an urbanism of entanglement: Site explorations in polarised Danish urban landscapes*. Arkitektskolens Forlag.

Twigger-Ross, C. L., & Uzzell, D. L. (1996). Place and identity processes. *Journal of Environmental Psychology, 16*(3), 205–220.

Developing an Indicator Evaluation System to Assess Cities' Identity

Peian Yao, Paria Bagheri Moghaddam, and Martina Corti

Abstract

As an ongoing subject, city identity has been researched in a range of disciplines and contexts, such as sociology, anthropology, urban planning, cultural tourism, design and art. The social and economic value of city identity as a strategic dimension of cultural and creative competition between countries is increasingly being recognized. (Anholt, What is competitive identity? In Competitive identity: The new brand management for nations, cities and regions, pp. 1–23, 2007). However, in the current search for strategies to make cities identifiable, liveable and inclusive, research projects have been undertaken in different disciplines, dimensions and fields of knowledge, yet there is no specific way to analyze and observe city identity, nor can research efforts from different disciplinary backgrounds lead to common discussions or collaborations. The paper aims to develop a method and approach to assess or measure elements of urban identity with the aim of developing a methodology for observing, analyzing and evaluating city cultural identity. Enable specialists from different disciplines to carry out a quantitative-based analysis of city identity. Against this backdrop, this paper will firstly define the cultural identity elements of the cities through the literature review, in order to develop an indicators system of city identity. Secondly, the appropriate evaluation criteria for the indicators are discussed based on the categories of the elements in the indicator system, including the qualitative and quantitative evaluation aspects. By doing this, the article aims at filling a research gap in current literature as it is the first attempt to systematically investigate the components of city identity on a wider scale, forming a framework for a structured insight into city identity. It contributes to a broader interest in the debate on the topic of urban identity.

Keywords

City space · Cultural heritage · City identity · Indicator evaluation system

1 Introduction

In the existing literature, research on city identity has been conducted both qualitatively and quantitatively. The main issues addressed in the qualitative literature include: exploring the meaning of city identity (Cheshmehzangi, 2020; Riza et al., 2012). How certain elements of urban space contribute to the formation of urban identity (ER, 2019; Onay, 2012; Samir & Arayici, 2015; Sofianto et al., 2021). Some studies investigate urban space, social well-being, and ethnic identity in the social and human sciences (Kotradyová, 2019). Quantitative studies on urban identity are mainly concerned with: comprehensive evaluation of indicators for a city, (Kiani et al., 2019; Martín et al., 2019). Statistical study through the collecting of data pertaining to city identity (Anastasiou et al., 2021; Sönmez, 2020). There are also studies on urban identity that cover topics such as city branding, city reputation and place marketing. Their research on urban identity content uses a combination of methods ranging from big data analysis to deep learning and other technological tools (Andéhn et al., 2014; Yan et al., 2021; Zhang et al., 2019). These have contributed

P. Yao (✉) · P. B. Moghaddam · M. Corti
Design for Spatial Relations Lab, DIDA Department, University of Florence, Florence, Italy
e-mail: peian.yao@unifi.it

P. B. Moghaddam
e-mail: paria.bagherimoghaddam@unifi.it

M. Corti
e-mail: martina.corti@stud.unifi.it

P. Yao
School of Industrial Design, Hubei University of Technology, Wuhan, China

M. F. Shahidan et al. (eds.), *Urban Narratives: Exploring Identity, Heritage, and Sustainable Development in Cities*, Advances in Science, Technology & Innovation, https://doi.org/10.1007/978-3-031-48517-6_2

directly or indirectly to the development of knowledge about city identity from different fields of study.

However, city identity has always been an abstract notion, with few concrete explanations in various fields and a lack of quantitative observation and evaluation methodologies. Regarding city identity, the design discipline focuses mostly on its cultural aspects. Due to the absence of a defined content, however, it is difficult to create a systematic reference framework of identity features to guide design practice, resulting in a lack of awareness of city identity by professionals, this increases the risk that cities will replicate one another.

In order to address this issue, the authors have sought to construct a system of indicators for the evaluation of urban identity, with the goal of achieving a thorough generalization of the subject-related factors based on previous studies. In order to determine the criteria for measurement and evaluation, a distinction is created between quantitative and qualitative analyses of the many forms of elements that comprise identity in urban space.

2 Literature Research

It all starts with the question of which elements of a city are considered to constitute a city's identity in past studies of urban identity. Which of these elements are more commonly mentioned, and which are less studied. To answer these concerns, the authors used a method of literature review to generalize the components of city identity by analyzing the existing literature on the subject. The result is a set of indicators using which professionals can obtain quantitative and qualitative observations on urban identity.

The authors searched the bibliographies of Google Scholar, ScienceDirect and CNKI for relevant references. Used keywords "city identity", "city image", "city branding", "city reputation", "place marketing", "cultural identity", search in CNKI using the conjunctions "and", "or". In order to broaden the scope of the search and to search more comprehensively for relevant factors, keywords such as "cultural tourism" and "urban heritage" were also used.

The final search included books, dissertations, journals and conference papers. A total of 197 references with the highest relevance were selected. The authors read further through this literature and eventually selected 137 useful articles. The authors further classified the obtained documents and the initial data to obtain a set of elements that influence the identity of the city, and sorted in descending order of mentions. The results are given in Table 1.

Based on the data, it is evident that the function of new media in the spread of city identity is receiving an increasing amount of attention. Common techniques of communicating city identity include short videos, online platforms, etc. However, they are not city identity components in and of themselves. The authors excluded 1 and 4 from the scope of the city identity evaluation indicators. Secondly, in the data obtained, the authors have analyzed the elements mentioned, which vary in type and form. There are both material and immaterial elements. Some elements have overlapping content, and some elements contain other elements. For example, 'Culture' is a large concept that encompasses other smaller items such as 'Music' and 'Art'. 'Public facilities' also includes items such as 'urban furniture' and 'sculpture'. In addition, 'Athletics', 'Gangs', 'Spice', etc., are elements of the identity of individual cities and are not representative, they can be classified as part of a larger category. They can be grouped into broad categories. The use of preliminary data does not allow for the development of a system of indicators for the evaluation of urban identity, so it is necessary to categorize and organize the preliminary data.

Depending on the content and form of the elements, the authors first categorized all elements according to their tangible and intangible forms. With regard to the scope of tangible elements in urban spaces, the authors refer to the definition of tangible cultural heritage from the RICHES resources website. The term "tangible cultural heritage" refers to a society's physical artifacts that are produced, preserved and passed down from generation to generation. It includes artistic creations, architectural heritage (such as buildings and monuments) and other tangible or physical products of human creativity that are significant culturally within a society (RICHES Resources, 2014). The authors classify tangible indicator elements as buildings, squares, streets, museums, theaters, transportation and landmarks, among others. The authors make reference to the concept described in the UNESCO Convention on Intangible Heritage. On its website, UNESCO defines intangible heritage as "the practices, representations, expressions, knowledge, and skills that communities, groups, and in some cases, individuals consider to be part of their cultural heritage" (Intangible Heritage Convention). It is stated that intangible culture is a living force that is "transmitted from generation to generation" and "continuously recreated by communities and groups" in response to its social and natural environment. Important component of community identity, intangible heritage "promotes respect for cultural diversity and human creativity" (UNESCO, 2003). Therefore, the authors categorize events, folklore, festivals, folk songs, religion, etc., as intangible indicators. There is also a category in between, which is not physical and cannot be touched. They can be perceived by the human senses and are classified by the authors as intermediate, such as taste, sound and dialect. The final system of indicators after the classification and hierarchy is given in Table 2.

In Table 2, the authors classify all the elements into several levels of indicators with encompassing relationships:

Table 1 Summary of urban identity elements from existing literature

No	Elements that related to the city identity	Authors of article and publish years	Research field	Mention times
1	New media and short video	Jing Li (2022), Wei Liu (2022), Ying Liu and Shuang Chen (2022), Jiaming Bao (2022), Rui Wang (2022), Zuo Wang, Huiwen Zhan and Cheng Xiang et al. (2022), Hao Zhang and Linghua Ran (2021), Shuyue Chen (2021), Yidan Tang (2022), Keyi Wei (2022), Xin Sun and Jing Zhang (2022), Hong Zhu (2022), Yawen Zheng (2021), Yue Zhang and Hailong Cai (2022), Jiaxuan Bai, Yuan Cheng and Binghui Sun (2022), Feng Yin (2022), Ling Wu, Chunyao Liu and Jun Du (2022), Zhigang Xia (2021), Jing Yang (2019), Jinping Zhao (2020), Aizhe Li and Xiaoming Chi (2019), Qi Zhou (2021)	Communication and media	22
2	Architecture	Linyan Wang (2014), Yingnan Liu (2020), Hongyu Li (2013), Yang Chao 2013), Müge Riza, Naciye Doratli and Mukaddes Fasli (2012), Zohreh Torabi and Yalda Sima (2013), Adham M. Hany Abulnour and Magdi Wasfi Wissa (2018) Hamed Hyab Samir and Yusuf Arayici (2020), Müge Riza, Naciye Doratli and Mukaddes Fasli (2012), Kunto Sofianto, Agusmanon Yuniadi and Agus Nero Sofyan et al. (2021), Abdulaziz Alzahrani (2022), Ali Salmo, Elena V. Scherbina and Lina Yaser Alibrahim (2021), Nik Mastura Nik Mohammad, Rohana Sham and Azlin Abdul Latif (2015), Imad M. Assali (2017)	Architecture design; sociology; history	14
3	Visual identity	Mengli Zhang (2014), Yang Jin (2020), Anqi Zhang (2020), Yuelin Zhou and Ye He (2021), Xu Han (2016), Xinyi Zhang (2021), Chengliang Ma and Huiliang Guan (2021), Xin Wang (2022), Xiangming Sun (2012), Jinhuan Wu (2019), Xin Zhao and Dangui Xu (2021), Juan Li (2022), Zhao Li (2019), Brian Alvin Hananto (2019), Yu. V. Gorgorova (2019), M. Alaa Mandour (2012)	Design; engineering	13
4	Film and TV, drama	Rui Yao and Wendong Kong (2022), Li Wen and Wanqing Yuan (2022), Xiaoyan Qin (2022), Chang Wang (2021), Zhihua Wang and Kaixue Xue (2022), Lu Li (2018), Xia Li (2022), Katie Milest (2008)	Communication and media; film; sociology	8
5	Culture	Dawei Zhang (2006), Yali Gao and Huadi Liu (2022), Wenting Lv (2022), Biqin Xia and Jiaxiang Wang (2021), Qingling Wang (2021), Lixin Han (2021), M. Alaa Mandour (2012)	communication and media; design; economics and management; arts	7
6	Cultural symbols	Xueqing Yan (2020), Xiaonan Zheng (2020), Zhexuan Wang (2020), Shuhan Yin (2021), Hao Yang (2022), Jing Liu and Wen Wangjun (2022), Jiuyang Lv and Geng Zhang (2022)	Communication and media; architecture design; design;	7
7	Museum	Huang Zhao (2022), Cenwei Kong (2020), Jie Wu (2021), Jiasheng Zhou (2022), Manli Zheng (2014)	Communication and media; design; environmental art design	5
8	Fashion and design	Juan Shan, Linjing Zhu and Yanchi Long (2021), Ping Xu (2007), Clarinda Rodrigues and Holger J. Schmidt (2021), Marina Parente (2016)	Marketing; design	4
9	Art in public spaces	Shuai Wang (2020), Yifan Zhu (2017), Lei Li (2014), Hao Wang (2011)	Art; art history; cultural studies	4
10	Landscape	Elif Sağlık and Abdullah Kelkit (2017), Annibale Salsa (2010), Yuan Feng (2018), Vera D. Damayanti, Hasti Tarekat Dipowijoyo and Jacqueline Rosbergen et al. (2019)	Architecture design; communication and media	4
11	Heritage	Liang Zhang (2022), Xiaohui Hou (2021), Mohamed Badry Kamel Basuny Amer (2018)	Architecture design; communication and media	3
12	Color	Sisi Wu (2014), Chen Xu (2021), Cristina Boeri (2019), Marco Ricciarini and Adelaide Tremori (2020)	Design;	3
13	Infrastructure	Xin Jin, Haochong Li and Zhihui Xi (2022), Yang Wei (2002), Dwi Lindarto Hadinugroho and Eunice Amanda Putri Matondang (2020)	Design; cultural studies; architecture design	3

(continued)

Table 1 (continued)

No	Elements that related to the city identity	Authors of article and publish years	Research field	Mention times
14	Folk music and music	Shuai Zhang (2020), Xuemei Luo (2019), Aksa Noya (2020)	Music	3
15	Square	Basak Damla Erdogan and Hatice Ayatac (2012), Nurettin ÖZGEN (2013)	Architecture design; geography	2
16	Traditional painting and art	Hailong Wang and Yafang Zhu (2019), Noa Yuval-Hacham (2019)	Art	2
17	Perception and memory	Maria Popczyk (2013), Damiano Razzoli, Fabrizio Montanari and Graziano di Paola (2020)	Anthropology	2
18	Materials	Paulina Dorota Lis (2014), Nilufer Saglar Onay (2012)	Architectural design	2
19	Food	Riccardo Pravettoni (2018), Peian Yao, Anqi Cheng and Stefano Follesa (2020)	Anthropology; design	2
20	Landmark	Pengfei Li (2022), Mohammad Ali Danesh Sedigh and Parnaz Goodarzparvari (2018)	Architecture design; design	2
21	Historic street	Xunfei Mu (2020), Zhaomo Zhang (2019)	Architecture design; communication and media	2
22	City furniture	Wei Hao (2019), Xinlu Song (2021)	Design	2
23	Events and activities	Yating Yang (2021), Ming Chen and Siyi Wang (2022)	Communication and media; economic management	2
24	Celebrities	Xiaolu Wang (2019), Eli Jamilah Mihardja,Prima Mulyasari Agustini and M. Bisyri (2019)	Art; sociology	2
25	Theaters	Li Yilun (2020), Sijie Qi (2022)	Arts industry and management; film arts	2
26	Rail transport	Xiaolong Huang (2021), Nianli Zhang (2021)	Communication and media; art	2
27	Handicraft	Marie Geneviève Cyr (2014), Birnaz ER (2018)	Fashion; art	2
28	Garden and park	Caimei Hong (2021), Lucie Poláčková (2014)	Landscape design	2
29	Bookstore	Xiaojing Jia (2019), Fangyuan Xiong (2019)	Communication and media	2
30	Sounds	Mengli Zhang (2014)	Design	1
31	Olfactory	Mengli Zhang (2014)	Communication and media	1
32	Mascot	Anjiao Xu, Yujie Jiang and Chun Zhong (2022)	Design	1
33	Road signs	Baopen Jin (2021)	Design	1
34	Literature	Kun Zhao (2012)	Literature	1
35	Religious events	Zohreh Torabi and Yalda Sima (2013)	Urban planning	1
36	Dialects	Manlu Wang (2021)	Linguistics	1
37	Road name	Ya'nan Zhu (2020)	Urban planning	1
38	Sculpture	Bowen Li (2019)	Art	1
39	Athletics	Yanjun Huang (2021)	Physical education	1
40	Linguistic landscape	Jiali Zhong and Meilan Liu (2022)	Linguistics	1
41	Gangs	Alistair Fraser (2015)	Sociology	1
42	Spice	Dewi Kumoratih (2022)	Design	1

Table 2 Indicators system for the evaluation of city identity elements

Indicator lever 1	Indicator lever 2	Indicator lever 3	Indicator description
Identity elements of city space	Tangible	City system	Architecture; squares; streets; museums; theaters; transport; landmarks; et al.
		City nature	Plants; landscapes; animals; greenery; et al.
		City infrastructure	Public space art; Street furniture; et al.
		City symbols	Patterns; materials; colors; skylines; architectural types and decoration; road signs; et al.
	Intermediate	City perception	The five senses (eyesight; aural; olfaction; tactility; taste)), et al.
		Historical and heritage	Handicrafts; dialects; road names; famous people; et al.
		Art and literature	Music; traditional painting; street art; poetry; literature; bookshop; et al.
		Life quality	Design; fashion; et al.
	Intangible	Traditional memories	Folk tales; festivals; ballads; et al.
		Life rituals	Lifestyle; poetic elements; religion; spirituality; et al.
		Events	Festive celebrations), religious events; significant events; et al.

the highest level is the level 1. Identity elements of city space, which is also the target level. The second level is the indicator level 2, which contains three broad categories: Tangible, Intermediate and Intangible. The third is indicator level 3, which contains the 11 categories of all the identity elements of the city space. Each level 3 indicator corresponds to the corresponding indicator level 2. Lastly, there is a description of the indicators corresponding to each indicator level 3, which includes specific things. It is the smallest individual component of the city identity evaluation system. Within a quantitative and fixed framework, these smallest individuals can be refined, extended or changed. There are no implications for the evaluation system.

The final indicator evaluation system can function as a standardized evaluation system for each city. The evaluation of these indicators is also an evaluation of the spatial identity of the city. It can provide data on the quality of the spatial identity of the city as reference data.

3 Quantitative and Qualitative Assessment of City Identity

Having obtained a system of indicators, the authors consider that different methods of assessment are needed for different types of indicators. Some need to be evaluated quantitatively and some need to be evaluated qualitatively. Quantitative evaluation includes: Quantity, Scale and Diffusion, Time. Qualitative evaluation includes: Distinctiveness, Aesthetic characteristics, Narrativity, Visual unity, Inclusivity. These are the various ways in which the identity elements of a city can be evaluated in a comprehensive way. The specific approach is summarized in Fig. 1.

3.1 Quantitative Evaluation Factors

3.1.1 Quantity

The assessment of whether a city has a strong identity starts with the number of identity elements in the city. The more identity elements a city has, whether tangible or intangible, the more likely it is that the city has a strong identity. The number of identity elements can be used as a reference for the magnitude of a city's identity. It represents the extent of a city's identity elements.

3.1.2 Scale and Diffusion

In addition to the number of identity elements, the size of the elements is also a reference factor, especially the size of the tangible elements. Large-scale identity elements are clearly more visually appealing than small-scale identity elements. The authors consider that there are three classes of urban space: small-, medium- and large-scale elements, which are characterized by: Large scale-with more recognizable value. Examples include: buildings, monuments, squares, streets, museums, gardens. It is noted that large objects in space are often the first elements that people notice. These large-scale features may be seen as the most intuitive conditions for the

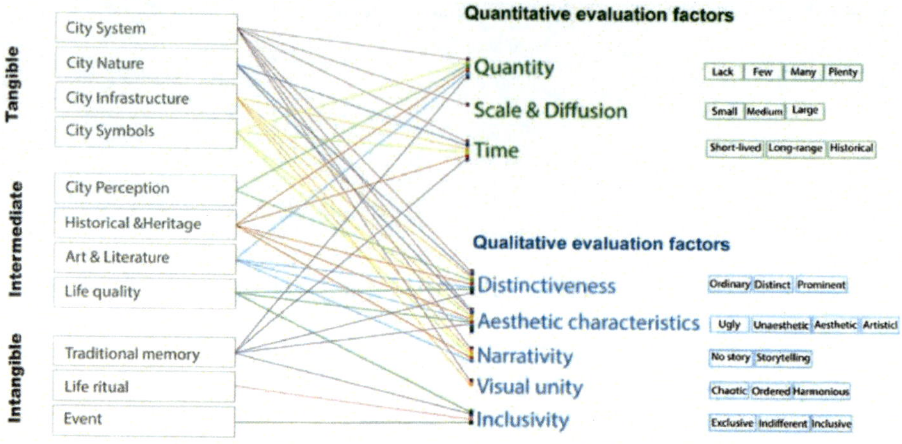

Fig. 1 Classification of city identity elements and their quantitative and qualitative evaluation factors. *Source* Author

character of urban space, shaping the character and quality of the urban landscape in the first place; Medium scale-more contact with people. For example, elements such as street furniture, street facilities, street decorations (shop signs, etc.), which are smaller than buildings. However, these are the things that urban dwellers come into meet with most often. They have the greatest impact on the human experience; small scale-providing a variety of rich perceptual experiences that add diversity to the space. For example, urban symbols, frontal patterns in spaces, decorations, signs, street names, street art. These subtle elements enrich the perceptual experience and reflect the cultural diversity of the space. They contribute to the narrative capacity of the urban space.

3.1.3 Time

The duration of the element in the urban space is also an important factor. Identity has both a fixed and a dynamic aspect. If identity only has a fixed architecture, then it has a consistent continuation throughout history. For example, architectural identity has both fixed and dynamic aspects in each historical period. People first perceive the actions and events provided by the forms and patterns of the physical environment. Then the features of shape, level and form capture people's attention and are perceived. The longer the duration can be considered, the more it affects urban identity and the less it affects it.

3.2 Qualitative Evaluation Factors

3.2.1 Distinctiveness

In addition to the quantity of urban identity elements, their uniqueness is also an important factor to assess. The uniqueness of identity elements distinguishes a place from other places. This uniqueness is formed by the involvement of culture, history and human activity over a long period of time. A city with a strong identity reflects a strong sense of individuality and self-confidence and is of great value.

3.2.2 Aesthetic Characteristics

Elements of urban identity, such as architecture, parks and public facilities embody aesthetic characteristics that often reflect the local history, culture and aesthetics, the artistry and charm of the space. The aesthetic characteristics of elements are reflected in, e.g., the form and decoration of buildings, sculptures, art installations or materials in public spaces, patterns, colors, landscapes, traditional costumes in spaces, etc.

3.2.3 Narrativity

The narrative nature of identity elements is inseparable from the memory of the everyday life of the inhabitants of a place. They are distinctive due to the differences of people. Spatial elements have a sense of authenticity in the world of human experience. Historically, cities have told micro-stories: place names, monuments, nameplates, murals, writing on walls, conversations about people, ordinary life and events. We attribute to narrative the ability to understand, interpret and represent, by giving form to reality, real or fantasy stories. Our attachment to place is actually achieved through the identification of those elements of affinity that involve our imagination.

3.2.4 Visual Unity

The strong visual coherence of the elements in urban space contributes to a strong city identity. Form and color elements in city space are symbolized in the human consciousness, thus creating a unified identity. Studies have shown that over 70% of the body's sensory receivers are concentrated in the eyes, and that humans use their eyes in conjunction with other organs to identify objective objects in most cases. To a large extent, visual perception factors such as field of view, perspective and visual balance will have a significant impact on the layout, size, proportion and color of the orientation facilities (Bao, & Wang, 2007). People always prefer to be in a comfortable and attractive space to perform various activities.

3.2.5 Inclusivity

The infectiousness of city spaces is reflected in the impact of festivals, events and rituals on people. The ability to include people in the city space, whether they are visitors or local residents. A positive experience for different cultural backgrounds, ages and social classes.

4 City of Florence as a Case Study

The article uses the evaluation of Florence's city identity as a case study to test the indicator system and evaluation method and to provide preliminary indications. Here, the article presents it not as an exhaustive survey but as an experiment. Florence's historic center was inscribed on the UNESCO World Heritage List on 17 December 1982. The history center is considered as "a unique artistic achievement of its kind, a masterpiece of work, the result of a continuous creation lasted for over six centuries", The Florentines have been able to preserve the old building processes by utilizing traditional building materials such as "pietra forte," "pietra serena," plasters and frescoes.

The purpose of this process was to experiment with the analysis of a city's identity in conjunction with the use of indicators and evaluation factors, and to provide insights into the feasibility of data-based urban identity research. Ultimately, the specific identity elements observed were content and evaluated as shown in Fig. 2.

The analysis shows that the identity of the city of Florence is the highest in terms of the evaluation factors Quantity, Scale and Diffusion and Time. They are also characterized by the highest in terms of Distinctiveness, Aesthetic characteristics, Narrativity, Visual unity and Inclusivity.

This process presents a preliminary approach to looking at the elements of identity in the city in a different dimension. The content of the observations will vary from city to city. The case of Florence's identity elements is not exhaustive and does not represent the result of a complete evaluation but rather a process that provides a method.

The results of this process analysis show that the indicators system and evaluation factors can be useful for systematic and in-depth observation and exploration of the elements of urban identity. Some possibilities for future research are offered.

5 Foundings

This is an attempt to analyze, in a broad sense, the framework system that influences the elements of urban identity. The results of the analysis indicate that this process is necessary. The index evaluation system of urban identity can be called an effective approach that helps researchers understand the identity of different cities and the factors that influence urban identity in more detailed and refined ways. It can also facilitate comparisons and promote models between different city identities, and it can unleash the enormous potential of cultural elements in cities.

Firstly, the indicator evaluation system provides a perspective on how people view urban identity. It can reflect how the diversity of elements in a city affects the quality of urban identity. Secondly, the indicator evaluation system also allows for continuing and expanding knowledge on, e.g., the quality of urban space, local attachments and cultural heritage. The theoretical framework of urban space as culture and collective memory is supported by the ways of seeing provided by the indicator evaluation system. Lastly, the process of observing the case cities indicates the operationalization of the framework for the indicator evaluation system in urban areas. The indicator system facilitated the observers' active exploration of the city's cultural context and provided a more comprehensive foundation for the analysis.

This process of analysis demonstrates that the system of evaluation indicators is a thorough and conclusive system of indicators. It is appropriate for addressing concerns pertaining to the quality of urban identity or urban cultural diversity. Due to the continuous progress of research in this field, it cannot be ruled out that the system may include more indicators that can be refined. Or due to the development of the research, some of the links between them may be adjusted. But any modification or updating of the content of the indicator descriptions will not affect the structure of the methodology. The indicator evaluation system can be utilized as a set of procedures or tools for organizing urban cultural identity investigation. It presents numerous opportunities for future investigation. It helps give a starting point and a foundation on which researchers from various research backgrounds can collaborate and analyze the subject.

6 Conclusion

Each city has its own specific and diverse identity components that constantly communicate stories to the audience. The elements of identity are diverse, and therefore the criteria for evaluation should be diverse as well. Without any method or approach to reveal and measure them, without a system of observation, evaluation or management, a large number of city identity elements cannot be properly observed. As a result, they are neglected, destroyed and eventually disappear. The authors argue for a data-based observation of the elements of city identity. This will lead to the establishment of a rational system for the conservation

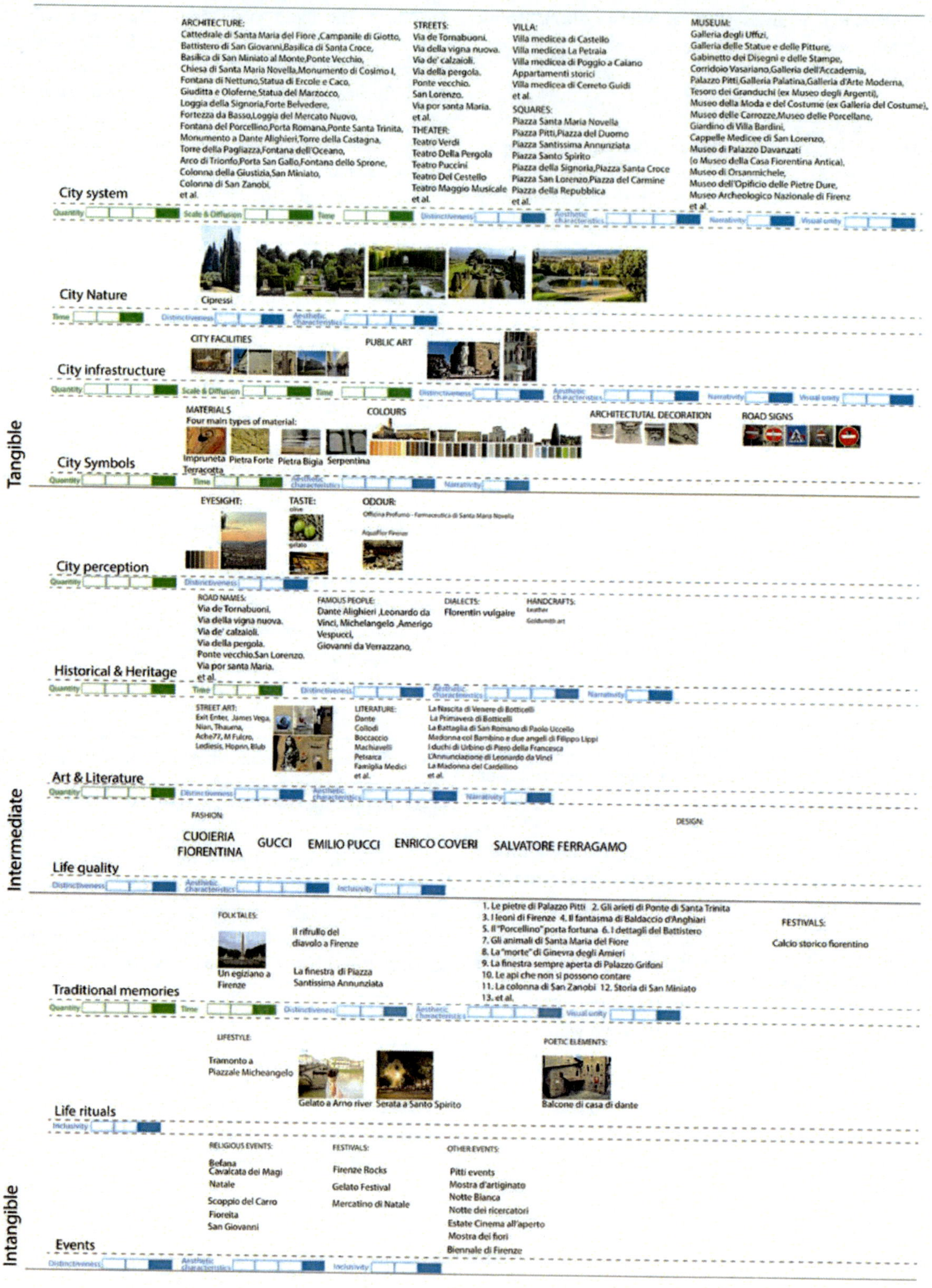

Fig. 2 Framework of city identity elements evaluation system. *Source* Author

and management of city identity, aiming to promote the conservation of the city's cultural diversity. This will enable a sustainable resource for urban development.

References

Anastasiou, D., Tasopoulou, A., Gemenetzi, G., Gareiou, Z., & Zervas, E. (2021). Public's perceptions of urban identity of Thessaloniki, Greece. *Urban Design International, 27*(1), 18–42. https://doi.org/10.1057/s41289-021-00172-8

Andéhn, M., Kazeminia, A., Lucarelli, A., & Sevin, E. (2014). User-generated place brand equity on Twitter: The dynamics of brand associations in social media. *Place Branding and Public Diplomacy, 10*(2), 132–144. https://doi.org/10.1057/pb.2014.8

Anholt, S. (2007). What is competitive identity?. In *Competitive identity: The new brand management for nations, cities and regions* (pp. 1–23). Palgrave Macmillan. https://doi.org/10.1057/9780230627727_1

Bao, S. D., & Wang, H.L. (2007). *Environmental signage guidance wayfinding design*. China Architecture & Building

Cheshmehzangi, A. (2020). Identity of cities and city of identities. *Springer*. https://doi.org/10.1007/978-981-15-3963-3

Er, B. (2019). Feretiko (Rize clohting) as a cultural identity of Rize City. *The Journal of International Social Research, 11*(59). https://doi.org/10.17719/jisr.2019.3244

Kiani, A., & Sardari, F. S. (2019). Evaluation and prioritization of urban identity elements using several decision models (Case study: Borujerd city). *Journal of Urban Social Geography, 6*(1), 91–108.

Kotradyová, V. (2019). Local identity in material culture as part of wellbeing and social sustainability. *Visions for Sustainability, 11*, 17–28.

Martín, J. M., Fernández, J. A. & Martín, J. A. (2019). Comprehensive evaluation of the tourism seasonality using a synthetic DP2 indicator. *Tourism Geographies: An International Journal of Tourism Space, Place and Environment, 21*(2), 284–305. https://doi.org/10.1080/14616688.2018.1505943

Onay, N. S. (2012). Stone as a determinant of architectural identity in the Florence of renaissance. *A|Z ITU Journal of Faculty of Architecture, 9*(2), 121–134.

RICHES Resources. (2014, November 27). *Tangible and intangible cultural heritage*. https://resources.riches-project.eu/glossary/tangible-and-intangible-cultural-heritage/

Riza, M., Doratli, N., & Fasli, M. (2012). City branding and identity. *Procedia—Social and Behavioral Sciences, 35*, 293–300. https://doi.org/10.1016/j.sbspro.2012.02.091

Samir, H., & Arayici, Y.(2015). *The Influence of modern architecture in transforming Iraqi city identity*. International Postgraduate Conference 2015, 10–12, Mediacity University of Salford.

Sofianto, K., Yuniadi, A., Nero Sofyan, A., & Gustaman, B. (2021). Symbol preservation and identity of Garut City. *Sosiohumaniora, 23*(1), 56. https://doi.org/10.24198/sosiohumaniora.v23i1.31553

Sönmez, B. E. (2020). A research on urban identity: Sample of Kadikoy District. *Journal of Contemporary Urban Affairs, 4*(1), 21–32. https://doi.org/10.25034/ijcua.2020.v4n1-3

UNESCO. (2003). *Text of the convention for the safeguarding of the intangible cultural heritage*. https://ich.unesco.org/en/convention

Yan, L., Wang, D., Zhang, S., & Ratti, C. (2021). Understanding urban centers in Shanghai with big data: Local and non-local function perspectives. *Cities, 113*, 103156. https://doi.org/10.1016/j.cities.2021.103156

Zhang, F., Zhou, B., Ratti, C., & Liu, Y. (2019). Discovering place-informative scenes and objects using social media photos. *Royal Society Open Science, 6*(3), 181375. https://doi.org/10.1098/rsos.181375

Compositional Approaches in Defining Istanbul's Urban Identity

Eliana Martinelli

Abstract

The goal of this study is to provide an interpretation of Istanbul's urban development by comparing compositional techniques in urban designs that have influenced the city's growth. The investigation is based on theories derived from Italian studies conducted over the course of the last four decades. In these studies, a compositional rather than merely historical perspective was used to analyze urban development and define an urban identity. The compositional research method, as well as the historical one, begins with the analysis of literature and archive materials. These materials, unlike the historical method, are analyzed with a critical stance toward the projects, interpreting, and understanding them through the drawings. When accompanied by a thorough historical analysis, it can provide a complete overview of the designer's work. The following study aims to demonstrate how the Ottoman city, despite its differences with North African and Middle Eastern medinas in terms of morphology and materials, can be considered an expression of the Islamic city. The primary research focus is on Mimar Sinan, an important figure in Ottoman architecture who worked from 1489 to 1558 and whose name is still widely recognized throughout Europe. The Turkish architect Turgut Cansever (1921–2009) provided new evidence that stems from further and more in-depth examinations of this topic.

Keywords

Architectural composition · Urban identity · Unity · Istanbul · Mimar Sinan · Turgut Cansever

E. Martinelli (✉)
Department of Civil and Environmental Engineering, University of Perugia, Perugia, Italy
e-mail: eliana.martinelli@unipg.it

1 Introduction: Research Background and Goals

"The cities and other forms of urbanization [...] united because they belong to the world of Islam are far from being a solid and unanimous matter, defined by precise chronological limits and territorial settings or by homogeneous ethnic contexts. On the contrary, they belong to a timeframe that corresponds to the last fourteen centuries of our history, which is unusually long for this kind of issue. Furthermore, they are widespread across a large area, extending to more or less significant portions of the Ancient World's three continents, and they express the recognizable contributions of a large number of civilizations, ideologies, and people" [Eng. trans.] (Cuneo, 1986, 5). Paolo Cuneo begins the first chapter of his book, *Storia dell'urbanistica. Il mondo islamico* with these words, emphasizing that the cities of the Islamic world, despite sharing similar physical characteristics, should not be considered in a singular manner. These cities are the materialization of various cultures and identities; they are the product of overlapping histories, transformations, and modifications that have resulted in the singularity of the urban areas.

The Ottoman city is not usually considered by historical literature as a representative example of an Islamic city since its development is not strictly related to the Sunnah, the system of practices that established the "tradition of life". This code of conduct, on the contrary, spontaneously determined the conformation of the North African and Middle Eastern medinas (Abu-Lughod, 1987; Burckhardt, 1985). From an architectural and compositional point of view, we can look at the Ottoman city as an expression of mystic Islam. The research aims at investigating this topic in detail by using specific bibliographical references and interpretative drawings.

This paper deepens and expands a study that was conducted over several years and was finally published in the book *Recomposing Unity. Turgut Cansever in Istanbul*

M. F. Shahidan et al. (eds.), *Urban Narratives: Exploring Identity, Heritage, and Sustainable Development in Cities*, Advances in Science, Technology & Innovation, https://doi.org/10.1007/978-3-031-48517-6_3

(Martinelli, 2022), the first Italian-English monograph about the work of the prominent Turkish architect Turgut Cansever. The study focuses on the development of Ottoman Istanbul as a great urban project that can be viewed as an interpretation of the existing Byzantine city and landscape. According to the author, the overall project depicts a true representation of mystic Islam. The work of two influential architects is taken into consideration: Mimar Sinan, the chief Ottoman architect, who lived for approximately one hundred years (1489–1588) and worked for the emperors Suleiman I, Selim II, and Murad II; and Turgut Cansever, a student of Sedad Hakkı Eldem, who was born in 1921 and worked between 1949 and 2009. The architectural approach of Mimar Sinan and, after four centuries, the theoretical design principles proposed by Turgut Cansever, which can be considered in continuity with Sinan's work, seek to find in Islamic principles and rules the expression of a form of modernity that stemmed and developed from the reinterpretation of the stratigraphic structure of the city.

2 Methodology

In architectural research, the "compositional perspective" leads the investigation and interpretation strategy. A "compositional approach" focuses on the methods and principles that can be used to design volumes and spaces while adhering to specific expression codes, features, formal logics, and settlement rules in order to modify the existing environment. Because of the compositional approach typical of Italian architectural design theory and practice, the theoretical sources include, in addition to Cansever's books, some writings by Italian architects such as Maurice Cerasi, Paola Sonia Gennaro, and Augusto Romano Burelli. The first concentrated on the Levantine and Ottoman cities, while the last two focused on Sinan's mosques and their implication in the urban layout.

Re-interpretative drawing is the fundamental instrument of compositional analysis. The research process consisted of re-elaborating many drawings detected in Istanbul's archives (at the SALT Research Center and the Cansever family's private archive), such as architectural sketches, technical drawings by the architects, and ancient and digital maps. Interpretative drawings have been used to provide a critical perspective on the city's evolution or the design ideas in the works of architects. They allow the architect-researcher to present an analysis while also extrapolating certain design concepts that can be implemented in future projects. This paper includes some samples of research re-elaboration drawings to demonstrate the method.

3 The Construction of Ottoman Istanbul

3.1 Istanbul as an Ideal City

The Turkish-Italian architect Maurice Cerasi wrote: "In Ottoman society's myths and nostalgia, there is a city that is a combination of all other cities, becoming the image and tangible version of the concept of urbanity: Istanbul. It is the city of material happiness: Sa'd- iābād. [...] Istanbul is the implemented version of the Ottoman city model" [Eng. trans.] (Cerasi, 1994, p. 229). *Sa'dabad* is also the title of a book written by the Turkish architect Sedad Hakkı Eldem (1977) about one of Istanbul's magnificent palaces, also known as the "palace of built happiness". Over the course of several centuries, the structure has been rebuilt three times. Each time, it was reborn from the ruins of its previous existence, persisting in people's collective imagination. As a result, it appears that both the building and the city have the capacity to regenerate; this common ability acts as the linking factor between the two.

Despite the fact that it lacks anything that might be deemed utopian or pre-ordered, Istanbul has the potential to be an ideal city. The Ottoman-ruled city does not aim to conform to any preset patterns or ideas that have been envisioned. The site's topography and character, including pre-existing forms and how they should be understood, have had a profound impact on its evolution. Because of this feature, certain Western historians, such as Armin von Gerkan (1924), refused to recognize Constantinople as a city and instead referred to it as a "densely populated landscape". Miniatures representing Constantinople in the fifteenth century and painted by Venetian painter Giovanni Andrea Vavassore (Fig. 1), as well as one painted by the Ottoman statesman Matrakçı Nasuh in 1537, are examples of image references that express this concept. The roads, which are difficult to distinguish in this scene, are not nearly as important as the urban open space, which is depicted as a continuous surface on which the structures are set. As seen in Byzantine mosaics or Ottoman miniatures, natural components, such as water and light, are directly included in the surroundings without having a perfect geometric order (Cerasi, 1988).

3.2 The Founding Criteria

An Ottoman city's founding criteria are substantially different from those of a Western city. Contrary to Western and Eastern classical traditions, the morphological origin of the Ottoman city is not based on the geometric rule that

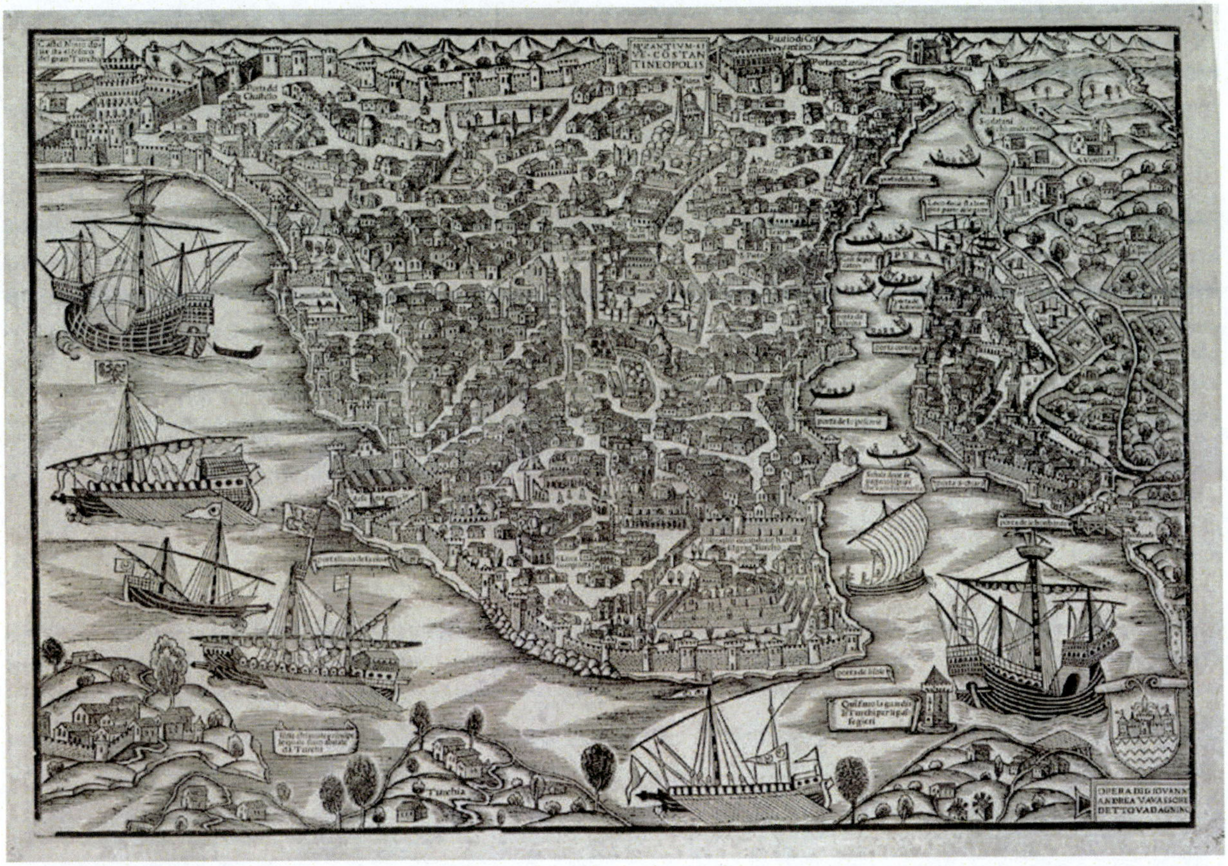

Fig. 1 Woodcut of Constantinople in the early Ottoman period by Giovanni Andreas Vavassore, circa 1535. Credits: Germanisches National museum, Nürnberg

governs the relationship between defined forms. It develops from more complex relationships, the most important of which are those with nature (Cerasi, 1988). Despite its magnificence, the Turkish Empire was not responsible for the development of any grand urban layouts based on celebration geometry, such as those seen in Central Asia. The Ottoman city, in particular, lacked a central point from which the surrounding road network spread outward.

The Divanyolu was the only urban axis implanted on the Byzantine Mese, Constantinople's original matrix. The imperial processions also traveled along this route. Every public space in Constantinople had a view of this arcaded street, which can still be observed in the contemporary city, albeit in a fragmented form (Fig. 2). Even after it was designated as the Ottoman city's urban axis, it maintained its character as a continuous public space determined by direction rather than geometry (Cerasi, 2004). The original street front was neither continuous nor compact; rather, gardens divided the buildings from one another, altering the width of the roadway and adding dynamism to the surrounding urban environment (Cerasi, 2005). Along with man-made structures, trees played a significant role in the overall composition of the landscape.

While the major road axis and public spaces remained those of the Byzantine Empire, neighborhoods were built around the new religious complexes, known as *külliyes*, in accordance with the curved or random structure of the Muslim city (Gennaro, 2022). Each *külliye* was made up of a mosque and many additional structures, such as a convent, a hospital, and a school, that served the religious, cultural, and social needs of the people who lived nearby. Paola Gennaro (2022) identifies Ottoman *külliyes* as "triggers of urban regeneration", which is consistent with the concept of urban colonialism. Her thesis is that, starting with Mehmed II the Conqueror, the Ottoman sultans envisioned the establishments of their *külliyes* as active grafts on an already degraded urban fabric, both architecturally and socially, in order to encourage and promote the development by putting pressure on new social activities dependent on these structures.

Even if it is the sole building rising in the urban landscape, the mosque does not represent the Ottoman city's

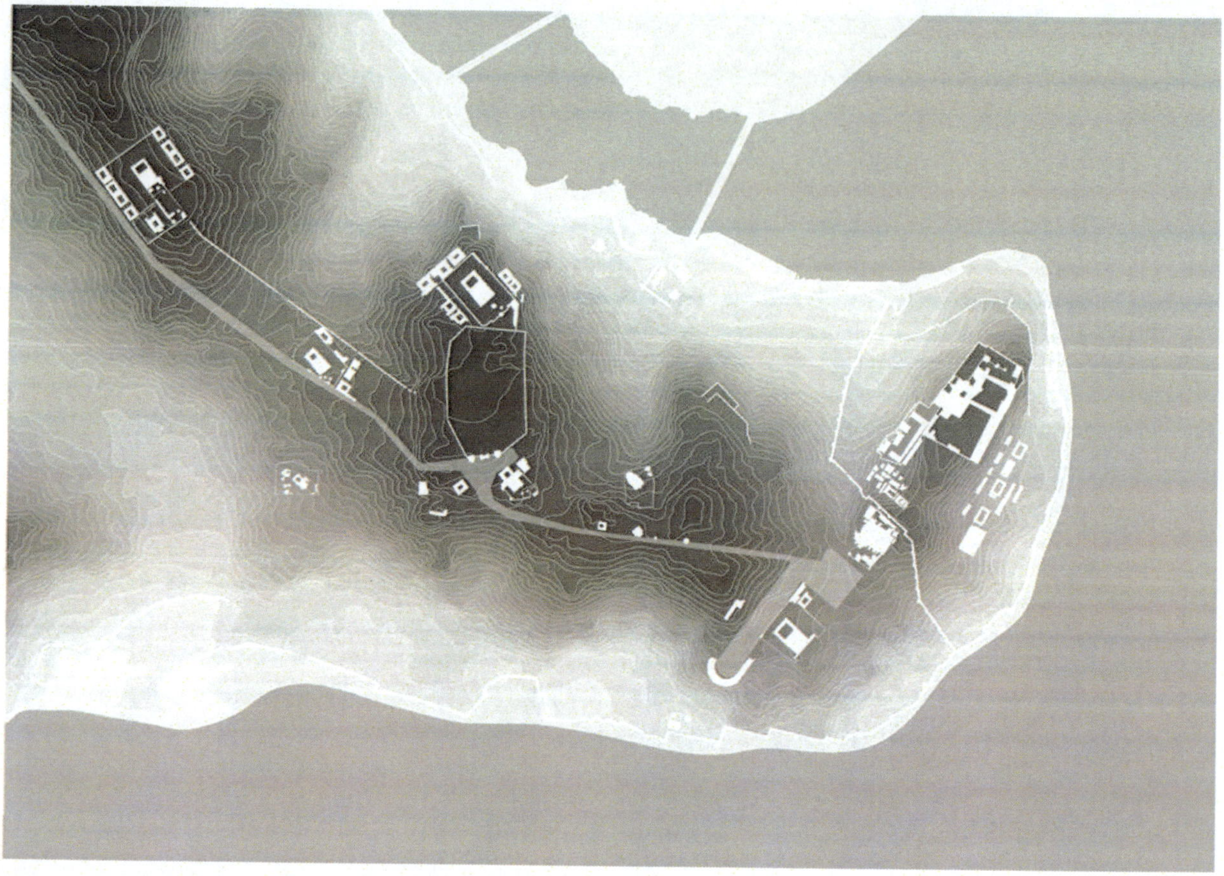

Fig. 2 Ancient Divanyolu in relation to the topography and monuments. Drawing by Martinelli (2022, p. 65)

center. Augusto Romano Burelli (1988) describes the prayer area as isotropic, with the physical core of the ritual substituted by a cosmic center that corresponds to the cupola and serves to unify the space. The mosque is the central focus of Mimar Sinan's architectural design, around which the associated buildings cluster. The architect built the city by defining the open spaces (*meydans*) that resulted from the disposition of the *külliye*'s buildings. Sinan's key instrument in the design process was the Turkish-Ottoman tradition of narrative sequences (Gennaro, 1992).

The Ottoman civilization was responsible for the subsequent urban restructuring by reinterpreting features of Byzantine architecture (Fig. 3). Turgut Cansever (1977) attributes the regeneration of the previously established heritage to an "Eastern" approach, which means making an effort to develop new connections between distinct entities in accordance with the Ottoman understanding of historical progression. Cansever believes that this operational method is capable of transmitting and retaining a strong historical consciousness. This strategy entails the development of means for determining relationships and interactions between various and opposing elements that lead to a sense of unity.

The use of counterpoint and the juxtaposition of complementary components are two essential compositional principles that we may discern in the ancient Ottoman city. "Counterpoint" can be defined as a narrative technique based on opposing but complementary motives. Combining different elements does not result in conflict; rather, it emphasizes their capacity for cohabitation and respect. Thus, rather than a distinct and finished form, unity emerges from the juxtaposition of opposing architectural elements. In the Ottoman city, there was a contrast between smaller wooden domestic structures that made up the neighborhoods and massive religious stone buildings like mosques and other *külliye* structures. The wooden dwellings had an ephemeral character and reflected nomad culture. There was also a contrast between the man-made buildings and the continual, overwhelming presence of trees. According to Islamic gnosiology, a complete comprehension of the world can only be achieved through the juxtaposition of complementary elements; if intuitive

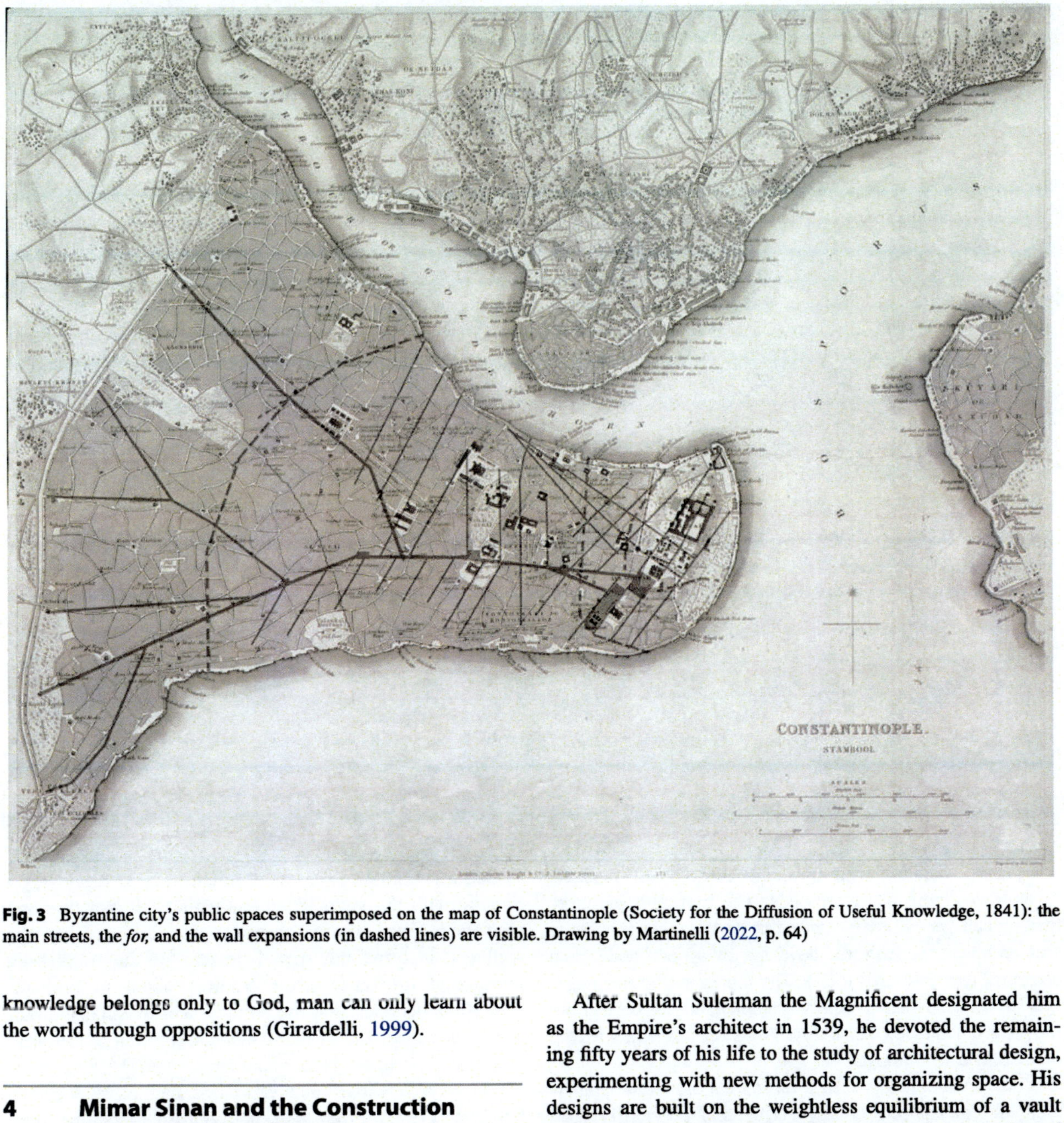

Fig. 3 Byzantine city's public spaces superimposed on the map of Constantinople (Society for the Diffusion of Useful Knowledge, 1841): the main streets, the *for,* and the wall expansions (in dashed lines) are visible. Drawing by Martinelli (2022, p. 64)

knowledge belongs only to God, man can only learn about the world through oppositions (Girardelli, 1999).

4 Mimar Sinan and the Construction of the Urban Landscape

Mimar Sinan was born an Orthodox Christian and joined the janissary corps after getting a military education. During that period, he and his unit participated in a number of different Ottoman Empire expansion campaigns. While he was traveling throughout the Mediterranean and the Middle East, Sinan had the opportunity to come into contact with a broad variety of architectural types, collecting a set of forms and figures that he would later use in his works (Gennaro, 1992).

After Sultan Suleiman the Magnificent designated him as the Empire's architect in 1539, he devoted the remaining fifty years of his life to the study of architectural design, experimenting with new methods for organizing space. His designs are built on the weightless equilibrium of a vault leaning upon a central organism connected to a series of smaller rooms (Gennaro, 1992).

According to Paola Gennaro, Sinan's network of buildings in Istanbul forms an archipelago of architectural islands nestled among the chaotic texture of the Ottoman city. Sinan did not plan a new city; he let the ancient city plan his own work (Goodwin, 1987). His urban project has resulted in the development of a real panorama, which can be completely perceived and enjoyed only from across the Bosporus. The overwhelming presence of mosques in this

Fig. 4 View of the Golden Horn and the Bosporus from the Süleymaniye complex madrasa by Mimar Sinan (1550–1557). Photo by Martinelli (2022, p. 67)

environment gave organization and order to the otherwise chaotic urban form (Fig. 4).

Sinan is credited with inventing the type of mosque, or at least giving it a clear code. The mosque's architecture, which was already existent in its vocabulary of parts and architectural features, develops in syntax and urban composition (Gennaro, 1992). The monument is no longer viewed as an isolated object, complete in its perfection; rather, it obtains new value through its relationship with the other *külliye*'s buildings.

Sinan methodically used a technique known as "imperfect" or "transposed axiality". This tool, previously used in other Ottoman architectural designs, consists of a deviation in the building's axial orientation (Cerasi, 1987, 1988) (Fig. 5). The mosque's directional axes are moved along its structure, or posed tangent to the external walls, while the other buildings are laid along with it. This translation improves angular vision and gives rise to the complex's asymmetrical construction. Imperfect axiality is not readable at the architectural scale, where symmetry is preferred; rather, it recurs at different scales of urban design. Sinan bases his *külliye*s on the systems of axial relationships between them and Hagia Sophia, his most important Byzantine period reference.

Selim II (1566–1577) tasked Sinan with restoring Hagia Sophia, with the goals of strengthening the structure and constructing two massive minarets. On that occasion, the architect analyzed the spatial devices of the basilica—a mosque at the time—while learning about its criticalities. He understood that the employment of imposing buttresses had made the structure heavier, limiting light access and conditioning the ornamental apparatus. The final result would be a "creative" restoration, as defined by Luciano Semerani (1991). From then on, the great Byzantine monument served as the fundamental reference in mosque designs, becoming the starting point for the development of Ottoman architecture.

Sinan's mosques are difficult to observe from a single point of view since they were created in accordance with the spatial sense of Ottoman cities. The main façade of the mosque is often partially hidden by a domed portico that is unusual in terms of scale and dimensions among the others in the courtyard. Each component is both hidden and visible, while the structural components are normally

Fig. 5 System of axial relations between the *külliyes* in the historical peninsula. Drawing by Martinelli (2022, p. 110)

concealed somewhere within the wall. The spatial perception is uniform and not hierarchical; for this reason, in Sinan's design, the light is studied to spread it homogenously. Drop shadows do not fragment the internal and external spaces because the disturbing elements, such as buttresses, are minimized, letting the solar light glide on the external spherical caps and spread across the space through the windows.

5 Turgut Cansever's Interpretation Through the Design

Turgut Cansever considers Istanbul the main reference for defining his own design approach. This perspective takes into consideration not only the typological and morphological elements of vernacular architecture but also a certain method of urban composition. Cansever offers a wide range of critical and interpretative readings on Islamic and Ottoman cities and architecture, always taking care to establish practical design principles. On numerous occasions, he stressed the similarities between Turkish-Ottoman spatial composition and that of Western modern art and architecture (Girardelli, 1999). His theoretical work, which has been published in a number of volumes, is still essential reading for Turkish architecture students. Cansever, in particular, deems Sinan the source of a modernity to take forward (Cansever, 2005).

According to Cansever, Sinan's compositional approach, which is characterized by repeated, distinct components (defined as "tectonics"), has its roots in nomad culture, in which individuals attempt to document the environmental changes that occur in their surroundings (Cansever, 2004). Sinan's mystic education had an impact on his work as well, since the life of a janissary is similar to that of a monk. Cansever (1997) believes that people's activities are

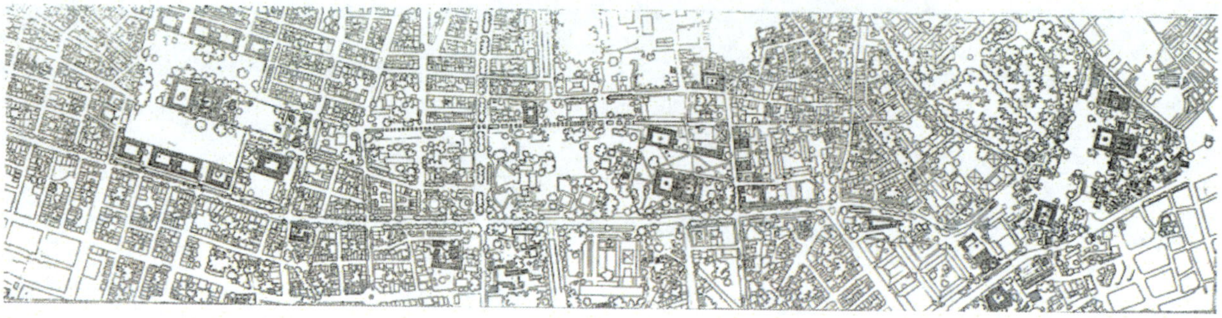

Fig. 6 Turgut Cansever's design sketch of the city section between the *külliyes* of Fatih and Beyazıt, 1963. This design exercise aims to demonstrate how to restore a human scale between the city and its monuments. Credits: Cansever Family Archive, digital copy obtained through SALT Research Istanbul (Martinelli, 2022, pp. 92–93)

mirrored in works of art as "forms of expression", and that the designer's faith and values impact the decisions that he makes.

Cansever investigates Ottoman architecture through the lens of Islamic ontology. If, as mysticism asserts, existence is a dynamic and ever-evolving process, then absolute urban schemes cannot represent the city itself. Cansever's concept of Islam is more than just a religious belief to celebrate through architecture; it is a worldview from which to extrapolate a design theory that combines and embraces diversities. Cansever's designs offer an architectural interpretation of Islam with the goal of integrating more ancient traditions into the contemporary world (Aksulu & Irklı Eryıldız, 2003).

Cansever's rhetoric on Islamic cosmology and the return to tradition, which is prevalent in his later works, diverts attention away from the true subject of his research, which is the awareness of site and construction (Bozdogan & Kasaba, 1997), expressed in two architectural notions dear to Cansever: type and building process (Martinelli, 2022). The first conveys the cultural relationship with the place through vernacular architecture, while the second allows architecture to be seen as the "art of building". Critics have frequently reduced his works to popular labels such as "critical regionalism" (Frampton, 1983), which can be defined as an approach that seeks to root architecture in the modern tradition while remaining tied to geographical and cultural context, attempting to mediate between global and local architectural languages. Critical regionalism, according to Frampton, implies a close relationship between architecture and nature based on geography, context, climate, light, and tectonic form. Cansever's architecture, on the other hand, is concerned with nature and places culture in a broader and more holistic sense. He looks at the city as a whole, not just portions of it. There is no longer a clear distinction between the new and the old, and the boundary between preservation and innovation is similarly blurred (Fig. 6). Each additional component contributes to the integration by providing an interpretation of the existing facts in relation to the contingent ones, as stated by the architect himself: "In the case of Istanbul and Ottoman city culture, the problem was the resolution of the dichotomies of the individual and universal, of the variety and the standardization, of the large and small, of the austere and gentle, of the eternal and the temporary, of the solid infrastructure and the independent superstructure, of the freedom and control of the personal experience and sensibility and the scientific objective research. All this in the contest of the attempted efforts for the resolution of historical contradictions between past and present cultures. Our present responsibilities are similar. […] The attitude of change for the sake of change and the separation of conservation from creation should be discarded. The fundamental principle in carrying out this task will be 'not to change until the old can be replaced by something better" (Cansever, 1977, 99).

In conclusion, Cansever's architectural interpretation does not admit dualism as an opposition between the eternal and the temporal, modernity and tradition, or regionalism and internationalism. Going beyond bipolar and mutually opposed categories can be considered a starting point for understanding a worldview that is not linked to Cartesian logic. Modernity is already within the tradition, and searching for a specific regional architecture would automatically exclude the complex cultural influences of Turkish architecture. Unity contains plurality; therefore, holding together a plural culture requires a unique cultural identity.

6 Conclusions

Since Islam is a non-figurative religion, the city, which was originally a catalyst for nomad populations (Marçais, 1928), actualized the faith by morphologically expressing the doctrine of Unity (*Tawhid*), an Islamic principle that asserts God's unity and oneness and does not recognize a distinction between the sacred and profane. As a result, Islam externalized the faith in new architectural forms (Kowsar, 1982). As illustrated in the previous sections and supported by the mentioned studies on architectural and urban composition as well as the interpretative drawings, the concept of Unity has been the core principle for the development of not just Arab medinas but also Ottoman cities. Here, unity is found not in the compact form of the medina but in the city's ability to embrace diversity in accordance with mystic Islam. In the urban cosmos, many forms and materials are gathered on the basis of specific types of relationships, such as "transposed axiality". This is because Istanbul, the city where East and West meet, was built over centuries through a series of stratifications while preserving a stark contrast with the different architectural cultures that were present at the same time. Mimar Sinan first attempts to define the compositional principles of Ottoman cities, using *külliyes* as "trigger points". More recently, Turgut Cansever, who had studied the work of Sinan in depth, wanted to transmit it to future generations by creating a new, hard-to-categorize concept of modern architecture. Both of these architects attempted to work in accordance with the idea of Unity, perceiving urban morphology as always evolving while not forgetting its past. Their design strategies were able to resist the city's unending alteration. This is a crucial idea for both comprehending Istanbul today and learning how to create its future architecture.

References

Abu-Lughod, J. (1987). The Islamic City. Historic myth, Islamic essence, and contemporary relevance. *International Journal of Middle East Studies, 2*, 155–176.

Aksulu, I., & Irklı Eryıldız, D. (2003). Local culture and historic experience for sustainable design. *Open House International, 28*, 67–73.

Bozdogan, S., & Kasaba, R. (1997). *Rethinking modernity and national identity in Turkey*. University of Washington Press.

Burckhardt, T. (1985). *L'art de l'islam. Langage et signification*. Sindbad.

Burelli, A. R. (1988). *La moschea di Sinan*. Cluva Editrice.

Cansever, T. (1977). The city center of Istanbul: Its past and its future problems. In *Urban Conservation in Europe and America: Planning, Conflict and Participation in the Inner City: Rome, 1975, Conference Proceedings* (pp. 91–100). Tipografia Olimpica.

Cansever, T. (1997). *İslam'da Şehir ve Mimari*. İz Yayınları.

Cansever, T. (2004). The architecture of mimar sinan. *Architectural Design, 6*, 64–69.

Cansever, T. (2005). *Mimar Sinan*. Albaraka Türk.

Cerasi, M. (1987). Place and perspective in sinan's townscape. *Environmental Design: Journal of the Islamic Environmental Design Research Centre, 1–2*, 52–61.

Cerasi, M. (1988). *Le città del Levante. Civiltà urbana e architettura sotto gli ottomani nei secoli XVIII-XIX*. Jaca Book.

Cerasi, M. (1994). Il giardino ottomano attraverso l'immagine del Bosforo. In *Il giardino islamico. Architettura, natura, paesaggio* (pp. 217–236). Electa.

Cerasi, M. (2004). *The Istanbul Divanyolu*. Ergon Verlag.

Cerasi, M. (2005). *La città dalle molte culture: l'architettura nel Mediterraneo orientale*. Scheiwiller.

Cuneo, P. (1986). *Storia dell'urbanistica. Il mondo islamico*. Laterza.

Eldem, S. H. (1977). *Sa'dabad*. Milli Eğitim Basimevi.

Frampton, K. (1983). Towards a critical regionalism. Six points for an architecture of resistance. In *The Anti-Aesthetic: Essays on Postmodern culture* (pp. 16–30). Bay Press.

Gennaro, P. S. (1992). *Istanbul. L'opera di Sinan*. Cittàstudi.

Gennaro, P. S. (2022). Colonie urbane: Le *külliye* ottomane come inneschi di rigenerazione urbana / Urban colonies: Ottoman *külliyes* as triggers of urban regeneration. *DAr Design Architecture Research, 2*, 24–39.

Girardelli, P. (1999). Ai confini del Mediterraneo. Radici storiche e tipologiche nei linguaggi architettonici contemporanei in Turchia. *Controspazio, 5*, 8–11.

Goodwin, G. (1987). Sinan and city planning. *Environmental Design: Journal of the Islamic Environmental Design Research Centre, 1–2*, 10–19.

Kowsar, M. (1982). Le città dell'Islam: identità di un passato? In *Architettura nei paesi islamici: seconda mostra internazionale di architettura* (pp. 28–35). Electa, Edizioni La Biennale di Venezia.

Marçais, W. (1928). L'islamisme et la vie urbaine. *Comptes rendus des séances de l'Académie des Inscriptions et Belles-Lettres, 1*, 86–100.

Martinelli, E. (2022). *Ricomporre l'unità. Turgut Cansever a Istanbul/ Recomposing Unity. Turgut Cansever in Istanbul*. Aión Edizioni.

Semerani, L. (1991). Sinan. Il restauro creativo. In *Passaggio a Nord-Est. Itinerari attorno ai progetti di Luciano Semerani e Gigetta Tamaro* (pp. 100–103). Electa.

Von Gerkan, A. (1924). *Griechische Städteanlagen*. W. De Gruyter and Co.

New Place Identity: Redefining Bangkok Old Town Area as the New Creative District

Thirayu Jumsai na Ayudhya

Abstract

In 2018, The Thai Government launched urban renewal projects in various Thailand provinces, including the capital city. In Bangkok, Charoen Krung district was selected by Bangkok Metropolitan Authority (BMA) as the new creative urban district in Bangkok. This paper aims to explore and evaluate the renewed urban identity of Charoen Krung district, whereby BMA's urban renewal project has been implemented. The project has been implemented for three years from 2018 to 2021, but there is currently no empirical research for the evaluation of the success in terms of a renewed urban identity. In 2021, this research was conducted with 77 research participants, including both local residents and visitors. Photo-elicitation and in-depth interviews were adopted within the qualitative research process, with the aim of understanding participants' given meanings; the renewed urban identity of place, to their everyday contextual experiences of the Charoen Krung district. The method triangulation, including (1) the on-site survey with researcher-produced-photograph (RPP), (2) participant-produced-photographs (PPP), and (3) in-depth elicitation interviews, with residents and visitors, was adopted. PPP and in-depth interviews were adopted for two groups of participants: (1) local residents and (2) visitors. Interview data were transcribed and interpreted with Interpretative Phenomenological Analysis (IPA) approach. Sub-themes and super-ordinate themes emerged as the reflective meaning of Charoen Krung's renewed place identity. Eventually, the meaning of creative district and specific identity has been differently perceived and defined by residents and visitors. In addition, the results and methods used in this research project are expected to be adopted and applied as practical guidance for evaluating the further evaluation of other urban renewal urban identity projects in other provinces of Thailand.

Keywords

Place identity · Sense of place · Renewed urban identity · Creative district · Bangkok creative district

1 Introduction

Charoen Krung district was established in the year 1864 in the reign of King Rama the Fourth of Thailand. The first road in Charoen Krung district was opened to the public on March 17, 1867, named Charoen Krung Road, meaning "prosperous city". The road created changes and evolutions in many dimensions for the district. The first change was that the public transportation and logistics were changed from traveling by boats in the Chao Phraya River or small canals traveling by rickshaw, horse-carriage, and car. Residential buildings, commercial buildings, and facilities were built along the 8575-m-long road to serve people's everyday life, so the road was considered the main factor drawing further developments to the district (Fig. 1). For about 150 years, urban contexts along Charoen Krung Road have been changed and evolved concerning complicated Bangkok's historical phenomena.

Around the 1870s, oversea trading companies settled their businesses along Charoen Krung Road, so this road became the new commercial hub for Bangkok city. A decade later, in the 1890s, Charoen Krung Road was packed with many types of buildings, including residential, commercial, and diplomatic buildings. One of the remarkable diplomatic buildings on Charoen Krung was Thailand's first British Consular Office. The building was designed

T. J. n. Ayudhya (✉)
School of Architecture, Art, and Design,
King Mongkut's Institute of Technology Ladkrabang,
Bangkok, Thailand
e-mail: thirayu.ju@kmitl.ac.th

M. F. Shahidan et al. (eds.), *Urban Narratives: Exploring Identity, Heritage, and Sustainable Development in Cities*,
Advances in Science, Technology & Innovation, https://doi.org/10.1007/978-3-031-48517-6_4

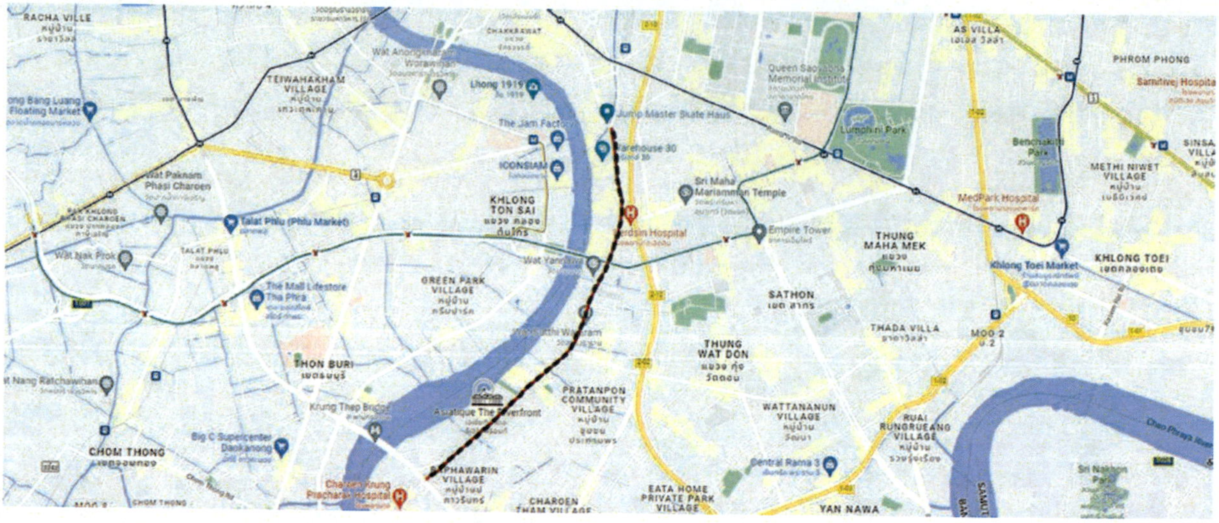

Fig. 1 Bangkok map, showing Charoen Krung Road as the dash-line (Google Maps, n.d.)

in an art deco architectural style. For nearly five decades ago, urbanization has been extended to other areas, in both the East-side and West-side of Bangkok. The Charoen Krung district has decreased its role as Bangkok's Central Business District, CBD. The local business sections gradually decreased, so many commercial buildings were closed down or abandoned. Many ruined buildings and built environments along Charoen Krung Road have been demolished due to the deterioration, and some remaining ones are in disrepair. In 2018, the Thai Government launched a re-urbanization project in many provinces. Charoen Krung district was selected by Bangkok Metropolitan Authority (BMA) as the new creative district in Bangkok. The main purpose of this urban renewed project was to revitalize existing dilapidated urban contexts in one of the oldest districts in Bangkok. For Charoen Krung urban renewal project, Bangkok Metropolitan Authority (BMA) initially began with the renovation of the historical Central Post Office building to be the Thailand Creative Design Center (TCDC).

The Central Post Office was built in 1943, two years before World War II, and also this building has been known as the landmark of the district. During World War II, the building survived several air strikes from the Alliance, shown in Fig. 2. In 2018, the renovation was completed, and since then TCDT started introducing several creative and artistic events to the district. One of the well-known international art events, named Bangkok Art Biennale, was held in this area, and this event helped to make the district well-known to domestic and international people. Then, in 2019, many creative and artistic events were drawn to public attention, so the district has become a new attraction point for many groups of people, including domestic and international visitors. Abandoned commercial and residential units have been renovated. New business trends with creativities, e.g., the art and home decoration shops, art galleries, and fusion coffee shops were opened in this district. A new beginning of business flourishing in the district was expected to boost the local economic growth and also build up a new significant place identity for Charoen Krung district. For nearly five years after implementing the creative district project in Charoen Krung district, every day of Charoen Krung local resident and visitors have been changed, but there are no pieces of evidence exploring the effects of this re-urbanized project on the new place's identity. This article aims to explore the emergent significant new place identity of Charoen Krung district, using a qualitative research approach; participant-produced-photograph (PPP) and photo-elicitation, in-depth interview, and interpretative phenomenon analysis (IPA). In addition, the process used in this research will be used as guidance for the re-urban projects in other provinces by The Thai Government.

2 Literature Review

2.1 Place Identity

Making sense of place emerges through individuals' perceptions and experiences with the meaningful places in which the people are emotionally connected (Oberlin & Gieryn, 2015; Westerholt et al., 2022). The two words, place and identity, are the connection in two ways: (1) the connection between an individual in the specific place the physical features of the place, and (2) the connection between an individual and the community (van Eijck & Roth, 2010).

Fig. 2 Thailand Creative Design Center (TCDC) in the historical Central Post Office building (photograph taken by author)

The strong place identity emerges through four main indicators: familiarity, belonging, involvement, and meaningfulness (Gross & Hochberg, 2016). A specific place evokes different meanings for different individuals. For example, at the moment of the first date for a young couple in a small garden on a corner of a big city, the place-making process emerges for this couple, considering this garden as their memorable place. For others, walking through this small garden at nighttime can make people feel dreadful because of the number of criminal cases in the garden. So, the small garden, in this case, plays a complex role in fully understanding the holistic nature of making sense of place. Place-based theories focus not only on making sense of place but also place attachment and place identity. Place identity is a multifaceted concept of contextual resonance in people's minds (Ruggeri, 2014). The place identity is also described as the ability of the particular locale to imprint memorable imageability in people's minds (Irwin, 1975). Also, it is the result of the continuous communication between an individual's self and the facet of experiences of the particular locale, including certain habits, rules, and norms related to spaces of multidimensional scales (Jones, 1998; Proshansky et al., 1983). Individuals' relationships bonded with the particular locale also contribute to building place identity (Reicher et al., 2006). This relationship is developed through the relationship between the settings in the everyday contexts and people (Gross & Hochberg, 2016). Developing place identity has been described as the gradient development that comprises individuals' emotional bond with places, including the sense of belonging, the place satisfaction and security, the place memorability and distinctiveness, and meanings embedded in a place (Southworth & Ruggeri, 2011). In place-based research, the identity of a place for a specific location, such as a tourist attraction site, is not limited only to local considerations and domestic values, but also extends to the level of the place branding (Ösgård & Spierings, 2021). For urban development from a tourism point of view, place branding is not a simple promoting issue of a city, such as creating weekend events, inviting people to the public concert, and opening street markets, but it may affect people's cognitive and emotional attitudes toward the city (Félonneau, 2004). On a global scale, the inter-urbanization aims to raise the place branding to the city branding producing the city's holistic images and symbols (Aydoghmish & Rafieian, 2022). With the inter-urbanization, some research found that the role of residents in place branding practice and place development was ignored. For city reimage and redevelopment projects aiming to attract visitors or tourists, the given place's identity can be notably different from the identity given by residents' projected images from their daily life and lived experiences (Ginting & Wahid, 2015). Making the creative identity for Charoen Krung area, Bangkok Metropolitan Authority (BMA) initially focused on how to build the place more attractive for both international and domestic visitors and tourists. A lack of paying attention to residents' place identity can affect the success of the re-urbanized project (Devine-Wright, 2013). In this research, the new place identity was investigated by qualitative triangulation approach through people's giving meaning to Charoen Krung district.

2.2 Place Identity and Photo-Elicitation

Photo-elicitation, known as photovoice, was initially utilized to collect the community's attitude toward the physical setting in the 1990s by Caroline Wang and Mary Ann Burris (Wang & Burris, 1997). It allows people to share and convey ideas about their lived experiences. It also focuses on a qualitative participant-action approach used to reflect a voice to chronicle their everyday surrounding contexts. Photovoice also enables participatory action among participants and communities. With photovoice, research participants can identify, represent, and enhance the contexts through a photographic technique to record and reflect the community's strengths and concerns. It encourages the research participants to raise critical issues through group discussion with photographs (Wang & Burris, 1997). This method relies on people's or research participants' roles as recorders. With photo-elicitation, people have the opportunity to record, reflect, and comment creatively. After photographic data collection, research participants are gathered for sharing their discussions, experiences, and attitudes about the studying issues. One of the distinctive advantages of this method is that it helps to overcome the language barrier between the research and the research participants) Torres et al., 2013). In addition, it also empowers participants with disabilities to comfortably reveal their everyday constraints (Macdonald et al., 2022). Photo-elicitation has unique advantages over other qualitative research methods, but its implementation in the real situation necessitates adaptations to existing contextual conditions (Falconer, 2013). In 2004, the survey showed that verbal methods such as journaling, interviewing, and questionnaire continue to be preferred methods among researchers studying people's emotional perceptions of ties to places. Small numbers of research combine the use of photovoice with other research methods, e.g., combining photovoice with a focus group (Hannay et al., 2013). Hannay et al. (2013) combined community-based participatory research methods, photovoice, and focus groups to explore Latina teens and their parents in identifying difficulties they encountered within their physical settings. Another example of research reveals that the exploration of the various emotions adolescents attach to the spaces they live and occupy was conducted in Udupi taluk, southern India. The researcher adopted photovoice and in-depth interviewed only one group of participants (Sekaran et al., 2020). For the research with a qualitative approach, using combining qualitative methods can provide comprehensive results and confirmedly verify emergent findings or themes (Adibelli & Korkmaz, 2022; Levin & Forward, 2021). In this research, applying photo-elicitation: (1) researcher-produced-photograph (RPP) and (2) participant-produced-photograph (PPP) were used in the elicitation interview as the interview medium with two groups of research participants; residents, and visitors.

3 Method

3.1 Triangulation in Qualitative Research

In qualitative research, the debate about reliability and validity has been continuously discussed. It focuses on how to ensure the rigor of the qualitative research process (Cypress, 2017). Triangulation, in qualitative research, is considered the use of manifold research methods or data resources to ensure a comprehensive understanding of an emergent phenomenon (Carter et al., 2014). There are four types of triangulation used in qualitative research, including (1) data triangulation, (2) methods triangulation, (3) investigator triangulation, and (4) theoretical triangulation (Denzin, 1978). Firstly, data triangulation is the use of various data sources, including differences in people, places, and times. With the use of different data sources, each data source can reciprocally compensate for the weaknesses of the other. Secondly, methods triangulation adopts multiple methods to study a particular situation or phenomenon, and this type can also reduce deficiencies and biases by using only one method. Thirdly, investigation triangulation uses more than one investigator, interviewer, observer, researcher, or data analyst, so this type can increase the ability to confirm research findings across different investigators, without their discussion and collaboration. Lastly, theory triangulation is the use of multiple theories to look at the situation or phenomenon with different perspectives, lenses, and sets of questions (Denzin, 1978; Patton, 1999). In this research, method triangulation was adopted, including (1) the on-site survey with researcher-produced-photograph (RPP), (2) participant-produced-photograph (PPP), and (3) in-depth elicitation interview with residents and visitors.

3.2 Photo-Elicitation (PE)

The qualitative research method that photographs are integrated into the inquiry process, Photo-elicitation, was initially published by John Collier in 1986 (Collier & Collier, 1986). Photo-elicitation adopts photographs as the medium that participants interpret their photographic media for the researcher during the interview (Loeffler, 2004; Turner et al., 2007). The photographic media is used to lead subsequent dialog and reflect the experiences of the participant (Torre & Murphy, 2015). For decades, photo-elicitation has been used widely in various research fields, such as psychology, anthropology, pedagogy, nursing, healthcare service, and education (Harper, 1998; Sensoy, 2011). Photographs can offer the possibility for the researcher to observe the research participants' life experiences, and they can create communicational links between strangers for a specific issue or subject. The advantage of integrating photographs into the interview process is that photographs provide an alternative representation to written and spoken language (Shaw, 2013). Using the photo-elicitation method in the interview process, compared with word-only-interview, was found that evokes more responses of individuals' in-depth thoughts, attitudes, and opinions from research participants during the interview. Photo-elicitation also allows the researcher obtains in-depth information with a lower number of participants than other methods (da Silva Vieira & Antunes, 2014). An example of individuals' determining the value of the urban forest was implemented in the city of Winnipeg (Sinclair & Larkin, 2009).

Two types of photo-elicitation widely adopted the inquiry approach with the photographic medium. The traditional photo-elicitation is the method in which the researcher takes the photographs used in the interview, the researcher-produced photograph (RPP). Another type, the auto-driven interview requires participants to take their photographs or asks participants to select photographs used for the interview. The traditional one, normally, is considered appropriate for the deductive research approach, but the auto-driven interview is useful for the inductive research approach (Clark, 1999; Frith & Harcourt, 2007). The auto-driven photo-elicitation, the participant-produced photographs (PPP), allows research participants to lead and teach the interviewer, with open expression and sharpens memory (Clark-IbáÑez, 2004; Hurworth, 2003; Taylor, 2002).

Photo-elicitation, especially participant-produced photographs, has been also adopted to study people's perceptions and experiences toward architecture, built-environment, landscape, and urban design and planning. In architecture, applying photo-elicitation with participant-produced photographs (PPP), Imamoglu (2000) studied architecture and non-architecture students' assessment of traditional and modern house facades in Turkish. Petermans et al. (2014) studied customers' experiences in a shoe and fashion retail environment in the center city of Belgium. The participants were asked to photograph inside or outside the environment during their store visit. In interior architecture, researchers attempted to explore customers' perceptions of the library spaces and to better understand customers' activities and behaviors within the library spaces during their use of the library (Haberl & Worthman, 2012). In another example, Adams et al. (2010) focused to elicit children's views about hospital environments at the Hospital for Sick Children (SickKids), Toronto. They explored how designers and patients understand the use of the hospital lobby, and their attempt was also to uncover the children's self-identifying and self-articulating place within contemporary pediatric hospitals. In urban design and planning, photo-elicitation interviews were used as a participatory tool to inform urban planning and decision-making processes on peoples' concerns (Bornioli et al., 2018).

Adopting photo-elicitation in the research of architecture, design, and other related fields can provide researchers with in-depth reflections on participants' though, experiences, and attitudes toward a specific phenomenon. There is a limitation to adopting photo-elicitation, especially the auto-driven one; participant-produced photograph (PPP), in everyday urban experiences and perception. In this research, with triangulation in the qualitative approach, photo-elicitation with researcher-produced-photograph (RPP) and participant-produced-photographs (PPP) were embraced as two methods in triangulation to collect individuals' place identity from the research participants; residents, and visitors, in Charoen Krung district. Implementing RPP in the Charoen Krung district context, the researcher went to the district, explored the context, and took photographs that represent the new identity of Charoen Krung district. Photographs taken in RPP and PPP are used in the in-depth interview with residents and visitors of Charoen Krung district.

3.3 Interpretative Phenomenological Analysis (IPA)

IPA mainly focuses on interpreting information, especially verbal information data, reflected from research participants' experiences or attitudes toward the particular phenomenon. Systematical interpretations with hierarchical evaluations are required to reveal essential data (Smith & Osborn, 2008). For IPA, abstruse meanings of a curtain situation can be retrieved through an understanding of people's making sense of their contexts (Smith et al., 2009). The connection, without interference, between the researcher's and participants' interview data, is a key success during the interpretative process. The two-stage interpretative process is used to expound on the emergent phenomenon; the researcher tries to make sense of people's making sense of their every day (Pringle et al., 2011). IPA process begins with a minimized preconception of what the researcher tries to explore, which can help the researcher avoid making a reckless decision. IPA allows the researcher to consider delicate data from case-by-case (Smith & Osborn, 2008). Data collection in IPA focuses on data that are likely to elicit research participants' detailed experiences reflecting the identity of Charoen Krung area. Purposive guiding questions were used to convey conversations to research participants. The initial set of questions was developed during the interview allowing possible issues and themes to emerge. The valued data, here, are unexpected stories, experiences, and concerns delivered by participants (Smith et al., 2009). Interview data from all participants are recorded in both audio and video formats to ensure that not only verbal data are collected but also participants' body language. In the first stage of IPA, reading and re-reading can help the researcher get familiar with emergent issues in all interview data. At this point, it also allows the researcher to revise previous interview questions and set up proper interview strategies for the coming interview activities. In the second stage, the free textual analysis is to explore interview content focusing on the use of language from all participants, and this can reveal the direction and connection among participants' stories. With IPA analytical processes, the researcher involves giving (1) descriptive comments on what participants responded to the phenomenon, (2) linguistics comments focusing on participants' expressive responses, (3) conceptual comments revealing abstract ideas within the interpretation; including participants' feelings, attitudes, and emotions (Smith et al., 2009). In summary, the key concept of IPA is to diminish exceeding details, keep the imperative complexity of emergent meanings, group similar patterns, and create sub-categories or themes, and this requires searching for reoccurrences among participants. The last stage of IPA is to categorize and build up super-ordinate themes, and it is considered the summative expression of the phenomenon (Smith et al., 2009).

3.4 Research Design

With the method triangulation approach, this research applied three qualitative methods to identify emergent findings and verify research results. Firstly, the aim of the on-site survey and researcher-produced-photograph (RPP) was to explore the imageability of Charoen Krung district with the researcher's perception and psychological reflection by taking sets of photographs in the studied context. The route of the on-site survey was selected according to the main roads and nodes in Charoen Krung district. One of the remarkable nodes in this district is Talad-Noi area, meaning "the small market" in English. Talad-Noi has been considered one of the areas with the highest density in both residential and commercial units. The surveying route started at Charoen Krung soi 30 to Talad-Noi area, on the north end of Charoen Krung road. The on-site survey was conducted during different periods in a day between January to April 2020. For example, in the first week of the month, the researcher went to the site on Monday, Wednesday, and Friday, and in the third week of the month, the researcher went to the site on Tuesday and Thursday. On the second weekend of the month, the researcher went to the site for the on-site survey on Saturday, and on the fourth weekend of the month, the researcher went to the site on Sunday. Visiting the site at different periods could help the researcher to increase the chances to discover possible emerging phenomena on the site. The researcher's sets of photographs were selected as the interview medium.

3.5 Fieldwork

For the interview, the researcher prepared sets of photographs for the in-depth interviews with research participants; (1) residents and (2) visitors. Research participants were asked to select available times for the interview at their convenience. Research participants were recruited via different approaches. The recruitment flyers were distributed on-site and posted on information boards, such as in coffee shops, restaurants, and bus stops. There were seventy-seven research participants, including both residents and visitors, who replied to be in the in-depth interview with researcher-produced-photograph (RPP). The interviews were proceeded by semi-structured with guiding questions. Researcher-produced photographs (RPP) were used as the medium of the interview. The interviewees were asked to reflect on their attitudes, feelings, and perceptions toward the new identity of Charoen Krung district.

In the second phase, there were fifty-four participants, including both residents and visitors, willing to proceed with participant-produced-photograph (PPP). Participants were asked to conduct PPP taking photographs reflecting their perceptions of the district's new identity and imaginability. Participants were allowed to use any type of camera, and most of them were using mobile-phone-camera. Participants were asked to select at least 15 photographs that mostly represent the new identity of the district. Participants' photographs were sent to the researcher via e-mail and prepared for the elicitation interview. Participants were also asked to complete a consent form allowing the researcher to use their photographs as an interview medium. The second phase of the research took nearly four months to be completed.

In the third phase, for both (1) researcher-produced photographs (RPP) and (2) participant-produced photographs, the series of elicitation interviews were conducted with semi-structured interview questions. The scope of questions used in the interview includes these three inquiries, (1) what is/are significant, remarkable, and distinguished aspect(s) of Charoen Krung district presented in the photograph, (2) among those aspects what is the best, and (3) why do you choose those aspect(s) as unique characters of the district. Interview conversations were recorded in audio format, and interview data were transcribed verbatim and analyzed with Interpretative Phenomenological Analysis (IPA). The interview transcriptions were placed in IPA analytical table and consequently interpreted into steps (Jonathan et al., 2009).

3.6 Data Analysis

The analytical process in IPA consists of several key steps. The first one is reading and re-reading. This step is conducted to ensure that participants are the focus of the analysis, and it allows the overall interview structure to develop while the researcher can gain an understanding of the connection of the whole interview section. The second step is the initial noting. In this step, the researcher examines semantic content and language at the exploratory interpreting level. The analytical process, here, is close to being a free textual analysis, with no restricted rules. The key focus in this step is to build comprehensive and detailed notes and comments on the original transcription. This step also includes different approaches, e.g., (1) descriptive comments; defining keywords, phrases, or explanations, (2) linguistic comments; paying focus on the way the content and meaning were presented in transcription, such as pronoun use, pauses, laughter, repetition, tone, and degree of fluency, (3) conceptual comments; dealing with the transcription data at a conceptual level. For the next step, developing emergent themes is one of the key steps in IPA. In this step, the researcher attempts to define the emergent themes by reducing the volume of details into keywords or phrases while maintaining complexity, connections, and patterns among the exploratory and initial notes. Searching for connections across emergent themes is another key step for IPA. This step involves the development of a diagram, charting, or mapping for connecting all emergent themes. At this level, the analyst is persuaded to explore and build up the organization of emergent themes. The representation of all three steps is required to analyze other cases. It is so important that the analyst should analyze other cases with its own approach. Finally, all emergent themes were categorized and brought to the highest level; super-ordinate themes. This step involves looking for patterns across cases. It also requires laying all tables and figures on a large surface and identifying them.

4 Findings

For more than one hundred years, Charoen Krung district has had its unique characteristics and sense of place. In 2018, Bangkok Metropolitan Authority (BMA) launched a plan to promote Charoen Krung district as the new creative urban. The main objective of this research is to explore the new place identity of Charoen Krung district through people's perceptions, lived experiences, and individual place recognitions. From IPA analytical process, all emergent themes were progressively interpreted at more abstract levels and succinctly named as super-ordinate themes. The process began with reading and re-reading interview transcripts to find verbal contents reflecting Charoen Krung place identity issues. All these contents were categorized and grouped in progressively abstract levels. All categories retrieved from all research participants. Thus, it is found that there are three super-ordinate themes in the identity of Charoen Krung district. The first one is varieties of integration. This super-ordinate theme represents people's sense of Charoen Krung district as the place where every variety of aspects has been integrated as one. For the second super-ordinate theme, the place for changes, people perceive all changes precisely, including culture, economical, social, and political changes happening in the area. Finally, the third super-ordinate theme, dynamic context, reveals the continuity of contextual changes in the area.

4.1 Varieties of Integration

Charoen Krung district has its intrinsic characteristics formed by complex layers of social, cultural, and political aspects. For decades, these aspects have been reflectively generated and accumulated by residents' lived experiences. These characteristics are gradually perceived by outsiders and become the unique identity of the place. The super-ordinate theme, "varieties of integration", were mentioned and revealed across cases. For Charoen Krung residents, when asked what things, issues, aspects, or phenomena represent ways of living in Charoen Krung, many participants, both residents and visitors, explicitly replies that mixed cultures and varieties of family lines lead their ways of living. Although Chinese culture can distinctively be perceived in this district, it has been blended with Thai living styles. For physical aspects and built environments, the integration of different architectural styles, Thai, Chinese, and Western styles, showed great harmony. The integration of various architectural styles was shown in everyday architecture, as follows:

...here, you can see the church is there, as my backyard view. My family has been living in this Chinese house for two generations. I've seen the roof through my windows since I was a kid. I often go to this church, especially when they have special events, Xmas and New Year...

...my friends are Christians and Muslims. I think this place is mixing us. We studied in secondary school together since the first grade...

...I always go to those Chinese grocery shops, and they also come here to order their lunch and dinner at my shop. They also love my menu...

4.2 Place for Changes

For decades, changes in Charoen Krung district have been influenced by tremendous factors, including cultural and social factors, local policy changes, and so forth. These changes have been willingly accepted by people, both insiders and outsiders. From the interview data and IPA analysis, there is no precise sign showing an individual's crucial conflicts among those changes, shown in Fig. 3. People's acceptance and sense of changes in this area can be perceived in the interview transcriptions shown in excerpts, as follows:

...for at least ten years, I've seen a lot of changes, especially physical environments. That pedestrian walkway was renovated at least five times. Lately, they placed new modern street furniture. I quite like it. The good thing is that we always have a new walkway...tidy and beautiful road...

...I've never seen the same view for a long time. I don't know why they have lots of renovations everywhere on this street. But, yeah we can live with it. New things always happen. We may face some difficulties during the construction period, but I'm ok after it has been done...

The senses of changes in this area can be seen across visitors' transcriptions, as follows:

...I love to visit Charoen Krung to see new events. Before Covid-19, new events continuously happened nearly every month. I love to see new things in this area...

Fig. 3 Example photographs showing individuals' crucial conflicts among place changing (photograph taken by author)

...It's so fun to see things around here be changed. You can see different cultural contexts just walking through all these small alleys. Turning left and right, your views keep changing, modern buildings located next to the Chinese temple, then look over there you will see the tip of the church's roof...

...I went here several times with my gang. This coffee shop keeps changing the decoration of the façade. Yeah, but I like it, so we always have a new corner for taking photos...

4.3 Dynamic Context

In the second super-ordinate theme, physical changes, physical aspects, and built environments may be easily perceived by people, both insiders and outsiders. The continuity of changes in all aspects is further interpreted on a more abstract level. The changes, here, have not been conceived as scattered fractions in the context but as the holistic phenomenon of changes. An example of continuity of change or dynamic context shown in Fig. 4. The 80-year-old warehouse was renovated to be the commercial units, including art galleries, shops, and café.

Excerpts from participants' elicitation interview transcriptions that reveal the sense of dynamic context, as follows:

...I think what is unique for this place is that they are all changes everywhere, from small pieces of pedestrian patterns, street furniture, graffiti on those walls, and so forth. I also would like to include people, and I can sense that there are so many groups of visitors coming here, especially before Covid-19. Three to four years ago, this area becomes in trend...

...people come and go. I can feel that new groups of visitors came here every weekend...

...I think everything around me here keeps changing. For example, many shops changed their business types to new trendy trading. In this area, only stable family businesses can be survived for years...

The three super-ordinate themes represent the current identity of Charoen Krung district through residents' and visitors' lived experiences within the context. The qualitative

Fig. 4 80-year-old warehouse was renovated to be art galleries and commercial units (photograph taken by author)

triangulation was applied as a methodological approach to verify the emergent identity of Charoen Krung.

5 Conclusion

In conclusion, Charoen Krung district is considered one of the oldest urban contexts and central business districts in Bangkok for several decades. This research aims to explore the identity of Charoen Krug district through people's lived experiences with the qualitative research triangulation approach. In this research, method triangulation was adopted, including (1) the on-site survey with researcher-produced-photograph (RPP), (2) participant-produced-photograph (PPP), and (3) in-depth elicitation interview with residents and visitors. For the first method of triangulation, the on-site survey and researcher-produced photograph were conducted to explore Charoen Krung's imageability via the researcher's perception and psychological reflection. Sets of researcher-produced photographs (RPP) were brought into the in-depth interviews with residents and visitors. There were seventy-seven participants willingly to be in the in-depth interview. They were asked to reflect their opinions on sets of photographs for what is/are the identity/identities of Charoen Krung. The interview data were collected and prepared for Interpretative Phenomenological Analysis (IPA). For the second method of triangulation, participant-produced-photograph (PPP), residents and visitors were asked to walk around Charoen Krung area and take photographs of what represents the new identity of the district. Each participant was asked to select at least 15 photographs used in the in-depth interview. Sets of participant-produced-photograph were sent to the researcher and prepared for the interview. The interview data from RPP and PPP methods were transcribed verbatim and analyzed with Interpretative Phenomenological Analysis (IPA). There are three superordinate themes as the result of IPA, (1) varieties of integration, (2) place for changes, and (3) dynamic context. Firstly, varieties of integration refer to characteristics and imageabilities of Charoen Krung in the way that different contexts, such as physical and social context, cultural context, diversity of people, and so forth, can be in harmony within the complexity of the district. Secondly, the place for changes implies the emergence of diversity in every dimension within the district, including cultural and social factors, local policy changes, and so forth. Lastly, dynamic context refers to the succession of changes happening in the Charoen Krung district. In this research, the triangulation used as researcher methodology is suggested to be a guidance to rechecking Royal Thai Government Project for creating or inserting a new identity in the current or old place.

References

Adams, A., Theodore, D., Goldenberg, E., McLaren, C., & Mckeever, P. (2010). Kids in the atrium: Comparing architectural intentions and children's experiences in a pediatric hospital lobby. *Social Science & Medicine, 70*(5), 658–667.

Adibelli, S., & Korkmaz, F. (2022). Pressure injury prevention practices of intensive care unit nurses in Turkey: A descriptive multiple-methods qualitative study. *Journal of Tissue Viability, 31*(2), 319–325. https://doi.org/10.1016/j.jtv.2022.02.001

Aydoghmish, F. M., & Rafieian, M. (2022). Developing a comprehensive conceptual framework for city branding based on urban planning theory: Meta-synthesis of the literature (1990–2020). *Cities, 128*, 103731. https://doi.org/10.1016/j.cities.2022.103731

Bornioli, A., Parkhurst, G., & Morgan, P. L. (2018). The psychological wellbeing benefits of place engagement during walking in urban environments: A qualitative photo-elicitation study. *Health & Place, 53*, 228–236. https://doi.org/10.1016/j.healthplace.2018.08.018

Carter, N., Bryant-Lukosius, D., DiCenso, A., Blythe, J., & Neville, A. J. (2014). The use of triangulation in qualitative research. *Oncol Nurs Forum, 41*(5), 545–547. https://doi.org/10.1188/14.Onf.545-547

Clark, C. D. (1999). The autodriven interview: A photographic viewfinder into children's experience. *Visual Sociology, 14*(1), 39–50. https://doi.org/10.1080/14725869908583801

Clark-IbáÑez, M. (2004). Framing the social world with photo-elicitation interviews. *American Behavioral Scientist, 47*(12), 1507–1527. https://doi.org/10.1177/0002764204266236

Collier, J., & Collier, M. (1986). *Visual anthropology: Photography as a research method*. University of New Mexico Press.

Cypress, B. S. (2017). Rigour or reliability and validity in qualitative research: Perspectives, strategies, reconceptualization, and recommendations. *Dimensions of Critical Care Nursing, 36*(4), 253–263. https://doi.org/10.1097/dcc.0000000000000253

da Silva Vieira, R., & Antunes, P. (2014). Using photo-surveys to inform participatory urban planning processes: Lessons from practice. *Land Use Policy, 38*, 497–508. https://doi.org/10.1016/j.landusepol.2013.12.012

Denzin, N. K. (1978). *Sociological methods: A sourcebook*. McGraw-Hill.

Devine-Wright, P. (2013). Think global, act local? The relevance of place attachments and place identities in a climate changed world. *Global Environmental Change, 23*(1), 61–69. https://doi.org/10.1016/j.gloenvcha.2012.08.003

Falconer, J. (2013). Photovoice participatory-action research design and adaptations for adverse fieldwork conditions. In J. Falconer (Ed.), *SAGE research methods cases*. SAGE Publications.

Félonneau, M.-L. (2004). Love and loathing of the city: Urbanophilia and urbanophobia, topological identity and perceived incivilities. *Journal of Environmental Psychology, 24*(1), 43–52. https://doi.org/10.1016/S0272-4944(03)00049-5

Frith, H., & Harcourt, D. (2007). Using Photographs to capture women's experiences of chemotherapy: reflecting on the method. *Qualitative Health Research, 17*(10), 1340–1350. https://doi.org/10.1177/1049732307308949

Ginting, N., & Wahid, J. (2015). Exploring identity's aspect of continuity of urban heritage tourism. *Procedia—Social and Behavioral Sciences, 202*, 234–241. https://doi.org/10.1016/j.sbspro.2015.08.227

Google Maps. (n.d.). *Bangkok Map, showing Charoen Krung Road*. Retrieved January 5, 2021 from https://goo.gl/maps/MBaHNvUDK1hv8HD87

Gross, M., & Hochberg, N. (2016). Characteristics of place identity as part of professional identity development among pre-service teachers. *Cultural Studies of Science Education, 11*(4), 1243–1268. https://doi.org/10.1007/s11422-014-9646-4

Haberl, V., & Worthman, B. (2012). Getting the picture: interviews and photo elicitation at Edmonton Public Library. *Library and Information Science Research E-Journal, 22*(2), 1–20. https://doi.org/10.32655/LIBRES.2012.2.1

Hannay, J., Dudley, R., Milan, S., & Leibovitz, P. K. (2013). Combining photovoice and focus groups: Engaging Latina teens in community assessment. *American Journal of Preventive Medicine, 44*(3, Supplement 3), S215-S224. https://doi.org/10.1016/j.amepre.2012.11.011

Harper, D. (1998). *An Argument for visual sociology*. Routledge.

Hurworth, R. (2003). Photo-interviewing for research. *Social Research Update, 40*(Spring).

Irwin, A. (1975). *The environment and social behavior.* Cole Publishing.

Imamoglu, Ç. (2000). Complexity, liking and familiarity: Architecture and non-architecture Turkish students' assessments of traditional and modern house facades. *Journal of Environmental Psychology, 20*(1), 5–16. https://doi.org/10.1006/jevp.1999.0155

Jonathan, S., Paul, F., & Michael, L. (2009). *Interpretative phenomenological analysis: Theory, method and research.* SAGE publication Ltd.

Jones, K. T. (1998). Scale as epistemology. *Political Geography, 17*(1), 25–28. https://doi.org/10.1016/S0962-6298(97)00049-8

Levin, L., & Forward, S. (2021). Explaining data analysis using qualitative methods. In R. Vickerman (Ed.), *International encyclopaedia of transportation* (pp. 107–112). Elsevier.

Loeffler, T. A. (2004). A photo elicitation study of the meanings of outdoor adventure experiences. *Journal of Leisure Research, 36*(4), 536–556. https://doi.org/10.1080/00222216.2004.11950035

Macdonald, D., Peacock, K., Dew, A., Fisher, K. R., & Boydell, K. M. (2022). Photovoice as a platform for empowerment of women with disability. *SSM—Qualitative Research in Health, 2*, 100052. https://doi.org/10.1016/j.ssmqr.2022.100052

Oberlin, K. C., & Gieryn, T. F. (2015). Place and culture-making: Geographic clumping in the emergence of artistic schools. *Poetics, 50*, 20–43. https://doi.org/10.1016/j.poetic.2015.01.001

Ösgård, A., & Spierings, B. (2021). 'It just didn't really happen': The lived space of entrepreneurial urbanism in Ørestad, Copenhagen. *Geoforum, 123*, 117–128. https://doi.org/10.1016/j.geoforum.2021.05.004

Patton, M. Q. (1999). Enhancing the quality and credibility of qualitative analysis. *Health Services Research, 34*(5 Pt 2), 1189–1208.

Petermans, A., Kent, A., & Van Cleempoel, K. (2014). Photo-elicitation: Using photographs to read retail interiors through consumers' eyes. *Journal of Business Research, 67*(11), 2243–2249. https://doi.org/10.1016/j.jbusres.2014.06.012

Pringle, J., Drummond, J., McLafferty, E., & Hendry, C. (2011). Interpretative phenomenological analysis: A discussion and critique. *Nurse Researcher, 18*(3), 20–24. https://doi.org/10.7748/nr2011.04.18.3.20.c8459

Proshansky, H. M., Fabian, A. K., & Kaminoff, R. (1983). Place-identity: Physical world socialization of the self. *Journal of Environmental Psychology, 3*(1), 57–83. https://doi.org/10.1016/S0272-4944(83)80021-8

Reicher, S., Hopkins, N., & Harrison, K. (2006). Social identity and spatial behaviour: The relationship between national category salience, the sense of home, and labour mobility across national boundaries. *Political Psychology, 27*(2), 247–263. https://doi.org/10.1111/j.1467-9221.2006.00005.x

Ruggeri, D. (2014). The 'my mission Viejo' project. Investigating the potential of photovoice methods in place identity and attachment research. *Journal of Urban Design, 19*(1), 119–139. https://doi.org/10.1080/13574809.2013.854693

Sekaran, V. C., Bailey, A., Kamath, V. G., Ashok, L., & Kamath, A. (2020). 'This is the place where I can be alone, no tension:' Photovoice evidence for adolescent perceptions of their microsystem and psychological adjustment. *Asian Journal of Psychiatry, 51*, 102021. https://doi.org/10.1016/j.ajp.2020.102021

Sensoy, Ö. (2011). Picturing oppression: Seventh graders' photo essays on racism, classism, and sexism. *International Journal of Qualitative Studies in Education, 24*(3), 323–342. https://doi.org/10.1080/09518398.2011.561817

Shaw, D. (2013). A new look at an old research method: Photo-elicitation. *TESOL Journal, 4*(4), 785–799. https://doi.org/10.1002/tesj.108

Sinclair, A. J. F., & Larkin, M. (2009). Elicitation of urban forest values from residents of Winnipeg, Canada. *Canadian Journal of Forest Research, 44*, 922–930.

Smith, Flowers, & Larkin. (2009). *Interpretative phenomenological analysis*. SAGE.

Smith, & Osborn, M. (2008). *Interpretative phenomenological analysis*. SAGE.

Southworth, M. F., & Ruggeri, D. (2011). *Beyond placelessness: place identity and the global city* (A. L. S. Tridib Banerjee, Ed. 1st ed.). Routledge.

Taylor, E. W. (2002). Using still photography in making meaning of adult educators' teaching beliefs. *Studies in the Education of Adults, 34*(2), 123–139. https://doi.org/10.1080/02660830.2002.11661466

Torre, D., & Murphy, J. (2015). A different lens: Changing perspectives using Photo-Elicitation Interviews. *Education Policy Analysis Archives, 23*(111). https://doi.org/10.14507/epaa.v23.2051

Torres, M. E., Meetze, E. G., & Smithwick-Leone, J. (2013). Latina voices in childhood obesity: A pilot study using photovoice in South Carolina. *American Journal of Preventive Medicine, 44*(3, Supplement 3), S225-S231. https://doi.org/10.1016/j.amepre.2012.11.020

Turner, M., Tamborini, R., & Limon, M. S. (2007). The moderators and mediators of door-in-face requests: Is it a negotiation or helping experience? *Communication Monographs, 74*(3), 333–356.

van Eijck, M., & Roth, W.-M. (2010). Towards a chronotopic theory of "place" in place-based education. *Cultural Studies of Science Education, 5*(4), 869–898. https://doi.org/10.1007/s11422-010-9278-2

Wang, C., & Burris, M. A. (1997). Photovoice: Concept, methodology, and use for participatory needs assessment. *Health Education & Behavior, 24*(3), 369–387. https://doi.org/10.1177/109019819702400309

Westerholt, R., Acedo, A., & Naranjo-Zolotov, M. (2022). Exploring sense of place in relation to urban facilities—Evidence from Lisbon. *Cities, 127*, 103750. https://doi.org/10.1016/j.cities.2022.103750

Architectural Identity of Benghazi City Between Tradition and Modernity Case Study–Urban Centre of Benghazi

Aziza A. Safour and Eman M. Elmazek

Abstract

Architectural character or identity is defined as a set of aesthetic characteristics and values that are expressed by a building and reflect its identity or that which distinguishes it. Urban character is a set of complex attributes within the character of the area and includes physical and non-physical components. Hence, architectural and urban character are considered as a reflection of community cultural identity, while place has a distinctive role in their formation. In Libya, there is a variety of architectural and urban compositions, and their characteristics differ according to place and time. Also, the elements of these formations interact with the nature of the place and its social and economic characteristics. In Benghazi, while there is variety in the architectural and urban heritage, its modern architecture and urban character are the outcome of global architecture, which has contributed to a loss of the city's identity. This research aims to review various views from architects and urban planners regarding architectural and urban character and examines the suitability of this for the local environment. It also reviews opinions on the modern character and new trends for architecture in Benghazi. The study uses a descriptive approach, analysing the opinions of a number of academics, architects and planners. The study outcomes point to the suitability of the inherited architectural character within the environment of Benghazi, while finding a lack of homogeneity between the modern architecture and the local environment, and that the modern architecture has failed to achieve sustainability.

Keywords

Architectural character · Urban character · Architectural heritage · Modernity

1 Introduction

There is no doubt that each city has a special character which distinguishes it from other cities, and architectural and urban heritage is one of the most important aspects of the cultural heritage that every nation is proud of. Torabi and Brahman (2013) suggest that "*an architecture that has its own identity acts like an identity certificate for its homeland and reveals the thoughts of its people*". Besides, Nooraddin (2012) states that "*the local architectural identity of any society is an important life container which reflects among others its cultural values and meanings that evolve over time*". Architectural character or identity is defined as a set of aesthetic characteristics and values that are expressed by a building and reflect the identity that distinguishes it, while urban character is a set of complex attributes in the character of the area and includes the physical and non-physical components. Hence, architectural and urban characters are considered as a reflection of the community cultural identity and place has a distinctive role in their formation.

Mahmoud (2016) explains that in Libyan cities, there is a strong influence from the geography of place on the architecture and urban patterns. He also identifies a clear footprint of place characteristics for each geographical region. Besides this, there is a diversity of architectural and urban compositions and styles whose characteristics differ with place and time, and the elements of these compositions interact with the nature of the place and its social and economic characteristics. This variety of compositions form a significant architectural and urban heritage based on the different regions of Libya and reflecting the specificity of each, whether coastal, mountainous or desert. Milod (2019),

A. A. Safour (✉) · E. M. Elmazek
Architecture and Urban Planning Department,
University of Benghazi, Benghazi, Libya
e-mail: Aziza.safour@uob.edu.ly

M. F. Shahidan et al. (eds.), *Urban Narratives: Exploring Identity, Heritage, and Sustainable Development in Cities*,
Advances in Science, Technology & Innovation, https://doi.org/10.1007/978-3-031-48517-6_5

in his study concerning vernacular architecture in Libya, confirms that "*Libya has experienced different historic stages, such as the Amazigh, Phoenicians, Greeks, Romans, Muslims, and Italian colonization. It is distinguished by a historic continuity, which has enriched its historic and architectural features*". As well, in order to follow the diversity of the architectural and urban character in Libya, it can be stated that the historical areas in the city centres are the most real expression of the architectural identity and urban character. The urban patterns of the coastal Libyan cities are characterised by their compact fabric and the main streets are perpendicular to the seashore. Figure 1a, b shows some examples of coastal Libyan architectural compositions which reflect the local features of the cities of Benghazi and Tripoli. Furthermore, the organic plan of the coastal cities is characterised by the clearness of the commercial streets, which extend from west to east, and this organic fabric is expressed by courtyards and squares.

However, the urban and architectural composition of the desert areas provides a wonderful model for compatibility with environmental conditions, in terms of the design of the buildings individually and through their relationship with other buildings and with the urban fabric as well. The philosophy of formation is commensurate with the social, urban and natural environment of the region and Fig. 2a, b illustrates some examples of the desert cities' compositions, where, Fig. 2a showing sample of the covered street in Gadamis old city, while Fig. 2b presenting example of the conical dome architecture in Ojalah.

Another significant example is the composition of mountain architecture, which is called passive architecture, and it is characterised by the drilling of houses into the mountains. This was done as a result of the climate's impacts and treatments, as well as a means of defence. In this style, all functional spaces are carved around a courtyard and accessed through a tunnel connecting the courtyard to the outside. Figure 3a–c provides an example of the passive spaces and houses in the city of Gerian, in Libya.

This research aims to explore the various views of architects and urban planners, reviewing their opinions regarding the architectural and urban character and examining the suitability of this character for the local environment. It also examines opinions regarding the modern character and new trends in architecture in Benghazi. Also, the research seeks to review the main characteristics of the architectural and urban composition of Benghazi old town, as one of the main coastal cities and most well-known Mediterranean Sea cities, and to enhance knowledge with reference to the traditional architecture in Benghazi by defining its main features and highlighting its characteristics. Furthermore, the study explores new directions and trends in modern architectural and urban character in the historical area of Benghazi, which is considered to be a product of the architecture of globalisation.

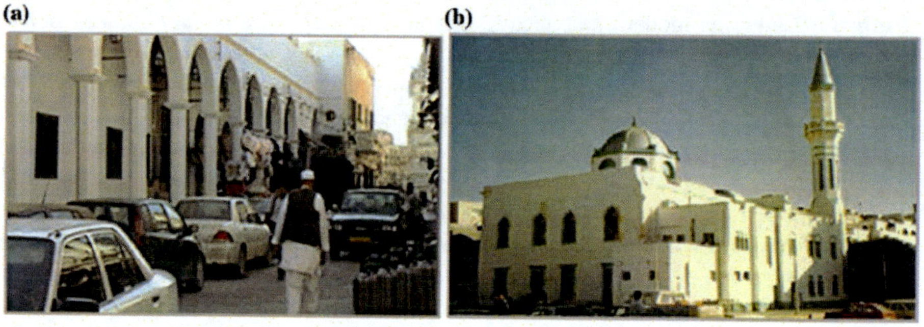

Fig. 1 **a** Commercial street in Tripoli (Safour & Elmazek, 2018). **b** Old Mosque in Benghazi (Safourand & Elmazek, 2018)

Fig. 2 **a** Public square in Gadamis (Safour, 2013). **b** Old Mosque in Ojalah (Safour, 2013)

(a) (b) (c)

Fig. 3 **a** Patio of drill house (independentarabia, 2022). **b** entrance of the drill house (independentarabia, 2022). **c** living room of the drill house (independentarabia, 2022)

2 Architectural Identity

Torabi and Brahman (2013) explain that identity is the ability to distinguish and identify one element from another, while Ettehad et al. (2014) state that "*architecture is scientific-artistic activities to create space and organize it; crystallisation of the culture of a community over time; and clear mirror of the society in different periods*". As well as this, Torabi and Brahman (2013) report that "*an architecture that has its own identity acts like an identity certificate for its homeland and reveals the thoughts of its people. All major and lasting works of architecture have been formed based on ideas of people living in that historical period. In fact, the architectural identity is formed based on orderly thoughts and organized action. This harmony between thought and action forms architecture with identity*". Consequently, architectural identity is an expressive characteristic of an entity or a group of attributes which reflects the essential reality of the architectural output, emanating from itself, from spirituality and the roots of its formation. There are three architectural features of identity: authentic, because of its connection to place; familiar and understandable, due to its association with prevailing customs; and the identity of a building which makes it distinct from others (Elsharif, 2019). Architectural character or identity is defined as a set of aesthetic characteristics and values that are expressed by a building and reflect its identity and what differentiates it. Urban character on the other hand is a set of complex attributes contributing to the character of the area and includes physical and non-physical components. Figure 4a–c illustrates the urban spaces and architectural landmarks of Benghazi which play an essential role in creating the image of the city and its history.

Moreover, architectural character has time and spatial dimensions, and both of these have an echo in the art of architecture. The time dimension is represented in the impact of contemporary changes in the human, social and cultural life of society. As for the spatial dimension, this is represented in the land and the conditions of each region that impose their presence on the architectural work.

Furthermore, attention is given to strengthening local urban identity as a matter of great importance. This interest is due to the many positive aspects of such work, starting with strengthening community identity and social cohesion among the population, raising the sense of belonging to a place and being connected to it, and leading to achieving economic positives resulting from the tourism and recreational activities that flourish in traditional places (Iballouz, 2020).

(a) (b) (c)

Fig. 4 **a** Government palace in Benghazi (Manzoni, 2009). **b** old fish market in Benghazi (Almarsad, 2022). **c** old parliament in Benghazi (Manzoni, 2009)

2.1 Determinants of Architectural Character and Urban Character

Lynch) 1960) in his study about the image of the city stated that the idea of the city's image can be explained by the perception of people for the built environment elements and the way that they adapt and interact with the city by creating mental maps based on five elements; paths, edges, districts, nodes, and landmark and Fig. 5a, b shows samples of landmarks, nodes and paths that involved in creating the mental map of Benghazi city.

Architectural character or architectural style can be defined as a natural result of several common and interacting factors. Some of these are melted into the crucible of the full use of the building, the construction methods and construction materials, the nature of the regions or the region and then its traditions and customs, in addition to social, economic, cultural and spiritual factors and the level of wealth. The architectural character of the building can be determined by its features, represented in the scale of the building, its proportions, the width of the building, the ratio of solid to open, the distribution of vertical and horizontal lines, density and quality of details, colours, building materials and texture. Figure 6a–c presents clear pictures of architectural details for squares and buildings from Benghazi.

However, urban character is determined by the urban formation, urban fabric, urban spaces, skyline, visual axes, roads, corridors and intersections. Mansour (2015) states that *"the main components of the old Arab-Islamic cities are the grand mosque which played the role of the heart in the city, as it was the religious, scientific, and social centre, and all the other public buildings were located around it. Markets: were typically located around the grand mosque, each one having its own kind of merchandise, and distributed very carefully in terms of place, size, activity, and craft"*.

Natthakit and Heath (2021), in their study of the urban heritage identity of Chiang Mai Old City, Thailand, argue that the distinctiveness and appearance of a place can be considered as essential aspects in shaping the identity of that place, and these aspects are what allow people to distinguish one place from another, essentially, when they are characterising urban heritage. Also, the authors state that the physical environment and the variety of human activities or events and cultures constitute the identity of a place, and through these elements, people define the identity of an urban environment. They conclude that urban identity is a vital issue, based on the understanding of spatial theories and the relation between socio-environmental values and the essence of place.

(a) (b)

Fig. 5 **a** Tree square in Benghazi (libyaakhbar, 2016). **b** Minaret of Benghazi (Iamviolet, 2017)

(a) (b) (c)

Fig. 6 **a** Landmark buildings in Benghazi CBD (Libyaalmostakbal, 2023). **b** one of the CBD squares (Manzoni, 2009). **c** municipal square of Benghazi (Libyaalmostakbal, 2022)

The architecture in Benghazi has gone through many historical stages, from its origins to the present. The history of the city reveals that the first appearance of Benghazi was during the Greek period, between 525 and 515 BC, as Hesperides, which means Golden Apple Gardens. After the Greek times, the city came under Roman rule as Berenice. At the end of the seventh century, the city was renamed during Islamic rule as Berniq. From 1450, the city was called the Marsa ibn Ghazi. In 1579, during the Ottoman era, a castle was built near the sea. Later, during the Italian occupation that started in 1911, the city expanded and developed to become an administrative and economic centre. As architecture always follows certain known orders, the architecture of Benghazi reflected most of these stages of political, economic, and cultural existence. The next Fig. 7a–c explains the development of Benghazi, illustrating the city at different times, from the ancient period, to the master plan set out in 1930, during Italian rule (Safour, 2013).

The new plans developed for Benghazi during the 60s, 80s and in 2009 started with the so-called first generation plan in 1968. Twelve years later, the government commissioned four consulting firms, including Doxiadis, to prepare a set of general and comprehensive plans to regulate urban growth and development, and this was called the second generation plan, covering the time from 1980 to 2000 (DOX-LIB, 1989). The third generation plan was to cover the period 2005–2030, and started in 2005: however, this third master plan was never implemented because it had no detailed implementation plans. Currently, the master plan of Benghazi covers an area of around 31,000 hectares, and the gross residential area is about 7900 ha. The gross residential density of Benghazi is about 100 people per hectare (UPA, 2009). Figure 8a–c shows the land-use plans for Benghazi, which include the distribution of different uses during the three consecutive generation plans. According to the General Authority for Information in 2012, the population of Benghazi was 700,000 inhabitants, with a growth rate of 2.117% (BSS, 2012). The city council of Benghazi (COB) has now started to rehabilitate the destroyed areas affected by the conflict, and the strategy of the COB is to focus on the rehabilitation of all sectors of the city with a comprehensive plan for 20 years ahead.

2.2 Identity of the Historical Centre of Benghazi

The historical centre of Benghazi is one of the most sustainable areas of the city, and Fig. 9 illustrates the mix of architectural compositions showing different historical stages. Also, the centre achieves a large number of sustainable environment foundations, such as: its sustainable urban composition in terms of compact pattern and high density; its mixed social composition; and vital life through the diversity of residents. Moreover, the multiple occupation of space (mixed use of land), in which apartment housing and service offices are located on the top of local shops, revives street life, reduces residents' need for various means of transport and encourages walking. Figure 9a–c shows the diversity of architectural compositions, with different historical stages; the local, the Ottoman and the Italian style.

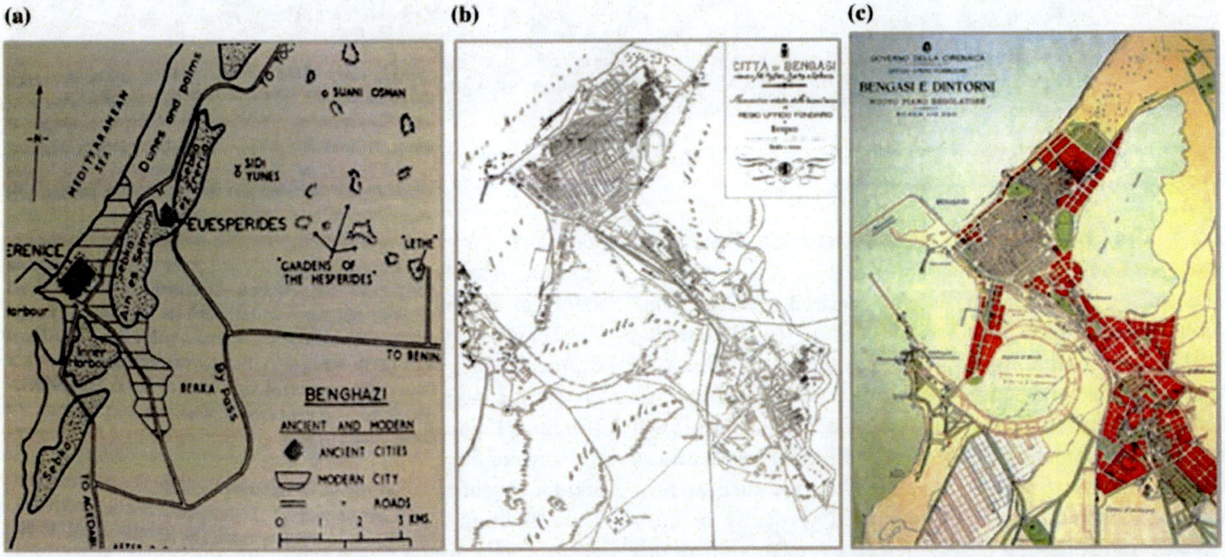

Fig. 7 **a** Benghazi, Ancient and modern (Bulugma, 1972). **b** Benghazi, 1927, City survey (Manzoni, 2009). **c** Benghazi new master plan, 1930 (Manzoni, 2009)

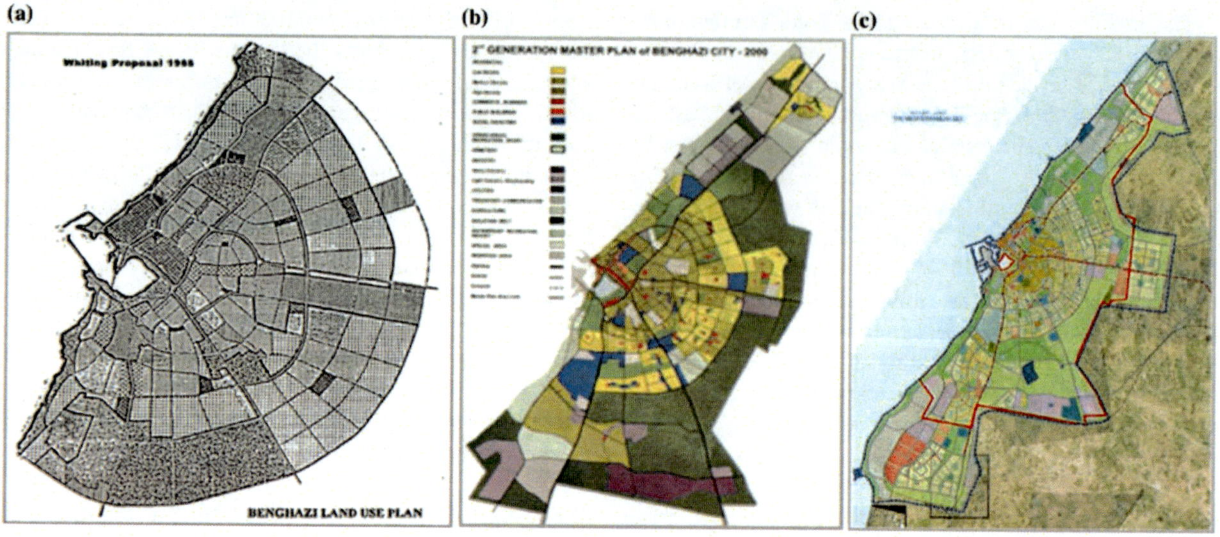

Fig. 8 **a** 1st gen. plan 1968–1988 (UPA, 2009). **b** 2nd gen. plan 1980–2000 (DOX-LIB, 1989). **c** 3rd generation 2005–2025 (UPA, 2009)

Fig. 9 **a** Local style (Safour, 2013). **b** Ottoman style (Safour, 2013). **c** Italian style (HPCU, 2010)

2.2.1 The Characteristics of the Historical Centre of Benghazi

The old town of Benghazi is characterised by lively souk activity and mixed land-use. According to UN-Habitat (2014), mixed land-use occurs where there is a variety of land-uses in a neighbourhood or urban area, including residential, commercial, industrial uses, etc. In other words, it is a mix between commercial uses, such as for retail, offices and entertainment, and non-commercial uses such as residential. This mix could be compacted vertically or horizontally. In addition, it enables relocation of more than one use in an integrated way to support sustainable forms of transport and gives priority to walking and biking. Mixed land-use refreshes the economy and increases the sense of safety in any neighbourhood by increasing the number of passers-by in public areas. The old town of Benghazi had different house types: the typical Arab house with a small *houch*; the large patio *houch*; and the mid-rise multi-storey building. Figure 10a–c describes the organic fabric of the old town of Benghazi, the main commercial axis of the *souk* and the layout of the patio *houch*.

In addition, in the historical area of Benghazi, workplaces can be found as a room inside the house, at a nearby shop, or at the nearest square. This was the lifestyle mostly followed and which succeeded in the old traditional communities. This pattern emerged for many reasons, one of the most important of which is the lack of shopping points or lack of a satisfying quality or typology for shops. Another apparent reason is the absence of urban regulations, which left enough space for spontaneity in land-use to address daily needs. In fact, most neighbourhoods in Benghazi old town stand as a good example of mixed land-use, and commercial and residential uses are integrated within the same street in a vertical way most of the time, although the lack of effective public transport affects the traffic and space for street parking needs. This should be considered as a model of mixed land-use that fits into the social and economic local environment. Figure 11a, b confirms the beautiful mix between different uses and styles.

Moreover, social mixing is an effective tool to avoid social exclusion and is one of the main principles of a sustainable neighbourhood. Therefore, it is a tool to achieve social integration and solidarity. A social mix can be achieved by interspersing different social levels in the same neighbourhood to ensure that all can access equal urban opportunities. This happens through a comprehensive neighbourhood that includes all types of housing. The social mix guarantees a healthy social network that influences the whole city in general and this factor and mixed building use support each other. In the old town of Benghazi, Alsherif Street stands as a good example of social mix (Safour & Elmazek, 2018). As Fig. 12a–c shows different social levels are mixed together due to the diversity of housing and services.

2.3 The Advantages of Preserving Architectural and Urban Character in Benghazi

Safour (2013) reports that there are a number of benefits to be achieved from preserving the architectural and urban character of historical areas, as this:

preserves the continuity of architectural and urban character of the historical areas;

protects the distinctive visual character of the areas and protects their value;

preserves the continuity of elements of heritage, and inherited values;

avoids inconsistencies in the relationship between architectural heritage and newly planned areas;

and prevents visual pollution and conflict between buildings.

It should be mentioned that the architecture and urbanism during the Italian period dealt with the local architectural heritage in a way that took into account homogeneity, integration and continuity. Also, this architectural output was in line with existing formations in a way that achieved the desired beauty. Figure 13a–c demonstrates the beauty of the

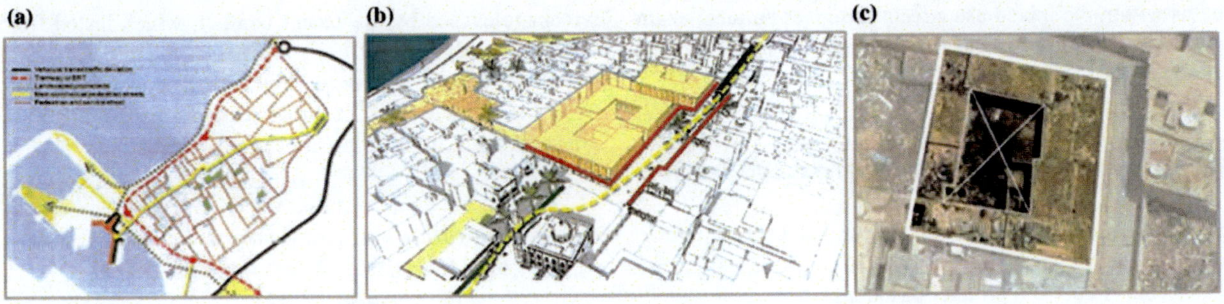

Fig. 10 **a** Organic fabric of Benghazi (HPCU, 2010). **b** commercial axis (HPCU, 2010). **c** patio *houch* (HPCU, 2010)

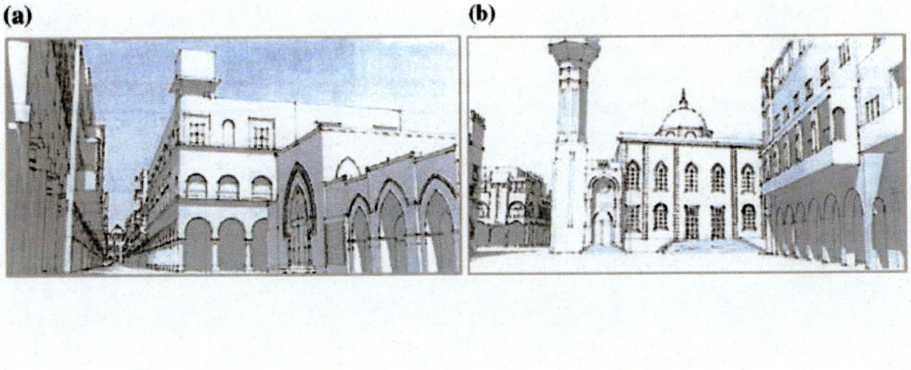

Fig. 11 **a** Mixed use between houses, shops and offices (HPCU, 2010). **b** harmony between different styles (HPCU, 2010)

Fig. 12 **a** Mid-rise multi-storey building (HPCU, 2010). **b** small patio *houch* (HPCU, 2010) **c** large patio *houch* (HPCU, 2010)

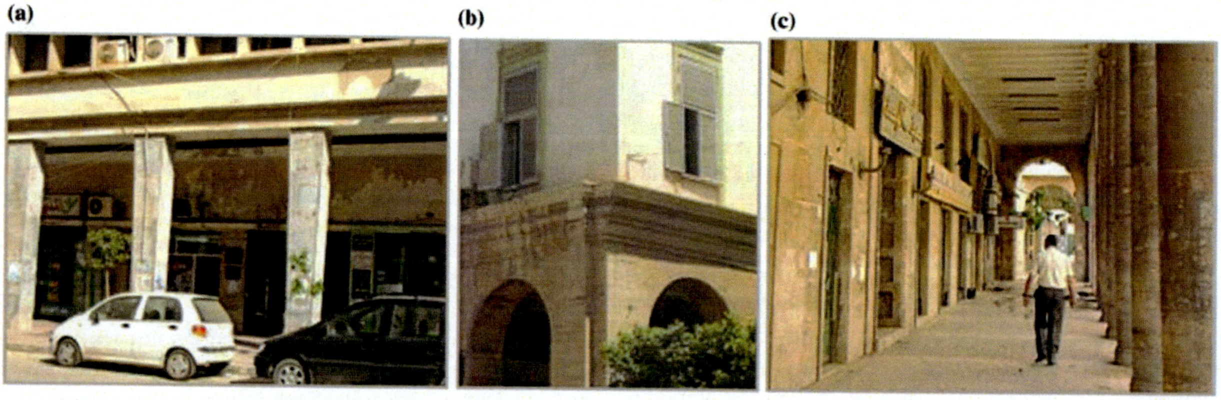

Fig. 13 **a** Mixed use with covered lane (HPCU, 2010). **b** corner of arched corridor (HPCU, 2010). **c** long arched corridor (HPCU, 2010)

baths within the mixed-use streets and gives an ideal example concerning the sustainable urban pattern in the old area of Benghazi.

3 Modern Trends in Benghazi Old Town

Despite the diversity of the architectural and urban heritage in Libya, today, the modern architectural and urban character is considered as a product of the architecture of globalisation, and can be found in all Libyan regions, whether on the coast, or in the desert or mountains. This architecture of globalisation has contributed to a loss of identity and of the distinctive architectural and urban character of the place. For instance, the old town of Benghazi was atypical Arab centre, which, due to the superficiality of Ottoman regime, remained almost unaffected by the Ottoman's architecture style. Later, the old part of the city was expanded vertically because of the limited ground space, and this expansion led to buildings seven storeys high, which became a common feature of most of the city's streets (Bulugma, 1972). Currentlys, Benghazi is suffering from a loss of identity as a result of the architecture of globalisation and the wreckage caused by the recent conflict, which lasted from 2014 until July 2017. During that time, many parts of the city were largely destroyed. Figure 14a–c shows the new trends in architecture and urban patterns in the historical area of Benghazi for different uses of building.

4 The Views of Academics and Consultants on Architectural Identity in Benghazi

This research has reviewed various views of architects and urban planners who participated in a focus group workshop taking place in April 2021 at the University of Benghazi. The study examined their opinions concerning the architectural and urban character of the city of Benghazi, and reviewed its suitability for the local environment. Their opinions were also sought regarding the modern character and trends of architecture in the city. The methodology used followed the descriptive approach, analysing the opinions of a number of academics from the Architecture Department at the University of Benghazi, the Architects of the Old Cities administration and planners from Benghazi Urban Planning Authority.

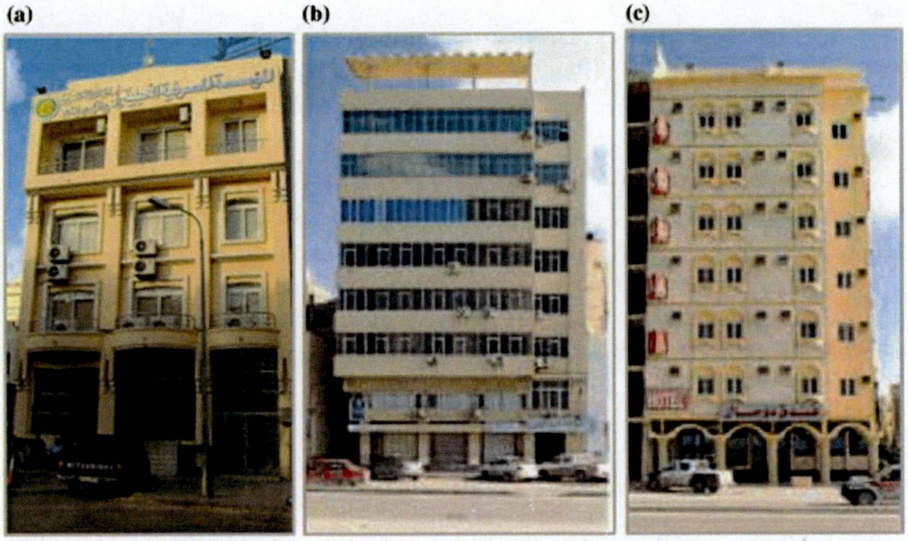

Fig. 14 **a** Mixed-use building (HPCU, 2010). **b** administrative building (HPCU, 2010). **c** waterfront Hotel (HPCU, 2010)

The participants state that the inherited architectural character shows that it is suited to the environment of Benghazi, while there is a lack of homogeneity between the modern architecture and the local environment, and moreover, that the modern architecture has failed to achieve sustainability. The participants argue that authorities and academics should pay greater attention to the preparation of standards to preserve the identity of architecture and urbanism. Also, they suggest that new constructions should have to respect the old alignment of the wall with the street, the ground surface, and respect the authorised maximal height. Also, volumes must be simple and cubic, with the use of inner courtyards to create cross-ventilation and bring about the necessary open spaces for sunlight access. Moreover, new façades should have to respect the principle of vertical and horizontal alignment of openings in the old building style.

Besides these points, the participants confirm that there are number of architectural details related to the local heritage which need to be considered such as: the simple semi-circular arch; the key of the arch; the simple form of the openings and doors; and also, the height of the buildings and the unity and continuity of the street elevations. Moreover, the organic fabric of the historical area needs to be considered in terms of sustainable solutions. In addition, the human scale and human baths should be taken in account. The simple style of public buildings such as mosques also requires to be regarded as an approach for new buildings. In addition, the participants suggested that for new extensions of Benghazi, a framework or guideline needs to be established that can be used to organise the existing buildings and maintain the overall style of the existing and new areas.

Another significant finding is that the participants are suggest that future studies should pay greater attention to the techniques used by the Italian architects and planners who worked on the development of Benghazi during the Italian occupation, in which the planners and designers succeeded in the integration of the old local part of Benghazi and the new Italian part, such as in the link between the city hall square, the long arched corridor (Almokhtar Street) and the old souk. The most significant landmarks of the Italian development are the city hall of Benghazi and the twin-domed cathedral, which are considered as the town's most prominent Italian-era features, as well as the public theatre. On the other hand, some of the participants indicated that the Italians had failed to change and hide the local town's features, with narrow and tortuous streets, single-storey buildings and an absence of open spaces. Figure 15a–c shows some of the landmarks of the Italian part of Benghazi.

5 Conclusions

Through a discussion of architectural and urban identity and its relationship to place, it was concluded that architectural identity is a part of the identity of the community, and expresses its culture and image. As a result of the prevalence of different modern trends, and the domination of globalisation architecture over local architecture, disharmony with the local environment has arisen. Also, Mansour (2015) confirms that "*it is clear that the Islamic Arab city lacks characteristics under these different physical, social, cultural and political pressures, which makes it in an urgent need to stick to its identity, not only in the form and image,*

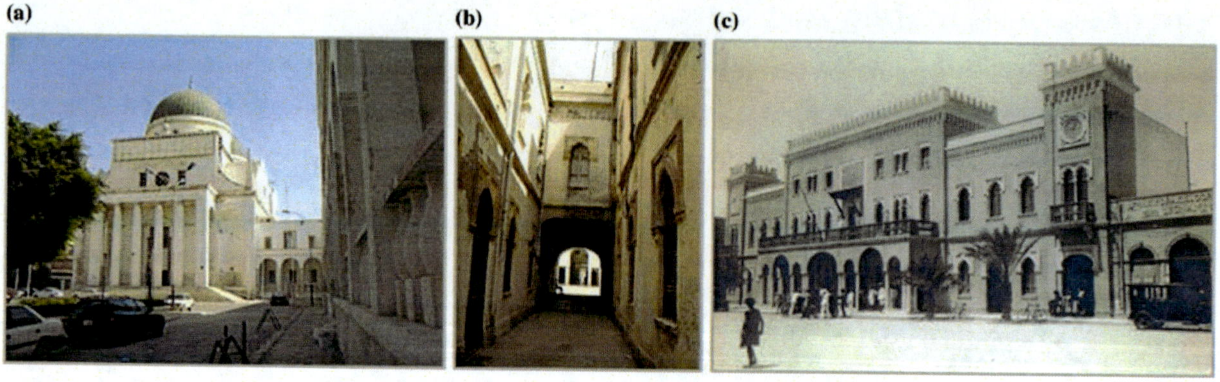

Fig. 15 **a** Cathedral building (Gasparoli, 2009). **b** houses and covered street (Gasparoli, 2009). **c** City Hall of Benghazi, 1925 (Manzoni, 2009)

but rather a sense and interdependence between man and the city". The current study finds that the inherited architectural character shows suitability for the environment of Benghazi, while there is a lack of homogeneity between the modern architecture and the local environment, and this form of architecture has failed to achieve sustainability. Research must be done to explore the interactions between local architecture, heritage and modernist architecture. Authorities and academics should pay particular attention to the preparation of standards that can be used as a guideline to preserve the identity of architecture and urbanism in Benghazi and other Libyan cities.

Acknowledgements Most of the pictures of the historical area and building in this research have been taking form Report 5.4 on Benghazi city centre's Architectural and Urban Charter, which has not been approved since July 2010. Because of that, I would like to acknowledge the people who were working on that report, as they saved many images of old buildings that were later destroyed during the conflict which occurred from 2014 to 2017.

References

Almarsad. (2022). *Almarsad*. https://almarsad.co/2022/03/22/. Accessed: April, 2023.

BSS. (2012). *Statistical book*. Bureau of Statistics and Sensus.

Bulugma, H. M. (1972). *Benghazi through the ages* (2nd ed.). DarMaktabata Al-Fikr.

DOX-LIB. (1989). *Benghazi Region, Baladiyat of Binghazi*. Final report on the master plan. Doxiadis Corporation. Urban Planning Agency.

Ettehad, S., Reza, A., Azeri, K., & Kari, G. (2014). The role of culture in promoting architectural identity. *European Online Journal of Natural and Social Sciences, 3*(4) Special Issue on Architecture, Urbanism, and Civil Engineering, ISSN 1805-3602.

Elsharif, O. (2019, March 4–6). Sustainability of Tripolitanian architecture in light of the identity crisis and the loss of architectural features. In: International Conference on Technical Sciences(ICST2019).

Gasparoli, P. (2009). *Benghazi; Diagnosis and restoration techniques on the surfaces of the Modern building*. Politecnico di Milano for Finmed.

HPCU. (2010, July). *Benghazi city center's architectural and urban charter. Phase 5* Report 5.4, Higher Popular Committee for Utilities. Benghazi Libya.

Iballouz, O. (2020). *The role of urban design in preserving the local urban identity in Al-Joufregion*. SFUDJP 2020, Jouf University, Saudi Arabia.

Iamviolet. (2017). *Iamviolet*. Iamviolet.arablog.org.2017/09/26/قرانم-شيبيرخإ-يديس. Accessed: July, 2023.

Independentarabia. (2022). Independentarabia.node/322701/تويب-لظلا-نم-جورخلل-ىعست-ةيرامعم-ةفحت-ةيبيللا-نايرغ-يف-رفحلا. Accessed: July, 2022.

Libyaakhbar. (2016). *Libyaakhbar*. https://www.libyaakhbar.com/wp-content/uploads/2016/02/0/page/1854. Accessed: April, 2023.

Libyaalmostakbal. (2022). *Libyaalmostakbal* . https://www.libya-al-mostakbal.org/top/37926/اذلكاقرة-اللليبية-ميدان-البلدية-خيرات-واحداث.html. Accessed July, 2023.

Libyaalmostakbal. (2023). *Libyaalmostakbal*. https://www.libya-al-mostakbal.org/.10/48016/شنيب-تدعو-خبراء-الآثار-والمهندسين-بنغازي-إلى-إيقاف-تدمير-سينما-برنيتشي.html. Accessed July, 2023.

Lynch, K. (1960). *The image of the city*. MIT Press.

Mahmoud, A. (2016). *Architecture of place in the Libyan city through geography and historical dimensions an approach to urban identity in Libyan City*. PhD Dissertation, Alexandria University, Alexandria, Egypt.

Mansour, H. (2015). *The lost identity of the city: The case of Damascus*. CITTA 8th Annual Conference on Planning Research AESOP TG Public Spaces & Urban Cultures Event.

Manzoni, M. (2009). *Modern architecture in Libya, preservation of a shared heritage*. Politecnico di Milano for Finmed.

Milod, M. (2019). *The vernacular Architecture in Libya: A case study of vernacular dwellings in the Nafusa mountain region*. PhD. Dissertation. Salford University, Manchester, UK. 2019.

Natthakit, P., & Heath, T. (2021). Defining the distinctiveness of urban heritage identity: Chiang Mai Old City, Thailand. *Social Sciences, 10*, 101. https://doi.org/10.3390/socsci10030101

Nooraddin, N. (2012). Architectural identity in an era of change. *Developing Country Studies, 2*(10), 2012.

Safour, A. (2013, April). *Urban and architectural character in Benghazi: between the heritage and the modernity directions*. Architectural Day Conference. Missrata University, Libya.

Safour, A. & Elmazek, E. (2018, March). Rehabilitation of Benghazi city—Ideas to become more sustainable. *Libyan Journal for Engineering Research (LyJER), 1*(1).

Torabi, Z., & Brahman, S. (2013). Effective factors in shaping the identity of architecture. *Middle-East Journal of Scientific Research, 15*(1), 106–113.

UN-Habitat. (2014). *A new strategy of sustainable neighbourhood planning: Five principles.* United Nations Human Settlements Programme.

UPA. (2009). *Third Generation Studies Benghazi region, Baladiyat of Binghazi.* Final report on the master plan. Urban Planning Agency, Benghazi, Libya.

Discovering Perceived Images of Reused Industrial Heritage from User-Generated Photographs: Three Mega-Event-Reinforced Industrial Heritage Transformation Cases

Huishu Deng

Abstract

Cities are increasingly reusing industrial heritage as part of cultural and creative regeneration strategies. However, designers and decision-makers face the challenge of determining which features and elements of industrial heritage are more perceived and preferred by the general public and will shape the future character of the site. The availability of large numbers of user-generated photographs (UGPs) on social media provides a new human-centered and grassroots lens for the above issue. This article proposes to analyze the content of UGPs to measure common regularities in people's perceptions of industrial heritage sites. The article focuses on three cases of event-driven industrial heritage transformation: Shougang Steel Factories in Beijing as the Big Air venue of the Beijing 2022 Olympic Winter Games, Parco Dora in Turin as the site of the Kappa FuturFestival since 2012, and Cockatoo Island in Sydney as the site of the Sydney Biennale since 2008. The analysis is conducted in three steps: collecting UGPs from social media sites with a time range from pre-event to post-event to compare the transition of perceived images; clustering the photographs based on visual similarity using deep learning models; and extracting commonly perceived image features, including landscape scenes, specific buildings/facilities, and certain elements.

Keywords

Perception · User-generated photograph · Industrial heritage · Urban transformation

1 Introduction

The human-oriented intangible factors such as emotional attachment and shared memory are increasing their role in the transformation of industrial heritage. Halbwachs (1950) has long demonstrated the link between the built environment and the collective memory, which involves people's emotional attachment to places. Today, voices of local bottom-up memory are a central point of industrial heritage studies as it critically addresses the idealized representations of a collective past showing a living dynamic evolving with time (Harvey, 2008). The transformation of industrial heritage accelerated and strengthened through events is a usual practice in recent years, which is conducted with the positive intention of defining a new area image and offering a better quality of life to the general public. However, such practice is facing the critiques of detachment between local people and industrial heritage sites that cause the inactivation and decline of the project (Müller, 2015). In recent years, decision-makers have begun seeking ways to build people's connection with places and reinforce place attachment which involves the elaborate interplay of emotion, cognition, and behavior in reference to a place (Morgan, 2010) and highlights the multiple-dimensional and interdisciplinary property of analysis.

Therefore, the inevitable question is, "what" are the common visually perceived and emotionally attached elements of a place? Many researchers address the question through numerical/literal indicators of a place, such as size, color, accessibility, and function, which have shown apparent correlations with the intensity of place attachment in some research (Casakin et al., 2015; Mandal, 2016). In architecture, the analysis of two-dimension functional layouts has been applied (Saymanlier et al., 2018). However, the graphic feature of place is absent in current studies on the measurement of place attachment (Raymond & Gottwald, 2020), but theoretically, they have been defined as a trigger of the sense of place by understanding the place

H. Deng (✉)
Heritage, Anthropology and Technologies Group, College of Humanities, EPFL, Lausanne, Switzerland
e-mail: huishu.deng@polito.it

M. F. Shahidan et al. (eds.), *Urban Narratives: Exploring Identity, Heritage, and Sustainable Development in Cities*, Advances in Science, Technology & Innovation, https://doi.org/10.1007/978-3-031-48517-6_6

through various graphic scenes and highlighting the symbolic meanings of these scenes. For instance, the snow-capped peaks or old-growth forests and the meaning of wildness are found to have substantial effects on raising a sense of attachment (Freestone & Liu, 2016; Ryden, 1993; Smith, 2018).

This research uses the photograph taken by ordinary on-site visitors (user-generated photographs) as the direct medium for studying place attachment, since the action of taking a photograph is not only triggered by the immediate environment but by many aspects of place attachment: attention, perceptions, preferences, memories, opinions, and more (Dunkel, 2015). By analyzing the contents of user-generated photographs (UGP) in three mega-event-reinforced industrial heritage transformation cases, this research aims to address two questions: (1) Before and after the transformation triggered by mega-events, what are the changes and consistency in the perceived images? (2) In the three cases with different cultural backgrounds and similar physical environments, are there common highly perceived images, and if so, what are they? What are the characteristics of the scene or formal composition? To what extent are they perceived?

2 Related Work

2.1 User-Generated Photograph and the Perception of Space

User-generated photograph (UGP) comes from regular people who voluntarily contribute graphic information and relative data (e.g., tags, spatio-temporal data, etc.) that then appear before others. UGP is unsanitized by regular media outlets and offers a bottom-up perspective (Krumm et al., 2008).

Many researchers have used UGP to analyze people's perceptions of space. Since 1960, in his influential book "The Image of the City" (Lynch, 1964), Kevin Lynch started to use user-generated content—the mental maps drawn by on-site visitors to study individuals' memory and navigation process through public space. Meanwhile, the anthropologist John Collier (Collier & Collier, 1986) first used photography as a tool for studying human perception. Handing out cameras to participants or self-directed photography (Dakin, 2003; Markwell, 2000) became generally accepted techniques for assessing the perceived/preferred public space to compensate for the inadequate levels of precision, reliability, and validity of expert assessments.

In recent years, benefiting from the proliferation of crowdsourcing photograph platforms (e.g., Flickr, Instagram, etc.) and the photograph-taking lifestyle in this information era, the availability of collecting a large number of UGPs has been opened. Numerous researchers in the fields of cultural heritage, tourism studies, and landscape and urban planning have applied the automatic recognition and categorization of UGPs content to study the relationship between people's perceptions and the characteristics of the city. For instance, in cultural heritage studies, researchers have defined photograph-sharing social media as a people-centered heritage platform that enables the creation of shared heritage and collective memory. The image content, text, geolocation, timestamp, and social network structure are used as multimodal datasets to reveal people's understanding of cultural heritage values and attributes (Bai et al., 2022; Ginzarly et al., 2019). The places presented in UGPs are also used to compare with the authorized heritage place lists to discover the preferred and neglected/forgotten heritage sites (Farahani et al., 2018). In tourism studies, the statistical measurement of the number of photographs in different scenes (e.g., architecture, plants, water, mountains, food, culture, shopping, etc.) and their spatial distribution is often used to analyze tourists' perceptions of destination images and their behavioral patterns (Stepchenkova & Zhan, 2013; Vu et al., 2018; Zhang et al., 2020). In the field of landscape and urban planning, similar methods have been used to study the common regularities of the esthetic appreciation of landscapes (Tieskens et al., 2018). Dunkel (2015) developed a visualization system to present the perceived environment, where the label of the photograph is attached to the corresponding position on the map and the size of the label is used to indicate the intensity of the perception. These studies demonstrate the possibility of using UGPs for spatial knowledge discovery in a variety of domains, especially cultural heritage and spatial experience, which is the main area of interest in this research.

2.2 The Technology of Image Content Analysis

The CV technology, built on artificial intelligence systems and deep learning algorithms, can extract "information" from images and supports the automatic image classification and content recognition of a large number of images. In the field of architecture and urban study, the commonly applied CV technologies include similarity clustering, content recognition, and semantic segmentation.

Similarity clustering can cluster pictures into groups based on their visual similarity. Images with high similarity will be closer in the coordinates, while images with low similarity will be far away. Previously introduced researches in the field of cultural heritage mostly used this approach, but they based the similarity clustering on the text tags set by the uploaders of the photographs, rather than directly on the visual content of the photographs themselves. This approach is limited when the photograph tags are not available. Papadopoulos et al. (2011) developed a novel scheme for automatically detecting landmarks and events in image collections based on both visual and tag similarity clustering, which enables the visual content of photographs to be used directly as the basis for clustering.

Content recognition employs the pre-trained model to automatically identify the main content in the photographs and classify them. The tourism research mentioned above all uses this technology. The key to this technology lies in how to label the pre-trained dataset. Based on different research purposes, the way of labeling is also different. Currently, the mainstream generic labeled datasets (e.g., ImageNet, Google's Open Images, etc.) mainly label "what object is it" lacking the labels for more complex spatial composition.

Semantic segmentation also applies pre-trained models but can recognize and segment multiple visual elements (e.g., vegetation, building, water, sky, people, car, etc.) in an image (Liu et al., 2016). Ma et al. (2021) use the ratios of visual elements on street photographs to calculate pre-defined spatial indices (greenness, openness, enclosure, walkability, and imageability) of the streets in an urban district.

The object of this research is industrial heritage, which, due to the specificity of its spatial characteristics, clearly does not lend itself to a generic labeled dataset, which would over-simplify much of the useful information. Therefore, instead of using the latter two methods, this research uses the similarity clustering method. The visual similarity of the photographs will be used as a basis for clustering, rather than the tag of the photograph, which will allow us to reach a wider group of users.

3 Data and Methodology

3.1 Cases

While intangible value, such as past productive memory, emotional attachment, and industrial esthetic, has been emphasized in the regeneration of industrial heritage, hosting mega-events adds another layer of value to the same site. At the built environment level, the mega-events reinforce and accelerate the urban transformation, including the reuse/beautification of existing industrial infrastructure, the creation of new landmarks, event venues, and sign/poster systems, and the area renovation by adding natural elements and leisure facilities. It increases the hybridity of the industrial heritage site and extends the issue of perception of industrial heritage to broader discussions on spatial experience. At the image level, the mega-events have been used as a vehicle for redefining the new branding of the city, for instance, from an industrial city to a creative/cultural/healthy city. The added new brandings are generally associated with the theme of events and followed by related consumption or cultural activities.

Hence, three industrial heritage transformation cases are selected, which are reinforced by different types of mega-events. The Shougang Park was a large compound of the steel industry at the side of the Yongding River in the western part of Beijing, China. The former steel production was shut down in 2008, and since then, the gradual reuse has started. Since 2017, with the official announcement that the Shougang compound would be transformed into Big Air venues for the Beijing 2022 Olympic winter games, the process of renovation was accelerated and expanded. After Beijing 2022, the Shougang compound is now a public leisure and sporty park open to citizens, and it also hosts temporary events such as trade fairs and exhibitions. The Parco Dora is located along the bank of River Dora in the northern part of Turin, Italy. It is an industrial zone gathering the construction of the Fiat Ferriere Piemontesi steel and sheet metal works and the Michelin tire factory, which was shut down in the 1980s and left as an abundant area in the city. The renewal of Parco Dora launched in 2004, and since 2012, it has been the location for Kappa FuturFestival, an annual music summer festival lasting three days. It is also a leisure park on regular days, including a skate park and playgrounds. The Cockatoo Island is a UNESCO World Heritage site located at the junction of the Parramatta and Lane Cove River in Sydney Harbor, New South Wales, Australia. It has a complex history of a former convict penal establishment, a naval ship dockyard, an industrial school for girls, and a reformatory. From 1857 to 1991, it functioned as a dockyard and shipbuilding facility. Since 2008, it has been a significant venue partner of the Biennale of Sydney, and after that, it has been used as a venue for multiple art events, including comedy, music, and film festivals. It is also a leisure site for picnic and camping activities.

As given in Table 1, the three cases share a similar industrial type involving large-scale production facilities, factories, and machine complexes, and similar land uses that comprise events venues, leisure facilities, and green space. By studying these three cases with spatial similarity, this paper aims to find the common regularity of perceptual experience in the renewed industrial heritage site.

3.2 Data Collection

The data used for this research are from public photograph-sharing platforms, including Flickr, Instagram, and Weibo in China. For each case, around 3000 user-generated photographs are collected under several criteria: (a)

Table 1 Introduction of cases

Name	Location	Area (km^2)	Industrial type	Mega-event type	Renewed land use
Shougang Park	Beijing	8.63	Steel industry	Olympic games	• Event venues • Leisure promenades around the lake • Sport/cultural consumption • Museum • Green park
Parco Dora	Turin	0.37	Steel industry	Music festival	• Event venues • Leisure promenades and bridges along the river • Sport facilities • Green park • Church
Cockatoo Island	Sydney	0.18	Shipbuilding industry	Art Biennale	• Event venues • Leisure promenades along the canal • Historic relics • Green park

The photographs are uploaded with the location tag of the case; (b) the photographs are uploaded by ordinary users instead of regular media or official account; (c) the selfies or portraits without enough environmental content will be excluded from the dataset; (d) the time when the first mega-event was held is used as the dividing point of data, in order to compare the changes between pre-event and on/post-event period (Table 2). The data are collected through Python by invoking Flickr API, Instagram Basic Display API, and Weibo (微博) API, all available photographs meeting the above criteria have been collected. This research emphasizes the visual content of photographs, which are collected and analyzed in detail, but other information like geotags, comments on this photograph, and public user profiles are also available for future research.

3.3 Data Analysis Method

The data analysis method is combined with two phases: photograph clustering based on visual similarity through unsupervised learning algorithms; the manual labeling of each clustered photograph group based on the spatial features.

Firstly, a Python program is constructed and tested, aiming to cluster a large number of photographs based on their similarity. The technical key points are the selection of the clustering algorithm and the convolutional neural networks (CNN) algorithm for vectorizing the picture. After several tests, we selected K-means clustering as the clustering algorithm, one of the simplest and most popular unsupervised machine learning algorithms. We also employed multiple well-known CNN algorithms such as ResNet, SENet, WRN, and VGG, and the VGG19 presented the best results in the test.

Secondly, the photograph clustering process goes through several rounds: (a) the original photograph data is clustered into 24 clusters through the Python program. In this process, the photographs without other similar photographs will be excluded, and only 50–70% of photographs will remain (Table 2); (b) the cluster including only one or two types of photographs will be manually grouped, and the

Table 2 Data sources and photograph amount

No.	Location	Platform	Time	Photograph amount of original dataset	Valid photograph amount	Ratio (%)
S1	Shougang Park	Weibo	On/post-event	4438	2349	52.90
S2	Shougang Park	Weibo	Pre-event	142	87	61.30
S3	Shougang Park	Flickr	Pre-event	585	495	84.60
P1	Parco Dora	Flickr	On/post-event	2226	2023	90.90
P2	Parco Dora	Instagram	On/post-event	1410	732	51.90
P3	Parco Dora	Flickr	Pre-event	469	390	83.20
C1	Cockatoo Island	Flickr	On/post-event	2413	2035	84.30
C2	Cockatoo Island	Instagram	On/post-event	624	318	51.00
C3	Cockatoo Island	Flickr	Pre-event	2235	1577	70.60

cluster including multiple photograph types will be automatically clustered again into 12 clusters, then manually grouped; (c) the process goes until all similar photographs are grouped.

Thirdly, all photograph groups are labeled and classed based on their contents. Appendix A shows the tags of three cases. The photographs are classified into five classes: architectural elements, industrial elements, cultural elements, consumption elements, and landscape elements. Each class includes several subclasses that illustrate the primary object in the photographs. However, one kind of object will be photographed from different perspectives. These perspectives are crucial information for this research. They need to be recognized in order to understand not only what objects are more perceived by visitors but also how visitors appreciate the holistic composition of space. Therefore, different perspectives of one object are also labeled in Appendix A, such as the "Orange Pillars-array" and "Orange Pillars-view on footbridge."

4 Results

4.1 Changes and Consistency of the Perceived Images Between Pre-event and On/Post-event Period

By comparing the overall statistical data of the three cases, the changes in the content of the photographs before and after the mega-event can be analyzed from the macro level.

It can be seen from Fig. 1 that, on the one hand, although all three cases are places with hybrid elements, no matter before or after the event, the industrial elements are the most attractive and perceived elements for visitors; on the other hand, the employment of mega-event has indeed changed the perceived image of the original industrial heritage site to a certain extent.

Specifically, in the case of Shougang Park, the Olympic venues embedded inside industrial fabric effectively attract the attention of visitors. The consumption business model adopted by Shougang after the Olympic Games also operates efficiently and gains a certain influence. In the case of Parco Dora, the massive increase in cultural elements, especially graffiti, reflects that the host of well-known music festivals annually gathers a large number of artist groups and makes it become an art destination. This shows that street culture and industrial heritage have merged through pop activities to form an impressive composite image. In the case of Cockatoo Island, the natural landscape elements on the island were mainly perceived before the mega-event. And the use of factories and historical buildings as venues for the biennale makes visitors pay attention to the exhibits while also increasing their attention to industrial heritage and historical relics.

4.2 The Common Perceived Images of Industrial Heritage

As mentioned above, the three industrial heritage cases are similar in the built environment. Therefore, by comparing them horizontally, the common regularity of people's perception of the renewal of industrial heritage sites can be explored. In Appendix A, the photographs with a proportion over 2.5% and have commonalities across different cases are extracted and analyzed. The scenes they showed included: macro scale-skyline with landmarks; medium scale-special geometric form and iconic objects; micro scale-machine and cultural symbols.

4.2.1 Skyline with Landmark

As shown in Fig. 2, the scene where massive landmarks stand out in the horizontal skyline is preferred by visitors. The landmarks of industrial heritage represent the character of one site, and they are the major physical carrier of productive histories, such as the cylindrical facility in Shougang Park and the crane in Cockatoo Island.

4.2.2 Special Geometric Form

The special geometric forms such as curves, vertical lines, and an array of repeating objects are well perceived by visitors in all three cases, which correspond to the classic

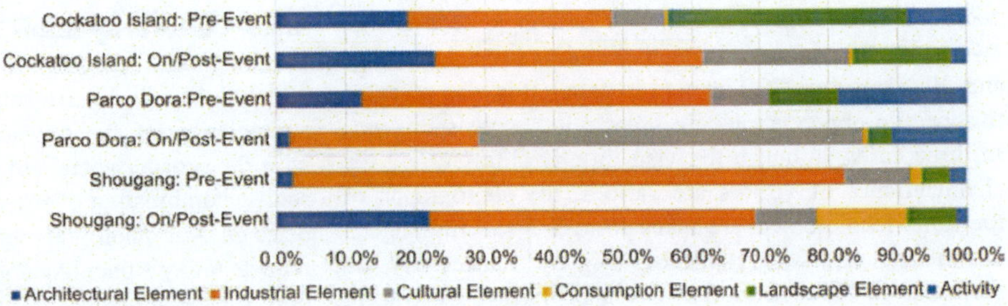

Fig. 1 Proportion of perceived elements in three cases. *Source* Drawn by author

Fig. 2 UGP examples of the scenes "skyline with landmark". *Sources* Shougang 1 (© Ju-cao-ye-jia-you-la, 2020, CC BY 2.0), Shougang 2 (© Jiang-mo-xiang-jiang, 2019, CC BY 2.0), Shougang 3 (© M, 2020, CC BY 2.0), Cockatoo 1 (© Grant Eaton, 2012, CC BY-NC-ND 2.0), Cockatoo 2 (© Davies, 2009, CC BY-NC 2.0), Cockatoo 3 (© Yap, 2008a, 2008b, CC BY-NC 2.0)

spatial composition theory and gestalt psychology. As shown in Fig. 3, the curve shape comes from the bridge arches that can be seen in industrial heritage sites or arched holes embedded in industrial facilities. The vertical linear objects such as chimneys are also well perceived by visitors. As shown in Fig. 4, the array of cooling towers and Olympic venues with similar figures, pillars, and machines make deep impressions on visitors. It is worth noting that they are not in the original production state. The designer reorganizes and arranges the original industrial facilities, or designs the new facilities into a similar form to add the array. This approach may strengthen the power and visual impact of industrial objects, making them become more perceptible images.

4.2.3 Iconic Objects in Iconic Perspective

We find that the iconic objects, such as Olympic venues, cooling towers, cranes, and factory buildings, have a high awareness of visitors throughout the industrial heritage site. By analyzing UGP content in detail, it is also revealed that people appreciate iconic objects in a consistent way. In three cases, although the visit time is different and the characteristics of visitors are various, the photographs of the same iconic object are taken from 3–5 specific angles, which are surprisingly similar (Fig. 5). Particularly, in the cases of Parco Dora and Cockatoo Island, there were similar factory buildings and both were used to host mega-events, visitors took a large number of similar photographs, although in different cultural contexts. These photographs show the attraction of the roof structure of the factory building to human perception.

4.2.4 Machine and Industrial Texture

Machine originals are one of the most perceived elements. They are regarded as "exhibits" connecting past memories. The preference for machine originals shows the importance people attach to past memories. In addition, a large number of photographs capture fragments of machines or facilities, focusing on the texture of industrial materials. This, to a certain extent, reflects the esthetic preference for "industrial style" (Fig. 6).

4.2.5 Cultural Symbols

Symbols with pictorial or written information are also widely perceived elements (Fig. 7). This connects the appreciation of the physical space of industrial heritage with the broader social culture. For example, graffiti represents a combination of street rebellion culture and industrial heritage. And photographing road signs and building names reflects today's phenomenon of people "checking in a location" on social media.

Fig. 3 UGP examples of arc-shape and linear objects. 2022 *Sources* Shougang 1 (© SuetVi, 2020, CC BY 2.0), Shougang 2 (© aqi1978,, CC BY-NC 2.0), Parco Dora 1 (© Vellut, 2020a, 2020b, CC BY 2.0), Parco Dora 2 (© raffaele sergi, 2016a, 2016b, CC BY 2.0), Cockatoo 1 (© Etheredge, 2012, CC BY-NC 2.0), Cockatoo 2 (© Wrigley, 2010, CC BY 2.0), Shougang 3 (© ID4533847, 2016, BY-NC-ND 2.0), Shougang 4 (© Xuangan, 2017, CC BY 2.0), Cockatoo 3 (© Wong, 2010, CC BY-NC-ND 2.0)

5 Discussions and Conclusions

In a quantitative way, this research verified from the bottom-up perspective that the holding of mega-events and the urban transformation triggered by them could play a role in changing the image of the industrial heritage site. It can be concluded through empirical evidence that mega-events have added more diverse images to industrial heritage, enriching people's presence and experience.

Further, by extracting image content from a large quantity of UGPs, highly perceived spatial elements can be discovered. On the macro scale, the representative giant industrial facilities are perceived as landmarks (e.g., cylindrical facility, cooling tower, and crane), which shape the

Fig. 4 UGP examples of array of repeating objects. *Sources* Shougang 1 (© Bei-zhai-han, 2022, CC BY-NC-ND 2.0), Shougang 2 (© Jiang-mo-xiang-jiang, 2019, CC BY 2.0), Parco Dora 1 (© raffaele sergi, 2016a, 2016b, CC BY 2.0), Parco Dora 2 (© Galli, 2013, CC BY 2.0), Cockatoo 1 (© Tudugalle, 2012, CC BY 2.0), Cockatoo 2 (© NAPARAZZI, 2011, CC BY-SA 2.0)

Fig. 5 UGP examples of iconic objects and their iconic perspective. *Sources* Shougang 1 (© Samnice, 2022, CC BY-NC 2.0), Shougang 2 (© RSC Han, 2022, CC BY 2.0), Shougang 3 (© Samnice, 2022, CC BY-NC 2.0), Shougang 4 (© Jiang-mo-xiang-jiang, 2019, CC BY 2.0), Parco Dora 1 (© diegofornero (destino2003), 2012, CC BY-ND 2.0), Parco Dora 2 (© Romero, 2017, CC BY 2.0), Cockatoo 1 (© BillOggPhotography, 2012, CC BY-ND 2.0), Cockatoo 2 (© Carmona, 2013, CC BY-NC-ND 2.0)

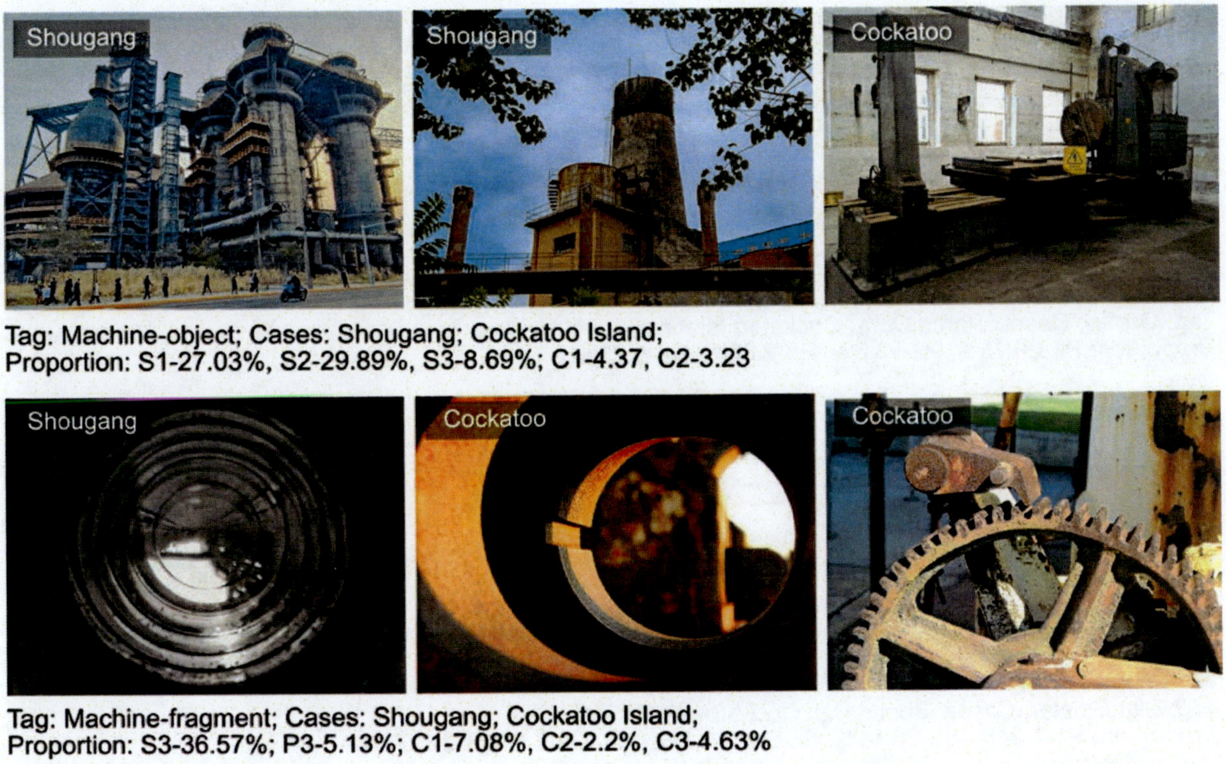

Fig. 6 UGP examples of machines and industrial texture. *Sources* Shougang 1 (© DYD, 2022, CC BY 2.0), Shougang 2 (© zyh, 2016, CC BY 2.0), Cockatoo 1 (© Yap, 2008a, 2008b, CC BY-NC 2.0), Shougang 3 (© Siren, 2014, CC BY 2.0), Cockatoo 2 (© Cingano, 2014, CC BY-NC 2.0), Cockatoo 3 (© Sanderson, 2016, CC BY-NC-ND 2.0)

memorial skyline of the industrial site with the surrounding horizontal environment. On the medium scale, the iconic objects hidden in the industrial complex are revealed, which are the physical carrier of the industrial history and the esthetic taste of "industrial style"; meanwhile, the perspectives of how people appreciate iconic objects appear to have a high degree of consistency, and this characteristic can be used for the adaptive reuse of iconic objects. On the micro scale, pictorial information (e.g., machine fragments) or written information (e.g., remaining signs on relics) are also well noticed and recorded, which link industrial heritage with a broader cultural context. These analyses based on digital tools can be utilized by the designer and decision-maker of industrial heritage renewal projects, aiming to establish a closer perceptual attachment between visitors and the physical space and connect on-site experience with past memories.

The limitation of this research is that there is no further distinction between the characteristics of photograph uploaders. First, the identities of mainstream users of different photograph-sharing platforms are different. For instance, Flickr users usually have more photography expertise or experienced photography enthusiasts; Instagram and Weibo users do not have this characteristic–they upload photographs to record and share their day-to-day lives. Second, a further breakdown of users' age, gender, and cultural background may also lead to more targeted results. Nevertheless, the methods and computer script constructed by this research can be applied to more cases and groups of people in future research to further reveal "what" in space can be more perceived and preferred by people and quantitative measure the degree of perception.

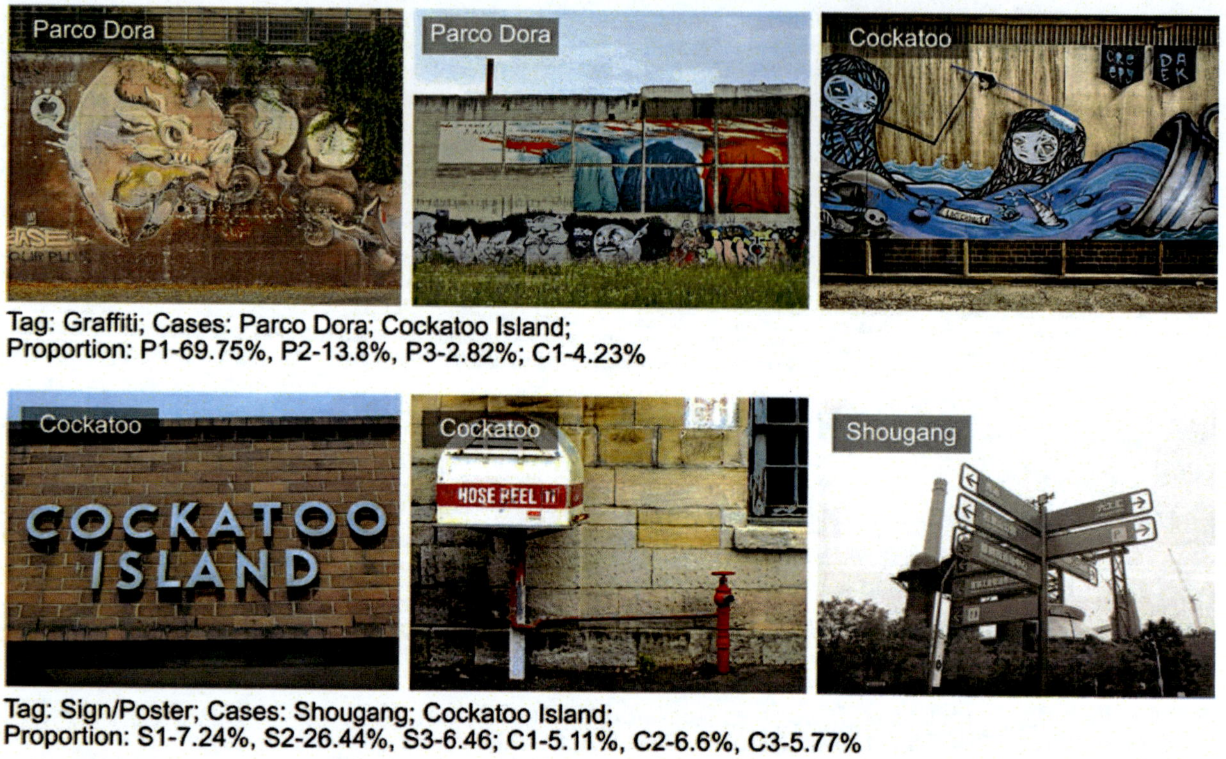

Fig. 7 UGP examples of cultural symbols. *Sources* Parco Dora 1 (© Vellut, 2020a, 2020b, CC BY 2.0), Parco Dora 2 (© Siclari, 2020, CC BY-NC-ND 2.0), Cockatoo 1 (© AegirPhotography, 2011, CC BY-NC 2.0), Cockatoo 2 (© Miguel, 2011, CC BY 2.0), Cockatoo 3 (© Ostergaard, 2014, CC BY-NC 2.0), Shougang 1 (© Feng, 2021, CC BY-NC 2.0)

This research can contribute to the future urban regeneration of industrial sites in two aspects. On the one hand, for policymakers and local administrations, the methodology adopted in this research allows the assessment of existing cases of industrial heritage transformation from a human-centered perspective. Spatial features with higher/lower perceptions can be extracted and evaluated quantitatively, which allows decision-makers to understand the use of industrial heritage sites and to plan the future use of the sites in a targeted way: applying higher perceived spatial features to similar types of industrial heritage transformation projects in the future; or improving forgotten and neglected sites. On the other hand, in terms of public action, this research not only helps the public to have a comprehensive understanding of the tangible features of industrial heritage and the intangible activities that take place within it, but also allows the public to contribute to this information and knowledge. Using the same methodology, UGPs can be obtained not only from tourists and social media, but also from the local community. This can facilitate the formation of participatory "crowdsourced knowledge," allowing the voices of local communities to be more collectively articulated and heard.

Acknowledgements This research was funded by the SNSF project funding "Uses of Cultural Heritage at the Beijing Winter Olympic Games of 2022" (2020-2024).

Appendix A. Labels and Proportion of UGP Group

Class	Subclass	Proportion (%)			Subclass	Proportion (%)			Subclass	Proportion (%)		
		S1	S2	S3		P1	P2	P3		C1	C2	C3
Architectural element	Bridge-arch	**3.53**	1.15		Bridge	0.44	2.05	**3.33**	Bridge-arch	0.98	**3.77**	0.51
	Bridge-historic	0.64										
	Historic building	0.72	2.30	1.82					Historic building	**5.55**	***10.69***	**2.54**
	Colorful pavement	1.06	1.15		Stair	0.25	0.82	0.51	Stair	0.34	0.63	0.38
	Footbridge	0.55			Church	0.69	0.82	**8.46**	Corridor	0.10	0.63	0.13
	Red slope	1.87							Architectural fragment	***14.84***	***13.52***	***15.35***
	Olympic venue -with array of cooling towers	**6.39**										
	Olympic venue -object	**7.11**										
Industrial element					Branch canal	0.49	0.14	*3.59*	Canal	1.52	1.89	2.35
					Main canal	0.69	0.68	0.77				
	Factory-façade	1.28		*3.23*	Factory-front façade	0.40	0.82	0.77	Factory-beam	0.39	1.89	0.63
					Factory-side façade	0.64	2.19	1.54	Factory-façade	*3.73*	*4.40*	*3.68*
					Factory-under roof	*3.16*	***11.89***	**8.21**	Factory-under roof	*3.05*	**7.55**	2.16
	Machine -in skyline	*2.55*	*4.60*									
	Machine-object	***27.03***	***29.89***	**8.69**					Machine-object	*4.37*	1.26	*3.23*
	Machine -fragment			***36.57***	Machine-fragment	1.53	0.41	**5.13**	Machine-fragment	**7.08**	2.20	*4.63*
	Frame	2.38	1.15	***13.54***	Frame	0.49	1.37	*4.62*	Frame	1.97	1.57	1.71
	Cooling tower	**7.83**	***3.45***	1.62	Cooling tower	1.19	2.46	1.54				
	Silos	1.45	1.15	1.21	Silos	0.84	0.82	1.79				
	Chimney	*2.94*		**6.87**					Chimney	2.36	*3.77*	1.78
	railway-arch/track	0.94		***15.15***	Columns-array	**5.68**	***14.62***	***12.56***	Crane-object	**9.39**	**6.29**	**7.04**
	Railway-train	0.60			Columns-view on footbridge	0.74	*2.60*	0.77	Crane-in skyline	*2.70*	*3.14*	1.08
	Tower-with Olympic rings	0.30			Columns-arch	0.64	0.82	*3.08*	Tunnel	2.26	*3.46*	1.20
					Columns-fragment	0.84	0.14	1.54				
					Columns-concrete/graffiti	*3.26*	*5.05*	*4.10*				
					Tube	0.15		0.26				

(continued)

Class	Subclass	Proportion (%)			Subclass	Proportion (%)			Subclass	Proportion (%)		
		S1	S2	S3		P1	P2	P3		C1	C2	C3
Cultural element					Graffiti	***69.75***	***13.80***	*2.82*	Graffiti	*4.23*	0.94	
	Sign/poster/moscat	**7.24**	***26.44***	**6.46**	Sign/poster		2.32		Sign/poster	**5.11**	**6.60**	**5.77**
	Olympics sign/moscat	1.83										
					Sculpture	0.40	0.55	**6.15**	Sculpture-array	*3.29*	0.31	1.08
									Exhibits	***10.42***	*4.09*	0.95
Consumption	Shop/brand	**8.60**	*3.45*		Shop/brand	0.10	1.64		Shop/brand	0.15	0.31	0.19
	Food/drink	*4.34*	**6.90**		Food/drink		0.82		Food/drink	0.25	0.63	0.25
Landscape element	Skyline	1.32	1.15	1.62	Skyline	0.84	0.96	*3.08*	Skyline	**9.29**	**6.60**	**5.20**
	Vegetation-object	*3.07*	**6.90**		Vegetation-object	0.54	2.19	0.77	Vegetation	1.03	*2.83*	1.52
	Vegetation-fragment	0.77	*3.45*	1.41	Vegetation-fragment	0.20	*2.87*	*4.36*				
					Lawn	0.59	0.82	1.79	Lawn	0.20		0.25
	Mountain	2.21	2.30		Stone	0.30	0.14		Stone	0.54	0.63	1.46
									Boat/car/flight	1.47	**5.35**	**5.83**
									Animal	1.47	1.57	***20.61***
Activity	Skateboarding/biking	0.68			Skateboarding/biking	*2.82*	**6.01**	**9.74**	Camping	0.69	*2.52*	0.25
	Camping	0.21			Events	0.49	***13.80***	*4.36*	Events	0.15	0.31	**7.42**
	Dance/run	0.21	*4.60*		Dance/gymnastics/parkour	0.79	*3.69*	0.77	Ball sports	0.69		
	Draw	0.04			Ball sports	0.44	1.23	*2.56*	Stationary activity	0.34	0.31	0.25
	Others (ballgame, paddle)	0.30		1.82	Stationary activity	0.15	0.55	0.26	Others (biking, swimming, performing)	0.05	0.31	0.57
					Games	0.15	0.00	0.00				
					Others (drawing, performing)	0.30	0.96	0.77				

The proportion of each group is the ratio of the number of pictures in this group to the total number
The 2.5–5% group is marked with Italic; 5–10% group is marked with bold; and > 10% group is marked with bold and italic

References

AegirPhotography. (2011). *2011 Outpost project* [Photograph]. Flickr. https://flic.kr/p/aCh7QY

aqi1978. (2022). *New Shougang bridge* [Photograph]. LOFTER. https://aqi1978.lofter.com/post/1d1c1006_1c6ec69a4

Bai, N., Nourian, P., Luo, R., & Pereira Roders, A. (2022). Heri-graphs: A dataset creation framework for multi-modal machine learning on graphs of heritage values and attributes with social media. *ISPRS International Journal of Geo-Information, 11*(9), 469.

Bei-zhai-han. (2022). *Beijing Shougang Park* [Photograph]. LOFTER. https://biaozhongyingjie.lofter.com/post/1f90b503_2b4607a9f

BillOggPhotography. (2012). *Cockatoo Island* [Photograph]. Flickr. https://flic.kr/p/byLLWi

Carmona, P. (2013). *Abandoned warehouse* [Photograph]. Flickr. https://flic.kr/p/gRZk8W

Casakin, H., Hernández, B., & Ruiz, C. (2015). Place attachment and place identity in Israeli cities: The influence of city size. *Cities, 42*, 224–230.

Cingano, S. (2014). *Untitled* [Photograph]. Flickr. https://flic.kr/p/nXMQoU

Collier, J., & Collier, M. (1986). *Visual anthropology: Photography as a research method*. UNM Press.

Dakin, S. (2003). There's more to landscape than meets the eye: Towards inclusive landscape assessment in resource and environmental management. *Canadian Geographer/le Géographe Canadien, 47*(2), 185–200.

Davies, S. (2009). *Untitled* [Photograph]. Flickr. https://flic.kr/p/6jR8tH

diegofornero (destino2003). (2012). *Parco Dora* [Photograph]. Flickr. https://flic.kr/p/coZ4Aq

Dunkel, A. (2015). Visualizing the perceived environment using crowdsourced photo geodata. *Landscape and Urban Planning, 142*, 173–186.

DYD. (2022). *Untitled* [Photograph]. LOFTER. https://edward-d-d.lofter.com/post/1df8aac4_2b5d48191

Eaton, E. (2012). *The cranes of Cockatoo Island* [Photograph]. Flickr. https://flic.kr/p/bcTG2g

Etheredge. D. (2012). *Cockatoo Island 004* [Photograph]. Flickr. https://flic.kr/p/bpwwW7

Farahani, L. M., Motamed, B., & Ghadirinia, M. (2018). Investigating heritage sites through the lens of social media. *Journal of Architecture and Urbanism, 42*(2), 199–211.

Feng, L. (2021). *Record life-Shougang* [Photograph]. LOFTER. https://linfeng830.lofter.com/post/1f4fc41c_1cc7d9365

Freestone, R., & Liu, E. (Eds.). (2016). *Place and placelessness revisited*. Routledge.

Galli, L. (2013). *Parco Dora: Foresta Meccanica* [Photograph]. Flickr. https://flic.kr/p/ejihjs

Ginzarly, M., Roders, A. P., & Teller, J. (2019). Mapping historic urban landscape values through social media. *Journal of Cultural Heritage, 36*, 1–11.

Halbwachs, M. (1950). La memoria colectiva y el tiempo. In *El conocimiento de la memoria colectiva* (pp. 103–137). UAT.

Harvey, D. (2008). The right to the city. *The City Reader, 6*(1), 23–40.

ID4533847. (2016). *Untitled* [Photograph]. LOFTER. https://creep-1.lofter.com/post/452e57_baa7a7d

Jiang-mo-xiang-jiang. (2019). *Untitled* [Photograph]. LOFTER. https://jiangmoxiangjiang.lofter.com/post/1d7fb9ac_1c73af47a

Ju-cao-ye-jia-you-la. (2020). *Shougang Park* [Photograph]. LOFTER. https://jucaoyejiayoula.lofter.com/post/1fee00b1_1cacf9702

Krumm, J., Davies, N., & Narayanaswami, C. (2008). User-generated content. *IEEE Pervasive Computing, 7*(4), 10–11.

Liu, L., Zhou, B., Zhao, J., & Ryan, B. D. (2016). C-IMAGE: City cognitive mapping through geo-tagged photos. *GeoJournal, 81*, 817–861.

Lynch, K. (1964). *The image of the city*. MIT press.

M. (2020). *Untitled* [Photograph]. LOFTER. https://linkinparkeminem.lofter.com/post/1e77a28e_1caaa3b4c

Ma, X., Ma, C., Wu, C., Xi, Y., Yang, R., Peng, N., Zhang, C., & Ren, F. (2021). Measuring human perceptions of streetscapes to better inform urban renewal: A perspective of scene semantic parsing. *Cities, 110*, 103086.

Mandal, A. (2016). Size and type of places, geographical region, satisfaction with life, age, sex and place attachment. *Polish Psychological Bulletin*.

Markwell, K. W. (2000). Photo-documentation and analyses as research strategies in human geography. *Australian Geographical Studies, 38*(1), 91–98.

Miguel, J. (2011). *Cockatoo Island* [Photograph]. Flickr. https://flic.kr/p/aCfwJ8

Morgan, P. (2010). Towards a developmental theory of place attachment. *Journal of Environmental Psychology, 30*(1), 11–22.

Müller, M. (2015). The mega-event syndrome: Why so much goes wrong in mega-event planning and what to do about it. *Journal of the American Planning Association, 81*(1), 6–17.

NAPARAZZI. (2011). *Cockatoo Island* [Photograph]. Flickr. https://flic.kr/p/aF3xyy

Ostergaard, P. (2014). *Cockatoo Island* [Photograph]. Flickr. https://flic.kr/p/mQC8k5

Papadopoulos, S., Zigkolis, C., Kompatsiaris, Y., & Vakali, A. (2011). Cluster-based landmark and event detection for tagged photo collections. *IEEE Multimedia Magazine, 18*(1), 52–63.

raffaele sergi. (2016). *Parco Dora* [Photograph]. Flickr. https://flic.kr/p/Nmwmnk

raffaele sergi. (2016). *Parco Dora B&W* [Photograph]. Flickr. https://flic.kr/p/Grizki

Raymond, C., & Gottwald, S. (2020). Beyond the "local": Methods for examining place attachment across geographic scales. In *Place attachment* (pp. 143–158). Routledge.

Romero, F. (2017). *Torino—Parco Dora* [Photograph]. Flickr. https://flic.kr/p/21B8Edj

RSC Han. (2022). *Untitled* [Photograph]. LOFTER. https://rschan.lofter.com/post/4d0fc4a0_2b68911be

Ryden, K. C. (1993). *Mapping the invisible landscape: Folklore, writing, and the sense of place*. University of Iowa Press.

Samnice. (2022). *Beijing 2022* [Photograph]. LOFTER. https://samnice.lofter.com/post/11568_2b608892b

Sanderson, I. (2016). *Cockatoo Island* [Photograph]. Flickr. https://flic.kr/p/GuTEgo

Saymanlier, A. M., Kurt, S., & Ayiran, N. (2018). The place attachment experience regarding the disabled people: The typology of coffee shops. *Quality and Quantity, 52*(6), 2577–2596.

Siclari, S. (2020). *Parco Dora pt.6 - La memoria* [Photograph]. Flickr. https://flic.kr/p/2jv3nSc

Siren. C. (2014). *Grey memory. Shougang* [Photograph]. LOFTER. https://sirencforever.lofter.com/post/366893_10fe8c5

Smith, J. S. (Ed.). (2018). *Explorations in place attachment* (p. 1). Routledge.

Stepchenkova, S., & Zhan, F. (2013). Visual destination images of Peru: Comparative content analysis of DMO and user-generated photography. *Tourism Management, 36*, 590–601.

SuetVi. (2020). *Curtain of night* [Photograph]. LOFTER. https://suetvi.lofter.com/post/4b514a9c_1cac3ac81

Tieskens, K. F., Van Zanten, B. T., Schulp, C. J., & Verburg, P. H. (2018). Aesthetic appreciation of the cultural landscape through social media: An analysis of revealed preference in the Dutch river landscape. *Landscape and Urban Planning, 177*, 128–137.

Tudugalle, H. (2012). *Cockatoo Island 4* [Photograph]. Flickr. https://flic.kr/p/csYjEw

Vellut, G. (2020). *Parco Dora (area Vitali) @ Turin* [Photograph]. Flickr. https://flic.kr/p/2ijKAMC

Vellut, G. (2020). *Pig graffiti @ Parco Dora (area Ingest) @ Turin* [Photograph]. Flickr. https://flic.kr/p/2ijHDmG

Vu, H. Q., Luo, J. M., Ye, B. H., Li, G., & Law, R. (2018). Evaluating museum visitor experiences based on user-generated travel photos. *Journal of Travel and Tourism Marketing, 35*(4), 493–506.

Wong, J. (2010). *Camping ground at Cockatoo Island* [Photograph]. Flickr. https://flic.kr/p/8ctBc5

Wrigley, G. (2010). *Harbor Bridge from Cockatoo Island* [Photograph]. Flickr. https://flic.kr/p/8Nbebg

Xuangan, Y. (2017). *Untitled* [Photograph]. LOFTER. https://yangxuangan.lofter.com/post/38f084_101a3d33

Yap, B. (2008). *Cockatoo Island* [Photograph]. Flickr. https://flic.kr/p/55WkcM

Yap, B. (2008). *Cockatoo Island* [Photograph]. Flickr. https://flic.kr/p/565UZc

Zhang, K., Chen, Y., & Lin, Z. (2020). Mapping destination images and behavioral patterns from user-generated photos: A computer vision approach. *Asia Pacific Journal of Tourism Research, 25*(11), 1199–1214.

zyh. (2016). *Once the Shougang* [Photograph]. LOFTER. https://20000819.lofter.com/post/32d06a_b7c8400

Transitional Arcades. Spatial Observations Between Italy and Wales: A Phenomenological Approach

Mickeal Milocco Borlini, Andrea Califano, and James Acott-Davies

Abstract

This study uses a phenomenological paradigm to study arcades between Wales and Italy to understand the evolution and direction of these public spaces/places and provide a framework to examine the bodily relationship between the urban role of arcades and galleries. The paper reviews the state-of-the-art and historiography of the architecture investigated, reviews the literature and its origins, and discusses a phenomenological and philosophical approach. The paper suggests that the intangible dimension of existence and experience is the key to understanding these places. By examining how we experience and perceive them, we can better understand their role in shaping our built environment and cultural heritage through time. Through the lens of phenomenology, we can gain insight into their historical context and better understand their intrinsic spirit. Furthermore, to contextualise the identity and spatial aspects of arcades, we draw upon the theorists M. Augé, T. Boettger, C. N. Schulz, W. Benjamin, and B. Rudofsky, as well as artistic reflections on Futurism and Cubism. The paper also questions whether traditional arcades have lost their intrinsic value of place or has their spirit of place has been transmogrified into a non-place through globalisation. Overall, the paper offers a rich and nuanced perspective on the cultural and historical significance of arcades, galleries, and porticos.

Keywords

Welsh arcades · Italian galleries · Architectural phenomenology · Mallisation critique

1 Introduction

In the following paper, a unique analysis of the arcades between Wales (1858) and Italy (1877) through the phenomenological lens helps gain a comprehensive understanding of the evolution and direction of these public spaces/places.

The first part of this text reports the state of the art and a historiography of the architecture investigated. The following section reviews the literature and its origins. At the same time, the concluding paragraph expresses our phenomenological and philosophical approach, drawing from influential authors such as C. N. Schulz, W. Benjamin, C. Rudofsky, and many others.

Briefly, phenomenology is a philosophical method that emphasises the study of conscious experience and subjective perception. It provides a framework to examine the role of arcades, galleries, and porticos in shaping our experience of the built environment. We can understand their cultural and historical significance by focusing on how we perceive and interact with these structures.

For example, as we will discuss in due course, the Galleria Vittorio Emanuele II in Milan, Italy, is an example of an arcade that offers a unique phenomenological experience. This grand shopping arcade, built in 1877, features a soaring glass dome, intricate ironwork, and luxurious storefronts. As visitors move through the space, they are enveloped by the sounds of footsteps, chatter, and music, creating a deep and immersive sensory experience.

From a historical perspective, arcades, galleries, and porticos have played an essential role in urban architecture and cultural development. They have provided sheltered spaces for commerce, socialising, and cultural exchange. They are

M. M. Borlini (✉) · J. Acott-Davies
Cardiff School of Art and Design, Cardiff Metropolitan University, Cardiff, Wales, UK
e-mail: mmiloccoborlini@cardiffmet.ac.uk

J. Acott-Davies
e-mail: st20027115@outlook.cardiffmet.ac.uk

A. Califano
Dipartimento Storia, Disegno e Restauro Dell'Architettura, Sapienza Università di Roma, Rome, Italy
e-mail: andrea.califano@uniroma1.it

M. F. Shahidan et al. (eds.), *Urban Narratives: Exploring Identity, Heritage, and Sustainable Development in Cities*, Advances in Science, Technology & Innovation, https://doi.org/10.1007/978-3-031-48517-6_7

often associated with the rise of modernity and consumer culture, which led to the most contemporary shopping mall (with their advantages and disadvantages).

A phenomenological approach to arcades, galleries, and porticos offers a rich and nuanced perspective on these structures' cultural and historical significance. By examining how we experience and perceive them, we can better understand their role in shaping our built environment and cultural heritage through time. Hence, our approach to the paper is to not consider a building as an object or terms such as mass, volume, and bulk. However, instead, we suggest that the intangible dimension of existence and experience is the key to understanding these places.

Through this phenomenological approach, we aim to discuss whether these arcade spaces are in a crisis of losing their identity in the contemporary world; we consider their modern counterpart (the shopping mall) with the same phenomenological approach: investigating the essence, leaving the conclusions open to further investigations. However, concepts such as place and non-place (Augé, 1995) help inform the discussion of whether traditional arcades have lost their intrinsic value of place or have their spirit of place (Norberg-Schulz, 1980) has transformed into a non-place through globalisation (mallisation).

The human relationship with space and its lived experience is crucial to understanding this discussion. Place and space must be perceived through the body, as our perception and experience of space are in constant flux, not fixed or absolute. Hence, we consider these places and their development over time, delving into their origin and transition into a modernist era. These places become grounded within their historical surroundings, and questioning their origin leads to better comprehending their intrinsic spirit (Norberg-Schulz, 1980).

2 Architectural Background

> By the end of the eighteenth century there were exchanges and colonnaded shopping pavements, a growing population with enough wealth and a desire for luxury goods, unconsidered trifles and personal adornment. It was an age when people loved to promenade, to keep in fashion, to be seen, and when window shopping was as attractive a diversion as visiting pleasure gardens and assembly rooms. (Mackeith, 1982 p. 26)

United Kingdom. As Mackeith (1982) reports in her research, we can state that arcades were born in response to the increase in population and their archetype was determined by a combination of developers, financial institutions, and classical architecture scholars, which directed where the arcades were built and whether they are a lasting commercial success. According to the same author, the retail industry remained the same until urban expansion was made possible through financial prosperity and improved transportation. The expanding market of goods and early beginnings of globalisation allowed a more significant consumption of goods; hence, Mackeith (1982) informs that the enthusiasm of new developers wished to create a shopper's 'paradise', a 'bazaar' of miscellaneous goods as well as a meeting point in the late nineteenth century.

Wales. Considering these historical elements, we can focus on the Cardiff arcades. Cardiff is known as the 'city of arcades' due to its high concentration of these city passages in a small urban area. The first arcades of Cardiff were built in 1858 as a regeneration process towards the first shopping paths in-between the urban fabric. These 'urban corridors' have a recurring typological element, as a portal is usually fenced (to be open during working hours). Generally, these are two stories with rows of commercial units, and the glass roof connects the two inner facades; generally having a linear plan in which the elevations all share a typological character (Fig. 1a, b).

(a) (b)

Fig. 1 **a** St Mary Street entrance to the Royal arcade in Cardiff city centre. Nelson, 2022. **b** View along the Royal arcade in Cardiff from the St Mary Street entrance. Nelson, 2022

Italy. Within the Italian territory, it is impossible to identify a city where the 'covered shopping street' system has represented a generative sign of the urban landscape, becoming an intrinsic typological character. This phenomenon is directly connected with the historical construction period and the type of commercial street spreading in Italy. The Italian galleries, born mainly in the second half of the nineteenth century, are characterised as monumental architecture, a city element with a pre-eminent urban function and a commercial one. Therefore, they are very different from the first Parisian passages and born exclusively as sheltered spaces to favour private commercial interests. The individual galleries in the major Italian cities are characterised as large-scale architecture (or the intent of acquiring the value of urban centrality), therefore, inheriting some characteristics from the European panorama aimed at transforming the cities in which they were born. The Galleria Vittorio Emanuele II (Milan) is probably among possibly the most representative of this reworking of the covered road and, for this reason, it represents one of the most popular on the European scene (Gioeni, 1995). The project was selected after three different competitions which, initially, the idea of renewal did not include a gallery but an arcaded street. Instead, the neighbourhood and narrow medieval streets would suit the size of the modern city. Giuseppe Mengoni designed the Milanese gallery in 1865 as an urban system that connects Piazza Della Scala with Piazza Duomo on the central axis; as a result, its entrances constitute a new architecture of the urban fronts. Mengoni's gallery is therefore constituted as a real designed space, both concerning the choice of the site, which, from now on, will always be related to crucial areas of the city (Fig. 2a, b). Furthermore, concerning the physical definition of the gallery and, with gutting and reconstructions, heights bold and colossal, they act not only on the new factory but on the city itself. The façade system, reminiscent of the triumphal arch scheme, is far from the entrance to the nearby Cristoforis gallery. Another example of this architectural typology is the Galleria Umberto I in Naples. Built between 1887 and 1891 to renovate the city and reclaim the infamous and unhealthy blocks around via Toledo, the Neapolitan gallery has the same cross-shaped plan with a central dome as the Emanuele gallery, but its weighted orientation connects via Toledo, Piazza Trieste e Trento with the San Carlo Theatre and the Maschio Angioino. The theme of the monumental facade becomes an even more complex element in the Neapolitan case where, at the entrance through a triumphal arch, a curved portico is added, which, with its colonnade, dialogues with the opposite theatre. Finally, other important examples can be found in Turin where the French influence had introduced the architecture of the passages as early as 1856. There are still three existing galleries today, while in Rome with the Colonna Gallery (1922) (today Galleria Alberto Sordi), located in the historic centre, blends some aspects of the Parisian passages with all the Italian architecture that precedes it.

3 Etymology

> [...] A vaulted place, open at one or both sides; an arched opening or recess in a wall. Obsolete.
>
> [...] 'A continued arch' (Johnson); a passage arched over; a walk formed by a succession of arches having a common axis and supported on columns or shafts. Also applied to an avenue similarly arched over by trees or shrubs; and extended to any covered avenue, esp. one with rows of shops or stalls on one or both sides.
>
> [...] Architecture. A series of arches on the same plane: 'a series of arches, either open or closed with masonry, supported by columns or piers.' Parker Gloss. Arch. 'In mediæval architecture, an ornamental dressing to a wall, consisting of colonnettes supporting moulded arches.' Gwilt. (= French arcature.) (Oxford Dictionary, 2022)

These current definitions provided by the Oxford Dictionary (2022) define what an arcade is. It is to be noted that the first lemma is specified as obsolete. However, the common denominator between these three definitions is

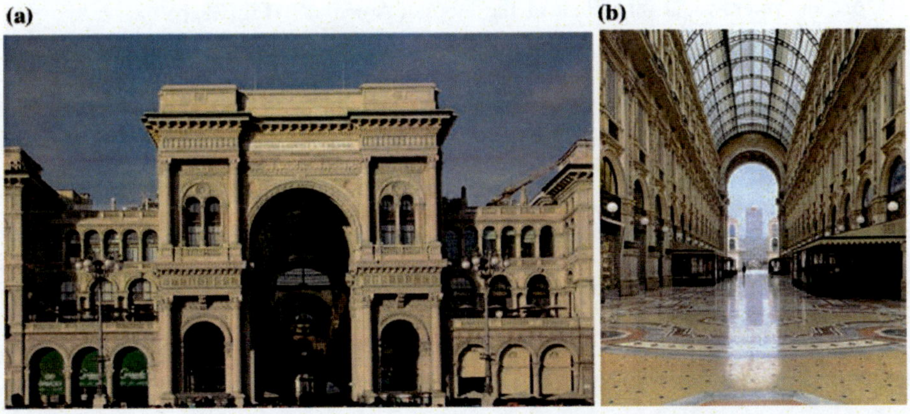

Fig. 2 **a** Galleria Vittorio Emanuele II, Milano, veduta dal sagrato del Duomo. Paolobon140, 2019. **b** Galleria Vittorio Emanuele II. MassimoZ75191, 2020

Fig. 3 **a** Interior of Galleria Umberto I, Naples; view from the entrance heading North (towards S. Brigida street)-Ober, 2020. **b**, Wyndham Arcade, Cardiff. Caulfield, 2017

always that of a covered path with ingress on either side. In the words of Rudofsky in 'Streets for People; a primer for Americans' (1981), we perceive that it is necessary to clarify the etymological meaning in the European panorama of the word 'arcade' in relation to the 'porch' (porticos). What emerges is a slight linguistic confusion, a loss in translation, which he utilises in a non-proper way, especially in Europe, which redefines the meaning of the terms according to the place of belonging. The author's analysis from the mid-twentieth century expresses that the word 'gallery' is misunderstood in its English transposition (Rudofsky, 1982); this means that according to the localisation/country in which we are (where we are situated), we may find big differences in the architectural meaning of the same word. Thus, the word gallery can mean 'porch', 'arcade' and 'square' at the same time; this is wrong typologically and architecturally speaking.

Rudosfky recognises the typological differences between the various lemmas by defining the 'Porch: a Porticus is "a colonnade a covered ambulatory, which consists of a roof supported by columns placed at regular intervals, and usually joined to a building even if it sometimes appears as a separate structure" (1982, p. 60 trans.).

'Porch' (porticus) etymologically is close to the word 'Door' (*porta*) and 'Port' (*portus*, entrance).

'Arcade' refers to an inhabitant of the utopian Arcadia, which is directly connected to 'life in the open air'; therefore, its correspondence in an architecture that is between closed (blind) walls is paradoxical. The only reference to the natural aspects can be found in the use of glass in the roofs and ceilings: it permits the natural light to pervade the space and therefore allows the user to recognise the passage of time.

Thus, the porticus mostly faces an open space, instead, the gallery envelops in an introverted way. The difference is that the porticus, in the act of protecting from the sun and bad weather, is defined not only as an urban element of aggregation but also with its own—still current climatic function (Rudosfky, 1982). However, the arcade is a completely closed element with a glass cover and does not fulfil this function. The porticus is open to all intents and purposes and, above all, it is a place where various social functions took place from the Greek city (stoá) to the contemporary one. The advantage of the Porch is its indeterminacy from a functional point of view, making it adaptable to any activity that can be carried out on the street, regardless of the historical period.

The arcade, on the other hand, was born as a place that must respect the closing or opening hours of the activities within its borders. We need to consider the size (width) of an Arcade, forced between two walls, that doesn't leave enough space to spread activities other than the 'transitional promenade' (depending on its shape). Specifically, if we think of a UK arcade, generally we will find a space that is only a few metres wide; instead, when we look at the Italian galleria style, such as Galleria Vittorio Emanuele, given the considerable size it was/is possible to opt for different types of social and cultural activities, including large-scale ones (as a concert, for example).

The gallery can be considered a typological adaptation of the Porch to the nineteenth-century city, but in this evolution, it has acquired from the point of view of the specificity of the 'object', probably losing its urban relationship (Fig. 3a, b).

4 Theories, Aesthetics, and Phenomenology

Following the previous historical excursus, it is now possible to state that arcades, galleries, and porticos are architectural features that have been prevalent throughout history and have played a significant role in shaping urban spaces. Reviewing these architectural elements with a phenomenological approach emphasises the importance of conscious experience and subjective perception the importance of conscious experience. It is subject to the cultural and historical significance of arcades, galleries, and porticos. It can help us better understand their role in shaping our built environment towards their most contemporary 'relatives', the shopping malls.

One notable example of this approach is in the work of philosopher Maurice Merleau-Ponty. In his book, 'The Phenomenology of Perception', Merleau-Ponty explores the relationship between perception and the built environment, arguing that the experience of space is inherently embodied. The author argues that arcades and galleries offer a unique sensory experience shaped by the physical environment and the movement of the body through space (Merleau-Ponty, 2012). Since, in the phenomenological sense, a spatial experience is a reciprocal relationship to the body, it's essential to understand Merleau-Ponty's concept of 'betweenness', where he refers to how our perception is situated between our body and the external world. In context to this discussion of the arcades and galleries, we understand the complex and layered spatial experience of the interplay between the individual and the built environment, as well as the social and cultural interactions that, within them, this notion of 'betweenness' emphasises the relational and contextual nature of our perception and how our bodily presence shapes our understanding of the world, which is paramount in understanding whether arcades become a non-place. Merleau-Ponty's idea of 'betweenness', in context to Marc Augé's concept of non-place (1995), suggests that our perception of non-places is mediated by our bodily presence and the context in which we experience them. In non-places, the individual can feel disconnected from the environment, resulting in alienation or disorientation. While arcades built before modernity do not have the inherent globalised features compared with their contemporary counterpart—the mall—perhaps these non-places, and their disorientating experience have been culturally transformed into what arcades originally were built for; a meeting place (Fig. 4a, b).

In addition, to better understand and discuss the phenomenological approach in architecture, we can recall Pallasmaa (2005), in the book 'The Eyes of the Skin: Architecture and the Senses', who explores the importance of sensory experience in architecture. Using a phenomenological approach, Pallasmaa argues that the built environment must engage all senses to create a meaningful and embodied user experience. Architectural theorist Till (2009) applies a phenomenological approach to the study of urbanism, exploring how the experience of the city is shaped by contingency and unpredictability. Till argues that a focus on the subjective experience of the city can help architects and urban planners better understand the needs and desires of their users. At the same time, social scientist and philosopher Spurlig (2010) discusses the relationship between phenomenology and the social sciences. Spurlig argues that a phenomenological approach can help us better understand the social and cultural significance of the built environment and how it shapes our experiences and perceptions. It highlights the importance of considering how architecture affects us personally and is a vibrant tool for discussing the embodied experience. These phenomenological approaches offer a valuable/critical perspective on architecture, emphasising the importance of conscious experience and subjective perception. These references demonstrate how this approach has been applied to explore the relationship

Fig. 4 **a** Die Gartenlaube, Illustrirtes Familienblatt, Ernst Keil, Leipzig. (Multiple Authors, 1867, p. 2019). **b** Napoli: interno della Galleria Umberto I. (Strafforello, 1894)

between the built environment and human experience and its potential to shape the design reflection and use of architecture.

> The people who first built a path between two places performed one of the greatest human achievements (Simmel, 2022).

Furthermore, through the theories expressed by M. Augé, T. Boettger, C. Norberg-Schulz, W. Benjamin (e.g. passages de Paris—place/non-places), and B. Rudofsky, we can focus on spatial and sociological aspects of the arcades. There are three elements to consider when reading the Italian arcades and galleries' spaces (or places). First, studying how they are inserted in their sociocultural and artistic context is necessary. Norberg-Schulz (1980) helps us with *Genius Loci*, where he not only reminds us to be respectful of the context, but also reminds us of its importance as a discriminating element and a mirror of the evolution of society.

For clarity, in this context, it should be remembered that when the arcades were born, they related to the spirit of the place at the time of their construction, and the same applies to the project that was made of them. We deduce that the spirit of the place changed after the arcade was built and inserted into its urban and social context; consequently, the contemporary city has changed again and repeatedly the spirit of the (urban) places; therefore, all the elements are contained within it, such as the arcades.

Rudosfky (1982) argues that there is an involution in the use of arcades/galleries; tourism transformed these spaces into an "incomprehensible street landscape" (1981, p. 82). We state that gentrification (globalisation) and emptying cities have amplified this phenomenon, transforming the city centre into a commercial gallery/mall with an open sky (Fig. 5a–c).

We consider the experience of the Cardiff arcades conflicting, thus, inspired by Marc Augé's words, simultaneously a place and a non-place. The author states that non-places are characterised by a repetition of the ritual of passage; when the repetition is too strong and sudden, the user completely loses the sense of place (as in most contemporary commercial malls or airports). We must question the value and the concept of space (or place) of these architectures that have the passage and 'movement of people' as their main features.

Intertwining the words of Augé (1995) with those of Norberg-Schulz (1980, 2000), we can state that these places find meaning only when connected with their sites' cultural and social environment; in other words, with their spirit. Regardless of the use of these in-between architectures, we must look at their architectural, social, and cultural function.

5 A Phenomenological Passage: Paths and Streets

> In the summer, the Gallery is almost completely abandoned by the Italians. In their place, those particularly pathetic vagrants that are modern tourists make their way as storm troops or as stragglers in an incomprehensible road landscape (Rudofsky, 1982 p. 82 trans.)

What is a road? The arcades and galleries are, in fact, an evolution of the road. They are a transformation and a reinterpretation of the concept of the street, an urban device and

Fig. 5 **a** Passage des Panoramas, Paris. Patafisik, 2019. **b** Lexicon shopping malls. Slaughter, 2018. **c** Via de' Tornabuoni. Pollastrini, 2020

architectural element that unites two (or more) points in space. This element is defined by a connection between two points, and ad Lynch (2006) recalls a place in and on which sociality takes place. Thus, the street has a dual soul: one defined by a quick connection and the other by social and commercial aspects that create new meeting places. In the gallery/arcade, the connection *passage* is always slow, so the second aspect is favoured over the first. So, it is easy to understand that when experiencing this transitional space, the crossing time must be considered.

The main difference is that the gallery/arcade is made only for pedestrians; the road instead is made for fast mobility, so we deduce that there is a solid multi-temporality aspect. In the galleries, the temporality is unique. That is considering all users to have a similar minimum crossing time. The 'slow' mobility favours the creation of space for social interaction.

"In Milan we spent most of our time in the vast and beautiful arcade, or gallery or whatever it's called. […] Tables everywhere on these marble streets, people sitting to eat, drink and smoke, crowds of people walking, this is a gallery" (Twain, 1880, p. 119 trans.) (Fig. 6).

However, by their very nature, the arcades encourage the commercial *spectacle*. As Augé (1995) and Benjamin (1999) recall, the commercial gallery can be a meeting place in two different historical periods. However, a commercial aspect will always influence the possible sociality connected to it. This involves the trivialisation of the concept of the street and, thus, the lack of leisure, where the user's experience is subject (primarily) to consumption times. The urban aspect is inherent in the meaning of gallery; this is not always valid in the arcade or passage. The passage was born as a space exclusively linked to commerce, while the gallery instead possesses characteristics of urban connection (Dameri & Dellapiana, 2011).

Fig. 6 Octagone of Galleria Vittorio Emanuele II, Catalogue # 4608-Brogi, 1880

The phenomenological approach is also demonstrated in the work of Walter Benjamin. In his essay 'Paris, Capital of the Nineteenth Century', Benjamin explores the cultural significance of arcades in Paris during the nineteenth century. He argues that the arcades served as a microcosm of modernity, offering a space for commercial and social exchange while also reflecting the fragmentation and alienation of modern urban life (Benjamin, 1999). For example, Elliot B. and Frisby D. report from W. Benjamin critiques Parisian arcades (2011; 2004) that the arcades reflected the bourgeoisie's consumerism in contrast with antiquity and modernity. Defining the arcades as the 'in-between' point between the two because a representative of the commercial evolution and as Elliot (2011) states: "[The] externalised drawing[s] room of the bourgeoisie and [...] an extrusion of the accursed interior into the metropolitan exterior". According to the Authors, Benjamin refers to them as labyrinths (2011; 2004) and icons of passage and transitions (see M. Augé). In fact, according to Benjamin in Frisby (2004, p. 275): "The ultimate and innermost affinity of modernity and antiquity reveals itself in their transitoriness" representative of the city of the first part of the twentieth century. So we have again the concept of passage, simultaneously expressed by M. Augé. Furthermore, we relate to Elliot (2011) words that arcades were, and still are, "dominated by a sense of the nineteenth-century structures as places of collective forgetfulness and alienation"; this is a very similar concept to M. Augé's research.

Fig. 7 La galleria di milano (draft). Carrà, 1912

We question whether the latter notion is relevant and if 'time' has modified these places. Therefore, to answer these questions, we suggest that arcades (and galleries) lie as a passage that insinuates themselves between the urban fabric. Phenomenologically, the threshold of the arcades in question allows seamless locomotion through the urban fabric and a dualistic perception of two spaces; a melding of interior and exterior and it is defined by spatiotemporal motion. According to T. Boettger, arcades are: "inherent in all the meanings, whether they manifest themselves in terms of space or time, is the common element of a transition, a threshold, a process, a measured route or something that passes. Something happens—the movement becomes a measured experience" (Boettger, 2014, p. 30).

The passage between the two buildings' trilithic shape by glass allows the feeling of being enclosed. However, simultaneously our vision is confronted with an exterior view, an ambiguity of space and time. The arcades' threshold or 'in-betweenness' has the danger of disorientation. Alienating as in a labyrinth—the viewer's bearings and consumption. "Benjamin ascribes significance to the fact that the first arcades constructed in the early nineteenth century were panoramas, 'views of the whole', dedicated to simulating for the viewer a situation of being present at another place in time" (Elliot, 2010, p. 92). Suppose we compare this dualistic nature of an arcade's nature of presenting the consumption of two or more spaces/places to the occupant. In that case, we can compare the experience of space and time to the paintings of the Futurists (Fig. 7). The paintings of Boccioni depict: "Time and motion are at the heart of Futurist art theory. It also follows that Bergsonian philosophy plays a far more central role in Boccioni's Futurist doctrine" (Henderson, 2013, p. 204). Hyperspace philosophy's interpretation of the fourth dimension is connected to time and motion; a visual depiction of space serves as a valuable tool, aiding in comprehending and enhancing the *Flaneur*'s experience by clarifying spatial understanding (Benjamin, 1999). We need to consider whether arcades are a place or a non-place, as to *feel* in a *place*, we need to be between two 'boundaries' (*margins*) to feel protected as if we were in an interior place, with the feeling to 'know' that place. This could happen by recognising a place as 'familiar' or making us feel comfortable.

Considering the bourgeoisie's consumerism and the alienating feeling of the arcades, we can compare them to the most contemporary shopping malls, according to the previous authors. The connection between consumerism and shopping malls (non-places) lies in the fact that many of these globalised spaces are emblematic of consumer culture. These are often designed to be generic and interchangeable—the opposite of the arcades—lacklustre in personality, sharing similar a familiar tectonic across different locations. These sites of consumerism allow consumers to feel familiar and comfortable in any mall they visit, allowing the Flaneur to drift endlessly.

In this sense, consumerism is intimately tied to the creation and proliferation of non-places. The desire for uniformity, efficiency, and convenience that underlies consumer culture is reflected in the design and function of non-places, which are intended to be easily navigable, comfortable, and unremarkable. However, this homogeneity also contributes to a sense of placelessness and disconnection from history, culture, and community, which can be seen as a negative aspect of consumerism and non-place.

6 Conclusion

In conclusion, we can read the old arcades and the contemporary malls with a phenomenological approach. For example, when walking through an arcade, we consider the experience of this void and time as directional (orientational), as opposed to the modern interpretation of an arcade—a shopping centre—where the superimposed layers present to the occupant a lack of direction; non-spatiotemporal, an acceleration of history and time. In the phenomenological sense, space is not directional in and of itself, as its orientation of viewing the space is through the observer. The bodily experience gives meaning to the idea of enclosure and betweenness, and this embodied approach brings meaning to the experience of the body in the arcades. As we have suggested through Merleau-Ponty's idea of 'betweenness', in context to Marc Augé's concept of non-place (1995), Malls (arcades' contemporary counterpart), perhaps these non-places by definition have undertaken a transformative action through globalisation, where these sites through its excess of time and acceleration of history and its disorientating experience have been culturally transformed into what arcades originally were built for, a meeting place, a place that is more relevant to the globalised world. Unlike their historical counterpart, malls have no sense of direction, but in contemporary times, like futurism, an excess of time results in a stimulating, dynamic experience of movement in time and space. The most significant difference between the sensation of entering the arcades discussed, and their modern counterparts is the sensation of being circumscribed in an interior space. Arcades have an ephemeral boundary, whereas the feeling of an interior space is less prominent than the performative action of the door creating an enclosure in a typical shopping centre. The identity of an arcade is the threshold space between existing exterior walls where space becomes a visible body with the soft, inviting boundary of an arcade. Thus, an arcade's interior space and exterior world are still being determined. We conclude that the structure of an arcade is directional; thus, the occupants can orient the space and transition into this space/place with ease. In the words of Boettger: "The flâneur can let himself drift and be tempted without being 'misled'" (Boettger, 2014, p. 30).

This study opens up to further investigations and critiques on 'urban experiences' in globalised city centres. In fact, we assist to a 'mallisation' of the historical centre, desensitising the occupant from leisure. Nowadays, the balance between *otium* and *negotium* overlaps with the business model (Cacciari, 2004). According to Augé, super-modernity has produced anthropological non-places, contrarily to modernity, which have not integrated the new spaces with the ancient places. The shopping centres (Busa, 2006) and the designer outlets of the Italian context created new spaces alienated from the sociality and historicity of the places to create 'commercial passages'. In contrast, the passages of the streets of Cardiff were converted into becoming places to be. Therefore, we question whether arcades can become places again and whether the globalised world has recontextualised or its intrinsic value of the spirit of the place has transitioned into a non-place (mall); a place to be.

Acknowledgements All Italian direct quotations have been translated (trans.) into English by the authors, unless stated otherwise.

Thanks to Sara Edwards for English proofreading.

References

Augé, M. (1995). *Non-places: Introduction to the anthropology of supermodernity*. Verso.

Benjamin, W. (1999). *The arcades project*. Harvard University Press.

Boettger, T. (2014). *Threshold spaces: Transitions in architecture: Analysis and design tools*. Birkhäuser.

Busa, A. (2006). Vita e morte delle grandi gallerie commerciali italiane. In *Amendola G. La città Vetrina, Liguori Editore, Napoli* (pp. 39–56)

Cacciari, M. (2004). *La città*. Pazzini editore.

Dameri, A., & Dellapiana, E. (2011). Gli abbellimenti interni dall'Europa a Roma. Architettura e progetto urbano nelle gallerie e nei passages. In A. Lo Monaco, & V. Nicolucci (Eds.), *La Galleria di Piazza Colonna*. Allemandi.

Elliott, B. (2010). *Benjamin for architects* (1st ed.). Routledge. https://doi.org/10.4324/9780203833872

Frisby, D. (2004). Walter Benjimin's arcades project: A prehistory of modernity. In M. Hvattum, & C. Hermansen (Eds.), *Tracing modernity: Manifestations of the modern in architecture and the city* (pp. 271–290). Routledge.

Gioeni, L. (1995). *L'affaire Mengoni: la piazza Duomo e la Galleria Vittorio Emanuele di Milano, i concorsi, la realizzazione, i restauri*. Guerini.

Henderson, D. L. (2013). *The fourth dimension and non-Euclidean geometry in modern art* (2nd ed.). MIT Press.

Lynch, K. (2006). *The image of the city*. MIT.

Mackeith, M. (1982). *History and conservation of shopping arcades*. The University of Edinburgh.

Merleau-Ponty, M. (2012). *The phenomenology of perception*. Routledge.

Multiple Authors. (1867). *Die Gartenlaube, Illustrirtes Familienblatt*. Ernst Keil.

Norberg-Schulz, C. (1980). *Genius loci: Towards a phenomenology of architecture*. Academy Editions.

Norberg-Schulz, C. (2000). *Architecture: Presence, language and place*. Skira Editore.

Oxford Dictionary. (2022). [on-line] at: https://www.oxfordlearners-dictionaries.com/definition/english/arcade [July 2022].

Pallasmaa, J. (2005). *The eyes of the skin: Architecture and the senses*. John Wiley & Sons.

Rudofsky, B. (1982). *Streets for people; A primer for Americans*. Editori Laterza.

Simmel, G. (2022). *Essays on art and aesthetics*. University of Chicago Press.

Spurlig, L. (2010). *Phenomenology and the social world: The philosophy of Merleau-Ponty and its relation to the social sciences*. Routledge.

Strafforello, G. (1894). *La patria, geografia dell'Italia/Parte Quarta—Italia Meridionale, Napoli*. Unione Tipografico-Editrice.

Till, J. (2009). *Architecture and contingency: On the urbanism of exceptional times*. Routledge.

Twain, M. (1880). *A trump abroad*. American Publishing company.

Revisiting Russo-Japanese Heritage in the Contemporary Chinese Context: The Identity Crisis of Dalian in Postcolonial Cultural Placemaking

Yufei Li

Abstract

The paper examines the identity crisis of Dalian, a postcolonial city in Northeast China, in the dilemma of heritage-making and cultural placemaking. Born in the turbulent transition between the nineteenth and twentieth centuries, Dalian experienced the intertwined power wrestling among East Asian countries, resulting in the complex layering of its urban image and cultural identity across the city's changing colonial regimes, from Imperial Russia to Imperial Japan, before being taken over by the People's Republic of China. The city's image reflects a multicultural pastiche following the tangled trajectory of its past. However, to fit the historical complexity into a socialist framework, political sensitivity often outweighs heritage awareness in the emergence of postcolonial nationalism. The paper explores Dalian's approach to cultural placemaking in its recent revival as an 'Cultural City of East Asia'. Through historical analysis and on-site fieldwork, the research identifies the city's hidden fusion and recognition of multicultural heritage, which has remained sensitive and not yet been officially advocated. Following comparative studies, the paper investigates the remaking of Russian and Japanese cultural streets in Dalian in light of recent controversies, which further intensifies the debate about adopting postcolonial heritage in the contemporary Chinese context.

Keywords

Architectural history · Postcolonialism · Urban planning · Placemaking · Colonial heritage · Culture

Y. Li (✉)
Department of Architecture, University of Cambridge, Cambridge CB2 1PX, UK
e-mail: yl667@cam.ac.uk

1 Introduction

A city's identity is fluid and unsettled, being defined and redefined constantly throughout its history of development and reformation. The image of the city is ever-changing, comprises not only the concrete structures of buildings, roads and landmarks that patched and filled the map, but also the tales of the past and the visions of the present. Arjun Appadurai argues that the sense of place is a product of history (Appadurai, 1996). The city's present identity is shaped by its past, in which we see the complex layering of the city's trajectory connecting the local meanings of place to the interactions and mingles of the vast world. Following those pathways across time and space, histories become memories, and imaginations constitute allusions.

If every recognition of identity would possibly start with a name, Dalian marks its identity as complicated enough for interpretation with its transition of names. Its current name Dalian (大连) in Chinese came from the same Chinese/Kanji characters as its Japanese name Dairen (大連)—which was a transliteration of Dalniy (Дальний, meaning 'remote'), a Russian name of the city at the very beginning of its founding (Jiang et al., 1984). Dalian did not grow as a traditional Chinese city. It was a fishing village located at the tip of the Liaodong Peninsula in Northeast China—or Manchuria as once known by the rest of the world. Due to its sensitive location being a gateway to the Bohai Sea and the Manchuria continent with an unfrozen coastal line, the city was born in the spreading and rivalry of imperial powers in the Far East in the late nineteenth century. Instead of organic growth, it underwent a sudden burst of planned development under colonisation, first by Imperial Russia (1898–1905), and then by Imperial Japan (1905–1945).

Like its name, the colonial traces did not fade into none, but linger in the city's contemporary urban scape. As Doreen Massey believes, a place is not cut off completely from its past, never has it lost connections with other places (Massey, 1995). Although the colonial past of Dalian—and

M. F. Shahidan et al. (eds.), *Urban Narratives: Exploring Identity, Heritage, and Sustainable Development in Cities*, Advances in Science, Technology & Innovation, https://doi.org/10.1007/978-3-031-48517-6_8

Manchuria to a further extent—is often referred to as 'negative legacies' for the region in the post-war accounts, colonial influence, as an inseparable part of the place's past, is still inevitable to be witnessed and addressed in the modern Chinese context. A daily ride on the tram in Dalian's city centre still displaces one to a scenic collage where modernity meets heritage, and the locality meets the otherness. From Dalniy to Dairen to Dalian, patches of historical remnants transcend time and are scattered in the form of distinctive cultural constructions among the urban landscapes.

In David Lowenthal's discussion about nostalgia, historical events constitute memories and act as the bridge connecting the past and the present, often in certain tangible forms for us to revisit constantly, such as in the form of landscape (Lowenthal, 1975). History is rooted geographically within a place where the past encounters the present in constructing a site of memories (Azaryahu, 2003). It reveals collective memories associated with the place, the landscape, as well as the city, and thus becomes an important material in establishing the local identity. The city of Dalian changes patently in its outer features—in its name and regime, while its inherent identity undergoes a delicate transformation that weaves in the everyday urban fabrics, which absorb and transform its multi-colonial cultures through different approaches. Meanwhile, the sensitivity towards 'negative' colonial legacies has potential impacts on shaping the current image of Dalian and interfering with the city's attempts at cultural placemaking.

The paper argues that the controversy brought by colonial heritages in the postcolonial environment results in the city's identity crisis, which is represented by both unintentional integration and vigorous denial of the colonial cultures. The paper first seeks a historical analysis of the initial founding of the city of Dalian from its Russo-Japanese colonial past, and by conducting fieldwork in Dalian via the lens of the locals of Northeast China, it observes the inherent approaches of the city in addressing the colonial heritage and postcolonial influence in both its physical and cultural constructions.

2 Dalian in Its Colonial Context

As a modern city born in the turbulent transition between the nineteenth and twentieth centuries, Dalian experienced intertwined power wrestling among East Asian countries, forming the complex layering of its urban image and cultural identity across the city's changing regimes throughout the First Sino-Japanese War (1894–95), Russo-Japanese War (1904–05) and the Manchukuo period (1932–45). It grew from a fishing village, Qingniwa, to a commercial town under Russian-European modern planning in the hope of building a 'Paris of the Far East', and was further transformed into 'the most Japanese city in Manchuria' under Japanese colonisation. After Japan's defeat, Dalian was taken over by the Soviet Union in 1945 and passed to the People's Republic of China in 1950, drawing a clear line between the socialist 'New Dalian' and its colonial past (Hess, 2011).

2.1 Russian Dalian, 1898–1905

Imperial Russia's initial interest in Dalian was strategic; the location granted the potential to develop both continental connections to Russia via a southern sub-branch of the Chinese Eastern Railway (CER) and maritime transport with its potential to develop an ice-free port (Qu, 2001, pp. 46–51). However, over the seven years of urban planning and intensive construction of the new city, Russia's vision grew into a grander ambition of establishing a port city to compete with other Chinese port cities such as Tianjin, Shanghai, and Hongkong, which were raised rapidly under other European colonial powers around the same period. As Franck Billé and Caroline Humphrey mentioned, Dalian's booming urbanisation during its Russian period signals 'ambitions that were novel and audacious, but also chimerical' (Billé & Humphrey, 2021, p. 258).

Without surprise, European approaches to planning and architectural design were favoured in laying out the foundation of the new 'Russian' city. The plan of Dalian exhibits a Baroque axial system which was influenced by the spatial arrangement of European cities, especially those of St Peterborough in Russia and Paris in France (Qu, 2001, p. 53). Meanwhile, as an attempt at modern urban planning, it also brought in inspiration from the Garden City Movement that thrived since 1898. As seen in a Russian plan of Dalian dated around 1899, urban constructions were centred around different nodes of the city, which were in form of squares and traffic roundabouts and interconnected by radial roads and boulevards. Five districts were planned in three major clusters, intercepted by parks and greenbelts as elements advocated in the garden city (Qu, 2001, pp. 52–53; Tang, 1995, p. 45) (Fig. 1).

A Russian Administrative District was set to a close approximation to the city's primary developments of the wharves and the railway station. Its importance was accomplished by the immediate construction of colonial institutional buildings such as the Chinese Eastern Railway Office/Dalniy City Hall (1898), Chinese Eastern Shipping Association (1902), and a lavish housing district in a combined Russian and German style as accommodations for officials. A wooden Russian Bridge stretched from the front of the shipping association southwards over the railways, and led the way to the business and financial district. At the heart of the plan, ten streets radiated from Nikolayevskaya

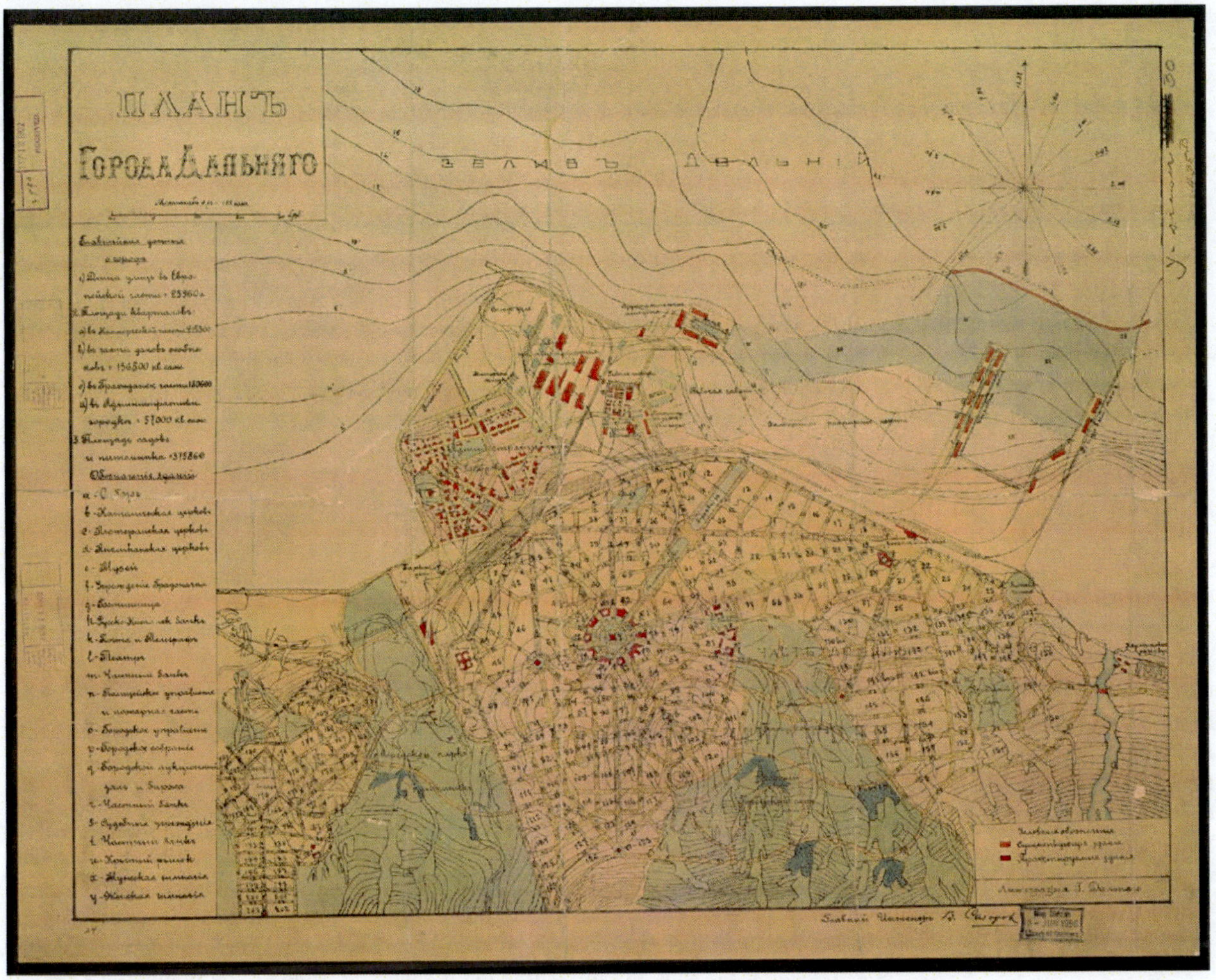

Fig. 1 Earliest masterplan of Dalian showing the urban zoning and axial system adopted in the planning. *Source Plan goroda Dal'niago*, Sakharov (1899), Geography and Map Division, Library of Congress. LC call number G7824.D3 1899.S3

Square (Nichola Square, 1899), with ten individual building blocks facing the concentric greenery, congregating financial and authoritative power in a prominent spatial configuration. Incorporating with the European city centre were two European residential districts set to the south and east of the city, in contrast to a Chinese Town to the western corner segregated by the West Park (Liu et al., 2021, pp. 4–5; Qu, 2001, p. 52; Xiao, 1988).

With a group of selected planners from Russia and Germany, 'Dalniy' was not designed as anything closer to a Chinese city from its planning stage but as another annexed territory like Vladivostok that Russia anticipated engaging in over a long period. However, apart from the major constructions in the Administrative District, most of the planning only began to take vague shape (Denison & Ren, 2016). The ambition was cut short by Russia's defeat by Japan in the Russo-Japanese War. Following the Treaty of Portsmouth, the tip of Liaodong Peninsula was handed over to Imperial Japan, who also took over the half-accomplished colonial plan of the Russian Dalian.

2.2 Japanese Dalian, 1905–1945

As an outstanding example of colonial development, Japanese Dalian appeared frequently in various cultural products under the propaganda policies of the Japanese colonial authorities. In a film by the South Manchuria Railway Company (SMR), a cruise ship set off from Kobe, Japan, and arrived at the Port of Dalian. The welcoming crowd stood in the upper corridor of the arrival hall, joyfully waving to the visitors walking across the bridge. People stepped out of the hall and began speculating on a modern, stylish urban life in Dalian before taking the 'Asia' train that sent them to the northern hinterland of Manchuria (Mantetsu Film Production Studio, 1937). As a port city, Dalian was once visitors' first arrival point via the sea and the last piece of the continent under the Japanese regime before stepping into the rest of Manchuria. By incorporating colonial expansion into infrastructural development, similar to Russia's approach to the CER, the SMR set up its railway network from Dalian and spread northward, making Dalian a popular destination for overseas travellers from the 1920s to 1930s.

With inputs from Japan, the city experienced a radical transformation from a Eurocentric establishment to an experimental spot of early Asian modernisation. The Lvshun-Dalian area formed the Kwangtung Leased Territory of the expanding Japanese Empire. While Dalian, once a Russian-European focal spot, was urgently awaiting an update on its cultural identity following the shift in power.

A round of decolonising Russian culture started first with names of places. The change from Dalniy to Dairen was among one of the slightest alterations. In the first few years of Japanese occupation, most places underwent the renaming process from Russian to Japanese wordings. Representative examples include the change of Administrative District to Russia District, Russia Bridge to Japan Bridge—rebuilt entirely in reinforced concrete, and Nichola Square to Grand Square (*Ōhiroba* in Japanese). The main commercial streets were combined under the name of Naniwa Machi—as *Naniwa* being the ancient name of Osaka, a renowned commercial capital in Japan.

Interestingly, after the Meiji Restoration of 1868, the trend of Westernisation flourished in Japan. At the time of Japan's colonisation in Dalian, western cultures, designs, and products symbolised an advance in civilisation, being the main reason behind Japan's inheritance of both the planning and architectural achievements left by its colonial predecessor. The Russian-completed buildings, mainly within the old administrative district, were re-inhabited by Japanese institutes. While those European buildings were only completed to the north of the railways by Tsar Russia before 1905, Japan took over the ambition and completed the European-inspired architectural cluster around the Grand Square with new buildings from 1908 to 1936. The SMR architects, many with a Western architectural background, adopted the European neo-classical styles with simplified ornaments and moulding, and some even combined Japanese cultural symbols such as *chidori-hafu* (plover gable) and *yamaboko* ('mountain halberd', a decorative wooden float) into design decorations, creating an early 20th-century eclecticism in an East Asian approach (Qu, 2001, pp. 165–170). The pastiche of architectural styles resulted in a myth of identities in Dalian's city centre, where the distinction between Russian and Japanese constructions was blurred through Japan's Westernisation and their attempts at duplicating, adopting, and modernising European classicism.

Further simplified modernism was implemented in everyday urbanscapes, especially the commercial streets, making the image of Dalian catch up with the rapid pace of urbanisation as seen in the Japanese-produced travel brochures and documentaries (Manchuria Railway General Bureau, 1938; Lvshun Fortress Command, 1939; Akutagawa, 1940). On shopping streets like Naniwa Machi, there were Japanese stores ranging from double-floored individual shopfronts to multi-level malls. Concrete facades, glazed windows, overhanging flat eaves, electric lighting, neon signs, and billboards formed the bustling streetscape similar to Ginza in Tokyo. On top of the modern glamour, archways with hanging lanterns went across certain sections of Naniwa Machi, imposing a strong Japanese identity with replications of traditional signage (Fig. 2).

More than mere decorations, the Japanese were migrating cultural landscapes to their colonial continent by recreating iconic Japanese sceneries in Dalian for the homelike hospitality of their migrants. In the aerial perspective of Dalian drawn by Hatsusaburo Yoshida, trees in the parks, gardens, mountain alleys, and lanes are all coloured in pink, indicating an exaggerated view of cherry trees in Spring (Fig. 3). The Longwangtang Cherry Yard in Dalian, constructed in 1926, was renowned for ten aged Japanese cherry trees and one rare century-old 'Star Magnolia' transplanted from Osaka. The cherry garden played the role of a main cultural venue for *hanami*, the traditional Japanese custom of people gathering and appreciating the transient seasonal blossom. The cultural scene was not complete without a view of ancient temple roofs. Major Buddhist temples in Japan, such as the Higashi Hongan-ji and Nishi Hongan-ji, set up their overseas branches in Dalian with similar appearances as their Japanese counterparts. With the Tang-imitating Zen style, the temples showed their uniqueness and proximity to traditional Chinese culture that contrasted with the Western-induced modernism in Dalian, further complicating the city's identity with the multicultural collage.

3 Retracing the Colonial-Influenced City Identity in Contemporary Dalian, 2021

At the time of drafting the paper, Dalian was successfully elected as the 'Cultural City of East Asia' of 2024 by the Ministry of Culture and Tourism of China for its

Fig. 2 Night view of Naniwa Street, as portrayed in the tourism brochure of Dairen. *Source The Ginzo of Dairen, Night View of Naniwa Machi*, Manchuria Railway General Bureau (1938). In the public domain

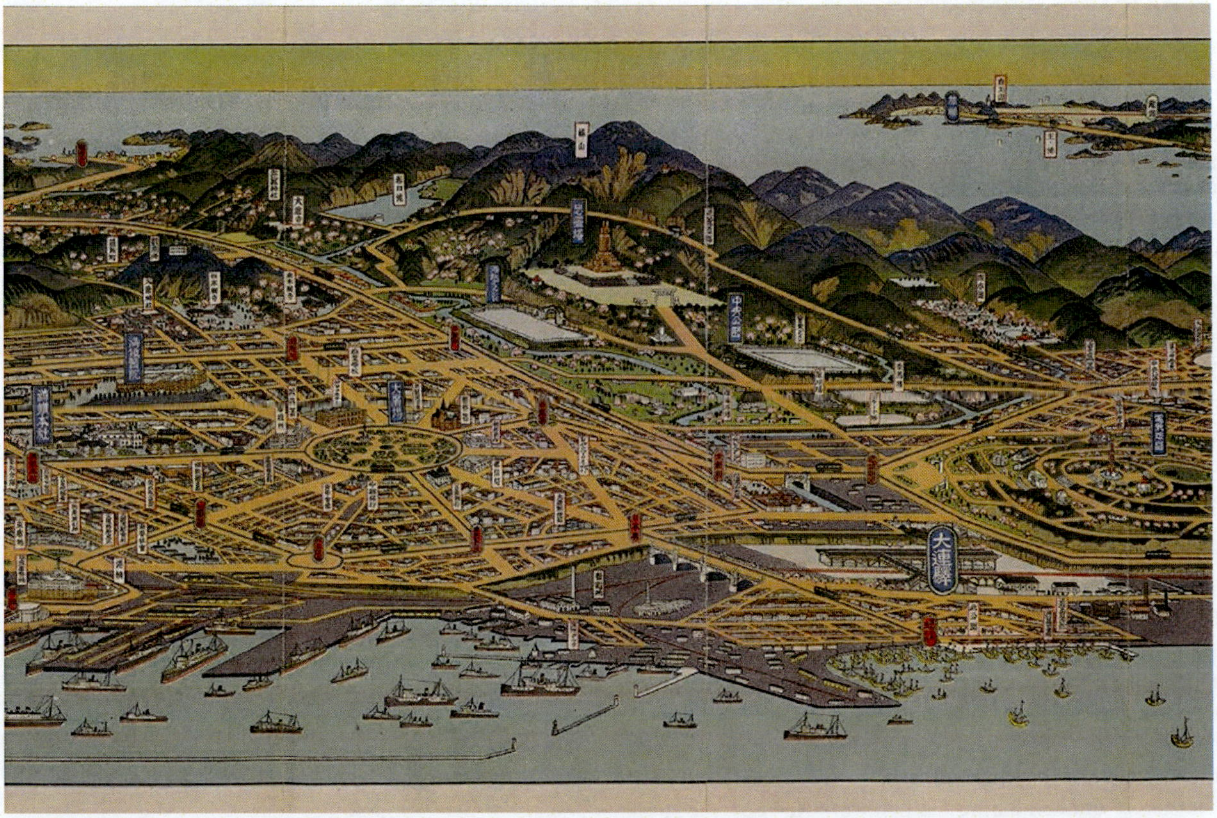

Fig. 3 Aerial perspective of Dalian (part) in Japanese painting. *Source Dairen*, by Yoshida (1929), Dairen: Japan Tourist Bureau Dairen Branch. In the public domain

Fig. 4 Itinerary map of the Dalian field trip with places mentioned in this article. *Source* Map by the author, 2022

placemaking efforts in relation to its construction of the cultural-historical urban image and economic contribution to the greater region of China-Japan-Korea (Peng, 2022). Do the Japanese and Russian colonial legacies still have a place in Dalian's contemporary urban landscape? How does the city manage the exotic image in a Chinese framework? What controversies and mixed feelings do the citizens of Dalian hold regarding the complexation of place identities in relation to colonial cultures? With these questions in mind, the study conducted a five-day field trip to Dalian to trace the status quo of Dalian's postcolonial heritage landscapes and cultural placemaking.

As the research was conducted in accordance with China's COVID-19 restrictions on international and intercity travel, the field trip was carried out remotely with the help of two local participants from 19th April to 24th April 2021. They followed the author's planned map and itinerary to Dalian's old city centre and recorded the image of Dalian in extensive detail through film and photography. Due to public health restrictions and social distancing requirements, interactive activities or interviews with local residents were omitted during the fieldwork. Instead, the focus was on the observational image and experiential space of the material urbanscape following the route of the designed trip (Fig. 4).[1] Formulating an urban tour through the design of itineraries and routes became part of the research process, as Sarah Pink suggests the opportunity to understand the making of place via making the touring routes (Pink, 2008, p. 179). In the people-centred perspective of approaching and experiencing the places, participants sense the flow of urban life in their neighbourhood surroundings where 'identity is constantly being made and remade' (Friedmann, 2010, p. 162).

During the field trip, the following key aspects related to the postcolonial construction of Dalian's urban identity caught our attention and stood out for detailed interpretation in this paper. First, official instructions of place decolonisation stayed in parallel with the continuation of the urban hierarchy brought by colonial heritage. Second, an exaggeration of exoticism was present in the use of the Russian

[1] The fieldwork was initially conducted as a part of the author's PhD research project *Atlas in Motion: Visualising Manchuria in Moving Images*, see a seminar talk of the same title by the author at UC Berkeley (Li, 2021). Unless referenced with sources, photos in the following sections were taken during the field trip by Yanchun Yan and Xiaowei Li, with complementary photos provided by Li Wang and Qu Qu for the purpose of this research.

colonial heritage in Dalian's 'romanticised' placemaking. Third, in comparison, there was the attitude of hesitant, avoidance and repellence in the conservation and tourism recreation of places with a strong presence of Japanese culture. However, hidden colonial cultural influences still blended into the identity of Dalian and became part of the everydayness in the depoliticised urban life.

3.1 Decolonisation Versus Inheritance

Decolonisation set the tone for the new Dalian after it rejoined the Chinese regime. Benefiting from its advanced modernisation due to rapid development as a colony, Dalian soon became a 'model city' of the People's Republic of China (Hess, 2011). Closely associated with this title was the raging decolonisation and renewal of urban identity based on the change of power.

The most straightforward observation was again from the change of place names. Just as the Japanese erased the Russian names from the map, new Chinese names of streets and districts were used to disperse their former Japanese identity and match the core value of socialist ideology. While walking on the streets of Dalian, changes in names clearly traced the city's shifting power authorities. Names inclined towards a centralised power of imperialism were the focus of the renaming process. The city's Ōhiroba Grand Square, formerly Russian's Nichola Square, was once again renamed Zhongshan Square to commemorate Sun Yat-sen (Sun Zhongshan in Mandarin spelling), the forerunner of the Chinese revolution. Central Park was changed to Laodong Park, with *laodong* carrying a more proletariat meaning of 'labour'. Places with blunt Japanese wordings were also changed to Chinese. The commercial street Naniwa Matchi, once named after a Japanese city, was renamed Tianjin Street, with Tianjin being a major port city in China. Yamagata Dori, a street named after Japanese imperial leader Yamagata Aritomo (1838–1922), was changed to Zhongshan Road and Renmin Road, with *renmin* meaning 'people'. Japan Bridge was rebranded with a more revolutionary name of Shengli Bridge, with *shengli* ('victory' in Chinese) marking the winning over the imperial invaders.

Despite the turbulence of the renaming movements, Dalian inherited the spatial zoning and hierarchy of the colonial past to a large extent. The central districts of colonial Dalian sustained most of their functions and prominent status that defined the foundation of Dalian's current layout. City squares remained as places congregating capital and power. From Nichola to Ohiroda and to Zhongshan Square, the round plaza surrounded by concentric green bands, traffic roundabouts, and well-preserved historical buildings continues to exist as the focal point of the city. The Elder's Square, once the political centre of Kwantung Leased Territory, is still the seat of the main government bodies in Dalian today after it was renamed Renmin Square. The former high streets also stuck to their commercial nature regardless of the changes in architectural streetscapes. Those central spaces were built to be the icons of an advanced modern metropolis, an identity that the current authority would prefer to retain.

As an unwritten tradition, official and institutional (or rather valuable) buildings of the colonial period were passed on to those with similar functions in New China. Many official buildings used by authorities, institutes and banks during Dalian's Russian and Japanese periods were well-preserved and still in function, mostly substituted by their contemporary Chinese counterparts. Plenty of instances could be found around Zhongshan Square—Dalian Yamato Hotel, once an SMR hotel chain branch, transferred its ownership to a Chinese state-owned enterprise and has been operated as Dalian Hotel. The Japanese-operated Chosen Bank was converted to the Industrial and Commercial Bank of China, and the Yokohama Specie Bank was passed over to the Bank of China. Among the Zhongshan Square heritage buildings, a People's Culture Club of Dalian was constructed in 1951, as the first new-built around the square in the postcolonial 'New Dalian', functioning as a theatre to host important cultural activities of the city. The architecture was designed by a Soviet team headed by Belarusian engineers, in a similar neo-classical style that echoes other buildings around the square (Tan & Han, 2004, p. 595). The inheritance and assimilation of the previous architectural style in the colonial time indicates a continuation of Europeanisation that originated from but grew beyond the city's colonial culture.

3.2 Restructuring the Cultural Identities in European Spectacles

Under the skyline of glazing skyscrapers and high-rise concrete towers, visitors could frequently encounter a different sightline at the pedestrian, seeing ornamented stone facades and porticos. One may believe that the mixed architectural encounters of Dalian are closely associated with the city's colonial legacy. However, in addition to the preservation, use, and maintenance of historical buildings, Dalian retained a special tendency to imitate classical European features in the design of its new buildings. Official promotion of the European-continental style by the Dalian government and the political preference for Russian cultural elements have covered the city in an exotic shell. This trend started from colonial legacies but has gone off the track of inheritance, developing into an aesthetic means of shaping the city's attractive public image.

Fig. 5 Dalian Custom House showing a collage of a modernist tower with a classical portico. *Source Dalian Haiguan Dalou*, by Qu (2022). Reprinted with permission

The trend of assimilating European style into new buildings started to develop in the 1990s with a governmental preference. To transform Dalian from an industrial production 'model' in the planned economy to a new trading and tourism destination, Dalian conceived the vision of 'Europe in China' and 'Capital of Romance' in positioning its identity during urban placemaking. This new definition of the image and identity of the city was reflected in the planning policies and the provision of corresponding architectural features by the local authority.

In Zhongshan Square, Russo-Japanese colonial heritage architecture has become the 'name card' of modern Dalian. Pairing with such Europeanised street views was black steel street lamps in antique shapes, each with a Roman numeral clock embedded on its lamp posts. Though many tourists might see the lamps as a sign of the past era, these new European-style facilities were actually built as additional features in the square's renovation in 2000 (Tan & Han, 2004, p. 297), with clock features added in 2011. Since then, they have become the 'new heritage' that has contributed to local collective memories in Zhongshan District over the past ten years, as locals developed the habit of checking the time with the lamp post clocks (Jin, 2021).

A number of new 'European-continental buildings' have emerged in Dalian since the late 1990s. Some newly designed institutional buildings, such as the Dalian Custom Building, were required to change their modernist schematic design to fit Dalian's 'renaissance' of European classicism in the mid-1990s, as advised by the local council and planning offices (Qu, 2002). Following the official instructions to introduce a European-continental style at the preliminary design stage, architects of the Dalian Custom Building added a front extension with a portico and Doric columns to the original modernist tower block, making a strange collage of architectural styles (Fig. 5). The trend of Europeanisation has continued for two decades and extended its scope from authoritative preference to commercial. Replicated European landscapes also appeared in condominiums developments, as seen in a 'miniature of Venice' by the east port of Dalian. (Fig. 6).

A Russian Cultural Street was built as a government-oriented commercial tourist attraction upon the colonial legacies of the old Russian Administrative District. Completed in 2000, the street covers a length of 500 m, with a planning area of 50,000 square metres. Ironically, with limited renovations of only three selective historical buildings, most of the project was completed in the process of destructing and rebuilding. The Russian Cultural Street, though once the authoritative power centre of Russian Dalian, appeared as a mere cultural replica for its visitors during the field visit. The main street reflected the government's efforts to recreate a Russian scenic spot, though its authenticity and integrity as a historical district were questionable. Russian souvenirs filled the shopfronts. In front of the building facades, temporary stalls were arranged on both sides of the street, selling Russian dolls, chocolates, colourful wigs, necklaces and every Russian-style souvenir to meet visitors' expectations—though most of them came from the wholesale markets in Yiwu, China. The street was equipped with vintage European streetlamps, displaying advertising banners with the pattern of Russian dolls.

Commercial tourism reached an end at the North Plaza, a junction of Yantai Street and Tuanjie Street where the cultural street is located. The plaza witnessed the real destiny of some Russian colonial remains. Buildings were in a surprising degree of dilapidation with rare traces of maintenance or refurbishment. The old Dalniy City Hall, a signature building of the Russian regime, was left idle in peeled

Fig. 6 'Miniature of Venice', a condominiums development in Dalian named Haichang Dongfang Shuicheng (Haichang Oriental Water City). *Source Miniature of Venice in Dalian*, by Wang (2022). Reprinted with permission

Fig. 7 Russian Cultural Street (left) contrasts the North Plaza (right). *Source Field trip visuals of Dalian*, by Li and Yan (2021). Reprinted with permission)

walls and surrounded by construction site panels. In the middle of the plaza, only local seniors were chatting by the edge of a dried-out fountain pool, in stark contrast to the well-tended tourism in the main street (Fig. 7).

With political positivity towards Soviet and Russian relations in China, Chinese cities with close proximity or historical encounters with Russia, such as Dalian, use Russian elements as 'European inspiration(s) to foster a convivial environment' in their placemaking (Billé & Humphrey, 2021, p. 262). Russia is reduced to just a symbol, being copied, amplified, and consumed. The new cultural street was where mimicries of heritage overrode their historical relics, lightened the burden of the place's colonial past, and reproduced a simplified and generalised foreign scene to fulfil Chinese people's imagination of the West. The postcolonial urban space has been turned into a commercialised spectacle in its new identity, but hardly an inheritance or supplement to the culture it once indulged.

3.3 The Avoidance and Repellence Towards Japanese Cultural Features

In contrast with the celebrated mimicries of Russian-European cultural images were the diminished Japanese ones. In over 40 years of Japanese occupation, many places in Dalian once bore strong Japanese identities. However, such Japan-related cultural influences are often categorised as 'negative legacies' by Chinese postcolonial research and scholarly addresses of the Manchukuo period (Chen, 2017, p. 148), which extends to the wider range of the Japanese colonial period since it stepped into Dalian. While institutional buildings such as offices and schools stand still and continue to serve similar functions in contemporary Dalian, places upholding more cultural influences among the public have undergone a different route of readaptation in the recent two to three decades, wearing down their Japanese origins to merge into the Chinese identities.

As gathering venues for the public, commercial streets were often regarded as display windows of everyday local culture, with their features and built environment acutely reflecting the changing social context. In the Japanese tourism brochures dated the 1920s, two distinctive high streets, Naniwa Machi and Chain Street, were listed as representative shopping and leisure destinations for Japanese visitors (Manchuria Railway General Bureau, 1938; Lvshun Fortress Command, 1939). Being the most prosperous commercial streets in Japanese Dalian, there were hardly any Japanese traces left in their current street views despite the only inheritance of their commercial nature.

Naniwa Machi, currently known as Tianjin Street, presented a completely different look from its historical appearance. The street lanterns and archways, as prominent cultural symbols of Japan, were long gone after the colonial period. However, due to the affluent social and economic establishment of the street, most of the original architectures, street fronts and building layouts were in use by local merchants and companies until a comprehensive renovation took place from the late 1990s to the early 2000s (Tan & Han, 2004, pp. 117–134). In top-down orders by the government, the streetscape underwent complete unification. Individual buildings were demolished and replaced with uniform designs of long commercial blocks along the street front. Dalian's Europeanisation was prevalent during the renovation of Tianjin Street, which resulted in the application of the trending European-continental style with decorative gables and mouldings on the plain cement facades of the newly built shopping complex. Modernist malls covered with full-height glazing intercepted the vintage vibe in the neighbouring façade (Fig. 8). By omitting the place's historical and cultural origins, the street's Japanese identity was erased and replaced with two rather contradicting labels—'Dalian's century-old traditional commercial street' and 'modernist business pedestrian'—under the preference of political correctness.

Compared with the blandness of Tianjin Street, the depression of Liansuo Street, the former Chain Street, was much more startling during the visit. With its proximity

Fig. 8 Tianjin Street in 2021. *Source Field trip visuals of Dalian*, by Li and Yan (2021). Reprinted with permission

to the railway station and inexpensive daily commodities, Chain Street once competed with Naniwa Machi as one of the most popular merchants' districts for Japanese civilians during the colonial period. Unlike the fully rebuilt Tianjin Street, the current Liansuo Street retained the style of the 1930s—and the degree of disrepair comparable with that age. The low-rise townhouses were dilapidated, with stains of smoke and rain covering their yellowish surfaces that had weathered for decades. Unlawful extensions were built upon individual buildings as if they were out of the reach of any planning regulators. Hardware stores, diners, and cheap hostels occupied the shopfronts, with sundries, cargo boxes, flashing signs, and vehicles parked randomly along the alleys. Disordered electric wires cut the sky overhead (Fig. 9). Though Liansuo Street once shared similar renowned fame with other places mentioned above in colonial Dalian, it was only listed as a historical conservation area since 2020 with no regeneration attempt taking place before then (Xiu, 2020). Unlike Tianjin Street, which was expected to attract high-end customers, Liansuo Street was left to decay with a more sensitive and less appealing colonial identity, having functioned as a gathering area for working-class Japanese migrants before 1945.

The sensitivity in refurbishing and adopting Japanese elements mostly took on a passive approach as seen in Tianjin Street and Liansuo Street. The political criticism of Imperial Japan's colonial history and its notorious invasion of China during WWII was the main reason behind the dismissal of Japanese images in the city, compared with the celebration of the Russian culture, which blended its colonial intrusion with the post-war friendship between the late Soviet Union, current Russia, and the PRC. Such political tendencies influenced the public's opinion and general acceptance of Japanese culture in a broader Chinese context.

In contrast to the blossom of European-romanticised cityscape, there was the radical rejection of the newly invested Japanese cultural placemaking. Indeed, with connections built up since its colonial era, Dalian has maintained a close relationship with Japan through collaborations among companies and cities at an official level. Conversely, the hidden bias towards Russian over Japanese identity became apparent when anti-colonial nationalism waves hit the city's recent tourism placemaking of a Japanese Cultural Street. Located in Jinshitan national tourist resort in Dalian, a commercial street 'Tang · Little Kyoto' announced its closure after a one-week trial opening due to strong opposition from Chinese netizens.

The street was newly developed in 2021 as part of a real estate complex integrating business, residence, and tourism. Its selling point was combining the traditions and cultures of the Chinese Tang Dynasty and Japanese Kyoto by reproducing the Nineizaka district in Tokyo with imported Japanese buildings, shops, and restaurants. JCAP, a group of multiple Japanese architectural firms, supported the project with technical details in creating traditional Kyoto-style streetscapes (JCAP7, 2020). Roof tiles and doors were imported from Japan, and dimensions of buildings and alleys were collected in Nineizaka to achieve a similar scale to that of the real Kyoto (Watanabe, 2021).

Although the project was supported by the government of Dalian to explore a new way of investing in Japanese cultural tourism, the opening of the 'Little Tokyo' caused a sudden uproar of an 'anti-Japanese' wave among Chinese netizens. In contrast to the Russian Cultural Street, which has been operated for 20 years seamlessly, the 'Japanese

Fig. 9 Liansuo Street in 2021. *Source Field trip visuals of Dalian*, by Li and Yan (2021). Reprinted with permission

cultural street' was widely criticised across social media platforms for 'being a Japanese cultural invasion', 'showing ignorance of colonial history', and 'hurting the feelings of Chinese people' (Wan & Deng, 2021). Even People's Daily, one of the official news media of the Chinese government, quoted an online criticism, stating that 'we should not use Japanese culture to hype commercial interests in a city such as Dalian, which suffered greatly from Japanese colonisation' (Hu, 2021). Interestingly, such voices came mostly from outside of Dalian, while the commercial street was packed full of local visitors who did not seem to mind having another exotic photogenic spot for shopping and touring over the concerns of postcolonial nationalism. Nevertheless, the controversy ended with the street's closure and reopening after a year—under a new name of 'Jin Shi Wan Xiang (Golden-stone Alleys)' that omitted Kyoto from its name. The shops started selling international groceries from Russia, Mongolia, and South Korea instead of Japanese products to blend the original Japanese identity into a multinational hotchpotch (Watanabe, 2022).

3.4 The Depoliticised Colonial Legacies in the Everyday Urbanscape

Apart from the postcolonial controversies that Dalian encountered in its cultural placemaking, some places of colonial origins have remained inherent in the contemporary everyday urbanscape of Dalian. They blended into the city's identity seamlessly with a rather generalised approach or less observant cultural features present in the built environment to fit in the broader East Asian cultural scope, thus lightening the weight of historical and political bundles in colonial legacies.

As traditional Japanese architectural forms bear Chinese influence dating back to the Tang dynasty, certain Japanese temple buildings in Dalian have withstood waves of decolonisation, cultural revolution, and postcolonial nationalism, being subsequently redefined as places of Chinese tradition. The Higashi Hongan-ji (East Temple) in Dalian was reused as the Dalian Theatre of Chinese Opera, matching the traditional cultural performance with the Tang-infused building exteriors. In a mix-match of the East Asian identity, the place dismissed its Japanese origin but retained the representative form of a generalised Asian tradition (Fig. 10).

Among those less noticeable colonial legacies, the natural landscape has been the main contributor to the city's identity and has become an inseparable part of the current urban life. The Longwangtang Cherry Yard, as mentioned in Sect. 2.2, has been carrying the tradition of cherry-viewing for nearly a century, with cherries transplanted from Japan in the 1930s. From late April to early May, cherry-viewing at Longwangtang became an annual routine for local people and tourists from neighbouring cities. The Yoshino Cherries and Star Mongolia planted by Japanese colonialists in the 1930s were accompanied by more cherry trees imported from Japan over the expansions of the yard in the past decades. The scenic blossom was carefully maintained. At the time of the

Fig. 10 Old Higashi Honga-ji Temple was used as the Dalian Theatre of Chinese Opera before the latter moved to a new location in 2019. The building stayed idle at the time of the field trip. *Source Field trip visuals of Dalian*, by Li and Yan (2021). Reprinted with permission)

Fig. 11 Cherry-viewing at Longwangtang Cherry Yard, April 2021. *Source Field trip visuals of Dalian*, by Li and Yan (2021). Reprinted with permission

field trip in late April, the cherry yard was filled with friends and families gathering under the blossoming trees, posing for photos or simply enjoying a breath of spring (Fig. 11). Similar practices have recurred at the exact location over the span of a century, with only the terminology for appreciating the scene changing from *hanami* in Japanese to *shanghua* in Chinese—transcending historical and cultural backgrounds.

Nevertheless, the natural plants were imbued with multiple meanings beyond their role as tourist attractions. Associated with the city's history but lived across time, additional cherries were planted after Dalian's colonial period to show friendship during the 'Golden Age' of Sino-Japanese relations from the late 1980s to the 1990s. Multiple cherry gardens, including the ones in the 203 highland and the Laodong Park, were established during Japanese official visits to commemorate the Sino-Japanese War and as a grateful sign of peacetime between Japan and China. However, the cheerfulness and beauty brought by the blossom season diluted the particular semiotics and awareness related to Japanese imperialism or postcolonialism. A 'cherry avenue' in Dalian's Xinghai Square was planted primarily for aesthetic appreciation in 2008. More cherries were hiding among small urban parks, schools, and residential districts, standing individually or in small clusters, unnoticed, as part of the common urban landscape. Cherry trees, rather than serving their original purpose of evoking nostalgia for the Japanese migrants, now complement Dalian's modern identity of East Asian romanticism in an everyday natural aspect.

The depoliticisation of landscape legacies extends from natural plants to designated squares and parks once holding colonial significance and functionalities. Located in Dalian's city centre, Laodong (Labour) Park is a quiet oasis surrounded by high-rise commercial and residential towers. Against the backdrop of a modern city, the landscape, waterscape, and aged trees still vaguely convey the lingering charm of history. Indeed, the park's prominent location is closely related to its past, which was never politically neutral. The Russians planned it as a greenbelt to separate the 'Chinese Town' from the colonial urban area. The Japanese built a Loyalty Tower in the park to commemorate their soldiers who died in the war. Now a new stela engraved with 'labour creates the world (laodong chuangzao shijie)' in Chinese has replaced the militaristic monument to set the tone for the socialist theme of the park. Though a change of political features necessarily occurred in redefining the identity of such places in the postcolonial context, returning the space to its people has significantly affected the city's spatial practice, which links to life itself. The current Laodong Park is a shared open space serving local residents, especially the retired elderly who exercise karaoke and group dances in the park square daily (Fig. 12). The place has been integrated into a joyful leisure life. Even the older people did not care if it was once a colonial spot tangled with segregation, inequality, hatred, and war. Only a few century-old trees would invite visitors to stop occasionally and witness the long-gone past engraved on their heritage plaques; local Dalianers merely stayed for a shade of greenery.

4 Conclusion

In Dalian's postcolonial regeneration and decolonisation attempts, rejections and controversies of 'negative legacies' coexist with fusions and adaptations of multicultural heritages. Parallel to the nationalist rebellion rising from the outside against the so-called colonial cultural invasions in commercial development are the inheritance of urban

Fig. 12 People are exercising in Laodong Park, Dalian. *Source* screenshots of video recording, *Field trip visuals of Dalian*, by Li and Yan (2021). Reprinted with permission

hierarchies, the romanticisation of foreign-imported styles, and the depoliticised appreciation of natural heritage from the inwards of the city.

Political considerations, exotic fantasies, and the vigilance and sensitivity to colonial influences have extensively shaped Dalian's contemporary identity, from architecture to urban spaces, from the image of the city to the sense of place. Prominent adoration of Europeanisation and inherent Asian recognition deviated Dalian from its targeting identity of an East Asian cultural centre to a much more intertwined, undefined pastiche that grew out of its dual-colonial complexities and developed in the postcolonial context of Northeast China. The city's identity crisis stemmed from its reluctance to acknowledge its historical past, including certain colonial connections and the legacies upon which the city was built. Resistance came from multiple aspects as reflected in Dalian's recently initiated urban regenerations and cultural placemaking projects. In some cases, permitted cultural recreation has gone far towards Disneyfication instead of heritage preservation with insufficient historical respect. Such situations exist not only in Dalian, but also in some other Chinese cities with similar colonial contexts, such as Harbin and Shanghai.

The paper is an initial attempt to gain insight into the dilemma of Dalian's city identity from an architectural and urban perspective. With first-hand materials gathered from fieldwork, the study presents Dalian as a thumbnail of the under-studied postcolonial cities in China. Although colonisation in Dalian came to the past tense, its influence and impacts are still present in the city's current built environment and cultural identity—embraced in the urban skin, triggering unexpected debates on occasions, while bringing in uniqueness through a multicultural indulgence which provokes opportunities of placemaking. Dalian's urban identity, though rooted in the city's colonial history, went beyond its geographical scope and became the result of constant compromising and rebalancing between the city's imagination of its past and the rewriting of its presence in the contemporary context.

Acknowledgments Despite the strict restrictions on travelling in China due to the COVID-19 pandemic, the fieldwork of Dalian was completed remotely with generous help from Yanchun Yan and Xiaowei Li as research participants and contributors back in China, and with the support from Kettle's Yard Fund and Faculty Fieldwork Fund provided by the Faculty of Architecture and History of Art, University of Cambridge. I would also like to extend my thanks to Li Wang and Qu Qu for providing photos to complement the fieldwork findings of the research.

References

Akutagawa, K. (Director). (1940). *Japan-Manchu children writing envoy /Ni-man Tsurikata Shisetsu* [Documentary]. Mantetsu Film Production Studio/Mantetsu eiga seisakusho.

Appadurai, A. (1996). *Modernity at large: Cultural dimensions of globalization* (Vol. 1). University of Minnesota Press.

Azaryahu, M. (2003). RePlacing memory: The reorientation of Buchenwald. *Cultural Geographies, 10*(1), 1–20.

Billé, F., & Humphrey, C. (2021). *On the edge: Life along the Russia-China border*. Harvard University Press.

Chen, Y. (2017). The 'teaching of the eyes' and the imperial sight of the Japan exposition—Also on the 'Modern Manchuria'. *Cultural Vision*, 144–148.

Denison, E., & Ren, G. (2016). *Ultra-Modernism: Architecture and Modernity in Manchuria* (1st ed.). Hong Kong University Press; JSTOR. https://www.jstor.org/stable/j.ctt1k3s9c9

Friedmann, J. (2010). Place and place-making in cities: A global perspective. *Planning Theory and Practice, 11*(2), 149–165.

Hess, C. A. (2011). From colonial port to socialist metropolis: Imperialist legacies and the making of 'New Dalian.' *Urban History, 38*(3), 373–390. https://doi.org/10.1017/S0963926811000575

Hu, J. (2021). 'Dalian Japanese cultural street' ammounced closure. *People's Daily*. https://finance.sina.com.cn/jjxw/2021-09-03/doc-iktzscyx2057179.shtml

JCAP7. (2020). *Tang little Kyoto project phase 1 model houses completed*. Japanese Consortium for Asian Projects. http://www.jcap7.com/en/news/dalian-littlekyoto/

Jiang, Y., Tian, J., & Wang, D. (1984). *Dalian Wenshi Ziliao (Dalian literature and history information)* (Vol. 1). Dalian CPPCC Cultural and Historical Materials Committee Internal Publication.

Jin, D. (2021). *Clocks in Zhongshan square stopped working*. Dalian News Media Group. https://www.163.com/dy/article/GC700LJN0534B3JB.html

Li, Y. (2021). *Atlas in motion: Visualising manchuria in moving images* [Presentation]. Central Asia Working Group, Institute of East Asian Studies, UC Berkeley. https://youtu.be/IBGqSBnnH68

Li, X., & Yan, Y. (2021). *Field trip visuals of Dalian* [Photography and video recording]. Atlas in Motion: Visualising Manchuria in Moving Images Research Project.

Liu, Y., Dupre, K., & Jin, X. (2021). Dalian's urban planning and design in evolution. *Journal of Chinese Architecture and Urbanism*, 1–16. https://doi.org/10.36922/jcau.v3i1.1024

Lowenthal, D. (1975). Past time, present place: Landscape and memory. *Geographical Review*, 1–36.

Lvshun Fortress Command. (1939). *Views of Dairen*. Lvshun Fortress Command.

Manchuria Railway General Bureau. (1938). *Dairen*. Manchuria Railway General Bureau.

Mantetsu Film Production Studio/Mantetsu eiga seisakusho (Director). (1937). *The tour around Japan, Chosen and Manchuria/Nai sen man shūyū no tabi* [Documentary].

Massey, D. (1995). Places and their pasts. *History Workshop Journal, 39*, 182–192.

Peng, H. (2022). *Dalian was successfully elected as the 'Cultural City of East Asia' of 2024*. Dalian Daily. https://www.ln.chinanews.com.cn/news/2022/0921/327548.html

Pink, S. (2008). An urban tour: The sensory sociality of ethnographic place-making. *Ethnography, 9*(2), 175–196. https://doi.org/10.1177/1466138108089467

Qu, X. (2001). *Jindai Dongbei Chengshi de Lishi Bianqian (The historical transformation of modern Northeast Cities)*. Northeast Normal University Press.

Qu, Y. (2002). Brief discussion of the creation of Dalian custom house and Shuixian Garden. *ArchiCreation, 2*, 38–41.

Qu, Q. (2022). *Dalian Haiguan Dalou (Dalian Custom House)* [Photography]. Yingxiang Dongbei (Imagine Dongbei). https://www.douban.com/group/topic/264048311/

Sakharov, V. (1899). *Plan goroda Dalniago* [Lithographic print]. publisher not identified; Library of Congress. https://lccn.loc.gov/2019585138

Tan, J., & Han, L. (2004). *Dalian Shizhi: Chengshi Jianshe Zhi (Chronicle of Dalian: Urban Construction, 1840–1990)*. Chronicle Publication. https://fz.wanfangdata.com.cn/details/newLocalchronicle.do?Id=FZ010656

Tang, S. (1995). *Dongbei Chengshi Guihua Shi (history of urban planning of Northeast China)*. Liaoning University Press.

Wan, H., & Deng, X. (2021). *Betrayal? Populist? What's going on in Dalian's 'Japanese Cultural Street'*. Sohu News. https://www.sohu.com/a/www.sohu.com/a/488029797_118927

Wang, L. (2022). *Miniature of Venice in Dalian* [Photography].

Watanabe, N. (2021). *The largest Japanese cultural street opens in Dalian, China*. NIKKEI CN. https://cn.nikkei.com/china/cpolics-society/45801-2021-08-23-05-00-01.html

Watanabe, N. (2022). *Dalian restarts Kyoto-themed commercial street, downplaying Japanese tone*. NIKKEI CN. https://cn.nikkei.com/politicsaeconomy/politicsasociety/47521-2022-02-04-17-10-56.html

Xiao, Z. Y. (1988). Urbanization of Dalian and its urban morphology in 1898–1904. *Journal of Dalian University of Technology, 28*, 21–33.

Xiu, Q. (2020). Liansuo Street—Dalian street's largest central heating starts here. *Dalian Evening News*. https://dalian.runsky.com/2020-12/06/content_6097308.html

Yoshida, H. (1929). *Dairen* [Map]. Japan Tourist Bureau Dairen Branch.

An Attempt to Analyze Ceramic Assemblages and the Function of the Pit from Gudiashvili Square

Tatia Butsuradze

Abstract

Gudiashvili Square is located in Kala in the historical part of Tbilisi, the capital of Georgia. Due to the fact that the square is located inside the old defensive wall of Narikala Fortress, it has a huge archaeological and historical importance. At first this square appears on the Map of Kartli, created by Vakhushti Batonishvili in 1735 and is mentioned with the name of Bezhana Garden. Of particular interest is a pit containing a remarkable assemblage of ceramics, unearthed within a dwelling where only the floor has been preserved. Above this dwelling, a cellar featuring vaulted arches was constructed, believed by art critics to potentially date back to the XVII–XVIII centuries. The pit was excavated under the floor, and it is noteworthy that such pits are typically excavated within the interiors of ceramic workshops, often adjacent to ceramic kilns. For instance, in Kutaisi, four pits were discovered within the workshop interior, yielding a diverse range of ceramic materials that have been dated to the X–XI centuries. In the case of Gudiashvili Square, the dwelling's interior where the pit was found may have served as a workshop, subsequently destroyed or transformed into a discarded deposit. Both conjectures merit consideration. To shed light on these findings, comparative analysis employing the study of ceramic materials in conjunction with other artifacts from medieval sites in Georgia was conducted, enabling the determination of the pit's function and its chronological placement.

Keywords

Gudiashvili square · Medieval ages · Cellar · Pit · Ceramic assemblages

T. Butsuradze (✉)
Tbilisi State University,
Tbilisi, Georgia
e-mail: tatiabutsuradze@gmail.com

1 Introduction

Gudiashvili Square is located in Kala, in the historical part of Old Tbilisi, which was surrounded by the defensive wall. Due to its location, it is obvious that this square would have a great historical and archaeological importance and because of this fact archaeological discoveries were expected. In late medieval ages, this square was named as Bezhana Garden. *"Until 1828 this square was known with the name of Moghnisi (as Moghni, as Moghnini) on the behalf of the Virgin Armenian Church, located near to the square and till nowadays this square is called its second name-Kolortaghi-same as Klortaghi. In 1828 this square was officially named of Abas-Abad on the behalf of capturing Persian city- Abas Abad by the Russian Imperial General Paskevich. In 1923 the square was named Alaverdiani, on the behalf of Bolshevik Stefane Alaverdovi. The contemporary name- Lado Gudiashvili Square, was given in 1988 on the behalf of famous Georgian painter Lado Gudiashvili"* (Elisashvili et al., 2011, p. 3).[1]

With the name of Bezhana Garden, which is presumably the first name of Gudiashvili Square, it is marked on the Map of Kartli, which was created by Vakhushti Batonishvili in 1735 (Dumbadze & Guchua (Eds.), 1990, p. 459, N65-e) (Fig. 1). There is an important and interesting note about the name of Bezhana. According to Mamisa Berdznishvili, Bezhana could be Phirkhulaant Nzarbega's son, the uncle of Osepha, whom this garden belonged and was well-known citizen living during the reign of the King Erekle II in the XVIII century (Berdznishvili, 1965,

[1] Special thanks to Tsira Elisashvili, Lia Bokuchava, and the whole team of ICOMOS Georgia for their support and contribution during the works and the research.

M. F. Shahidan et al. (eds.), *Urban Narratives: Exploring Identity, Heritage, and Sustainable Development in Cities*,
Advances in Science, Technology & Innovation, https://doi.org/10.1007/978-3-031-48517-6_9

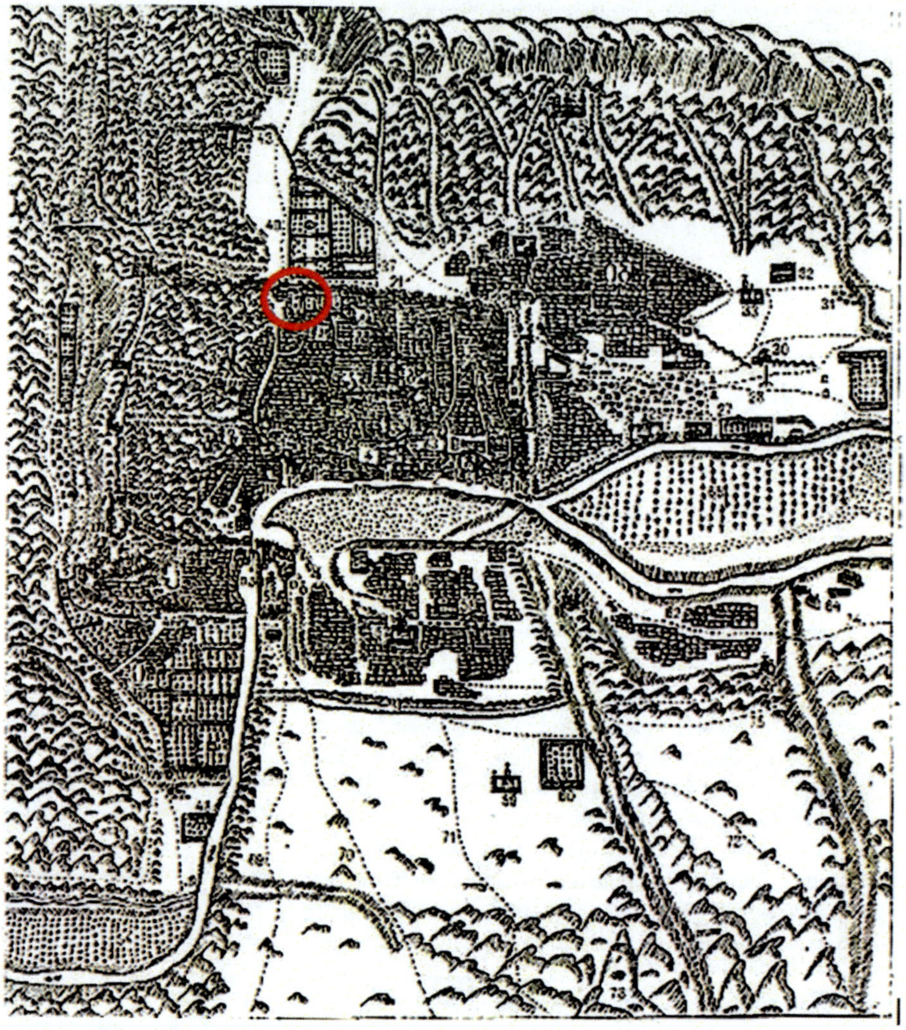

Fig. 1 Map of Kartli created by Vakhushti Batonishvili in 1735; with red circle location of Gudiashvili Square (Dumbadze & Guchua (Eds.), 1990, Pl. 50)

pp. 26–27). As it appears, the square is located near to the one of the gate of Tbilisi called "Kojori Gate", which was located at the place, where Leselidze Street (contemporary Kote Apkhazi Street) joins to the Liberty Square. This gate was located in the western part of the defensive wall of Tbilisi (Dumbadze & Guchua (Eds.), 1990, p. 243). Over the centuries, the square has undergone multiple reconstructions, finally assuming its contemporary layout in 1844, complete with surrounding streets, dead-ends and corresponds to the present-day situation (Suramelashvili & Chachkhunashvili, 2007, pp. 5–6) (see also Butsuradze, 2022, p. 1).

Gudiashvili Square occupies historically and archaeologically significant area, where archaeological findings are primarily associated with construction activities. It is crucial to highlight that conducting thorough archaeological excavations in Tbilisi poses considerable challenges due to the city's dense population and the complexities of urban development, particularly within the confines of Old Tbilisi.[2,3] Of particular interest is a specific building located at Beglar Akhospireli

[2] Rehabilitation-restoration project started in 2018 and is funded by Tbilisi Development Fund and Tbilisi City Hall. Archeological monitoring was being carried out from the beginning of the project (2018) by National Agency for Cultural Heritage Preservation of Georgia (by Z. Giorgadze, T. Niniashvili, T. Gurgenidze). From 2019 archeological monitoring continued by myself and Lana Chologauri. For nowadays archeological monitoring still continues by myself.

[3] Conservation plan of Gudiashvili Square and its surrounding area was fulfilled by ICOMOS Georgia in 2015. The first stage of restoration-rehabilitation projects were done by ICOMOS Georgia and the works were performed under their supervision. Special thanks to the team of ICOMOS Georgia for their support and contribution during the works and the research.

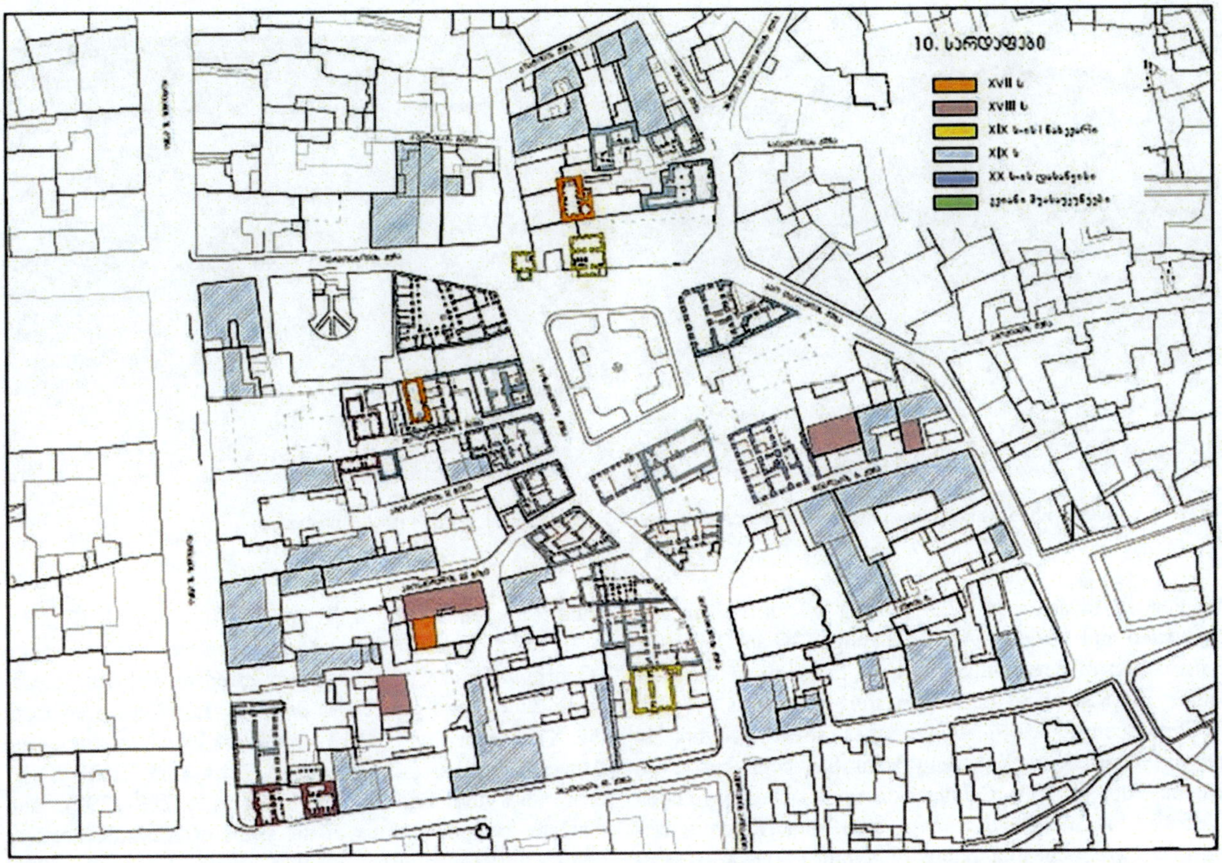

Fig. 2 Plan of Gudiashvili Square. Different period cellars indicated with different colors. Thematic plan by ICOMOS Georgia[6]

I dead end, building N2/4,[4] where cellar with vault arches has been uncovered and in the earlier dwelling's interior where the pit was found may have served as a workshop, subsequently destroyed or transformed into a discarded deposit. A wide range of ceramic materials from the pit consists of local and imported materials as well. Among them could be distinguished kitchenware, tableware, household, and fine ware. Most of them are wholly preserved and give possibility to identify and date. Here rises important questions—(a) what is the purpose of this pit? (b) why this pit is in the middle of the dwelling? and (c) is there a chance that this dwelling was used as a ceramic workshop? The main aim of this paper is to answer these questions, determine, date, and characterize materials from the pit with the help of comparative methodology, define the purpose of the pit, and preliminarily to give an idea about the importance of Gudiashvili Square during the medieval ages.

To have a general idea about Gudiashvili Square archaeology, should be remarked that archaeological remains discovered here encompass more than just ceramic assemblages, ranging in date from the IX to the XIX–XX centuries. Among the ceramic assemblages have distinguished more than 20 groups such as: bowls, sphero-conical vessels, tobacco clay pipes, oil lamps, salt holders, inkpots, stewing pots, tubs, pots, jugs, lids, vases, tripods, fragments of baskets, imported vessels (materials of western[5] and eastern world), defective materials, pitchers, as well as glass fragments, coins, clay toy, wooden bath shoe, building materials—bricks, tiles, tubes, etc. Except ceramic materials, the findings include cellars, wine cellars, and other dwellings with various functions. Maia Mania notes that following

[4] Construction works have been done by the construction company—IN-SI. Special thanks to: K. Babutsidze, G. Babutsidze, M. Muradashvili, E. Kekelishvili, L. Sadaghashvili, G. Mezvrishvili, M. Jafiashvili, R. Maghaldadze, and L. Maghaldadze, who were working on this building.

[5] About Bartmann jugs see Chologauri (2022), 28–44.

[6] Conservation plan of Gudiashvili Square and its surroundings, 2015, by ICOMOS Georgia. Note: Since 2015, ICOMOS Georgia has updated data, though it's not published yet. Courtesy to ICOMOS Georgia.

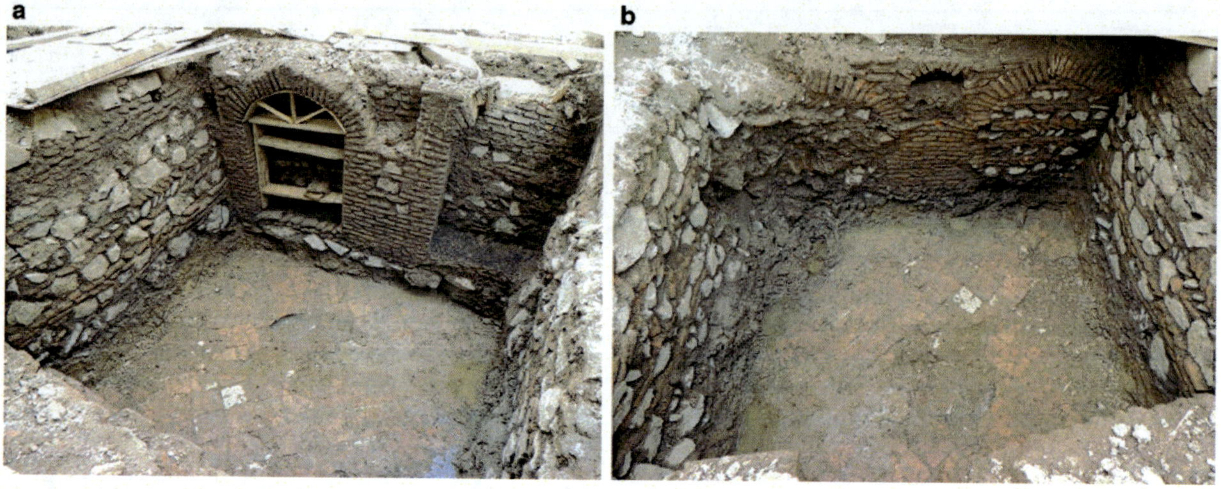

Fig. 3 a, b Cellar with the vault arches revealed at Beglar Akhospireli I dead end, building N2/4 (© T. Butsuradze)

the Persian invasion in 1795, during which Tbilisi was devastated and razed by Aga Mohammad Khan Qajar, the construction of new buildings in the first half of the XIX century often involved utilizing existing basements.

Consequently, the medieval ages street network in the old districts of Kala and Isani has been preserved (Mania, 2003, p. 150). Cellars with vault arches have been unearthed on Gudiashvili Square, with Maia Mania asserting that these arches undoubtedly belong to the XVII century, while semi-circular arches emerged concurrently in the XVIII century (Mania, 2003, pp. 152–153) (Fig. 2). Maia Mania also mentions that those cellars were used for storing food items and often contained refrigerating wells constructed with bricks. Some of these cellars also served commercial purposes. Wine cellars with buried Kvevri (*pithos*-large wine vessels) in the ground are common feature, and on occasion, cellars were even utilized as living spaces (Mania, 2003, pp. 151–152). As Gabriadze notes, refrigerating wells of this kind were typically found in the residences of feudal lords and other wealthy person's dwellings, storing snow and ice that were used for medicinal purposes by the doctors (Gabriadze, 2013, p. 149) (see also Butsuradze, 2022, p. 2).

2 Discussion of the Ceramic Assemblages from the Pit

As it is remarked above, the main focus centers on a specific vault arched cellar's interior. Inside, below the basement of this cellar a floor level was discovered, which belongs to a previous, older dwelling that as it appears, was demolished. The cellar by an art critic supposedly dates back to the end of the XVII and XVIII centuries[7] (Fig. 3a, b). It has a square shape and is constructed using so-called "Georgian Bricks", commonly referred to as bricks measuring 22 × 22 × 4 cm. These bricks were widely used from the XV century until 1812 in various locations, including Alaverdi Fortress (XVII century), Bochorma (XVII–XVIII centuries), the Fortress of Kolagori (1788–1799), and the later period sections of the walls of Narikala Fortress (XVII–XVIII centuries), among others (Jghamaia, 1980, pp. 80–81). Generally, the cellar is constructed by the combination of bricks and cobblestones. The eastern and western walls of the cellar feature vault arches. Throughout the interior, there is evidence of a burnt layer, suggesting that this cellar was destroyed during the invasion of Agha Mohammad Khan Qajar, who devastated and razed Tbilisi to the ground in 1795. A similar trace of a major fire has been observed in another cellar-ceramic workshop unearthed on Gudiashvili Square.

Below the basement of the cellar is a floor level belonging to the older dwelling. It is worth noting that this floor was constructed using Georgian bricks of a different size (than the cellar itself), measuring 25 × 25 × 3.5 cm. Such bricks were commonly used in the XII and XIII centuries, as seen in sites like Dmanisi (XII–XIII centuries) (Jghamaia, 1980, pp. 79–80). In many instances, these bricks were reused in later periods. The floor level extends outside the eastern wall of the later cellar and connects to a wall constructed with rock stones. Furthermore, it joins a platform built with larger bricks measuring 27 × 27 × 6

[7] Special thanks to Tatia Ghvineria for identification and dating the cellar; Moreover special thanks to Mate Akhalaia for his help.

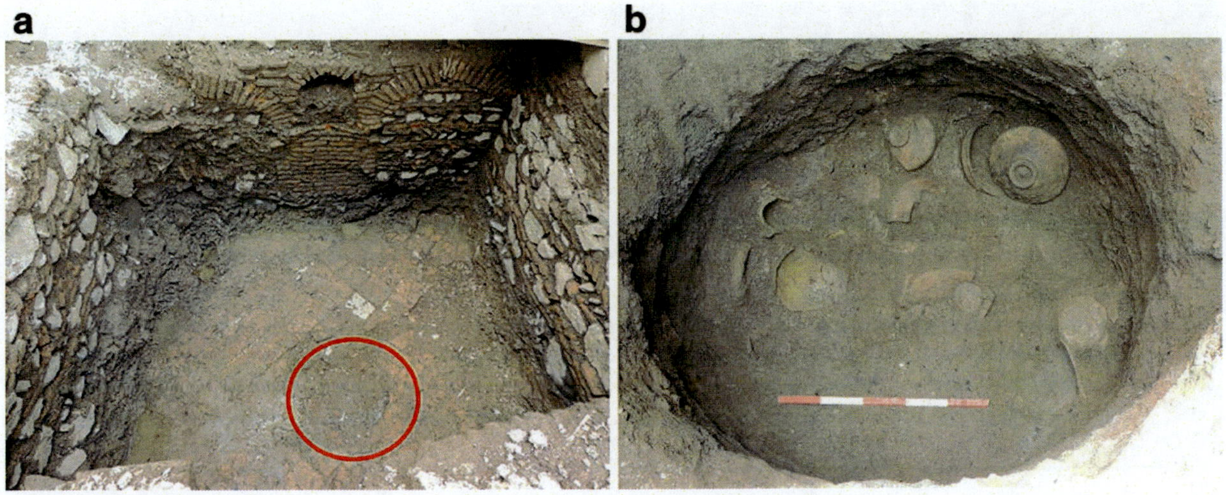

Fig. 4 **a, b** Pit full with ceramic assemblages excavated under the floor (© T. Butsuradze)

cm. Bricks of this size were used in both VIII and IX centuries (e.g., Basilica of Alvani, VIII–IX centuries) and the XII and XIII centuries (e.g., Geguti Palace walls, XII century; Timotesubani, XII–XIII centuries) (Jghamaia, 1980, pp. 78–79). But undoubtedly the floor level and platform belong to the same period and dwelling as well.

As it is mentioned above, pit, filled with ceramic assemblages, was excavated under this floor (diameter: 1.20 m; depth: 2 m). During the cleaning of the pit, fragments of charcoal and traces of ash were discovered, indicating that the inhabitants occasionally used it for firing purposes (Fig. 4a, b). It is noteworthy that the residents of Tbilisi would typically discard ceramic production remnants in the Mtkvari River or bury them in the ground (Dumbadze & Guchua (Eds.), 1990, p. 125). These pits are usually found in proximity to ceramic kilns and within workshops. For example, in Kutaisi, four pits were excavated within a workshop dating back to the X-XI centuries. These pits varied in size and shape and contained a large number of artifacts, which helped determine their function during the workshop's existence. The author suggests that these pits were dug during the workshop's operation, likely for soaking clay, and were subsequently filled with fragments of ceramic materials, including defective ones (Gabriadze, 2013, pp. 80–82). Of particular importance is the discovery of a ceramic workshop near 300 Aragveli Garden in Tbilisi. The remains of the workshop include kilns (IX and X trenches), trenches, dump deposits (VIII, XII, XIII, and XIV trenches), ceramic assemblages, remains of a pipeline (XVI trench), trenches near the city wall (XI and XIV), building materials, bracelets, and more. The author suggests that the workshop was active during the XII–XIII centuries (Lomtatidze, 1955, pp. 130–160). The ceramic assemblages from this workshop were extensively studied by Marine Mitsishvili (see Mitsishvili, 1979).

Similar pits found on Erekle II Square were excavated during an archaeological campaign in 1956–1957. For example, one pit (NK-I) had walls lined with bricks measuring 25×25 cm. The authors of the study indicate that, in this case, the pits were associated with the remains of ceramic workshops and served practical purposes (Grdzelishvili & Tkeshelashvili, 1961, pp. 16,22). They suggest that these pits had different functions, with deeper ones with unfired walls likely serving as wells, while others connected to ceramic production (Grdzelishvili & Tkeshelashvili, 1961, p. 28). The authors also divide these pits into chronological groups, suggesting they were used between the V and XVI centuries (Grdzelishvili & Tkeshelashvili, 1961, p. 31).

Regarding the pit revealed in Gudiashvili Square, no traces of a workshop were found nearby. However, considering that the older cellar was completely destroyed, leaving only the preserved floor level, it is possible that some kind of workshop existed here. It should be noted that there is a ceramic workshop with its kiln and additional structures on Gudiashvili Square as well, but this does not deny existence of another one. The ceramic assemblages excavated from the pit should be examined as a whole since they form a contextual grouping. Here are distinguished glazed and unglazed materials. We should divide materials into two groups—(a) local production and (b) imported pottery. But functionally those materials should be subdivided into four main groups, such as—(a) tableware—bowls, jugs, and salt holders; (b) kitchenware—pots, stewing pots, and ketsi (pan

Fig. 5 **a**, **b** Bowls with light blue glaze and scratched out wavy ornament from Gudiashvili Square (© T. Butsuradze); **c** Bowl with wavy ornament and green glaze from Tbilisi (Bakhtadze, 2013, p. 48, N40)

like vessel); (c) household ware—oil lamps; and (d) other materials—mainly fragments of different vessels.[8]

Discussing and analyzing of the materials should start from the tableware that comprise well-preserved bowls with different ornaments and glaze colors. In total from the pit are represented 34 well-preserved bowls. These bowls are primarily made from reddish clay and are initially covered with white engobe, upon which various ornaments are scratched out before being covered with a monochrome glaze. The rims of the bowls are vertical, the hems are sharply rolled down to the bottom, and the bottoms have a wheel-like shape.

Discussion should be started with two similarly ornamented and light blue glazed bowls. The wavy ornament is scratched out on the white engobe beneath the glaze. One of these bowls is completely preserved, while the other is fragmentary (Fig. 5a, b). According to Marine Mitsishvili, such pottery with monochrome glaze and scratched-out ornaments dates back to the eleventh–twelfth centuries. It is found in the Tbilisi Ceramic Workshop (the workshop excavated in 300 Aragveli Garden). They are also present in Rustavi Fortress and settlement, Dmanisi, Ujarma, and Gudarekhi (Mitsishvili, 1969, pp. 24–25). Mitsishvili also notes that various ornaments such as looped triangles, rings, and squares, are represented on bowls with a light blue glaze (Mitsishvili, 1979, p. 39). A bowl with a wavy ornament similar to these examples was also discovered in Tbilisi, but it is covered with a green glaze. Nodar Bakhtadze suggests that it dates back to the XI–XII centuries (Bakhtadze, 2013, p. 48, N40) (Fig. 5c).

[8] Ceramic assemblages were restored by Nino Kamkamidze.

Among the ceramic assemblages, three bowls with similar ornamentation were found (Fig. 6a–c). Two of them are covered with a light blue glaze and are ornamented with a single row of double rings, with a smaller ring scratched out inside the larger one. The third bowl is glazed with a brownish manganese glaze, and it is slightly deformed due to the effects of fire. According to Marine Mitsishvili, such ornamented bowls date back to the XI–XII centuries (Mitsishvili, 1969, pp. 24–25). She also notes that various ornaments such as triangles, rings, squares, and loopy triangles are scratched out on bowls with a light blue glaze (Mitsishvili, 1979, p. 39). Regarding bowls covered with a brownish manganese glaze, Mitsishvili remarks that a large number of such glazed bowls from the Tbilisi ceramic workshop are decorated with geometric ornaments like loops and rings (Mitsishvili, 1979, p. 46). Another bowl with similar ornamentation but covered with a green glaze was also found in another pit on Gudiashvili Square.

One of the bowls from the pit bears a scratched-out image of a snake/dragon in the combination with floral images (Fig. 7a). Bowls with images of animals and birds are also found on Gudiashvili Square, but they are covered with polychrome glaze instead of monochrome glaze. Ceramic materials with scratched-out ornaments and light blue glaze depicting birds and animals are found in Tbilisi Ceramic Workshop as well. Marine Mitsishvili, who studied the ceramic assemblages from this workshop, categorizes bowls with scratched-out images of animals, covered with light blue glaze, in a subgroup. She mentions that only five such ornamented pieces have been found, including bowls with calf images, a dog's head, unidentified animal's feet, and an uncertain animal image (Mitsishvili, 1979, p. 45). Vessels with bird images are also found in

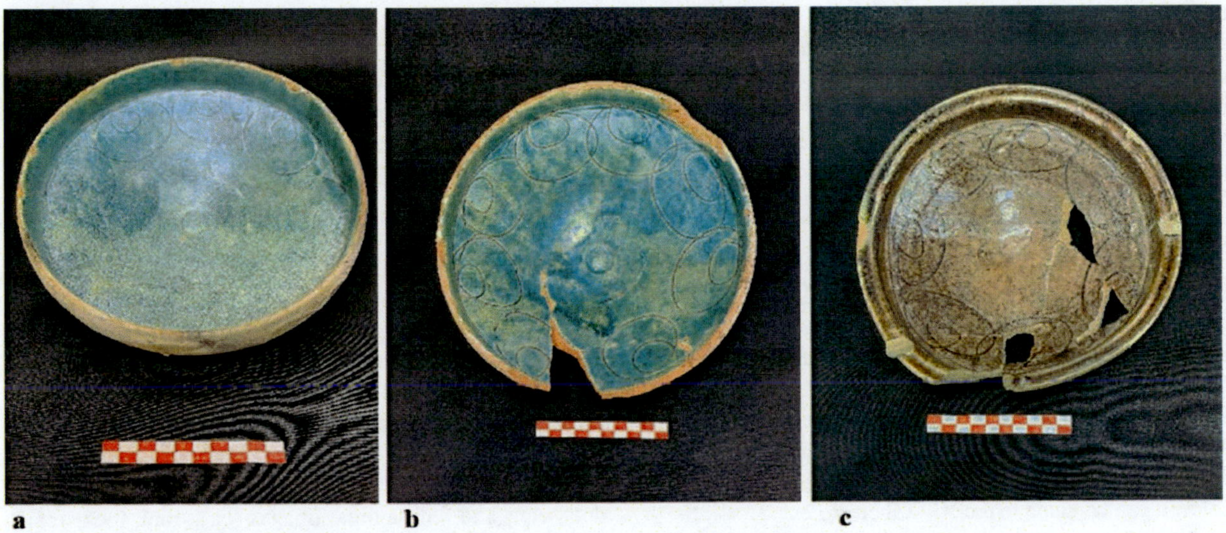

Fig. 6 **a–c** Bowls with single row of double rings from Gudiashvili Square (© T. Butsuradze)

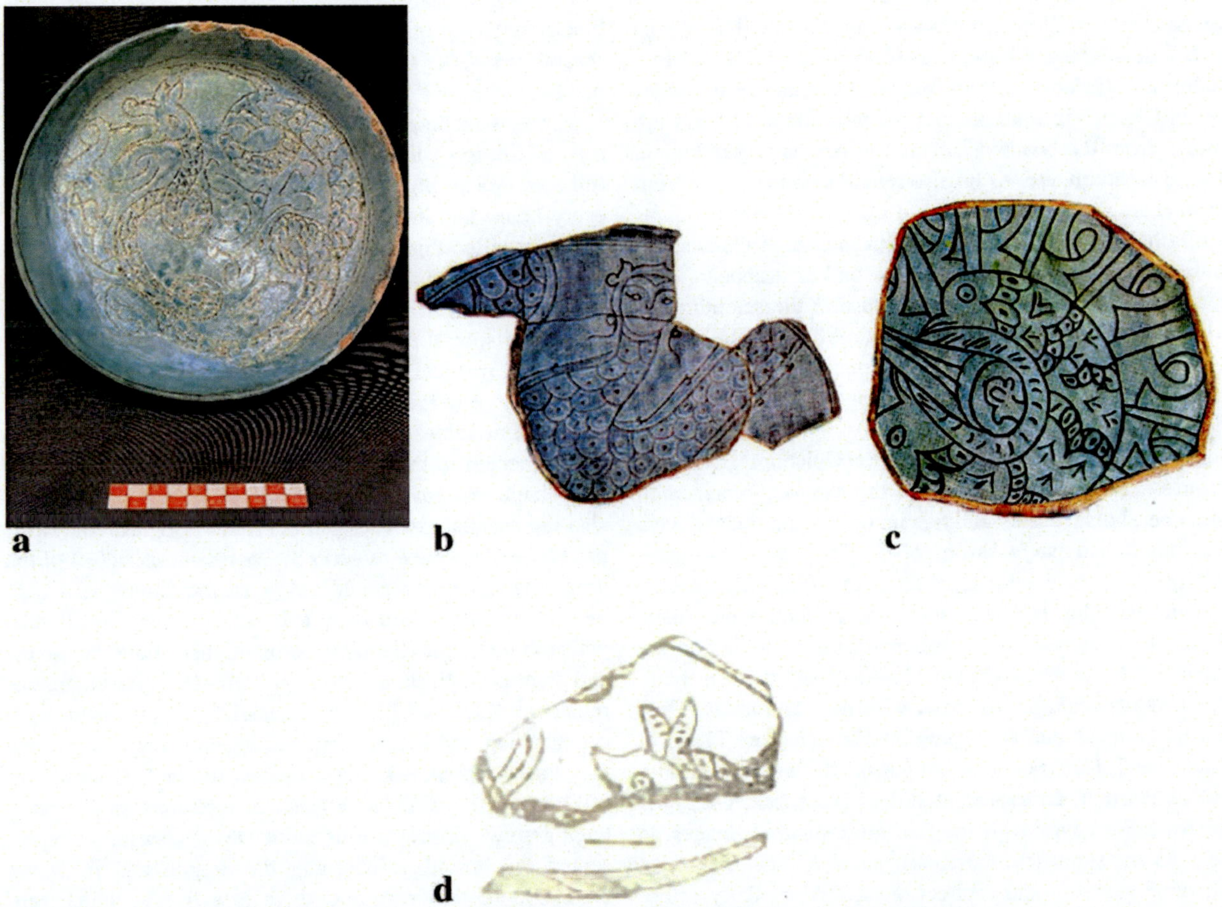

Fig. 7 **a** Bowl with dragon image and light blue glaze from Gudiashvili Square (© T. Butsuradze); **b, c** Bowls' fragments with bird images and light blue glaze from Tbilisi Ceramic Workshop (Mitsishvili, 1969, Pl. XIX–XX); **d** Bowl from Erekle II Square with the image of pig and covered with light blue glaze (Grdzelishvili & Tkeshelashvili, 1961, Pl. XXV-108)

the Tbilisi ceramic workshop, featuring images of bird-women, two-headed birds, birds with double endings, and more (Mitsishvili, 1979, p. 44). Mitsishvili remarks that bowls with scratched-out ornamentation depicting birds and covered with monochrome glaze, along with different ornamented bowls with geometric decor, date back to the XI–XII centuries (Mitsishvili, 1969, p. 28, Pl. XIX–XX) (Fig. 7b, c). Regarding bowls with animal images and light blue glaze, similar vessels were found in the garden located on Erekle II Square during the archaeological campaign in 1956–57 in pit N10. The authors attribute this pit to the third layer and date it to the eleventh–fourteenth centuries (Grdzelishvili & Tkeshelashvili, 1961, pp. 31,65- N108; Pl. XXV-108) (Fig. 7d). Although an exact parallel for the bowl with the dragon image was not found, based on the discussed examples, ornamentation principles, and glaze color, it can be dated to the XI–XII centuries.

According to the animal images, Vakhtang Jafaridze remarks that in ceramic decoration special place has animal and bird motifs. Bird motifs were widespread during XI–XIII centuries. He also mentions bowls with animal images, such as lions from Kaspi and Samshvilde. Bowls with animal images come also from Dmanisi, Gudarekhi, Ikalto, etc. (Jafaridze, 1956, p. 30–31). About bird images Levan Chilashvili remarks that different kinds of birds are known from the bowls discovered in Armenia, Azerbaijan, Iran, Byzantium, etc. As he also remarks on the bowls from Rustavi are represented images of animals and even humans (Chilashvili, 1958, p. 122–123). But in case of Gudiashvili Square there is sneak/dragon image and as it appears such image was not very common, because in the scientific literature such images are not discussed. What was the reason, why they have made snake/dragon image on the bowl with floral images as well, we can just suppose and find its roots in folk, mythology.

Snake images are represented on Colchian art-on the Bronze axes, girdles, fibulae. On the axes are represented dog, deer, horse, snake, and fish images. Leila Phantskhava has mentioned supposition of I. Meshchaninov, who remarks that on Bronze axes are represented fish, sneak, dragon, then dog and deer and those animals which have sneak body, opened gullet and paws ended with fins are identified as dragons (Phantskhava, 1988, p. 21). Leila Phantskhava remarks that axes with deer and sneak, deer and fish images that are represented together are found in Tsiossi and Tlia. For an example on the axes from Tlia and Chabarukhi are represented deer and sneak. On a girdle from Samtavro with the fish are two deer images as well. As she supposes in this case, sneak and fish represent the same meaning and connect to the fertility deity or are personification of it. As the author remarks, on the seals from Babylon near to the "life tree" is represented sneak (Phantskhava, 1988, p. 43). According to her, Colchian axes decorated with the images of sneak and fish are represented from several graves of Tlia, from the hoards of Chabarukhi and Tsageri and from Northern Caucasia—Kuban and Galiati. N18 grave from Tlia dates back to XII–X centuries B. C, N23 grave dates to VII century B. C. N39, 208, and 215 graves date back to XII–X centuries B. C. From those graves are coming axes with the images of sneak and fish. Hoard from Chabarukhi dates back to VIII–VII centuries B. C (Phantskhava, 1988, p. 66–67; PL. X-2; XVI) (Fig. 8a–c). Snake is also represented on fibula from Tlia—N202 grave (Phantskhava, 1988, p. 35, Pl. XXIII-2) (Fig. 8d). Moreover, snake images are represented on the silver bracelets from Pichvnari on the V century B. C. Greek necropolis. Those bracelets are decorated with geometric (triangles, circles, parallel lines, dots, etc.) and with animal images (calf, ram, ibex, snake, etc.) (Kakhidze, 2007, p. 189). Such golden bracelets with inwards back are known from Vani date back to V century and the first half of IV century BC, end up with the images of different animals (Chkhonia, 1981, pp. 47–49) are made locally with the influence of Achaemenid pattern and represents models of Colchian goldsmithing (Chkhonia, 2013, p. 117).

Nino Abakelia remarks that snake played an important role in Georgian tradition—in mythology, cult, ritual, art, folk, etc. According to the most common beliefs in Georgia, every house had its sneak-patron, which protected it and if someone killed this snake that meant the destruction of the family. The person who killed the snake was getting ill or sometimes died (Abakelia, 1997, pp. 5–7). She remarks that sneak had special role in folk medicine and in many cases appears in medical formulas (Abakelia, 1997, p. 13). From the history of religions it is well-known that fish and sneak have similar values. Sneak and dragon in Georgian tradition are connected to the water. Dragon is the holder of water, its resident and guardian (Abakelia, 1997, p. 14). Dragon is also represented in Caucasian fairytales (Gogiashvili, 2011, pp. 156–161). As she remarks in Georgian folk defeat of the dragon is not only metaphor. Against the dragon all weapons, which have been used and are described in all fairytales that exist in Georgia are mentioned: bow and arrow and sword (Gogiashvili, 2011, pp. 161–166). As the author remarks in Christian literature sneak h as metaphorical meaning and in the Christian legends warriors against the snakes represent struggle against the evil (Gogiashvili, 2011, pp. 167–168). As Gogiashvili mentions, in Georgian hagiography—in the life of Saint David Garejeli is represented the episode of burning the dragon and in David Gareja monastery complex, there is a fresco, which represents this fact (Gogiashvili, 2011, p. 167) (Fig. 8e). Dragon illustrations are also represented in Medieval Ages

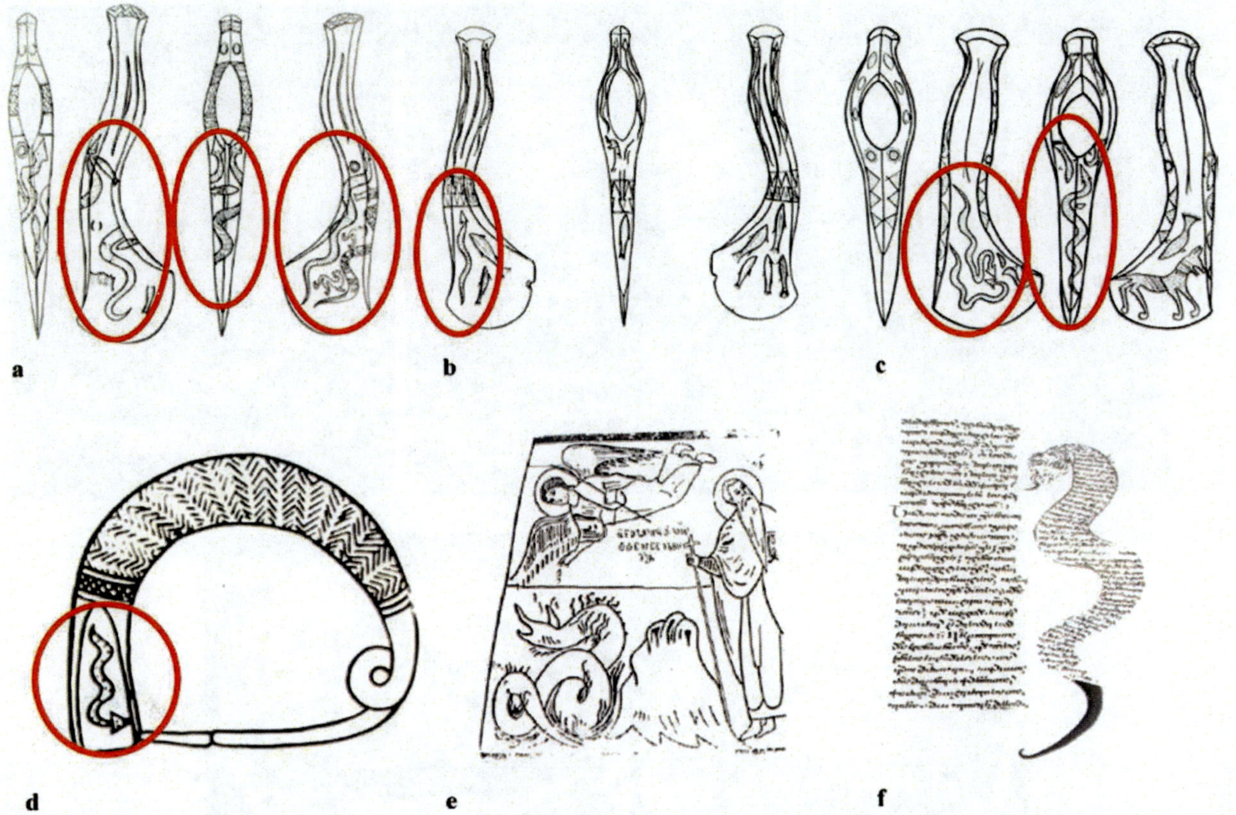

Fig. 8 **a** Colchian axe with snake image from Kuban (Phantskhava, 1988, Pl. X-2); **b** Colchian axe with sneak and fish images from Chabakhuri hoard (Phantskhava, 1988, Pl. XVI-1); **c** Colchian axe with dog, sneak, and fish images from Kuban (Phantskhava, 1988, Pl. XVI-2); **d** Fibula with snake image from Tlia (Phantskhava, 1988, Pl. XXIII-2); **e** Fresco from David Gareja Monastery complex represents the burning of the dragon (Gogiashvili, 2011, p. 167); **f** Image of dragon on Georgian manuscript "Climacus" (Gogiaschwili, 2006, p. 67)

Georgian manuscripts. As Elene Gogiashvili remarks XII century Georgian manuscript—"Climacus" by Saint John Sinaites is important because on the lace of the manuscript are represented drawings by the copyists and mainly are represented animals and mixed creatures images: fishes, crab, deer, man with goat horns, birds, and dragons. As the author remarks, this manuscript was rewritten by the calligrapher Nicholas Nikrai in 1160 in David Gareja monastery complex and all these images are connected to the content. The author remarks that *"XII century Georgian manuscript is kept in the National Center of Manuscripts (No. H-1669)"* (Gogiashvili, 2011, pp. 168–169) (for the image, see Gogiaschwili, 2006, p. 67) (Fig. 8f). It is noteworthy that in the dragon image's contour are written different kinds of sins (Gogiashvili, 2011, p. 170).

As Elene Gogiashvili remarks, in Christian symbolism dragon represents evil creature and in the fairytales where is combat against the dragon, the influence of Christianity is powerful (Gogiashvili, 2011, p. 175).

As it appears sneak/dragon had a special place and meaning as in Ancient Colchian handicraft monuments as in Georgian tradition-mythology, cult, folk, etc. The image of sneak/dragon on the bowl from Gudiashvili Square could be the continuation of this tradition that lasted over the centuries from the ancient times and in the Medieval Ages had special meaning as well.

Among the bowls from the pit are three bowls with "motion lines" ornamentation. These bowls have white engobe painted motion lines that are covered with light blue glaze. One of the bowls is fully preserved, while the other two are fragmentary (Fig. 9a–c). Similar ornamented bowls are known from the Tbilisi ceramic workshop, with a total of 43 items, including two semi-fabrics and 17 defective ones. These ceramics were found in different trenches and the workshop area but not in the XVIII trench (Mitsishvili, 1979, p. 29). Marine Mitsishvili dates such ornamented and glazed ceramics to the eleventh century (Mitsishvili, 1979, p. 36). It is noteworthy that Mitsishvili mentions two groups among materials covered with light blue glaze: those with scratched-out ornaments on the engobe surface and those with motion lines and other ornaments painted with white engobe on the fabric (Mitsishvili, 1979, pp. 38–39).

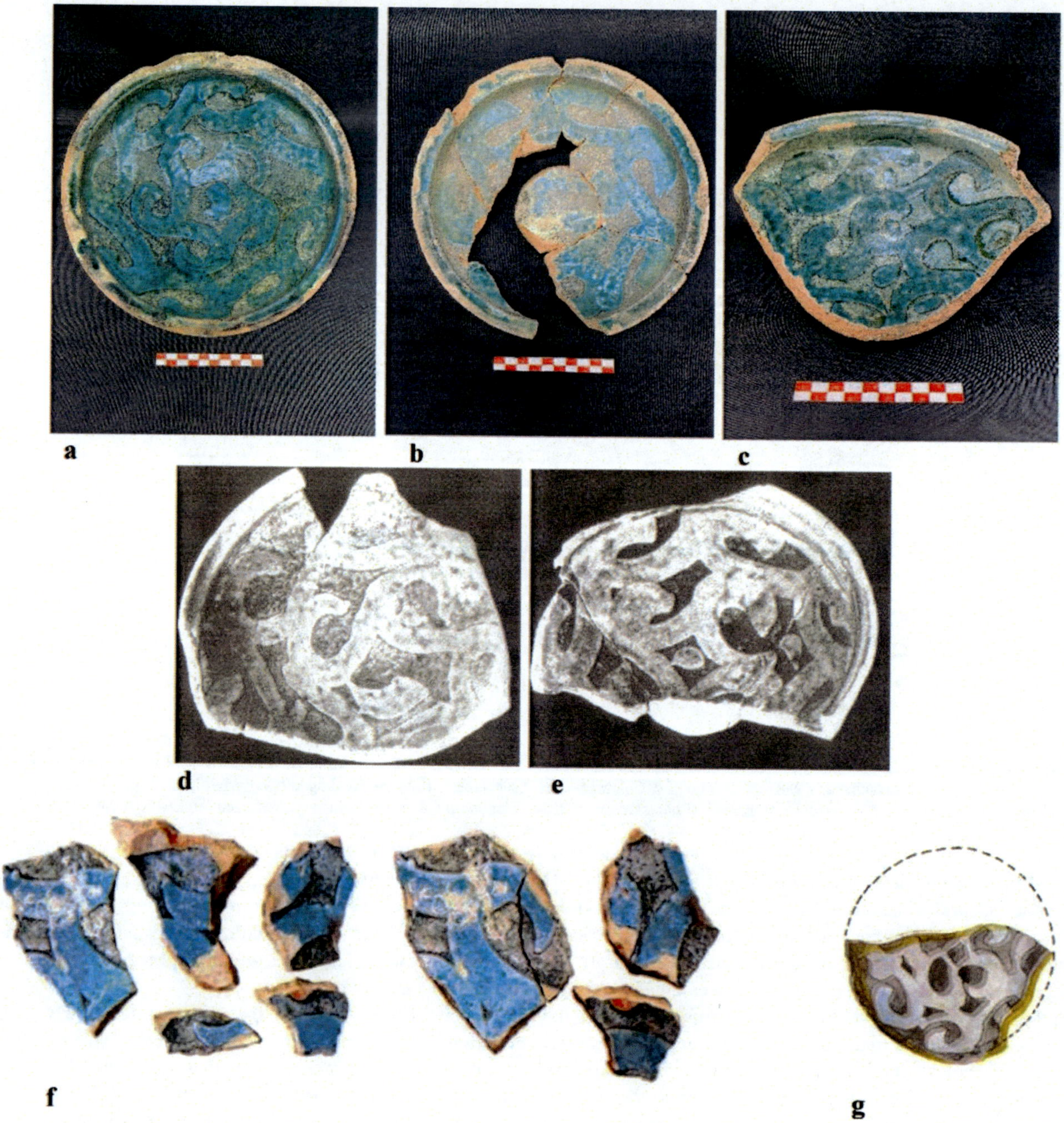

Fig. 9 **a–c** Bowls painted with white engobe and covered with light blue glaze from Gudiashvili Square (©T. Butsuradze); **d**, **e** Bowls painted with white engobe and glazed from Tbilisi Ceramic Workshop (Mitsishvili, 1979, Pl. I-1-2); **f** Bowls painted with white engobe and covered with light blue glaze from Samtavisi Monastery (Berikashvili, 2013, Pl. IV-1-2); **g** Bowl from Dmanisi (Kopaliani, 1996, Pl. XXV-5)

Mitsishvili remarks that the technique of fabric painting with white engobe is characteristic for the IX–X centuries and is continued to be used in the XI century, but with a "new" light blue glaze instead of green glaze. In Tbilisi ceramic workshop, white engobe painted and light blue glazed bowls (XI century) were produced alongside scratched-out ornamented and light blue glazed bowls. In the XII century, monochrome glazed ceramics (green, brownish manganese glaze, etc.) and scratched-out ornamented ceramics coexisted. In the last quarter of the XII century, polychrome glazed vessels appeared, and in the XIII century, vessels painted with a brush were produced (Mitsishvili, 1979, pp. 39–40). Mitsishvili notes that white engobe painted and light blue glazed vessels in the Tbilisi ceramic workshop are ornamented with so-called "motion lines" and, in very few cases, with different ornaments (Mitsishvili, 1979, p. 43; Pl. I-1-2) (Fig. 9d, e). Similar ornamented fragments were also found during the excavations of Samtavisi Monastery, which archaeologists date to the XI century, with five items represented (Berikashvili, 2013, p. 204, Pl. IV-1-2) (Fig. 9f). Another ornamented bowl fragment was found in Dmanisi in building N10 and is dated to the XI century (Kopaliani, 1996, pp. 85,125, Pl. XXV-5) (Fig. 9g).

In the pit, three bowls with a similar ornamentation were found. These bowls have a decoration consisting of pair lines forming two intersecting squares, creating a geometric pattern and covering them with light blue glaze (Fig. 10a–c). A similar ornamented bowl was discovered in Tbilisi ceramic workshop, where it is possible that two squares were scratched out and covered with light green glaze (Mitsishvili, 1979, p. 47, Pl. XI-1) (Fig. 10d). Another bowl with the same ornamentation is known from Rustavi, specifically from grave N21 in a medieval cemetery, which dates back to the XI–XII centuries (Pachikashvili, 2014, p. 143, Fig. 178) (Fig. 10e). According to the decoration style, Vakhtang Jafaridze remarks that geometric motifs were widespread from which are distinguished: double parallel lines, triangles, pentagon, and so-called "Sheild of David"—hexagon-compound of two equilateral triangles. Rarely are represented circles cursive to each other like chain, cross-like ornaments on which are scratched-out loops and sometimes are scratched-out zigzag lines and rectangles. All those geometric figures are made by hand. He also remarks that in the museum are preserved light blue glazed bowls with pentagon, bowl with triangle and covered with brownish manganese glaze. As Vakhtang Jafaridze remarks "Shield of David" was mostly spread motif in different fields of art as in Muslim as in Christian cultural world. Six-pointed star appears not only on the Georgian ceramic but also on the coins, such as anonym Qā'ānic coins from Tbilisi (1266–1280) (Jafaridze, 1956, pp. 25–26). In case of Gudiashvili Square there is no image of "Sheild of David" on the bowl, but are represented two intersecting pentagons and squares.

In the pit, five bowls with similar decoration were discovered. These bowls are covered with light blue glaze and feature scratched-out looped triangles and pair lines

a b c

d e

Fig. 10 **a–c** Bowls with crossed squares and covered with light blue glaze from Gudiashvili Square (©T. Butsuradze); **d** Bowl with crossed square and glazed from Tbilisi Ceramic Workshop (Mitsishvili, 1979, Pl. XI-1); **e** Bowl from Rustavi (Pachikashvili, 2014, p. 143, Fig. 178) ornamentation. They can be dated to the XI–XII centuries (Fig. 11a–e). A similar ornamented bowl was found on Rustavi fortress, as noted by Marine Mitsishvili (Mitsishvili, 1969, p. 88, Pl. XIV-1) (Fig. 11f). Other bowls with similar decoration were unearthed on Erekle II square in pit N14, which belongs to the third cultural layer and dates back to the XI–XIV centuries (Grdzelishvili & Tkeshelashvili, 1961, p. 65–N116, Pl. XXV-116) (Fig. 11g). Additionally, Nodar Bakhtadze mentions a similarly decorated bowl from Rustavi in his work (Bakhtadze, 2013, p. 54, Fig. 49) (Fig. 11h).

The next bowl to discuss features an image of circles filled with loops and triangle-like images filled with loops in between (Fig. 12a). Nodar Bakhtadze, in his work, has presented two similar ornamented bowls. One is covered with light blue glaze and was found on Rustavi settlement, while the other, covered with green glaze, comes from Tbilisi. The ornamentation on both bowls is the same, and Bakhtadze dates them to the XI–XII centuries (Bakhtadze, 2013, p. 50–51, N44-45) (Fig. 12b, c). Based on this, it is reasonable to date the bowl from Gudiashvili Square to the same period, the XI–XII centuries.

The other eight bowls covered with light blue glaze and featuring different ornamentations exhibit drop-like ornaments filled with net-type scratched décor, triangle ornaments filled with net-type decoration, rings-type ornaments

Fig. 11 **a–e** Bowls with light blue glaze and with the ornamentation of looped triangles and also pair lines from Gudiashvili Square (© T. Butsuradze); **f** Bowl with light blue glaze and with the ornamentation of looped triangles and pair lines from Rustavi fortress (Mitsishvili, 1969, Pl. XIV-1); **g** Bowl from Erekle II square (Grdzelishvili & Tkeshelashvili, 1961, Pl. XXV-116); **h** Bowl from Rustavi (Bakhtadze, 2013, p. 54, Fig. 49)

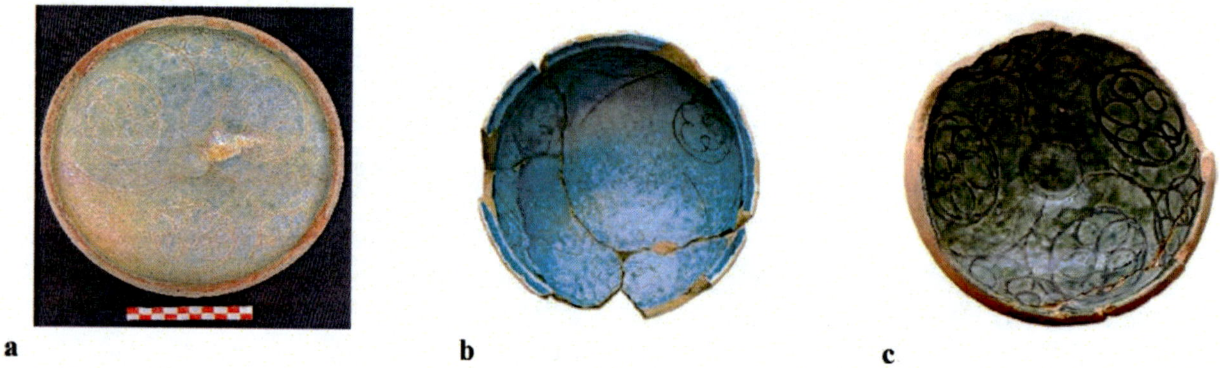

Fig. 12 **a** Bowl with light blue glaze and circles and triangles filled with loops from Gudiashvili Square (© T. Butsuradze); **b** Bowl from Rustavi Settlement (Bakhtadze, 2013, p. 50, Fig. 44); **c** Bowl from Tbilisi (Bakhtadze, 2013, p. 51, Fig. 45)

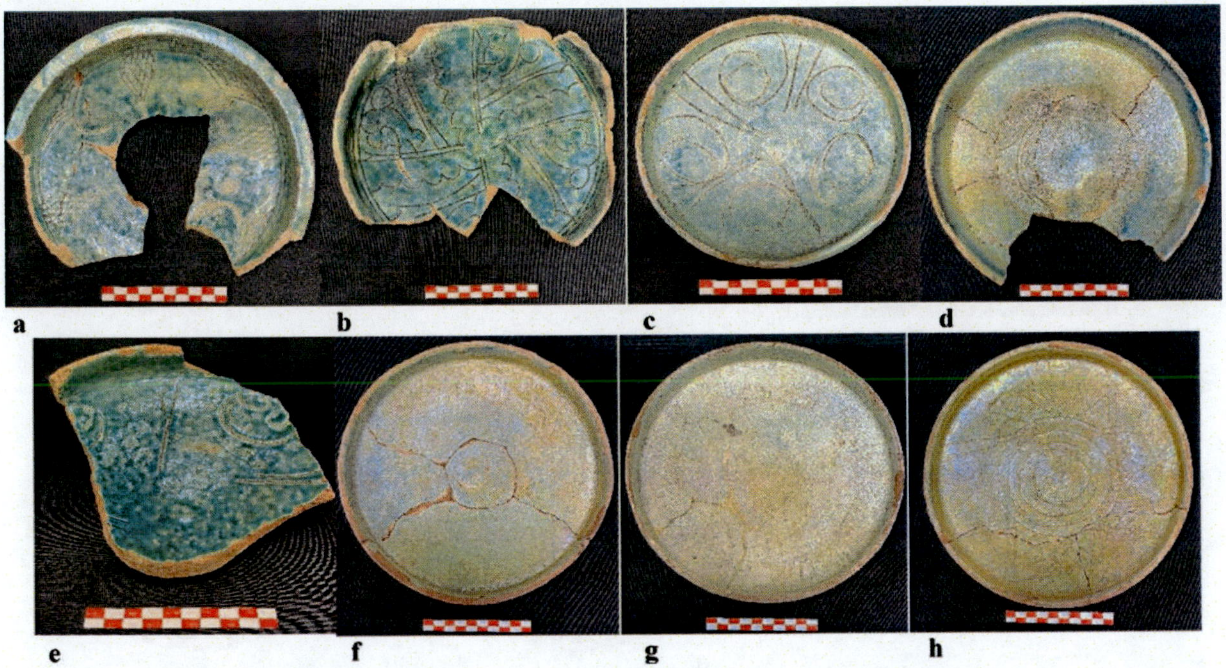

Fig. 13 **a–h** Bowls with light blue glaze and scratched-out ornaments from Gudiashvili Square (© T. Butsuradze)

with two pair lines, floral ornaments with vertical wavy and looped pair lines, rings with triangle-like décor in the center, looped circles, and more. Among these bowls, two of them are so iridescent that it is challenging to identify the color of the glaze, but it may be light blue. Based on the materials discussed earlier, including their glaze color and ornamentation style, these bowls can be confidently dated to the XI–XII centuries (Fig. 13a–h).

In the pit, a bowl with a brownish manganese glaze was also discovered. The bowl is decorated with a sun-like ornament consisting of double circles in the center and triangles above them, creating an imitation of rays (Fig. 14a). Marine Mitsishvili mentions that such ornamented bowls are known from various sites such as Rustavi fortress, Dmanisi, Ujarma, Gudarekhi, and the dump deposit of Tbilisi ceramic workshop (Mitsishvili, 1969, p. 25, Pl. XVI-1-2) (Fig. 14b). In Kutaisi, as Apolon Gabriadze notes bowls covered with brownish manganese glaze are abundantly represented. The author also mentions the presence of fragments of "greater" and "smaller" vessels with such glaze. It is worth noting that in western Georgia, such glazed pottery is rarely found. For example, a piece of Kotoshi (a vessel used for a cradle) with its inner side glazed with brownish manganese glaze was found in Batumi Fortress. Additionally, in Vani, on Tsikhesulori Fortress, a small piece of pottery covered with such glaze was discovered (Gabriadze, 2013, pp. 122–123, Pl. XXX-9-10-13) (Fig. 14c).

The bowl with a brownish manganese glaze and scratched-out ornament is mentioned by Vakhtang Jafaridze in his work. It features two pairs of horizontal lines that are scratched out and repeated three times before being covered with glaze. This bowl was found in Tbilisi. Another similar glazed bowl with scratched-out radial lines, resembling cross shoulders with looped circles between them, is preserved in the Tbilisi History and Ethnography Museum (Jafaridze, 1956, p. 59, Pl. XX-1-2) (Fig. 14d, e). Marine Mitsishvili mentions that 84 such glazed bowls were found in the Tbilisi ceramic workshop, with the majority of them decorated with geometric ornaments such as loops and circles. However, she also notes two exceptions, with one bowl featuring a scratched-out bird image and the other having a star point (Mitsishvili, 1979, p. 46). Gudiashvili Square also has representations of such glazed and ornamented bowls, including fragments from different areas.

Regarding other monochrome glazed bowls, there are two bowls covered with dark green glaze and scratched-out ornaments. The other two bowls were also covered with dark green glaze, but the ornamentation differs. One bowl has scratched out semi-circles created with pair lines, filled with net-like ornamentation. There is also a wavy ornament dividing the bowl into two parts. The second bowl has two scratched-out circles in the middle, above which oblique vertical and waved lines are present, creating a geometric decor (Fig. 15a, b). According to Marine Mitsishvili,

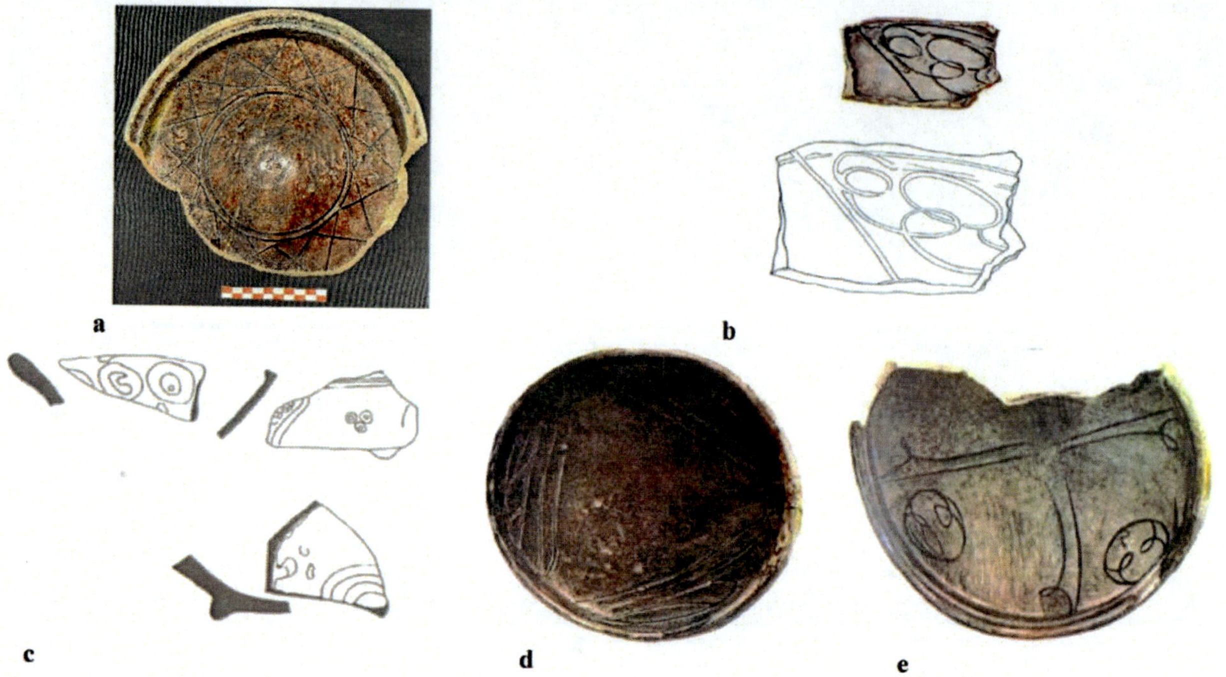

Fig. 14 **a** Bowl covered with brownish manganese glaze and scratched-out ornaments from Gudiashvili Square (© T. Butsuradze); **b** Bowl with brownish manganese glaze and scratched-out ornaments from Tbilisi (Mitsishvili, 1969, Pl. XVI-1-2); **c** Bowls with brownish manganese glaze from Kutaisi (Gabriadze, 2013, Pl. XXX); **d, e** Bowls with brownish manganese glaze from Tbilisi (Jafaridze, 1956, Pl. XX-1-2)

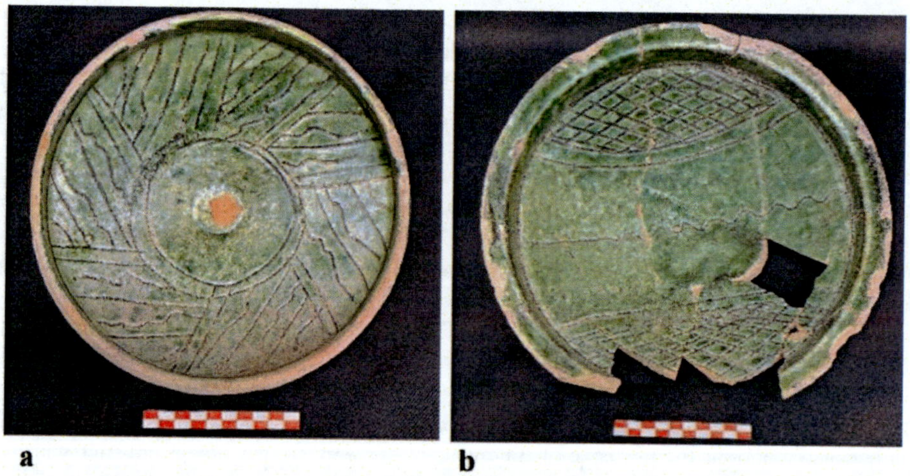

Fig. 15 **a, b** Green glazed and scratched-out ornamented bowls from Gudiashvili Square (© T. Butsuradze)

such glazed and ornamented bowls are abundantly represented in the Tbilisi ceramic workshop. Most of them feature scratched out four- and six-pointed stars, covered with light green glaze. Other bowls are decorated with looped lines and rings. The author mentions that such ornamented and glazed bowls were also found in the dump deposit of the workshop (Mitsishvili, 1969, p. 23). Mitsishvili emphasizes the difference between bowls produced in the workshop, glazed with light green glaze and adorned with faint ornaments, and bowls from the dump deposit, decorated with sharply scratched ornaments and covered with dark green glaze, similar to the bowls from Gudiashvili Square

a b c d e

Fig. 16 a–e Semi fabric unglazed bowls from Gudiashvili Square (© T. Butsuradze)

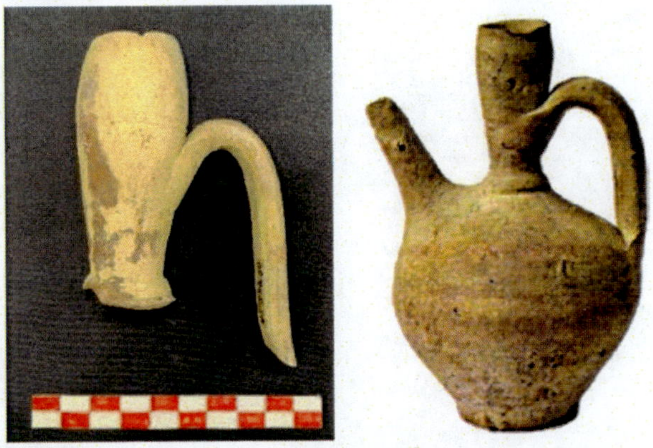

a b

Fig. 17 a Fragment of a jug with narrow spout from Gudiashvili Square (© T. Butsuradze); b Jug with narrow spout from Rustavi (Pachikashvili, 2014, p. 152, Fig. 197)

(Mitsishvili, 1969, p. 24). She also points out that this difference indicates that before the demolition of the workshop, bowls with light green glaze were produced, while earlier, bowls from the dump deposit were made and covered with dark green glaze, suggesting a time difference between the productions of these vessels. Mitsishvili further mentions that such glazed bowls are known from Rustavi fortress, but they are covered with light green glaze and scratched-out ornaments, while the bowls from Rustavi settlement are covered with dark green glaze and have colorful flecks (indicating incomplete dilution of the paint by the craftsman). A smaller number of such glazed bowls are found in Dmanisi, Ujarma, and Gudarekhi (Mitsishvili, 1969, p. 24).

In the pit, five unglazed bowls were discovered, which appeared to be semi-fabrics and prepared for glazing. Two of them were not even covered with engobe, while the other three were covered with white engobe and featured scratched-out ornaments such as looped circles with pair vertical lines, looped triangles with pair vertical lines, and looped triangles with looped drop-like ornaments. It seems that all of these bowls were intended to be glazed but were ultimately left unglazed (Fig. 16a–e). Similar semi-fabricated bowls were also found in the Tbilisi ceramic workshop, totaling around 250 pieces. They were covered with white engobe and had scratched-out ornaments, indicating their preparation for glazing. According to Marine Mitsishvili, based on the ornaments, these bowls were likely intended to be covered with monochrome glaze, but for some reason, they were not glazed (Mitsishvili, 1979, p. 47). Therefore, we can hypothesis that the semi-fabric bowls found in Gudiashvili Square were also intended to be covered with monochrome glaze, considering the style of ornamentation.

In the tableware are also unified two jug fragments from the pit, which is more or less possible to identify. The first fragment appears to be from a jug with a narrow spout, made from straw-colored clay (Fig. 17a). A complete jug of this type was found in Rustavi settlement and is dated to XI–XIII centuries (Pachikashvili, 2014, p. 152, Fig. 197) (Fig. 17b). It is plausible to suggest that the fragment from Gudiashvili Square belongs to a similar vessel and can be dated to XI–XIII centuries as well.

Another significant vessel found in the pit is made of white clay and appears to have been made in a mold and is imported vessel as well. It is partially preserved and could be identified as a jug or pitcher. The surface of the preserved fragment is divided into two parts. The lower part is decorated with vertically arranged alternating stripes of rings and rosettes, with a row of rosette ornamentation surrounding the entire surface. The upper part features two rows of rings encircling the surface, with images of lions and fishes arranged vertically above them. The fishes create sections in which lions, rosettes, rings, and a medallion are located. Above the lion, there is an image of a medallion that is poorly preserved and cannot be identified. The empty space between the two images is filled with rings and rosettes (Fig. 18a–c).

Although an exact parallel for this vessel could not be found, it is likely that it belongs to the Islamic period and could be an imported piece. Similar mold fragments with complex images are known from Nishapur. According to Wilkinson on the upper register is an Arabic inscription

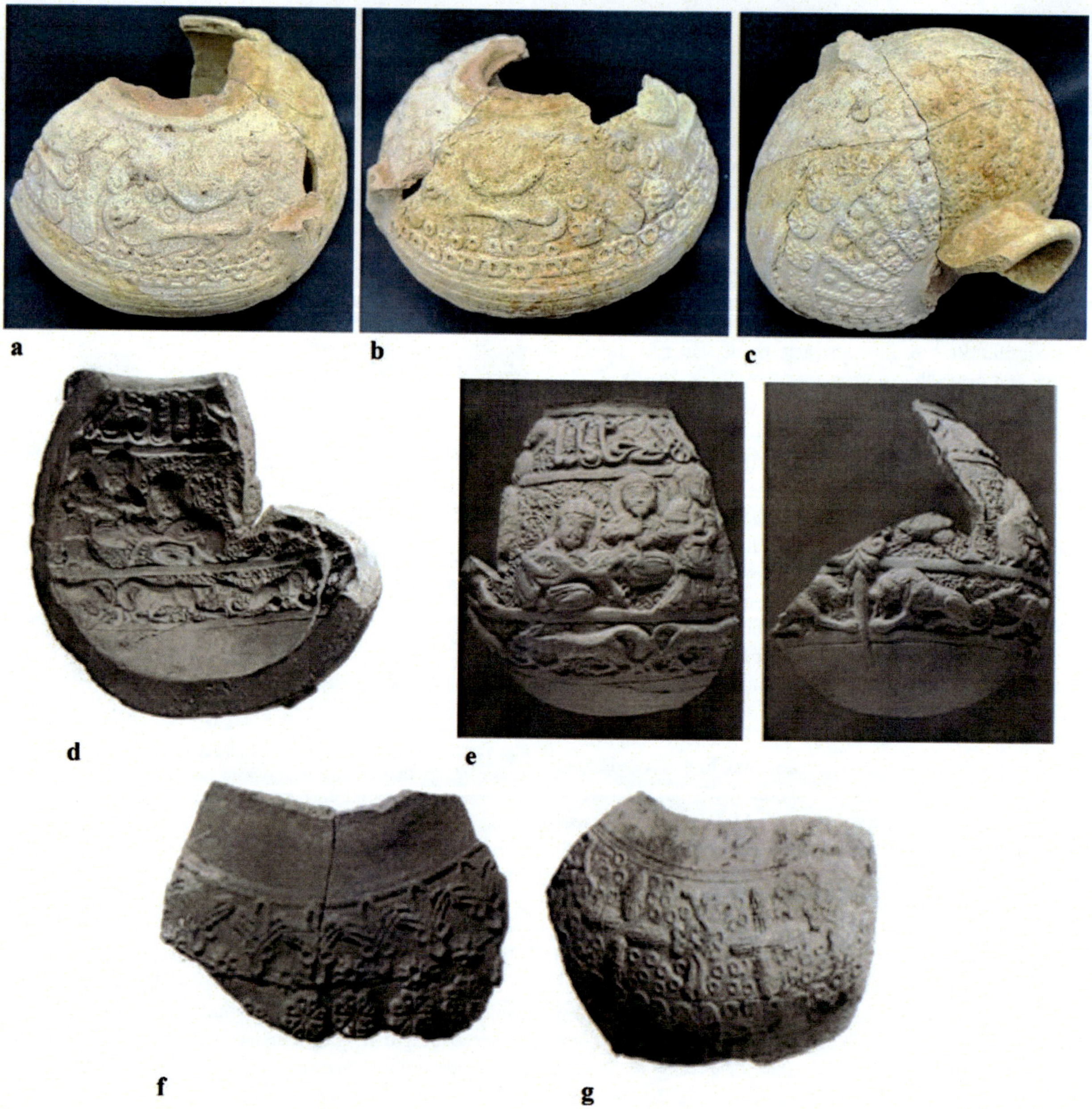

Fig. 18 **a–c** Jug or pitcher fragment from Gudiashvili Square (© T. Butsuradze); **d, e** Mold fragment from Nishapur (Wilkinson, 1973, p. 284, Fig. 57a–c); **f, g** Pitchers' fragments from Nishapur with rosettes and rings ornamentations (Wilkinson, 1973, p. 360, Fig. 187–188)

in Naskhi script; the middle register represents human figures *"appear against a background of elaborate curling foliations"*, and the bottom register represents different animals, possibly jackals and lions (Wilkinson, 1973, pp. 272–273, 284; Fig. 57a–c). It is worth noting that the images of lions on the vessel from Gudiashvili Square are more schematic compared with the delicate depictions on the Nishapur mold. Wilkinson suggests that the central register of the mold represents a literary subject, Farhad and Shirin. He also remarks that various elements of the mold were common in Iranian art of the XI, XII, and XIII centuries (Wilkinson, 1973, p. 272) (Fig. 18d, e).

Wilkinson categorizes this mold into the Alkaline-glazed ware, noting that it was made in Nishapur in the latter part of the XI century and throughout the XII century (Wilkinson, 1973, p. 263). He also remarks *"that glazed molded pottery with background foliations was made in Kashan as well as Nishapur is certain"* and *"it is not all*

Fig. 19 **a–c** Salt holders from the pit (© T. Butsuradze); **d** Salt holders from Tbilisi ceramic workshop (Mitsishvili, 1979, Pl. LIV-1); **e** Salt holders from Rustavi fortress (Pachikashvili, 2014, p. 144, Fig. 180)

impossible that this particular type of mold was imported into Nishapur" (Wilkinson, 1973, p. 273). As he remarks "*the absence of such pieces may also indicate that the areas excavated, including the East Kilns, were destroyed, not by the Mongols in the second decade of the thirteenth century, but by the disastrous earthquakes of 1145 and that subsequent to the quakes and the sacking in 1153, these areas were completely abandoned*" (Wilkinson, 1973, p. 263).

On the vessel from Gudiashvili Square are also images of rosettes and rings. Such rosettes are also known from Nishapur, for an example on the pitchers' fragments (Wilkinson, 1973, pp. 333, 360, Fig. 187–188). On one of the pitcher fragment, where are images of small rings, the author dates this fragments about twelfth century because of the double-knotted forms between the crosses and he mentions that it is characterized for this period (Wilkinson, 1973, pp. 333,360; Fig. 188) (Fig. 18f, g). Should be mentioned that studying and analyzing the scenes, details of this vessel from Gudiashvili Square, is still ongoing process. But roughly according to all above mentioned this vessel could be dated back to XI–XII centuries.

From the tableware should be distinguished three pieces of salt holders, from which one is wholly preserved (Fig. 19a, c). From them one is covered with brownish manganese glaze and the rest two fragments are covered with blue glaze. Salt holders are also very common type of vessels represented on every medieval age's sites. Marine Mitsishvili has remarked that in Tbilisi Ceramic workshop, in different trenches are represented 87 in total, covered with blue, light green and brownish manganese glaze and among them are distinguished semi fabrics and defectives as well (Mitsishvili, 1979, pp. 35,63; Pl. LIV-1) (Fig. 19d). Salt holders are known from Rustavi fortress and are dated back to XI–XIII centuries (Pachikashvili, 2014, p. 144, Fig. 180) (Fig. 19e).

Among the household ware are represented oil lamps, in total with five items. Four of them were covered with light blue glaze, while the fifth lamp was covered with brownish manganese glaze (Fig. 20a–e). These oil lamps, known as shell-like lamps, are characteristic of the medieval period. Tbilisi ceramic workshop appears to have produced a significant number of such oil lamps, with approximately 126 items discovered. All of them are shell-like lamps, except

Fig. 20 **a–e** Shell-like oil lamps from Gudiashvili Square (© T. Butsuradze); **f** Shell-like oil lamps from Tbilisi ceramic workshop (Mitsishvili, 1979, Pl. LIV-2); **g** Shell-like oil lamps from Tbilisi (Bakhtadze, 2013, Fig. 62); **h** Shell-like oil lamps from Dmanisi (Kopaliani, 1996, Pl. XXXIV-2); **i** Shell-like oil lamp from Rustavi settlement (Pachikashvili, 2014, p. 144, Fig. 181)

for one lamp that features a spout (Mitsishvili, 1979, p. 64, Pl. LIV-2) (Fig. 20f). Nodar Bakhtadze has published shell-like oil lamps with handles, which date back to the XII–XIII centuries and are covered with light blue glaze (Bakhtadze, 2013, p. 64, Fig. 62) (Fig. 20g). Shell-like oil lamps have also been found in Dmanisi Fortress, specifically on the territory of the gate. According to the author, these lamps are dated to the XI–XII centuries and are covered with green glaze (Kopaliani, 1996, p. 94, Pl. XXXIV-2) (Fig. 20h). They have also been discovered in Rustavi settlement, dating to the XII–XIII centuries (Pachikashvili, 2014, p. 144, Fig. 181) (Fig. 20i).

Another group of vessels are represented with kitchenware from the pit. Here were discovered about eight partly preserved pots, stewing pots and one ketsi-pan like vessel. In total nine kitchenware was identified and possible to find analogies and datable as well. Most of these items are fragmentary, making it challenging to identify or date them accurately. David Mindorashvili's work highlights the significance of unglazed vessels from the high medieval period (Mindorashvili, 2015, p. 129). He notes that the shapes of medieval pots are generally uniform but vary in size. The pots typically have handles, although there are some examples without handles. According to Mindorashvili, the ornaments on these pots often feature waved or linear girdles, as well as oblique and circular notches (Mindorashvili, 2015, pp. 131–132,143, Pl. I-22) (Fig. 21d).

One of the fragmentary pot fragments from the pit on Gudiashvili Square features a handle that is plastered onto the rim and body of the pot. A small dot is plastered on top of the handle, and the neck of the pot is decorated with a row of notched nail-like ornamentation. Below the neck, there is a grooved wave ornament. Another pot fragment from the pit exhibits two rows of notched nail-like decoration, with the handle plastered on the neck and body. A small dot is also plastered above the handle. This pot is also fragmentary. Additionally, a fragment of the pot's side and handle has been found (Fig. 21a–c).

These three pots are similar to a pot discovered in the N10 pit on Erekle II Square, which belongs to the third cultural layer and is dated to the XI–XIV centuries. However, the pot from Erekle II Square is decorated with two rows of notched nail ornamentation (Grdzelishvili & Tkeshelashvili, 1961, pp. 31,66-N131, Pl. XXIV-131) (Fig. 21e). Based on these similarities, it is reasonable to assume that the pot fragments from the pit on Gudiashvili Square could also be dated to the XI–XIV centuries and might belong to a similar pottery tradition. However, further study and analysis are necessary to provide more precise information about these fragments and their dating.

Among the kitchenware found in the pit on Gudiashvili Square, there is a fragmentary representation of Ketsi-pan-like vessel (Fig. 22a). David Mindorashvili observes that these vessels have a very thick fabric and are handmade. The rims of these vessels are usually smooth but sometimes decorated with wide notches. Some of these vessels come in large, medium, and small sizes, and occasionally feature a grooved cross in the center. Mindorashvili suggests that such decorated pottery might have been used for baking ritual bread (Mindorashvili, 2015, pp. 131, 61-Pl. IV-a, p. 100-Pl. I-8) (Fig. 22b, c).

Due to the fact that Ketsi-pan-like vessel is fragmentary represented from the pit, it is difficult to determine whether it had a cross in the center or to make definitive statements

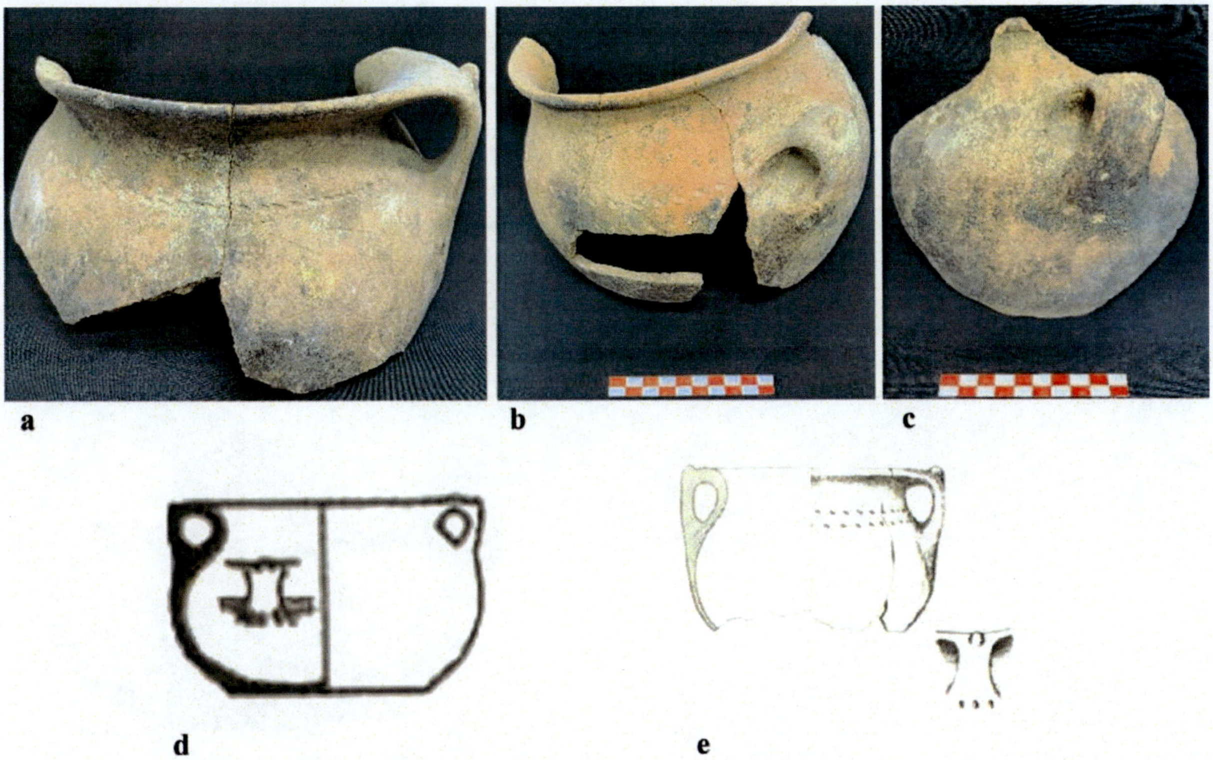

Fig. 21 **a-c** Pots fragments from Gudiashvili Square (© T. Butsuradze); **d** Pot from Rustavi (Mindorashvili, 2015, p. 143, Pl. I-22); **e** Pot from Erekle II Square (Grdzelishvili & Tkeshelashvili, 1961, Pl. XXIV-131)

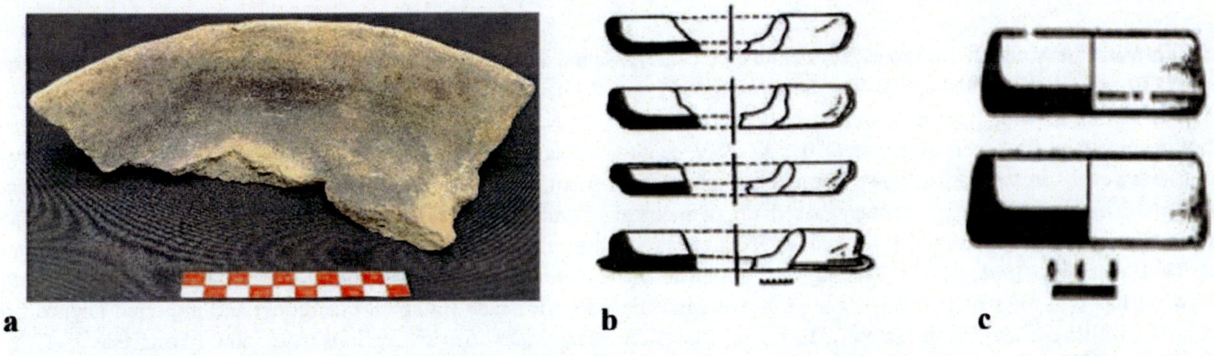

Fig. 22 **a** "Ketsi"-pan-like vessel from Gudiashvili Square (© T. Butsuradze); **b** "Ketsi" from Adjaristskhali Gorge Fortresses (Mindorashvili, 2015, p. 61-Pl. IV-a); **c** "Ketsi" from Tikheli (Mindorashvili, 2015, p. 100- Pl. I-8)

about its specific characteristics. Further analysis and examination are required to better understand this fragment and its original form.

In addition to the previously mentioned vessels, fragmentary representations of stewing pots were also found in the pit. According to David Mindorashvili, these stewing pots differ from regular pots in that they are smaller in size and have wide rims. They are also more decorated compared with pots. These vessels were designed to be hung over a fire, so they have upswing triangle-like rounded handles with holes in the middle and an oval-shaped bottom (Mindorashvili, 2015, p. 132). One of the stewing pots discovered in the layers of the pit had four upswung triangle handles with holes in the middle. The handles on both sides were decorated with a coniferous ornament, and the side of the pot was adorned with a notched girdle (Fig. 23a, b). Stewing pots have also been found on Erekle II Square, where archaeologists have classified them into five groups,

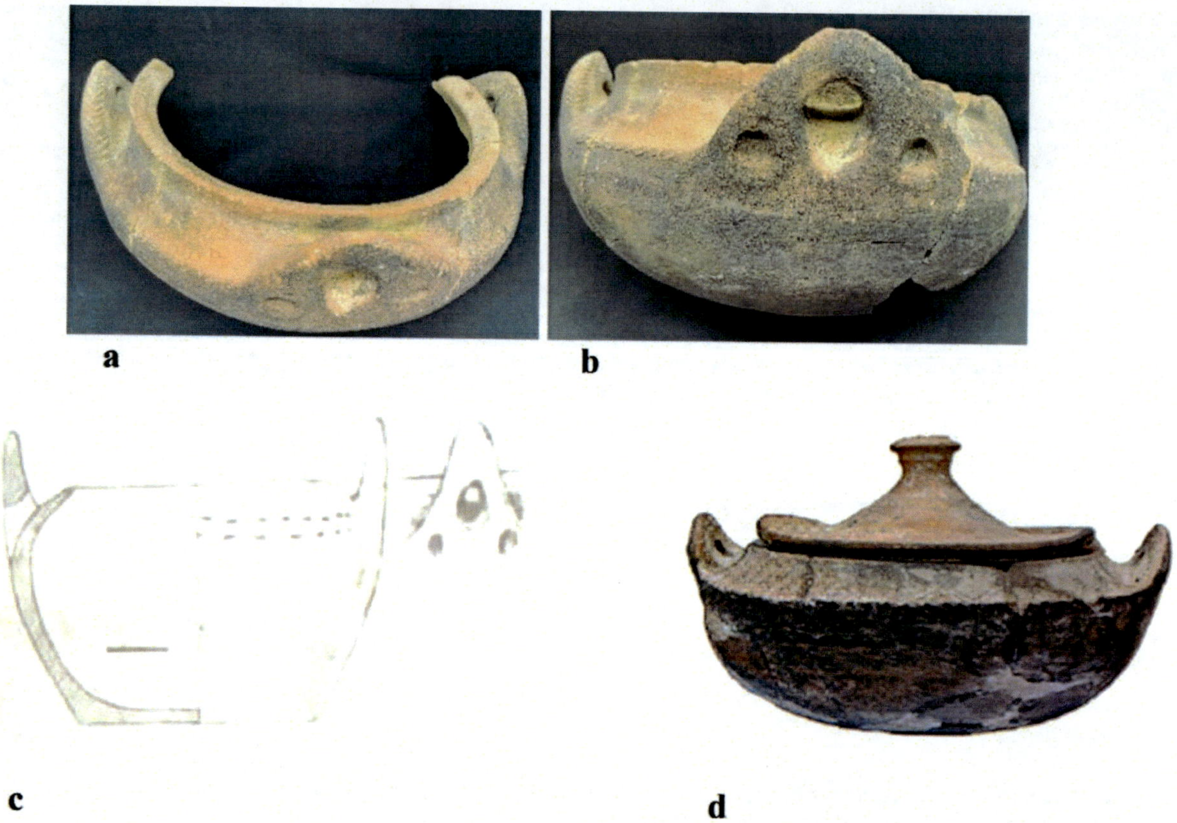

Fig. 23 **a, b** Stewing pot from Gudiashvili Square (© T. Butsuradze); **c** Stewing pot from Erekle II Square (Grdzelishvili & Tkeshelashvili, 1961, Pl. XXVII-177); **d** Stewing pot from Rustavi (Pachikashvili, 2014, p. 153, Fig. 199)

dating from the V to VII centuries, VII century, X century, and XII century (Grdzelishvili & Tkeshelashvili, 1961, p. 47). A similar stewing pot was found in N14 pit on Erekle II Square, which is generally dated to the XI–XIV centuries. However, archaeologists have more precisely dated this particular pit to the XI century based on two coins from Shirvan that are from the XI century (Grdzelishvili & Tkeshelashvili, 1961, p. 31). The stewing pot from the N14 pit has two rows of parallel linear girdles around the rim (Grdzelishvili & Tkeshelashvili, 1961, p. 69-N177, Pl. XXVII-177) (Fig. 23c). A similar stewing pot was also found in the Rustavi fortress and is believed to date back to the XI–XIII centuries (Pachikashvili, 2014, p. 153, Fig. 199) (Fig. 23d). It appears that this type of stewing pot also had lids, as fragments of lids have also been found in the pit.

Among the other fragments was found in the pit are pieces of faience, from which one is much more diagnostic one. Vakhtang Jafaridze notes that faience collections excavated in Georgia date from the XI to the XVIII centuries. He distinguishes two groups among the materials found in Georgia: a relatively small group of locally made faience and a larger group of imported faience mainly from Iran (Jafaridze, 1956, p. 36).

Nana Mamaiashvili, who studied medieval faience, mentions that during the high medieval period (XI–XIII centuries), Georgia experienced economic, political, and cultural development as a feudal state. Faience from this period has been found in Tbilisi, Dmanisi, Gudarekhi, Ujarma, and Rustavi of Kukheti. Among these findings, there are both locally made (in small quantities) and imported faience (in large quantities and various materials) (Mamaiashvili, 1976, p. 17). Mamaiashvili also notes that there is a wide variety of faience in Georgia dating to the XII–XIII centuries, and she categorizes them into several groups (Mamaiashvili, 1976, pp. 18,20).

Regarding the other pieces of faience found in the pit, there are four rim fragments, one handle fragment, and one bottom and side fragment, with the latter being the only one that is ornamented. This ornamented bottom fragment is made from white clay and is covered on both sides with blue glaze. The outer part is decorated with relief vertical stripes and oblique notches (Fig. 24a, b). Nana Mamaiashvili mentions that similar ornaments, including

a b c

Fig. 24 a, b Faience bottom fragment decorated with relief vertical stripes with oblique notches from Gudiashvili Square (©T. Butsuradze), c Faience fragments from Tbilisi (1) and Dmanisi (2) (Mamaishvili, 1976, Pl. IV-1-2)

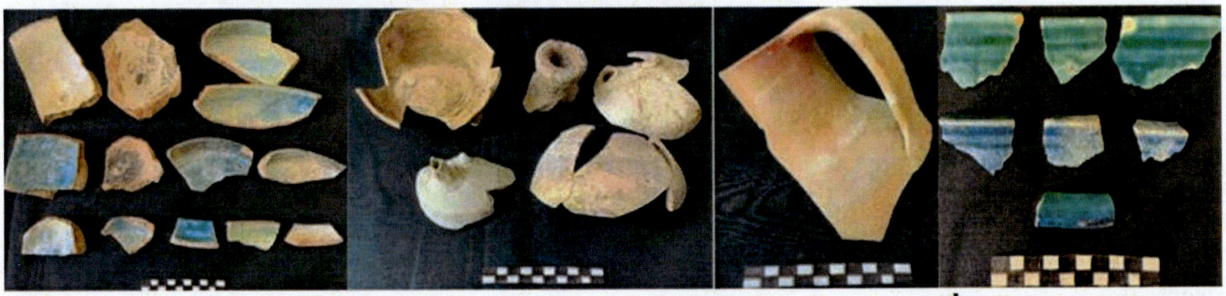

a b

Fig. 25 a, b Part of other ceramic sherds from the pit (©T. Butsuradze)

the so-called "tack" decoration, have been found in Tbilisi (during excavations in the 300 Aragveli garden), Dmanisi, and Rustavi (Mamaiashvili, 1976, p. 25, Pl. IV-1-2) (Fig. 24c).

The ornamented faience fragment found in the pit on Gudiashvili Square, featuring relief vertical stripes with oblique notches and "tack" type ornamentation, resembles locally made faience described by Nana Mamaiashvili (Mamaiashvili, 1976, p. 25). Mamaiashvili also describes a similar vessel found in Dmanisi with the same ornamentation, including a "tack" type ornament in the middle, and notes that such pottery is also known from Anisi. The vessel from Dmanisi is covered with blue transparent glaze (Mamaiashvili, 1976, p. 26, Pl. IV-4). Mamaiashvili suggests that pottery from Dmanisi and Anisi can be dated to the XII–XIII centuries based on the ornament type, noting that vessels with relief vertical stripes and "tack" type ornamentation were produced until the first half of the XIII century (Mamaiashvili, 1976, p. 26). Based on these observations, it is plausible to date the ornamented faience fragment from the pit to the XII–XIII centuries as well.

What about the last group distinguished among the materials from the pit, here were unified 83 pieces of different vessels, such as: bowls, pots, lids, faience, jugs, handles, bottom fragments, pottery sherds made of white clay, and other unidentified vessel fragments (Fig. 25a, b).

It is important to determine the MNV and percentage of those materials from the pit, which help to define mostly produced and used vessels. Total amount of ceramic assemblages are 136 items, from which with a big amount are represented bowls with different ornamentation motifs and glaze color. As it is shown of the diagram (Fig. 26), if we don't take into consideration different vessels' sherds (overall 61%), with a big amount are represented tableware, more precisely bowls (overall 29%), then kitchenware (6%), and household ware (4%).

3 Conclusion

The presence of these diverse ceramic assemblages indicates a wide range of tableware, kitchenware household ware, and other ceramic sherds with different decorations, both glazed and unglazed. Studying these materials makes it possible to date the pit to the XI–XIII centuries. This dating is supported by the presence of building ceramics such as bricks used during the medieval ages, also with bowls with monochrome glaze and scratched-out ornamentation,

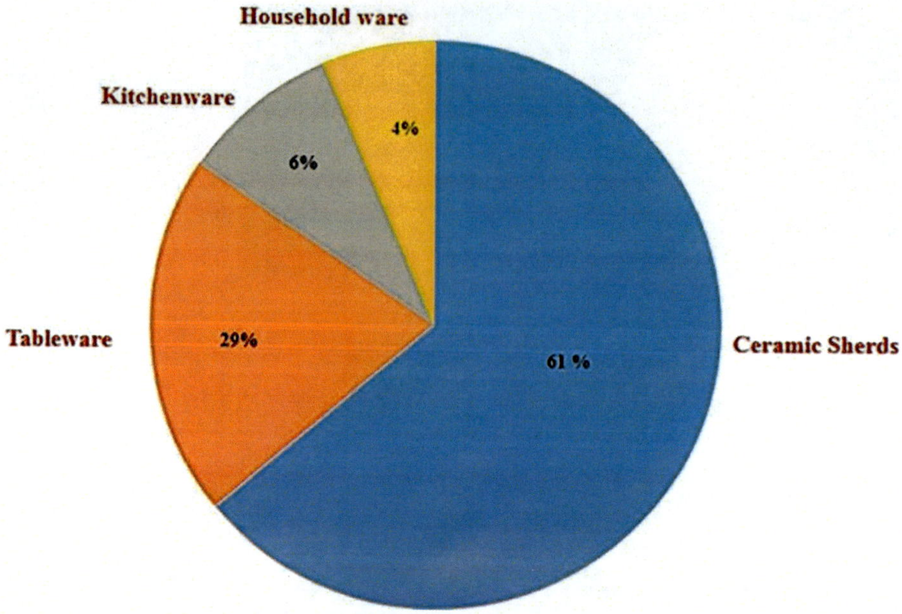

Fig. 26 Percentage of overall ceramic assemblages from the pit (©T. Butsuradze)

faience, and Islamic pottery. It is worth noting that a significant number materials indicating on both local production and import.

According to all abovementioned discussion, we can suppose that this pit, like others, excavated not only in Tbilisi (Tbilisi ceramic workshop, Gudiashvili Square), but in Kutaisi, was used in the workshops. According to this fact, this floor level belonged to before existed ceramic workshop, which was destroyed in later period and another cellar was constructed above it. Some semi-fabric bowls (five items) might indicate on this fact as well. The existence of the workshop itself indicates on the local production of the ceramic materials, which were commonly used by the residents of this square and moreover it indicates and proved the development of this handicraft. According to the fact that on Gudiashvili Square is another ceramic workshop, with kiln and other adjacent dwellings won't exclude the possibility of existence of another workshop. The fact that imported materials (ceramics made of white clay, such as jug with the lion, fish and other images, and other pottery sherds) indicates to the cultural and trade connections to the adjacent regions. According to the comparative analysis of above discussed materials, it is well seen what kind of vessels were used and spread during XI–XIII centuries and what kind of motifs were used for decoration of the those vessels. Discussion about this pit and materials have a crucial importance for studying medieval ages Tbilisi and specifically for Gudiashvili Square, because this square became archaeologically sensitive area not a long time ago and analyzing this information will help to determine the importance of Gudiashvili Square during the medieval ages and to define its role in the urban environment.

References

Abakelia, N. (1997). *Symbol and ritual in Georgian culture.* Georgian National Academy of Sciences, Institute of History and Ethnography named Ivane Javakhishvili.

Bakhtadze, N. (2013). *Ceramics in medieval Georgia.* Bakur Sulakauri Publishing.

Berdznishvili, M. (1965). External side of Tbilisi in XVIII century (according to the Georgian documentary materials). In *Georgian National Academy of Sciences, Institute of History, Archaeology and Ethnography named Ivane Javakhishvili.* Publishing House 'Metsniereba'.

Berikashvili, D. (2013). Medieval period glazed pottery from Samtavisi. In *Bulletin of the Georgian national museum. Series of social sciences* N4 (49-B) (pp. 199–211).

Butsuradze, T. (2022). Gudiashvili square-a model of urban archaeology. In *Conference abstract—International Conference on Cultural Heritage and New Technologies-CHNT 27, Vienna* (pp. 1–5). https://chnt.at/wp-content/uploads/2022/11/ID139_Butsuradze.pdf. Accessed June 18, 2023

Chilashvili, L. (1958). *City of Rustavi (historical-archaeological research).* Publishing House of Georgian Scientific Academy.

Chkhonia, A. (1981). Golden jewellery from Vani settlement. In *Vani VI-archaeological excavations.* Publishing House "Metsniereba".

Chkhonia, A. (2013). Ancient Colchis connections to the Achaemenid world (golden bracelets). In *Iberia-Colchis N9-researches on the archaeology and history of Georgia in the classical and early medieval period.* Georgian National Museum, Journal of Otar Lortkipanidze Centre of Archaeology.

Chologauri L. (2022). Bartmann jugs from Tbilisi-import and local imitations of the German stoneware in Georgia. In *The International Scientific Journal of Humanities-Contemporary Challenges of Archaeology and History, TSU-TI N1* (pp. 28–44).

TSU Press. https://tsuti.tsu.ge/index.php/journal/issue/view/1/1. Accessed May 31, 2023

Dumbadze, M., & Guchua, V. (Eds.) (1990). *Tbilisi from ancient times till the end of XVIII century* (Vol. I). Publishing House „Metsniereba".

Elisashvili, A., Amashukeli, T., Elisashvili, Ts., Tsotskhalishvili, M., Gordeladze, T. (2011). *Squares of Tbilisi-Gudiashvili Square*. Tiflis Hamkari-The Union of Tbilisi Caretakers.

Gabriadze, A. (2013). *History and archaeology of medieval ages Kutaisi (X–XV centuries)*. Publishing House „Motsameta".

Gogiaschwili, E. (2006). Die christliche Symbolik von Tierfiguren im georgischen Volksmärchen, dargestellt an einer Handschrift. In *Fabula 47, heft 1–2, Walter de Gruyter Berlin-New York* (pp. 65–78).

Gogiashvili, E. (2011). *Dynamics of mythical and religious symbolism into fairytale structure*. Ivane Javakhishvili Tbilisi State University, Faculty of Humanities, Department of folkloristic.

Grdzelishvili, I., & Tkeshelashvili, O. (1961). *Material culture sites of Tbilisi (according to 1956–57 archaeological campaigns)*. Publishing House of Georgian Scientific Academy.

Jafaridze, V. (1956). *Ceramic production in XI–XIII centuries Georgia (according to archaeological materials)*. Publishing House of Georgian Scientific Academy.

Jghamaia, J. (1980). *Building ceramic in feudal ages Georgia*. Publishing House „Metsniereba".

Kakhidze, A. (2007). *Pichvnari II- Antique world in Eastern Black Sea coast (The 5th century BC Greek necropolis at Pichvnari)*. Part I, Archaeological museum of Batumi, Oxford Ashmolean Museum.

Kopaliani, J. (1996). *Fortress of Dmanisi (historical-archaeological research)*. Publishing House „Sakartvelo".

Lomtatidze, G. (1955). Archaeological excavation in Tbilisi in the winter of 1948. In: *Materials for the archaeology of Georgia and The Caucasus* (Vol. I).

Mamaiashvili, N. (1976). *Faience in medieval ages Georgia*. Publishing House „Metsniereba".

Mania, M. (2003). Cellars of Old Tbilisi. In *Georgia antiquities N3*. National Research Center of Georgian Art History and Sites protection named Giorgi Chubinashvili.

Mindorashvili, D. (2015). *Archaeology of Georgia II-III- high and late medieval ages*. Publishing House „Batumi Shota Rustaveli State University".

Mitsishvili, M. (1969). *Glazed Pottery in Georgia in IX–XIII centuries*. Publishing House „Khelovneba".

Mitsishvili, M. (1979). *Glazed pottery workshop in medieval ages Tbilisi (IX–XIII centuries)*. Publishing House „Metsniereba".

Pachikashvili, N. (2014). *Ancient Rustavi*. Rustavi Historical Museum, Published with the Financing by Municipality of Rustavi City Municipality.

Phantskhava, L. (1988). *Artistic handicraft monuments of Colchian Culture*. Publishing House "Metsniereba".

Suramelashvili, M., & Chachkhunashvili, T. (2007). *Lado Gudiashvili Square N2 (Historical-artistic research of the building)*, Tbilisi.

Wilkinson, Ch. K. (1973). *Nishapur: Pottery of the early Islamic period*. The Metropolitan Museum of Art, Printed by the Press of A. Colish, Inc.

Shaping Cities and Formation

Legislation as an Important Element of the Management of the Casbah of Algiers: A Historical Review

Farah Hadji, Emad Mushtaha, Zaki Aslan, and Quenza Bougherira

Abstract

The historic center of Algiers (Casbah) presents a state of severe decay. Even that the state has financed projects for its preservation, and promulgated laws to frame these actions with adequate conservation process, the Casbah's buildings are still in danger of collapsing. This paper investigates the managing system of this heritage and its evolution through the public regulations and the legislative decisions. The management structure bodies of the Casbah have known a significant development since the promulgation of Law 98-04 regarding the preservation of cultural heritage. This law brought a new tool for the conservation and preservation of historic living cities that did not exist before 1998, what causes an unclear management strategy. This paper highlights anomalies concerning the management of the historic site of the Casbah before and after the promulgation of Law 98-04, and the efforts made by the state, as well as the results obtained so far after the approval of the new conservation instrument which is the PPSMVSS (Permanent Plan of Safeguard and Enhancement of Safeguarded Sectors) since 2012.

Keywords

Heritage · Urban heritage · Heritage management · Heritage law · Safeguarded Sector · Safeguard plan (PPSMVSS)

Abbreviation

ANAPSMH:	National Agency for Archeology and Protection of Historic Sites and Monuments (*Agence Nationale d'Archéologie et de Protection des Sites et Monuments Historiques*)
ANSS:	National Agency of Safeguarded Sectors (*Agence Nationale des Secteurs Sauvegardés*)
APC:	Municipality's Popular Assembly (*Assemblée Populaire Communale*)
APW:	Wilaya's Popular Assembly (*Assemblée Populaire de Wilaya*)
ATRC:	Atelier Technique pour la Revitalisation de la *Casbah*
BEREP:	Study Office for the Restoration and Preservation of Old Districts (*Bureau d'Etudes de Restauration et de Préservation des Vieux Quartiers*).
CNERU:	National Centre for Studies and Applied Research in Town Planning (*Centre National d'Etudes et de Recherches appliquées en Urbanisme*)
COMEDOR:	Standing Committee for the Study and Development of Organization and Planning of the agglomeration

Safeguarded Plan of the Safeguarded Sector.

F. Hadji (✉) · Q. Bougherira
Department of Architecture, University of Algiers 1, Alger Ctre, Algeria
e-mail: hadji3farah@gmail.com

E. Mushtaha
Department of Architectural Engineering, University of Sharjah, Sharjah, UAE
e-mail: emushtaha@sharjah.ac.ae

Z. Aslan
ICCROM, Sharjah, UAE
e-mail: aslan.zaki@gmail.com zaki.aslan@iccrom.org

M. F. Shahidan et al. (eds.), *Urban Narratives: Exploring Identity, Heritage, and Sustainable Development in Cities*, Advances in Science, Technology & Innovation, https://doi.org/10.1007/978-3-031-48517-6_10

of Algiers (*Comité Permanent d'Etudes et de Développement, d'Organisation et d'Aménagement de l'Agglomération d'Alger*).

DARQ: Neighborhood Planning and Restructuring Department (*Direction d'Aménagement et de Restructuration des Quartiers*).

D.C.W.A.: Direction of Culture of the Wilaya of Algiers (*Direction de la Culture de la Wilaya d'Alger*).

DEP: Direction of Publics Equipment (*Direction des Equipements Publics*).

ETAU: Central Study Office for Public Works, Architecture and Urban Planning (*Bureau Central d'Etudes et de Travaux Publics, d'Architecture et d'Urbanisme*).

OFIRAC: Office for the Intervention in and Regulation of Development operations in the Casbah of Algiers (*Office d'Intervention et de Régulation d'Opération d'Aménagement de la Casbah d'Alger*).

OGEBC: National Office for the Management and Exploitation of Protected Cultural Goods (*Office National de Gestion et d'Exploitation des Biens Culturels Protégés*).

OPGI Hussein Dey: Promotion Office for the Property Management of Hussein Dey (*Office de Promotion et de Gestion Immobilière de Hussein Dey*).

PDAU: Master Plan for Development and Urbanism (*Plan Directeur d'Aménagement et d'Urbanisme*).

POS: Land Use Plan (*Plan d'Occupation des Sols*).

S.S.: Safeguarded Sector (Protected Area) (*Secteur Sauvegardé*).

1 Introduction

The protection of cultural heritage has known a long process to arrive to what we know nowadays, process that keeps on evolving. During the last three decades, the management of cultural heritage has become an important approach to the preservation of the latter. Indeed, the UNESCO and its advisory bodies consider the heritage management system as a framework divided into nine components that are common to all heritage management systems. These components can be categorized under three categories (Fig. 1): three elements, three actions to undertake (or process), and three results (or objectives to achieve).

This has been reflected in the various doctrinal texts issued internationally (from Venice Charter in 1964, to HUL Recommendation in 2011), laws promulgated nationally, and in some cases even locally.

This paper focuses on the legal framework and their evolution regarding the management of the World Heritage site of the Casbah of Algiers.

To do so, an anthology of the national Algerian laws concerning the protection of the cultural heritage is explained shedding light on the management and texts having a direct relation with the protection of historic centers, more precisely, the living historic center of the Casbah of Algiers, set up as Safeguarded Sector (S. S.)

2 Algerian Laws Regarding Heritage Protection

The heritage protection in Algeria has known several steps, starting by the laws issued during the French colonization to a post-colonization state, to a new law more complete.

2.1 During the French Colonization (1830–1962)

During the French colonization, all the existing heritage in the Algerian lands was managed and used by the colons. Their interest was limited to the vestiges from the Roman period, and it was a justification of colonization, to prove that Algeria is part of France.

In 1887, the French state issued a law regarding the conservation of monuments and works of art of historical and artistic interest. It was only then that Algerian heritage experienced the first recognition of some of its monuments. The first list of historical monuments of Algeria was established this same year, having 7 monuments including three mosques of the Casbah of Algiers.

These events affected the collective memory of Algerian society, which led to the quasi-total neglect of the heritage inherited from the periods preceding and following the Roman era (such as Numidian period, Islamic period and Ottoman period).

After independence, the Algerian government reconducted the legislation governing the country at that time through Law 62-157 (OJ n°02, 1963), which led us to a period of pseudo-legislative independence where colonial legislation still governed the Algerian national heritage.

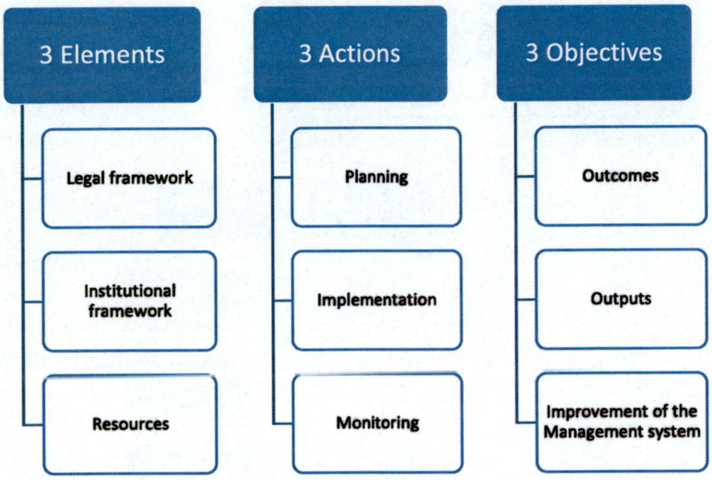

Fig. 1 Scheme of a heritage management system. *Source* The authors

2.2 Ordinance N°67-281 (1968)

In 1967, an ordinance was promulgated (OJ n°7, 1968; OJ n°26, 1981), related to the "*Excavations, and the Protection of Monuments and Historical and Natural Sites*" attempting to broaden the scope of protection of the Algerian heritage.

The ordinance had six titles and contains 138 articles, the 1st title includes general principles, the second title deals with excavations, and the 3rd is devoted to historic sites and monuments.

The 4th title deals with Natural Sites and Monuments by specifying their nature, definition, the way of classifying them and setting the conditions for the various actions and operations to be undertaken within their perimeters. The fifth and penultimate point deals with sanctions.

The last point is entitled "organizations," and closes the legal text by explaining the constitution of the National Commission for Sites and Monuments, its members, its missions, the frequency of its meetings, its skills, and roles, and adds details on the constitution of the Departmental Commission of Sites and Monuments.

Numerous Algerian specialists reproach to the Ordinance n°67-281 a lack of not taking in charge living urban historic centers (such as the Casbah, Medinas, etc.). Except of the expert Zadem (Zadem, 2008), who pointed out that in 1967 the essential texts in terms of development and town planning that we know now, had not been promulgated yet.

It was until 1998 (Fig. 3) that a new and more comprehensive cultural heritage protection law was enacted, including more typologies of heritage such as intangible heritage and subaquatic heritage.

2.3 Law 98-04 (1998)

This new law, related to the protection of cultural heritage, promulgated on June 15, 1998, aims to "*define the cultural heritage of the Nation, to enact the general rules for its protection, safeguarding and enhancement, and to set the conditions for their development work*" (OJ n°44, 1998).

The law is subdivided into 9 titles, making it possible to distinguish the definitions of different heritage categories, the different approaches to the treatment of the different types of heritage that the law undertakes to protect, as well as financial aid, sanctions and the various bodies that have the role of applying this law. The 9 titles are as follows:

- Title I: General Provisions (Art. 1–7)
- Title II: Protection of Immovable Cultural Property (Art. 8–49)
- Title III: Protection of Movable Cultural Property (Art. 50–66)
- Title IV: Intangible Cultural Property (Art. 67–69)
- Title V: Archeological Research (Art. 70–78)
- Title VI: The Organs (Art. 79–81)
- Title VII: Financing of operations for intervention and enhancement of cultural property (Art. 82–90)
- Title VIII: Control and Sanctions (Art. 91–101).
- Title IX: Final Provisions (Art. 102–108).

Article 8 of Law 98-04 defines immovable cultural property in three categories (Fig. 2):

- Historic monuments
- Archeological sites
- Urban or rural groups/ensemble

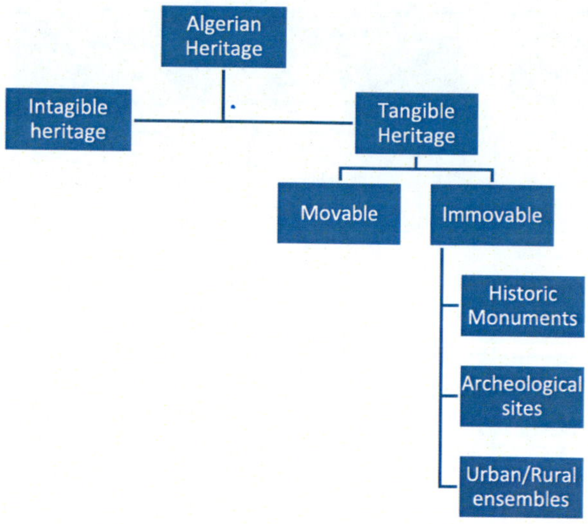

Fig. 2 Organigram of heritage types according to the Law 98-04. *Source* Hadji et al. (2023)

A specific protection regime is applied for each category of cultural property, according to Article 8 of Law 98-04:

- Registration on the list of additional inventory
- Classification
- Creation in a Safeguarded Sector.

The classification order is issued by the Ministry of Culture after consultation and opinion of the National Commission for Cultural Goods. It must determine the conditions for classification and set out the ensuing easements and obligations.

From article 8 to 49, five chapters detail the different methods of protection according to the type of property:

- Registration on the additional inventory of cultural real estate
- Classification of cultural real estate
- Safeguarded Sectors
- Expropriation for public utility
- Right of pre-emption

From article 41 to article 45, are detailed the definition of the Safeguarded Sectors, the process of creation of the S.S., and the ministries integrated in the process.

From article 50 onwards, the conditions for the protection of movable cultural property, intangible cultural property and scientific research are dealt with.

From article 79 to 81, it is treated of the organs and the national commission of the cultural goods and the modalities of its formation.

The financing of operations is dealt with in articles 82 to 90. Followed by the provisions of control and sanctions announced in articles 91 to 105.

And the end deals with the final provisions concerning the relationship with the environmental protection law and the repeal of Ordinance n°67-281, in the article 106–108 (Fig. 3).

3 Evolution of the Management Bodies of the Casbah of Algiers

3.1 The Management of the Casbah After1962

The management of the Casbah has been based on a non-constant supervision and this since independence, where we saw its supervision change from one ministry to another and from one organization to another, or even the creation of a new organization for its management.

In what follows, we will explain briefly these different changes of supervision with the change of the organizations which carried out the orders of the Ministries Supervising and the results obtained before the promulgation of the Law 98-04 relating to the Protection of Cultural Heritage.

3.1.1 Situation After the Independence (1962)

After the declaration of independence of July 1962, French legislation in terms of town planning was renewed with minor changes.

Fig. 3 Algerian heritage laws chronologically. *Source* The author

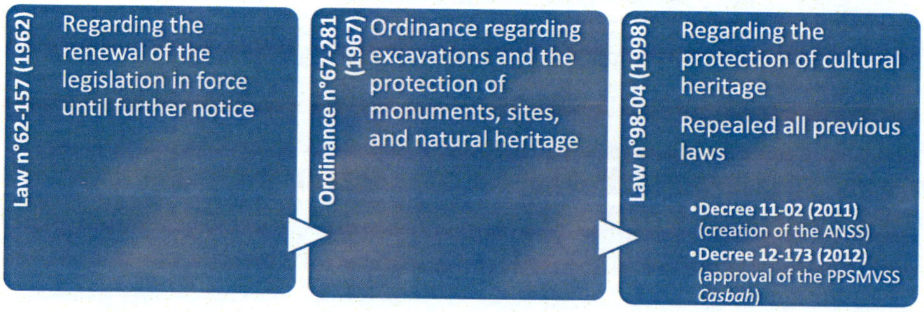

The Casbah was perceived as an increasingly overcrowded degraded district of the contemporary city. Modest actions were undertaken in the 1960s, including the initiative of a UNESCO expert (Lezine) in 1966 who outlined the main lines for the conservation, safeguarding and development of the Casbah. He recommended taking urgent measures concerning.

1. Legislative support.
2. The launching of studies and projects.
3. Emergency works.

3.1.2 COMEDOR (+ Ateller Casbah) (1968)

3.1.3 Executive Decree N°68-625: (1968)

At the end of the 1960s a Standing Committee for the Study and Development of Organization and Planning for the Agglomeration of Algiers (COMEDOR) was created (OJ n°79, 1968). It was only at the beginning of the 70s, that the COMEDOR; was laying, for the first time, the foundations of an action aiming at safeguarding the historic ensemble of the Casbah.

This safeguarding action was guided in such a way as to take into account all the socio-economic problems of the integration of the Casbah into the contemporary city, while inscribing this action in the problematic of the development and the planning of Algiers.

3.1.4 The Listing of the Casbah as National Heritage

3.1.5 Decree of September 12, 1973

In 1973, a decree (*arrêté*) was issued to start the procedures of the classification of the Casbah of Algiers as Historic Site according to ordinance n°67-281. (OJ n°91, 1973).

3.1.6 Ordinance N°75-22: (1975)

In the same concern of taking in charge this historic fabric, the ordinance of March 27, 1975 approving the General Orientation Plan of development and planning of the agglomeration of Algiers (P.O.G.) was promulgated (OJ n°46, 1975). This plan scheduled the upgrading and revalorization of the Casbah.

The P.O.G. specifies the two criteria which must govern this upgrading:

1. The entire site of this district with a high population concentration and low level of equipment must be considered globally as a historic site.
2. The Casbah has a predominant character as a residential area and must retain it.

In 1978, a UNESCO mission was sent to Algeria to prepare for the UNDP/UNESCO convention. This cooperation led to the development of the study entitled "*Project for the Revaluation of the Casbah of Algiers—Preliminary Development Plan*" completed in March 1981.

This project of a very high quality level was able to carry out an important part of the tasks assigned to it by the agreement. The basic studies concerning the history, the demographic and economic structure, the characteristics of the buildings, the circulation, the equipments and the infrastructures, have been made. A global vision of the Casbah has been identified, priority areas have been identified, and development proposals for the entire Casbah and for priority areas as well as architectural projects and sectoral recommendations have been developed.

3.1.7 The Action Plan—Programme of Priorities (1985)

Following the project of revalorization of the Casbah of Algiers and in accordance with its orientations contained in the proposals (completed in March 1981), detailed studies of an operational nature began to emerge and this from October 1981.

These studies resulted on a graphic file and a written document. The graphic file consisted of an inventory of typologies, state of the properties, economic activities, demographic data as well as urban and architectural analysis.

The written document consisted of an analysis of the situation at the historical level and presents a proposal for an urban planning regulation.

In June 1985, the government decided to create a working group who was responsible for drawing up a priority action program to be initiated in the Casbah. It was composed of representatives from: (ANAPSMH, 1992).

- Ministry of Town Planning, Construction and Housing.
- Ministry of hydraulics, environment and forests.
- Ministry of Culture and Tourism.
- Ministry of Planning and Regional Development.
- The Wilaya of Algiers.

The program resulting from this working group was very ambitious, and considered the production of a document that presented an assessment of the operations undertaken on the Casbah, and was to be presented as an action program aiming at taking in charge all the problems in the Casbah (such as the restoration of buildings, in an individual or concerted framework by islet, the restoration of monuments, resumption of all networks, construction of new equipment, de-densification of the Casbah through the relocation of families), and finally produce a reflection on the organizational framework likely to manage the complexity of these problems. This program included the financial evaluations related to the selected actions.

Of this very ambitious program, only a certain number of actions have been achieved: (ANAPSMH, 1992).

i. **OFIRAC**: The action plan proposal to create an office with the necessary prerogatives to ensure the project management of all projects located within the protection perimeter was crowned by the birth in the same year of the Office of Intervention and Regulation of Development Operations on the Casbah of Algiers (OFIRAC) (OJ n°31, 1985, p. 694)
ii. **Relocation**: The state of deterioration of the buildings until this period (1985) forced the relocation of 3324 families representing 23,268 inhabitants who occupied 490 homes.
iii. **Studies**: Within the framework of the action relating to the study of restoration of buildings, there were (in 1985) more than 250 houses covered either in surveys, or in complete or partial studies.
iv. **Bastion 23 and the Citadel of Algiers**: This committee also recommended the restoration of the Citadel of Algiers by the State, former seat of power before the French occupation and Bastian 23, a small urban complex constituting the last evidence of the pre-colonial fabric in the navy quarter, demolished in the 1940s.

OFIRAC was created to respond on site to questions that have remained unanswered (humanitarian, economical, legal, technical, etc.) for more than twenty years after independence (Lesbet, 1988).

However, the statute of the Office would have allowed it to demolish many buildings in the Casbah. Djaffar Lesbet enumerates the demolitions planned by OFIRAC in 1985, numbering 210 with a construction project of 3 three penetrating sections cutting through the old fabric, keeping Bastion 23 and the Citadel intact. (Lesbet, 1988).

3.1.8 The 1986 Master Plan of Development

Following the 1985 action program, a master development plan was created in 1986, in collaboration with a German team, to provide a coherent framework for all these actions. It takes up the general orientations of the 1981 "*Project for the Revaluation of the Casbah of Algiers,*" nevertheless introducing a few variants of strategies for taking in charge the historical urban fabric.

A draft town-planning regulation was drawn up in July 1986 to regulate all the interventions recommended in 1985 (in terms of restoration, renovation, restructuring or reconstruction).

3.1.9 The Master Urban Plan of Algiers P.U.D. (1987)

After the questioning of the General Orientation Plan for the development of the agglomeration of Algiers (P.O.G.), its scheduled revision was carried out by the *Center National d'Etude et de Recherche en Urbanisme* (CNERU) (OJ n°48, 1980) which resulted in 1987 in the elaboration of a Master Urban Plan (P.U.D.) for the city of Algiers.

The roles of this plan were the restoration and rehabilitation of the historic fabric as well as the de-densification by transferring the population to other neighborhoods.

The document confirms the vocation of the Casbah as provided for in the project to upgrade the Casbah of Algiers (1981) and recommends the development of specifications (*cahier des charges*) accompanying the development plans of the various zones. It was proposed to classify the perimeter of the Casbah as a historic site.

3.1.10 Executive Decree N°87-10: (1987)

In the same year, another body was created to strengthen the intervention system of the authorities on national heritage. It is the "National Agency for Archeology and the Protection of Historic Sites and Monuments" (ANAPSMH), which was created in 1987 (OJ n°02, 1987).

This public establishment is administrative in nature and endowed with legal personality and financial autonomy. Within the framework of the national cultural development plan, it is responsible for all actions of inventory, study, conservation, restoration, and presentation to the public of the national historic cultural heritage according to art. 3 of decree n°87-10 (OJ n°02, 1987).

It is to this Agency that was given the role of contracting authority in the major operations to safeguard the Citadel of Algiers and the Bastion 23.

3.1.11 New Legislation 1990–1991

Although the movement of the 88 s was known by the beginning of a difficult period in Algeria called "the black decade," and which lasted about ten years, the beginning of the 90 s was marked by the promulgation of several laws on land use. Planning and town planning, state law, law on land orientation, rules relating to expropriation for public utility, town planning and construction planning rule, modalities of instruction and issuance of building permit, certificate of conformity and the demolition permit, procedures for drawing up and approving the master plan for land use planning and development (PDAU), procedures for drawing up and approving the land use plan (POS).

This legal device having come to regulate the use of the grounds could also enrich the regulation as regards protection of the sites and historic buildings, and to generate the need so expressed by heritage experts for an urgent promulgation of a new law for the Protection of Heritage.

Indeed, a few years later, the President of the Republic, Liamine Zeroual, promulgated the law relating to the protection of Cultural Heritage on June 15, 1998.

3.1.12 The Classification of the Casbah as World Heritage

After 18 years from the opening of the procedures to list the Casbah in the National Heritage list, the Casbah is still not a National Heritage.

3.1.13 Decree of November 24, 1991

When the registration procedures on the World Heritage List started, the Equipment Minister (in charge of UNESCO Algerian delegation) noticed the incomplete classification of the Casbah as a National Heritage. To finalize these procedures, a decree classifying the Casbah of Algiers on the Algerian Official Journal had to be published. That happened in November 1991 (OJ n°60, 1991). Following this approach and other procedures as well as proposing a safeguarding plan to the World Heritage Committee, the Casbah was inscribed on the World Heritage List in December 1992.

3.2 The Management of the Casbah Today

Numerous Algerian heritage specialists (Khelifa, 1991; Ouagueni, 2011; Righi, 2014) and others agreed that Ordinance n°67-281, regarding the archeological excavations and preservation of historic and natural sites and monuments, is not a sufficient legislative text to take in charge the living urban historic sites.

The said ordinance did not only have a lack of precision regarding the instrument that must be used to intervene and manage the historic urban center, there was also a lack of directives and regulations aiming to take into account urban development near monuments and sites.

In what follows, the management of the Casbah of Algiers will be discussed: first in relation to the legislative texts concerning the cultural heritage, but also, in relation to the decisions taken by the Algerian State. It divided into four chronologically overlapping time parts, namely:

- The period after the Promulgation of Law 98-04 till creation of the Safeguarded Sector of the Casbah (1998–2005)
- The period after the creation of the Safeguarded Sector of the Casbah and of the OGEBC till the change in the applicant authority (2005–2016)
- The period after the Creation of the ANSS (from 2011 till today)
- The period of change of applicant authority between OGEBC and the Wilaya (since–2016)

3.2.1 After the Promulgation of the Law N°98-04 (1998–2005)

This phase represents the period from the promulgation of Law 98-04 on the protection of cultural heritage in 1998, until the creation of the Casbah of Algiers as a Safeguarded Sector in accordance with the implementing texts of this law in 2005.

At the time of the promulgation of Law 98-04 on the protection of cultural heritage in 1998, the Casbah of Algiers was under a different kind of supervision, in place since 1997, where it was separated from the different ministries. This supervision was the Governorate of Greater Algiers (OJ n°38, 1997). This new supervision took charge of studies and works for the rehabilitation, restructuring, restoration, and renovation of the urban fabric according to art. 14 of the decree creating the Governorate of Greater Algiers. (OJ n°38, 1997, p. 6).

Although the Law 98-04 relating to the Protection of Cultural Heritage was promulgated in June 1998, with a new system of protection for living historic urban centers, the application text having to give details regarding the development of safeguard instruments was promulgated only in October 2003.

In 2003, 5 texts of application of the said law were issued, implementing the modalities to the development of the national inventory, cultural goods, and the development of PPMVSAs (*Plan de Protection et de Mise en Valeur du Site Archéologique et ses zones de protection*) as well as PPSMVSS (*Plan Permanent de Sauvegarde et de Mise en Valeur du Secteur Sauvegardé*).

There have been a significant number of studies of "safeguard plans" that were launched, for most of them, by the Ministry of Housing, in haste and even, for some, before the promulgation of Law 98-04.

As for the safeguarding plan for the Casbah of Algiers, it was drawn up between 2000 and 2005, even before the creation of the Safeguarded Sector of the Casbah of Algiers, launched by the Wilaya of Algiers elaborated by its contactor the CNERU (Ouagueni, 2011). This action was a follow-up to the recommendations issued by UNESCO when the Casbah was proposed to be on the World Heritage List.

The Governorate of Greater Algiers was abrogated due to the fact that ordinance n°97-15 of its creation was declared unconstitutional by decision n°02/DO/CC/2000 of February 27, 2000 (OJ n°7, 2000, p. 3). So the Casbah regains its status as a municipality again under the supervision of the Wilaya of Algiers.

3.2.2 After the Creation of the Safeguarded Sector of the Casbah: Between D.C.W.A. and OGEBC (2005–2016)

3.2.3 Executive Decree N°03-324 (2003)

In 2004, the CNERU, after several years of studies, ended up producing a safeguarding plan that has been approved by the APW[1] of Algiers (Righi, 2014). In 2003, the Executive Decree 03-324 relating to the methods of establishment of the PPSMVSS was promulgated (OJ n°60, 2003a).

The Decree n°03-324 comes into application of Law 98-04 relating to the protection of cultural heritage. It stipulates that "*The development, instruction, content, implementation of the permanent safeguard and enhancement plan, as well as the safeguarding measures applicable before its publication and the conditions for its modification, revision and regular updating will be specified in a regulatory text*".—article 45.

The executive decree 03-324 announces the distribution of the roles of the organizations in charge of the Management of the Safeguarded Sectors. The role of project management in the development of Safeguarding Plans for the Safeguarded Sectors was assigned to the Direction of Culture of the Wilaya concerned (D.C.W.).

3.2.4 Executive Decree 03-322: (2003)

This executive decree requires that to be able to work on a listed cultural property, there must be a qualified architect in the team of the project management (OJ n°60, 2003b). It stipulates that work on cultural property must be entrusted to project managers meeting the conditions for *professional qualification*, as well as the obligation to entrust the studies and works to approved Design offices and qualified architects. The qualification of the architect is awarded by the Ministry of Culture under a certain circumstance with an experience in field.

3.2.5 Executive Decree N°05-173: (2005)

The Urban Historic Site of the Casbah of Algiers was created as a Safeguarded Sector in 2005 by the executive decree n°05-173 published in the Official Journal n°34. (OJ n°34, 2005).

In the meantime, in December 2005, the National Archeological Agency underwent on an administrative restructuration and this action resulted in the creation of a new body, the National Office for the Management and Exploitation of Protected Cultural Property (OGEBC) (OJ n°83, 2005; OJ n°13, 2012).

The CNERU has started a new study to bring the old safeguard plan proposed in 2004 up to standard with the provisions of Executive Decree 03-324. The study was launched in January 2007 with D.C.W.A. as the contracting authority in accordance with the requirements of the executive decree.

In February 2010, after 3 years of launching the study of the PPSMVSS of the Casbah of Algiers by the CNERU under the contracting authority of the D.C.W.A., "*a decision took place which transferred the contracting authority*" of the Emergency Works operation, from the D.C.W.A. to the OGEBC (Righi, 2012). Making the Casbah, once again, relive the instability of its management.

CNERU had completed the final version of the PPSMVSS in 2010 and had submitted it for approval. One year later, the safeguarding plan was adopted by the APW in April 2011.

3.2.6 After the Creation of ANSS (2011 Till Today)

3.2.7 Executive Decree N°11-01 (2011)

Decree n°03-324 was amended in 2011 by decree n°11-01 (OJ n°01, 2011a) when the National Agency for Safeguarded Sectors (ANSS) was created. Decree n°03-324 stipulates in article 17 that D.C.W. concerned is responsible for the implementation and management of the PPSMVSS. This article was amended in 2011 by decree n°11-01 stipulating the creation of a new entity that will be responsible for the implementation of the PPSMVSS and their application.

This modified article transfers the mission of implementing and applying the PPSMVSS from the Direction of Culture of the Wilaya to the National Agency for Safeguarded Sectors (ANSS).

3.2.8 Executive Decree N°11-02: (2011)

At the moment when the OGEBC holds the reins of emergency measures/works within the Safeguarded Sector of the Casbah of Algiers, on January 5, 2011, Prime Minister decreed, the creation of the National Agency for Safeguarded Sectors (ANSS) (OJ n°01, 2011b).

The missions of the ANSS mainly consist of the implementation of the various Safeguard Plans already approved, as specified by the amended article 17 of Decree 03-324 (mentioned above). Nevertheless, the Agency carries out other missions stipulated in article 4 of the decree of its creation and aims to make the application of the PPSMVSS easy for specialists.

The ANSS was created by decree in 2011, but did not become operational until 2013.

3.2.9 Executive Decree N°12-133: (2012)

It is the decree of approval of the PPSMVSS of the Casbah of Algiers (OJ n°18, 2012). This decree comes into application of Article 16 of Executive Decree n°03-324. The article 6 of the executive decree n°12-133 re-enforce the role of the ANSS of being the main body to apply the PPSMVSS.

[1] People's Assembly of the Wilaya of Algiers.

The OGEBC created the Cellule Casbah, in 2013, composed of 8 architects and a coordinator responsible for coordinating and transmitting the various decisions taken by the *steering committee* to the 8 architects of the Cellule Casbah.

For the launch of the "*Plan d'Attaque,*" the "Cellule Casbah" had started to draw up the specifications (cahiers des charges) for the launch of the studies of:

- The 07 major monuments;
- The 05 places of worship/Cult;
- The 03 collectors: Ourida Meddad Boulevard, La Lyre Street, Divan Street.

After the development of the famous Attack Plan, in April 2014, the OGEBC made a request to the Minister of Culture (at that time Khalida Toumi), this request emanates from the urgency of the operation which must be undertaken given the state of degradation of the Casbah building, and just before its approval, the minister was changed and replaced by Nadia Labidi, May 05, 2014. (Figs. 4 and 5).

Approval of the plan of attack and his agreement sees a temporary halt, until the new minister understands the procedures, the plans established and the process to be followed.

A few months later, and after eminent efforts on the part of the services detached from culture (OGEBC, ANSS) to explain to the new minister the state of progress and the urgency in which the Casbah support project found itself, the minister was changed again, replaced by Azedine Mihoubi from May 14, 2015, to April 1, 2019.

3.2.10 The Casbah Today: Managed by the Wilaya of Algiers (Since 2016)

In January 2017,[2] the supervision on the intervention undertaken inside the Safeguarded Sector of the Casbah was transferred from the OGEBC to the Wilaya of Algiers.

The restoration operation was supported with the various bodies of the Wilaya. Where the Directorate of Public Facilities (DEP) was responsible for work monitoring on some buildings and the Directorate of Development and Restructuring of Neighborhoods (DARQ) was responsible for other buildings restoration studies. As for the emergency works operation, it was taken over by the OPGI Hussein Dey.

The OGEBC became the technical assistant for those bodies (OPGI Hussein Dey and DEP) for the restoration studies launched in 2018. The studies carried out that year concerned the certain mosques of the Casbah, we can mention: the Safir Mosque, Sidi M'Hamed Cherif Mosque and Abdellah Mosque.

[2] Rapport de reunion inter-ministerielle.

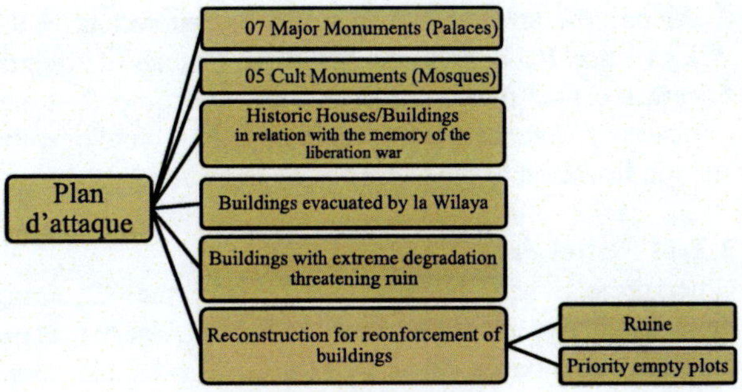

Fig. 4 Plan d'Attaque. *Source* The author

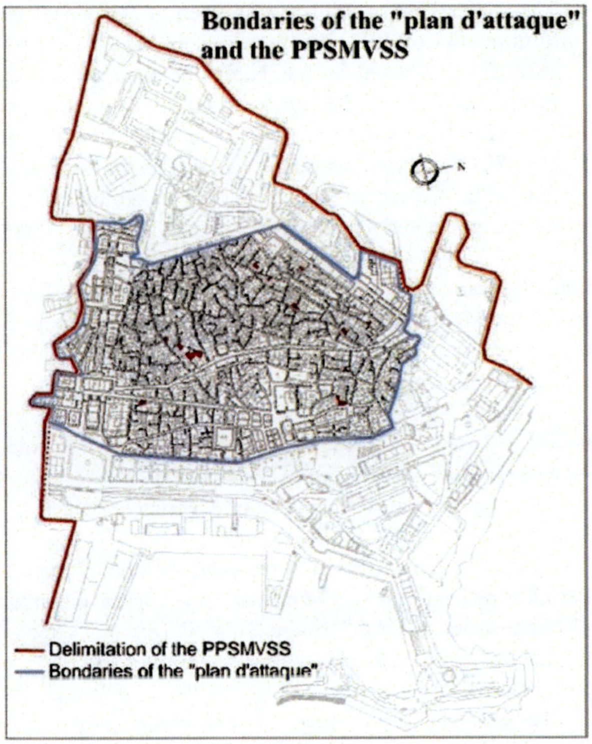

Fig. 5 Boundaries of the "plan d'attaque". *Source* Hadji et al. (2023)

In April 2018, the reception and reopening of the Ketchaoua Mosque at the lower Casbah took place with a presidential ceremony. The event was also an occasion for the presidential inauguration of the Metro-Museum station at the lower Casbah and the extension of the second metro line in the capital.

Although in 2021 the return to normal life takes place with difficulty and with great caution, there has been the acceptance of finished restoration projects.

Among the projects launched for the restoration by the Wilaya since 2017 inside the SS of the Casbah of Algiers, 4 were accepted during the year 2021, we name: 2, 4 and 6 Impasse Silène, as a set of three adjoining buildings, the historic Bouhired house located at 03 Impasse Lavoisier.

3.2.11 Other Projects

Other projects have been launched within the SS during these last 20 years, some launched by the Algerian State, others by collaborations between Algeria and other countries, or still, other projects launched by local communities, associations, and heritage lovers.

We will present in what follows the most significant projects during the two previous decades, in chronological order:

- The Algiers Metro project: was open at the end of the third quarter of 2011. And it passes through the Casbah of Algiers. Arrived at the border of the S.S. of the Casbah of Algiers, the underground construction work of the Metro encountered vestiges. In 2007, a mission by UNESCO experts took shape relating to the Safeguard Plan of the Casbah of Algiers and the archeological constraints generated by the Metro route. (UNESCO, 2008).

One of the solutions presented, and which was applied, was to flee the Metro to 30 m below the ground surface, to avoid any destruction of the archeological remains present and escaped inside the S.S.

- The "Turath" Program is a collaboration between Algeria and the European Union, which consists of assistance for the enhancement of Algerian heritage.

The program is made up of 4 components, one of them is the pilot project that was launched on two sites classified as National and World Heritage. This pilot project affected the Casbah and its defensive system, as well as the Tomb of Medghacen.

The project of the defensive system of the Casbah consisted in highlighting the two bastions still existing in the urban fabric of the historic center, and revitalizing the route connecting them. It turned out that the route connecting them passed by the mosques of the upper Casbah, and was thus called "the route of the mosques."

- The Ketchaoua Mosque, of great symbolic value representative of the Casbah, was closed in 2008, following the degradations occurred to it by the famous earthquake in 2003. The project of its restoration was being delayed and postponed constantly by in relation to the various changes in the management of the historic center of the Casbah. (APS, 2018)

It was finally restored jointly by the Algerian Ministry of Housing and the Turkish Cooperation and Coordination Agency (TIKA), following Algerian–Turkish agreements signed in 2013.

- Several exhibitions are organized each year to make society aware of the importance of safeguarding heritage, but also to express and display the conditions in which it finds itself.

We will mainly mention for this time frame the exhibition of "Casbah Behind the Legacy" that was an initiative of young Algerian artists who invested themselves in expressing their love for the Casbah and translating it into national and global awareness through social networks, which was inaugurated at the end of January 2018, which lasted for months.

- In January 2018, experts met under an international conference prepared by the Algerian Ministry of Culture jointly with UNESCO.

The conference aimed to address the issues in direct relating to the conservation and management of the Casbah, trying to propose a coordinated approach and an integrated one to allow an act of revitalization of the old city. It was entitled "*International meeting of experts on the revitalization of the Casbah of Algiers.*"

The event closed with a list of recommendations made by the head of the section for Arab countries at the World Heritage Center, we will quote:

- Broaden the view on the Casbah, and not be satisfied with technical problems;
- Evaluate past experiences, which will represent an in-depth analysis as well as an observation and an evaluation, certainly not an audit. An evaluation of the institutional, administrative, legal framework, etc.; as a self-criticism to improve oneself.
- Improve the PPSMVSS through an integrated approach;
- Cooperating with the Wilaya considering an Interministerial approach, to be able to get out of the vicious circle of rehousing, and stop making the Casbah as transit point for this vicious circle;
- Invest in preliminary impact studies, for all kind of projects (from large infrastructure projects to other less large), on the dynamics of the city and assessing the footprints and make sure that this dynamic is positive;
- Shed lightening on the various administrative procedures and consider a management mechanism;
- Ensure a monitoring mechanism at the national and international level;

- Improve the integration of all responsible actors (different ministries and different bodies of the Wilaya of Algiers) in the planning and implementation of actions;
- Consider public facilities in the Casbah and propose a mix of functions.
- A workshop was installed in Algiers ATRC,[3] where all the representatives of the organizations involved in the Casbah (ANSS, OGEBC, DEP, D.C.W.A., APCC)[4] were present, as part of a decentralized cooperation signed between the Wilaya of Algiers and the Ile de France in 2017. Following the meeting of experts in January 2018, they added several actions to safeguard and enhance the Casbah of Algiers (OUAR, 2019). One of the ATRC's missions was to organize a participatory worksite for the Transitional Urbanism project.

The Table 1 summarizes the experience of managing the Casbah of Algiers under the two laws (Ordinance 67-281 and Law 98-04).

We observe the presence of the CNERU lasted more than 10 years between elaborating different safeguarding plans and conforming them to the new legislation from 1998 to 2010. We also notice that the change of supervisory authority was more frequent under the Ordinance n°67-281, before the promulgation of Law 98-04. This can be explained with the fact that the ordinance did not have a specific supervisory authority. With the Law 98-04, the role of its application goes to the Ministry of Culture.

4 Conclusion

In the legislative and application texts of Law 98-04, we often find the same bodies/directorates which are in charge (by law) of the "Management" of the historic site of the Casbah, as well as the implementation and the application of its PPSMVSS, which are: The Department of Culture of the Wilaya of Algiers, and the Municipalities concerned. And since 2011, another organization has joined them, the National Agency for Safeguarded Sectors.

On the ground, things happened differently, and the organizations present on the ground were other than those dictated by the law and the legislative texts.

We therefore find the Wilaya (with its decentralized departments: the DEP and the DARQ), and the OGEBC since its creation in 2006, which are more present than the other organizations, with more qualified staff and longer experience.[5] However, the Department of Culture, which was to see its staff improved in accordance with article 17 of decree 03-324, saw this project repealed with the promulgation of decree 11-01 by assigning the mission to the ANSS.

Unfortunately, even the ANSS does not benefit from an improvement in its staff, despite what is stated in article 17.[6]

The precision that the Law 98-04 brought made it easier to identify the organizations that work on the Casbah, but still lacks precision regarding the instruments of conservation.

The PPSMVSS is considered as an instrument of safeguard and protection, a precision in the legislative texts allowing to know the mode of employment of its application is to be established.

One of the recommendations that we can propose is to elaborate a regulation text addressing a short, a medium, and a long-term management plan for the application of the PPSMVSS of the Safeguarded Sectors. This text must precise how to elaborate those plans, who is the management body in charge of their implementation and a monitoring committee detached from this management body to periodically assess the situation and the conduct of work.

[3] Technical Atelier for the Revitalization of the Casbah.

[4] **ANSS**: Agence Nationale des Secteurs Sauvegardés; **OGEBC**: Office national de Gestion et d'Exploitation des Biens Culturels protégés; **DEP**: Direction des Equipements Publique de la Wilaya d'Alger; **D. C. W. A.**: Direction de la Culture de la Wilaya d'Alger; **APCC**: Assemblée Populaire Communale de la Casbah.

[5] The staff of the OGEBC was mainly that of the Archaeological Agency after its restructuring in 2005, and therefore were already familiar with the different situations of urban and archaeological heritage management.

[6] Article 17 of executive decree 03-324, JO n°60 (2003a, 2003b), pp. 17–12.

Table 1 Evolution of the casbah's management under the Algerian laws regarding heritage protection (Ordinance 67-281 and Law 98-04)

	Year	Stakeholders/contributors/organism	Supervisory authorities	Objectives/projects
Before promulgation of law 98-04	1968	COMEDOR: Atelier Casbah	Presidency	1975: P.O.G. (general plan of orientation): comprend un plan de renovation et de restauration de la Casbah
	1973	Sous-direction des Beaux-arts avec Direction de la Culture	Ministry of information and culture	Procedures were open for the classification process of the Casbah of Algiers as a national heritage under the category of historical site
	1981	ETAU: Atelier Casbah+ PNUD/UNESCO	Ministry of public works and construction	1981: revaluation of the casbah of Algiers—preliminary development plan:
	1983	BEREP: Atelier Casbah	Ministry of housing and urban planning	Studies were carries out relating to the Casbah on behalf of the Wilaya by BEREP (OJ n°18, 1983), under the supervision of the ministry of housing and urban planning
	1985	OFIRAC	Ministry of urban Planning, construction and habitat	The action plan—programme of priorities
	1986	OFIRAC	Wilaya of Algiers	Master plan of development (and its regulation)
	1987	CNERU (OJ n°10, 1983)	Ministry of housing and urban planning	P.U.D.: master urban plan—city of Algiers
	1989	Ministry of Information and Culture	Ministry of information and culture	The minister of information and culture promulgate a first decree (*arrêté*) approving the classification of the Casbah as a national heritage
	1990	Algerian Delegation to UNESCO	Ministry of equipment	Procedures to register the casbah of Algiers on the world heritage list
	1991	Ministry of Culture	Ministry of culture	Publication on the official journal, the decree (*arrêté*) about the listing of the casbah as a natoinal heritage.
	1992	UNESCO/WH Division		The Casbah of Algiers is registered on the WHL.
	1998	CNERU and DARQ	GGA : great Governorate of Algiers	G.P.U.: grand project urbain
After promulgation of law 98-04	2000	CNERU	Wilaya of Algiers	Elaboration of a conservation plan
	2006	D.C.W.A. and CNERU	Ministry of culture	Permanent plan of safeguarding ad enhancement of the safeguarded sector of the casbah (PPSMVSS) elaboration
	2010	OGEBC and CNERU	Ministry of culture	Starting the implementation of the PPSMVSS of the Casbah (emergency works—the first phase)
	2013	OGEBC and ANSS	Ministry of culture	Elaboration of « *Plan d'Attaque* » by la Cellule Casbah
	2016	DEP and DARQ and ANSS	Wilaya of Algiers/ministry of culture	Implementation of the « *Plan d'Attaque* » and other emergency works
	2017	DEP and DARQ and ANSS and OGEBC	Wilaya of Algiers/ministry of culture	Restoration works were lunched in 5 buildings, as well as studies for restoration of numerous mosques, historic monuments and other buildings inside the S.S..

Bibliography

ANAPSMH (1992). *Elaboration du Plan de Sauvegarde de la Casbah d'Alger—Phase 1*. September 1992. Alger. Dossier de Classement de la Casbah d'Alger comme patrimoine mondial. Not published. ANAPSMH. Centre de documentations Bastion 23, Casbah, Algiers.

APS, Algérie Presse Services. (2018). "Le président Bouteflika inaugure la mosquée Ketchaoua restaurée et les extensions du métro d'Alger". In *APS, magazine en ligne, rubrique Algérie*. Last retrieved in world wide web November 2021. https://www.aps.dz/algerie/72371-le-president-bouteflika-inaugure-la-mosquee-ketchaoua-restauree-et-les-extensions-du-metro-d-alger

Hadji, F., Aslan, Z., Bougherira, Q., et al. (2023). Impact of legislative precision in the management of the *Casbah. Built Heritage, 7*, 8 (2023). https://doi.org/10.1186/s43238-023-00088-y

Khelifa, A. (1991). « Aperçu sur la législation et la réglementation Algérienne, dans la préservation du Patrimoine Monumental et sa mise en valeur ». In Préface du *Recueil législatif sur l'archéologie, la protection des sites, des musées et des monuments historiques*. ANAPSMH. Alger (pp. 7–10).

Lesbet, D. (1988). La Casbah d'Alger, entre Réhabilitation et Réanimation. In *Peuples Méditerranéens : Les urbanistes dans le doute. N°43*, Avril-Juin 1988. Les archives du Centre d'études Diocésain les Glycines—Alger.

Ouagueni, Y. (2011). *Algérie, la politique de sauvegarde et ses outils. In Algiers, from city to agglomeration*. Planum. Retrived in April 2016 in: http://www.planum.net/algiers-from-city-to-agglomeration

Ouar, L. (2019). Contribution à la revitalisation de la Casbah d'Alger, une clarification utile. In *MADINATI la revue de l'urbanisme, de l'architecture et de la construction: Le patrimoine/Avatars, dénis et espoirs, n°8, Décembre 2019*. ATP Editions. Oran, Algérie.

Righi, K. (2012). *Réflexion autour du rôle de la maitrise d'ouvrage dans l'élaboration d'un Plan Permanant et de Mise en Valeur d'un Secteur Sauvegardé (PPSMVSS) Quelques cas exemplaires et leur valeur d'application à la Casbah d'Alger*. [Unpublished Master's thesis] EPAU, Algiers.

Righi, K. (2014). *Réflexions sur l'histoire et l'actuel dispositif opérationnel de protection de la Casbah d'Alger*. In Doctoral *Séminaire of DREAM Laboratory* INAMA—ENSA-Marseille (pp. 6–15) [Dissertation proposal].

UNESCO. (2008). Brodovitch M., Souk F., Baccar M. Rapport de Mission Casbah d'Alger, effectuée en décembre 2007. Report. Février 2008. Alger.

Zadem, R. (2008). Contribution Pour une Mise en Œuvre des Plans Permanents de Sauvegarde des Ensembles Urbains ou Ruraux d'intérêt historique ou architectural [Conference presentation]. In *the 1st Euro-Mediterranean Regional Conference on Traditional Mediterranean Architecture, Present & Future*. Barcelona (pp. 174–176).

Algerian Laws and Texts Published in the Official Journal (OJ)

OJ n°02. (1963). *Law n°62-175, tendant la Reconduction jusqu'à nouvel ordre de la législation en vigueur au 31 décembre 1962*. Official Journal. Algeria.

OJ n°07. (1968). *Ordinance n°67-281 relating to excavations and the protection of historic and natural sites and monuments*. Official Journal. Algeria.

OJ n°79. (1968). *Ordinance n°68-625 creating COMEDOR* Official Journal. Algeria.

OJ n°91. (1973). *Decree of September 12, 1973 opening a classification authority among the historic sites of the Casbah of Algiers*. Official Journal. Algeria.

OJ n°46. (1975). *Ordinance n°75-22 approving the general orientation plan of development and planning for the agglomeration of Algiers*. Official Journal. Algeria.

OJ n°48. (1980). *Decree n°80-53 creating the CNERU: National Center for Studies and Achievements in Urban Planning*. Official Journal. Algeria.

OJ n°26. (1981). *Modification of the ordinance n°67-281*. Official Journal. Algeria.

OJ n°10. (1983). *Decree n°83-155 creating the CNERU: National Center for Studies and Applied researches in Urban Planning* Official Journal. Algeria.

OJ n°18. (1983). *Decree n°83-293 creating the B.E.R.E.P. (Bureau d'études, de restauration et de conservation des quartiers anciens)* and, *Decree n°83-297 dissolving the Central Office of studies and public works of Architecture and Urbanism (E.T.A. U.) and transfer of its assets, activities, structures, resources and personnel.*

OJ n°31. (1985). *Decree n°85-196 establishing the office for intervention and regulation of operations in the Casbah of Algiers (OFIRAC)*. Official Journal. Algeria.

OJ n°02. (1987). *Decree n°87-10 establishing the ANAPSMH: National Agency for Archeology and Protection of Historic Sites and Monuments*. Official Journal. Algeria.

OJ n°60. (1991). *Order (Arrêté) of 24 November 1991 classifying the Casbah of Algiers among the historical sites*. Official Journal. Algeria.

OJ n°38. (1997). *Ordinance n°97-15 establishing the Governorate of Greater Algiers*. Official Journal. Algeria.

OJ n°44. (1998). *Law n°98-04 relating to the protection of cultural heritage*. Official Journal. Algeria.

OJ n°07. (2000). *Decision n°02/D.O./CC/2000 relating to the constitutionality of Ordinance n°97-15 of May 30, 1997*. Official Journal. Algeria.

OJ n°60. (2003a). *Executive Decree n°03-322 includes the procedures for drawing up the general inventory of protected cultural property*. Official Journal. Algeria.

OJ n°60. (2003b). *Executive Decree n°03-324 2003 includes the modalities for preparing the permanent plan for the preservation and reclamation of the preserved sectors*. Official Journal. Algeria.

OJ n°34. (2005). *Executive Decree n°05-173 includes the establishment of the Safeguarded Sector of the Casbah of Algiers*. Official Journal. Algeria.

OJ n°83. (2005). *Executive Decree n°05-488 transforming legal nature of the National Agency for Archeology and Protection of Historic Sites and Monuments and changing its name*.

OJ n°01. (2011a). *Executive Decree n°01-11 Amending and completing Executive Decree n°03-324 of Shaban 9, 1424 corresponding to October 5, 2003 includes the modalities for preparing the permanent plan for the preservation and reclamation of the Safeguarded Sectors*. Official Journal. Algeria.

OJ n°01. (2011b). *Executive Decree n°02-11 corresponding to January 5, 2011. It includes the establishment of a national agency for the Safeguarded Sectors and defining its organization and functioning*. Official Journal. Algeria.

OJ n°13. (2012). *Executive Decree n°12-98 Amending and supplementing the executive decree n°05-488 fixing the statutes of the OGEBC*. Official Journal. Algeria.

OJ n°18. (2012). *Executive Decree n°12-133 Approving the PPSMVSS of the Casbah of Algiers*.

Urban Renewal of the Lost Heritage of the Casbah, Between Stakes and Challenges. Case Study of the Casbah of Algeria

Saouane Med Boudiaf

Abstract

The Casbah of Algiers is a beautiful and rich historic site that has been shaped by various eras and types of spatial occupation. However, the development of the peripheries has led to the marginalization and devaluation of historic cities, with modern cities becoming machines that produce urban voids in the historic centers. These abandoned spaces pose a problem of discontinuity and rupture with the urban dynamism, which needs to be addressed through requalification of urban voids. The Casbah of Algiers, a UNESCO World Heritage site, requires revaluation and reappropriation of its identity. This article explores these themes through a case study of the Casbah of Algiers, highlighting the obstacles that hinder renewal and give a new image to the city. In this procedural framework, we will focus on the most important neighborhoods and landmarks to identify the key issues, define jobs, and allocate a path for intervention.

Keywords

Urban renewal · Heritage · Casbah · Requalification · Revaluation

1 Introduction

Even with the diverse typological mix and urban spread that characterizes Algiers, it is important to note that the structural layout and typological expression of its neighborhoods, from the "old town" to the "new town," persist, with their physical and morphological characteristics generally preserved.

S. M. Boudiaf (✉)
Echahid Cheikh Larbi Tebessi University, Tebessa, Algeria
e-mail: saouanemohamed@yahoo.com

Algiers has grown from a delimited nucleus (the Casbah) to become a significant part of the city, extending along the riverside spaces of the bay, expanding toward the south and southeast, transforming a medium-sized medina into a Mediterranean metropolis in less than two centuries.

The text discusses Algiers' historical process and how its fundamental structure and urban patterns, along with other urban centers in the wilaya, reflect their heritage and urban planning, serving as a source of identity. The Casbah, a unique type of Medina or Islamic city located in a picturesque coastal area of the Mediterranean, is highlighted as a prime example of urbanization in the western part of the Mediterranean Sea and sub-Saharan Africa. The Casbah features a citadel, ancient Ottoman-style mosques and palaces, and the remnants of a traditional urban structure that has deep roots in a sense of community. Overall, Algiers' formation had discontinuities and contrasts, but its heritage and urban planning are still visible in its urban patterns (Benzenine, 2016).

The Algiers Kasbah was constructed according to the classic layout of an Islamic city, consisting of three main elements: a religious and cultural center, a commercial district, and a political–administrative area. The religious and cultural center includes the Great Mosque, accompanied by a madrasa—a Koranic school divided into the four schools of Sunni jurisprudence—and the hammam—a public bath that serves as a social hub. The commercial district is comprised of the Souk, a center of shops and craft workshops divided into sections, with the most valuable goods located closest to the mosque and general goods further away. The political and administrative center is housed within the fortified citadel. Lastly, the residential areas are organized into hawma, or neighborhoods, each with its own small mosque, school, and hammam.

The discourse surrounding the Kasbah brings to mind both past shortcomings or deficiencies, as well as the need to safeguard the Medina with its simple qualities. As the heart of the city's collective memory, the Kasbah represents

M. F. Shahidan et al. (eds.), *Urban Narratives: Exploring Identity, Heritage, and Sustainable Development in Cities*,
Advances in Science, Technology & Innovation, https://doi.org/10.1007/978-3-031-48517-6_11

an urban cultural heritage that encompasses architecture, history, identity, and powerful symbols. The social changes experienced by both its physical site and its population have led to degradation that is often denounced, requiring the mobilization of all institutional actors and civil society. Without continuous efforts toward preservation, development, and integration into the modern city, the Medina may continue to fade away.

The data reveal a distressing situation: 300 homes have been demolished, 450 are empty and boarded up, 200 show signs of significant decay, and only 50 have been restored so far. Additionally, only 250 homes are occupied by their original owners, while most are rented out or occupied illegally. In some cases, newcomers from rural areas have exacerbated the problem, as housing initiatives from the 1970s provided alternative accommodation for families in substandard homes, unintentionally discouraging the renovation of homes in the Casbah (Fabiani, 2021).

Regardless of the discontinuities and contrasts resulting from the historical process of its formation, the basic structure and urban fabric of Algiers, and even other urban centers located in municipalities integrated into the wilaya, remain identifiable and expressive in heritage and town planning, serving as an identity resource (Rafik, 2015).

The themes discussed in this article are examined within the context of a case study of the Casbah of Algiers in Algeria, which is a World Heritage site facing the threat of decay and loss. The objective of our study during this case study is to analyze the factors that contribute to the degradation of old buildings, as well as characterize the mortars used in these buildings, in order to propose respectful interventions and solutions. This article concludes that our participation in this process requires consideration of various parameters and requirements that account for the urban dimension of identification, as well as the significant obstacles that have impeded the renewal and revitalization of the Kasbah city. In a procedural framework, a second reading by the teacher is conducted, with a focus on the most crucial neighborhoods and landmarks to identify key issues. Based on this reading, we will define tasks and allocate responsibilities accordingly.

2 Methodology

The research will involve conducting an urban analysis to comprehend the formation and transformation of the medina's urban fabric, ranging from the macro-scale to the micro-scale, including the medina itself, neighborhoods, and individual houses. The main focus of the study is not solely on the medina or its structure, but rather on analyzing the factors that contribute to the deterioration of old buildings, which form the basis of the traditional building module in the urban structure of the medina. However, studying the medina is crucial in order to grasp the principles of traditional buildings in Casbah, both in their original and current state.

The research will heavily rely on fieldwork to investigate the changes made to dwellings over time and to understand the reasons behind these modifications. To obtain meaningful results, the work will be divided into two main components.

Architectural component: This will involve analyzing the design aspects, collecting various documents such as diagrams, photos, plans, sections, and facades.

Social component: This will entail gaining insights into the living conditions of the inhabitants and their perspectives on housing. Interviews will be conducted with families who still reside in Casbah.

2.1 The Casbah is a Unique Type of Medina City

The Casbah, also known as Kasbah or Qasbah, is primarily a constructed environment, an urban ensemble that carries the legacy of ancestral craftsmanship and architecture. It is recognized as an authentic heritage of architecture and urban planning.

The Casbah is a narrative, a tale imbued with emotions and sensations that are shared by the Algerian people, expressed through various forms such as construction trades, crafts, miniature art, illumination, and music. These emotions and sensations stem from the experience, whether positive or negative, of over three centuries of Ottoman rule when Algiers, known as el Djazair el Mahroussa, was a central hub of Muslim power in the Mediterranean. During this period, a rich culture emerged through trans-Mediterranean exchanges (ACHAB ep CHERNAI, 2012). The Casbah, with its purpose of the sea, now holds significance only within this historical context.

The Casbah, located in the old town of Algiers, was named after the fortress that overlooks it and was built by the Ottomans in the sixteenth century. The old town developed on the steep eastern slopes of a rocky hill that descends sharply toward the west, forming a triangular core carved into the hills facing the Mediterranean Sea by the Deys (see Fig. 1). The fortifications define the boundaries of the Casbah, shaping the architecture to grow vertically, resulting in a high-density settlement. The housing typology in the area consists of interconnected, white, geometric houses with rooftop terraces facing the bay, totaling around 1200 houses spread across 36 hectares of land (CNERU, 2012).

The urban layout of the Kasbah is in harmony with the site and its topography, as acknowledged by Ravereau and Le Corbusier. The site itself played a significant role in

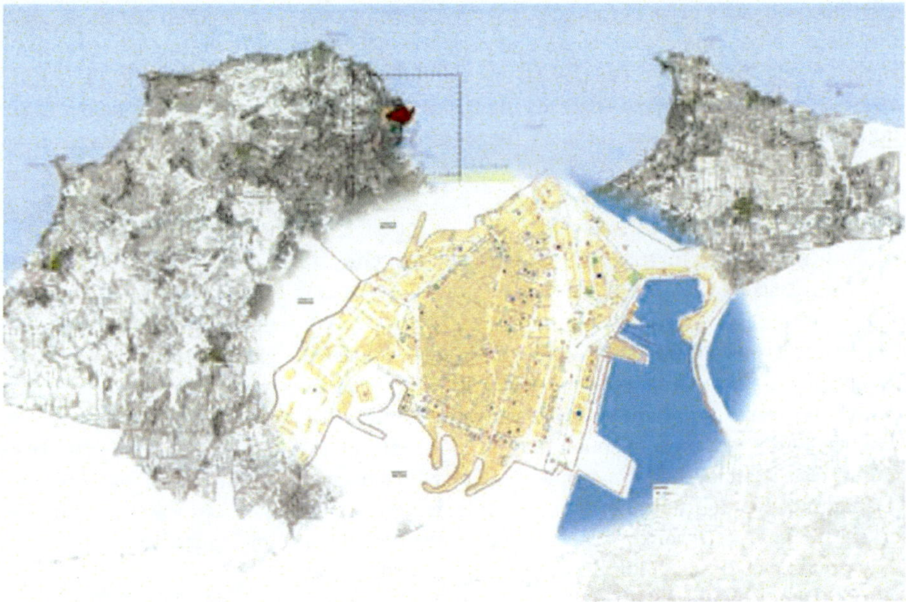

Fig. 1 Geographical of the Kasbah of Algiers. *Source* EXPO 2015 park

shaping the city. Le Corbusier praised the town planning as flawless (see Fig. 2), with houses positioned in a way that their terraces unfold like natural balconies, offering a view of the sea. The spatial organization of the Kasbah reflects the societal values, with the "hawma" or district considered as a semi-private space, while trading centers (souks), fountains, and places of power are considered fully public. Additionally, each district in the Kasbah has its own mosques and kouba (shrines) of local saints, although this arrangement is not uniform, with some houses overlooking wider lanes.

The site's inscription is upheld by the preserved attributes of its Outstanding Universal Value. The Kasbah stands as a testament to its remarkable authenticity, including its dense urban planning, use of traditional construction materials such as earthen bricks, lime rendering, stone, and wood, and its diverse functions as a residential, trading, and cult center. The continuation of traditional architectural skills, particularly in building trades and decorative elements, serves as a significant advantage in maintaining the site's Outstanding Universal Value (Dris, 2005).

Our research focuses on the exploration of the lower Casbah of Algiers, specifically the district of the navy, which serves as a significant testimony to our history. The ancient city of El Djazair, known today as the Casbah of Algiers, holds a special place as a site of historical importance, being listed as a universal heritage site and dominating the Mediterranean for three centuries.

As we reflect on the Casbah toward the end of this year, it evokes a sense of pride for those who have witnessed

Fig. 2 The Casbah of Algiers the certificate of authenticity Google photos 2022

more than 50 years of independence. It is a testament to the resilience of our first architects who successfully defended against foreign aggression. The Casbah represents the rhythm of power and a glimpse into Algeria's illustrious past, standing as the sole remnant of our capital's former glory.

2.2 Evolution of Life Styles and Transformations of Heritage in the Casbah City

The process of modernization does not entail wholesale acceptance of Western culture, but rather the reshaping of cultural systems. It is crucial to thoroughly examine the various forms of modernization before adopting them. Unfortunately, many non-Western societies have attempted to address the issues caused by Western-style modernization by simply applying foreign standards, rather than adapting them to their own rich cultural values and architectural traditions (Bouteche Besma, 2015).

Nevertheless, authenticity must be preserved within the constructed system of World Heritage, as it serves as a bridge between the past and the present. While this is often viewed positively, as it portrays the image of the Algerian state and its cultural heritage, it also leads to two false impressions. Firstly, the notion that the Casbah alone represents Algeria and that the poverty and deterioration of the site are reflective of the state as a whole, similar to the outsider interest in slums. Secondly, it creates a misleading perception that these conditions were characteristic of the past during the Ottoman Empire. By associating poverty and decay with the Casbah, the contrast with the modern European city below is accentuated, visually implying that the French colonization improved the city. This adds to the already misunderstood planning methods, overlaying yet another erroneous image to combat, pushing it further into the past. (Zeghiche, 2022).

The Casbah is a complex and beautiful space that deserves to be preserved, not only for its intricate urban planning and passive design, but also for the people who call it home. However, the cycle of hope and disappointment has become all too familiar, indicating that relying solely on the government may not be a viable solution. Moving forward, the preservation of the Casbah is a challenging task, but it is not a new issue, and the reasons for the need for change in the Casbah can be traced to the following factors:

- **The Utilization of Modern Materials that are not Compatible with Traditional Masonry**
  Referring to errors committed by the current inhabitants themselves, the individual explained that they have substituted traditional materials like earth and lime with inappropriate modern materials like cement, and have made unauthorized interior modifications without seeking approval from the relevant authorities (see Fig. 3). Additionally, he pointed out that wood doors and windows have been replaced with aluminum, and alterations

Fig. 3 Face-lift for the Casbah of Algiers Author 2022

to the facades of old buildings have weakened their foundations. In agreement, Othmane Bouras, a board member of the foundation, highlighted that over 80% of the houses in La Casbah are privately owned, totaling more than 4000 properties compared to only 20% that are habous (public properties), and he estimated that this represents a significant obstacle to the effective management of the city (CNERU, 2012).

- **Transforming Residential Properties into Commercial Buildings in Order to Enhance the Market Worth of Urban Lands**
  The escalating market value of urban land has led to a rampant demolition of old houses in the city center, deemed outdated. This phenomenon is becoming pervasive, met with apathy from both the residents and public authorities who fail to acknowledge these structures as valuable heritage worth preserving (see Fig. 4). Despite being modest and unassuming, these buildings hold significance as they embody the historical urban heritage of the city. They possess architectural and technical excellence that must be safeguarded and passed down as a testament to the past of this region. (Farrand, 2015).
- **Pathologies Linked to Human Intervention**
  After conducting an analysis on the conservation status of the Palais des Raïs, from its restoration to its current state, it has been observed that despite restoration efforts and periodic cleaning (see Fig. 5), the palace continues to undergo similar deteriorations, resulting in a degraded state. These deteriorations are primarily physical and chemical in nature. Being located in a coastal zone, the Palais des Raïs is affected by the deposition of sea spray, which particularly damages the facade coatings and even the masonry (Gabriel, 1901).

The presence of humidity in the walls, either through infiltration or capillary rise, is a significant factor in the degradation of the coating layers, and over time, it can even lead to the deterioration of the underlying structures. It is presumed that the use of contemporary materials that are incompatible with the old masonry may exacerbate the situation. Cement, introduced in the restoration and renovation work, quickly replaced lime due to its faster setting time and higher strength.

However, cement mortars are generally hard and brittle, and their low porosity leads to the stagnation of moisture inside the masonry, which adversely affects wall protection. To test this hypothesis, it is crucial to understand the physicochemical characteristics of traditional lime mortar and compare them with those of cement.

- **Spatial Configuration and Current Appropriation**
  The physical changes that the patio houses in the medina of Algiers have undergone are apparent when comparing their current state with descriptions from existing literature (see Fig. 6). However, at this stage, we are unable to verify the accuracy of these changes. To address this, we have utilized three indicators of space syntax—namely, topological relations, integration, and average depth—in analyzing the first corpus to extract the genotypic characteristics of patio houses in Algiers. The results of this analysis can serve as a reference for future comparisons of use (Arras, 2020).

Fig. 4 Casbah of Algiers: The soul of the city author 2022

Fig. 5 Greasy soiling due to atmospheric pollution author 2022

Fig. 6 Patio closed by a wall and a curtain author 2022

- **Pathologies of Materials and Structures**
  Seismic tremors can cause significant damage to the built heritage, leading to its collapse. Our country, like northern African countries such as northern Algeria, is situated in a high seismicity zone due to its geological location at the convergence of two tectonic plates, resulting in frequent earthquakes (see Fig. 7). Seismic risk is a major threat to the built heritage, but there are also other factors that can cause damage to materials and structures, such as water and humidity in various forms. Material degradation can occur due to chemical actions such as efflorescences, saltpeter, mushrooms, and molds, as well as physicochemical processes like freeze–thaw cycles, physical impacts, and increased mechanical

Fig. 7 Action of the earthquake on the foundations author 2022

action due to factors such as overexploitation and reduced soil carrying capacity, including earthquakes (Mouloud, 2007).

- **The State of Conservation of Old, Public Facilities of Yesteryear and Public Fountains**
  The history of Algiers' old city, known as the Casbah, has been the subject of increasing research. While previous studies have mainly focused on the streets, houses, palaces, bathhouses, and mosques, little attention has been given to the public fountains that line the pathways of the Kasbah. Although mentioned in some research, these hydraulic structures have not been thoroughly analyzed in terms of their historical and architectural significance.

During the Ottoman period, these fountains were numerous and played an important role in the urban and social fabric of the Casbah. They were often located at road junctions and were recessed or leaned against the facades of mosques and zaouiyas (Islamic religious schools) (see Fig. 7).

Today, only a limited number of these fountains remain, with around twenty still in existence, but only nine are functional and actively used by the citizens. However, their preservation has been neglected, and only sporadic interventions by local residents, such as repairs to taps or occasional painting, have helped maintain their functionality, but sometimes at the cost of altering their original architecture.

Beyond the heritage issues and the looming threat of erasure of the past, this pattern of (re)building the city on itself can be seen as a form of "ordinary" urban renewal driven by citizen participation, which is part of the city-making process. However, the land reclaimed from demolition and reconstruction projects is often earmarked for high-value private developments, while public projects such as social housing are relegated to the outskirts. This situation raises questions about the contradictions between the discourse promoting urban development to reduce social inequalities and protect the environment, and the rentier and speculative urban practices that often disregard these concerns. (Slimani, 2015).

Prior to commencing the studies, an assessment of the condition of the buildings was conducted, resulting in the formulation of an emergency action plan aimed at mitigating the degradation of this heritage site (see Fig. 8). The objective is to establish regulations that will prescribe land use, easements, and proposed operations within the framework of conservation and development of the protected area.

The interest of this type of analysis (typo morphology) is to identify the relationships between the different components of the urban fabric, as well as the contradictory or converging relationships that are established between the shapes of streets, buildings and plots. The urban fabric is therefore an integral part of the urban form, which is itself an integral part of the urbanized site. (Ministry of Culture & Arts, 1992).

The emergency measures are implemented to halt the processes of degradation, allowing for further investigation and analysis of the safeguard plan project. These measures are of a conservatory nature, aimed at stabilizing the affected areas in both physical and human aspects (see Table 1). By identifying the degradation phenomena, their

Fig. 8 Ayn Ali Medfaa street in Algiers, the Casbah Google photos 2022

causes, and their overlapping effects, an alarming diagnosis has been established. This diagnosis has also identified cases of extreme urgency that require immediate attention from the state, with the involvement of all relevant sectors and key stakeholders in managing the affected territory.

2.3 Urban Renewal of the Casbah, Between Stakes and Challenges

Urban renewal is a process of transforming deteriorated neighborhoods and cities to improve their living environment and integration. Our research focused on the city of Algiers, known for its nineteenth to twentieth-century architecture. We examined the dynamics of urban forms, occupation patterns, and relationships between urban morphology and housing types.

In Algiers, urban renewal has been carried out gradually, not by a government-initiated policy, but through voluntary participation of residents who may not always be aware of the implications. These transformation efforts, including house demolitions and reconstructions, have contributed to the changing face of the city center. In the 2000s, the public authorities in Algiers initiated significant changes in urban production methods due to various urban dysfunctions. This involved seeking expertise from foreign engineering firms to develop urban projects with international recognition and establishing local public structures for urban production (Bouteche Besma, 2015).

The main challenge is to create favorable conditions for continuous urban requalification. This involves finding the right balance between on-site and off-site rehousing at the local and agglomeration scale. In the past, urban planning in Algeria focused on extending the city beyond its fringes to address urgent demands for housing. However, recent economic and social changes have led to a shift in approach, with the public authorities implementing reforms aimed at revitalizing the economy, increasing housing supply and employment opportunities, and undertaking restructuring efforts. This new socio-economic context has resulted in unprecedented actions and interventions in existing urban areas, reflecting the evolving nature of the Algerian city and the emergence of new complex issues (Humanite, 2018).

The Kasbah of Algiers has had a significant influence on architecture and urban planning in North Africa, Andalusia, and sub-Saharan Africa during the sixteenth and seventeenth centuries. This is evident in its unique habitat and dense urban stratification, which represents a synthesis of various traditions including the local Muslim way of life. Despite undergoing changes and facing seismic hazards, the Kasbah of Algiers has largely retained its integrity. Its esthetic characteristics, materials, and architectural elements continue to express the values that led to its classification as a World Heritage site in 1992. The residential function of the Kasbah has helped maintain the viability and integrity of the site. Restoration efforts within the framework of the Safeguard and Enhancement Plan, in line with local and national standards, have contributed to the preservation of the site's integrity. However, there are threats to its integrity due to over-densification, uncontrolled interventions, earthquakes, fires, landslides, and floods (Arras, 2020).

As more structures in the Kasbah deteriorate, neglect spreads, with few residents having the means to repair their houses or prevent the collapse of neighboring structures.

Table 1 Conservation state of the building

Neighborhood	Plot	Empty	Building	Old building 1st degree old	2nd degree old	3rd degree old 4th degree	4th degree vintage	Ruin	Walled	Closed	Constillicit	Renovated	Works in progress	Travaux al'arret
Red sea	253	28	199	33	50	15	38	25	05	09	10	22	07	10
Amar Ali	917	62	855	99	171	167	80	104	18	74	30	74	22	16
Sidi Ramdane	203	39	164	04	16	31	50	14	03	26	03	03	14	02
Amar El Kama	330	11	319	20	40	58	130	16	02	29	11	07	04	03
Souk El Djemaa	204	29	174	10	17	14	74	06	11	17	02	13	03	10
Lallahoum	96	63	33	02	02	02	18	–	01	–	–	03	–	05
Marineand the harbor	82	Place	78	0	01	14	63	–	–	–	–	–	–	–
The citadel	16	02	13	–	01	03	07	–	–	–	–	01	01	–
The peripheral zone	88	06	82	10	11	24	36	–	–	01	22	–	01	–
Total casbah	2189	240	1816	178	273	328	496	165	40	155	78	123	52	46
% total casbah			83	10	15	18	27	10	2	8	4	6	2	2

Source Synthesis drawn up on the basis of the CNERU surveys. Data are constantly being updated as part of the implementation of emergency measures

Since 2012, the Algerian government has shown significant interest in the Kasbah of Algiers, adopting the Permanent Plan for the Safeguarding and Enhancement of the Safeguarded Sector of the Casbah of Algiers. This plan, financed with 90 billion dinars over ten years, aims to restore the Kasbah to its original state, with a focus on historical sites, while also finding a solution that balances the preservation of its cultural and historical value with the needs of its long-term residents. The plan also includes the reconstruction of approximately 400 collapsed buildings to strengthen the overall urban fabric, as empty plots pose a great danger to neighboring houses in the interconnected fabric of the Kasbah. Additionally, the plan addresses sanitation, infiltration, drainage, and water and energy supply issues without compromising the city's esthetics and heritage (Bachar, 2022).

Despite the historical discontinuities and contrasts in its formation process, the base structure and urban matrices of Algiers, including the Kasbah, reflect the city's evolving nature and the complexities of urban renewal efforts.

3 Conclusions

Any object that has a mark in time refers to a historical or cultural period of a place; it has a heritage dimension. It is also a witness of a period of evolution of a place and, therefore, of humanity. It is the common identity factor between generations that have occupied the same territory, the only link that remains but evolves over time.

Cities have experienced unprecedented urban growth. Their mode of growth is based on urban expansion rather than building the city upon itself. In order to combat urban sprawl and strengthen sustainable development, we speak of "remaking the city on the city." Cities that have a historic center or an old fabric offer an opportunity for the city to reconcile between the two entities: the new town and the historic town. The urban renewal of old fabrics is moving toward sustainable urban development and responds to the new challenges of twenty-first-century cities.

The Casbah, a mythical city and a city unlike other Algerian cities, the Casbah of Algiers, where each corner reflects an identity and each space testifies to the history of a past community, is distinguished by its urban singularity and architecture that lies in the harmony of its volumes and its integration into a steep site facing the sea.

Despite all these advantages, the Casbah of today is in danger; it suffers from the effects of time, neglect, and lack of maintenance. Independently of the discontinuities and contrasts resulting from the historical process of its formation, the basic structure and urban matrices of Algiers, and even of other urban centers located in municipalities integrated into the wilaya, remain identifiable and expressive in terms of heritage and urban planning, prevailing as an identity resource. Among the set of actions to be implemented, it is important to identify urban renewal not only at the level of the central fabrics of Algiers but also at the level of other centralities that can be reported in terms of heritage value, architecture, and urban planning. The rehabilitation of spaces and structural urban components will be carried out according to an order of priorities for the progressive application of the mechanisms and instruments suitable for this purpose, ensuring the cohesion of the urban heritage components, namely systems, fabrics, elements, sets of elements, or existing structures. Urban renewal operations will be inseparable from the housing component, given that the elements or sets of existing heritage value themselves constitute an important resource for filling in the gaps in this sector.

On the evaluation of aid for disadvantaged neighborhoods: "urban policy must be based both on the necessary urban renewal and on a specific policy of social interventions in the broad sense. Because beyond the dignity of the habitat, the effectiveness of the republican promise and, therefore, part of the future of our country are at stake in these neighborhoods; to be born, to have grown up, to live somewhere on the national territory cannot seal a social destiny."

From now on, we no longer speak of "renovation" but of urban "renewal." This favors a global approach to the neighborhood, and not an intervention mainly focused on the built environment. We are moving away from a logic of "urban repair" of the living environment, to move toward a trivialization of neighborhoods in their environment, and their inclusion in broader development dynamics. Also, new principles give a stronger ambition to this program: [No further text provided] (Sato, 1973).

- Working and studying old fabrics means that we must take into consideration all the elements that constitute the memory and value of heritage. This approach is important because it allows for the establishment of an extensive and significant collection during an intervention, in order to better integrate and successfully carry out a project.
- The requalification of the Casbah district fits perfectly into this logic, as its integration is based on the referential elements of the district, including its history, monuments, and architecture, with the aim of enhancing the district.
- The demolition of social housing complexes is not essential; decisions will be made according to the local context. Another novelty is that agreements will be signed at the intermunicipal level and formalized in a single document, with the involvement of inhabitants at all stages, from design to evaluation. The environmental

quality of the works will determine a portion of the subsidies, and the development of private housing and economic activity to promote social and functional diversity will be taken into account.

In the context of spatial transformations in an Algerian city that is being rebuilt without prior design and regulated by the market and private initiatives following public action, questions that public authorities will need to address relate to responsibilities and the roles of stakeholders: Who should do what and how? If the old buildings in Casbah show signs of dilapidation and could present an opportunity for land recovery in an urban environment, what strategy should be adopted? What should be the consideration of social, economic, historical, and heritage dimensions in a process of qualitative change? How should private actors and inhabitants be involved in such a process? How should tasks be defined for each stakeholder in a coordinated partnership approach, considering the diversity of territories to be renewed, scales to be articulated, and social situations to be addressed?

Bibliography

ACHAB ep CHERNAI, S. (2012). *Development of a technical guide for the rehabilitation of heritage (Habitat) from the Ottoman period*. Mouloud Mammeri University.

Arras, K. (2020). *Réhabilitation et revitalisation de la Casbah d'Alger: Le Wali d'Alger s'engage à rattraper le retard des travaux de Restauration*. Retrieved from reporters. https://www.reporters.dz/rehabilitation-et-revitalisation-de-la-casbah-dalger-le-wali-dalger-sengage-a-rattraper-le-retard-des-travaux-de-restauration/

Bachar, K. (2022). *Dynamics of urban renewal through the transformations of residential heritage in the high Algerian steppe plain*. Retrieved from openedition: https://journals.openedition.org/tem/8794

Benzenine, A. (2016). *Renewal of urban empty centers background: Redesignation of the Bab Zir District*. Abou Bekr Belkaid University Of Tlemcen.

Bouteche Besma, D. L. (2015). *Urban marginalities: An urban and social phenomenon that persists and ends* (pp. 61–74). Sciences & Technologie.

CNERU. (2012). *Safeguard and site enhancement plan Sauvegarde* (Presentation Report Phase Iii-Final Edition). Algaer.

Dris, N. (2005). *Habiter le patrimoine: monde en marge et identité urbaine*. openedition (pp. 93–104).

Fabiani, R. (2021). *Algeria: the restoration project of the Casbah of Algiers. Récupéré sur Med-Or Leonardo Foundation*. https://www.med-or.org/en/news/algeria-il-progetto-di-restauro-della-casbah-di-algeri

Farrand, A. G. (2015). *The Casbah of Algiers: An imperiled heritage*. Retrieved from Ibn Ibn Battuta: https://ibnibnbattuta.com/blog/casbah-algiers-imperiled-heritage

Gabriel, C. (1901). *Corpus of Arabic and Turkish inscriptions from Algeria*. E. Leroux.

Humanite. (2018). Retrieved from: Casbah of Algiers: open letter to Jean Nouvel. https://www.humanite.fr/en-debat/alger/casbah-dalger-lettre-ouverte-jean-nouvel-665443

Ministry of Culture and Arts. (1992). Preliminary draft of the general plan for the safeguarding of the casbah of algiers..development. Alger.

Mouloud, D. (2007). *Methods of strengthening old buildings*. University Publications Office.

Rafik, M. (2015). Benchmarks around the safeguard plan for the Casbah of Algiers. Retrieved from babzman. https://babzman.com/reperes-autour-du-plan-de-sauvegarde-de-la-casbah-dalger-partie-1/

Sato, J. D. (1973). *Revaluation of the Casbah of Algiers*. Unesco.

Slimani, J. (2015). De la rénovation urbaine au renouvellement urbain : glissement sémantique ou nouvelle approche ? Après-demain (pp. 11 à 13). Retrieved from : From urban renovation to urban renewal: Semantic shift or new approach?: https://www.cairn.info/revue-apres-demain-2015-1-page-11.htm

Zeghiche, B. K. (2022). The renewal of the Algerian city by the demolition-reconstruction of old buildings. From sanction to the search for legitimacy, Case of the city of Annaba. Retrieved from openedition. https://journals.openedition.org/cybergeo/26597

Morphological Changes of Urban Nodes and Their Effect on Urban Mobility and Transportation Modes in Heliopolis

Mahmoud Amgad, Omar M. Galal, and Ayman Wanas

Abstract

The transformation of urban forms and the organization of automobility networks were the main factors affecting urban transportation and mobility throughout the years. Moreover, the relationship between movement and urban forms shapes the city's urban mobility. The research significance lies behind documenting and analyzing the situation in Heliopolis urban nodes after the major additions in flyovers and intersections in the favor of vehicular modes of commuting in 2019 so that it provides an opportunity for future studies in this field. However, these intrusions neglected the vital role of urban nodes and their effect on mobility in Heliopolis. This study's objective is to investigate the morphological changes in urban nodes resulting from Heliopolis' new urban development and review it according to previous urban theories to clarify its impact on the city's urban mobility and transportation modes. The study will be conducted through a comparative analysis between Heliopolis urban nodes before and after the aforementioned development. Moreover, the study will apply a morphological analysis to investigate the transformation of the urban nodes (Heliopolis Square and El-Hegaz Square) physical form by abstracting the urban elements to clarify the transformation of the street patterns, geometrical form, and movement in the squares. Thereupon, the paper expounds on whether these intrusions enhanced or diminished the ability of urban nodes to serve the city's urban mobility. The results of this review demonstrate that this new development affected positively vehicular movement in the selected nodes but affected negatively walkability. Furthermore, this development removed the tramlines which were one of the identity elements and a traditional mobility mode in Heliopolis. Consequently, the results also show that there is a mutual relationship between mobility and identity. This provides new opportunities for future scientific researches to understand to what extent identity affects mobility.

Keywords

Urban morphology · Urban mobility · Urban developments · Identity · Heliopolis

1 Introduction

Over the years, cities reflected the culture and identity of people. Moreover, cities' urban development has always been connected with the development of cultural, economic, and social aspects. Thus, studying cities' urban development has attracted researchers from different fields, aiming to understand the factors affecting city development. Thereupon, the process of urban planning and design kept developing to enhance the city experience and make cities keep up with the rapid development of technologies and ideologies.

The city of Heliopolis is one of the significant urban experiments in the history of Egypt. It introduces a new urban pattern and planning ideology providing a unique character and identity to the city. Located in the north part of Cairo, Heliopolis was built in 1906 due to *Baron Empain's* vision of orienting the new urban development in the desert instead of the riverbank (Ilbert, 1985). The planning of Heliopolis was profoundly influenced by Sir *Ebenezer Howard's* principles and the '*Garden City*' planning.

M. Amgad (✉) · O. M. Galal · A. Wanas
Department of Architecture, Arab Academy for Science and Technology, Cairo, Egypt
e-mail: mahmoudamgad18@aast.edu

O. M. Galal
e-mail: omar.galal@aast.edu

A. Wanas
e-mail: awanas@aast.edu

M. F. Shahidan et al. (eds.), *Urban Narratives: Exploring Identity, Heritage, and Sustainable Development in Cities*, Advances in Science, Technology & Innovation, https://doi.org/10.1007/978-3-031-48517-6_12

According to the Central Agency for Public Mobilization and Statistics, the bulletin inventory of roads and bridges 2017/2018 states that from 1970 to 2018, Cairo has 15 vehicles bridges for the general authority of roads and bridges, 77 vehicles bridge for road and localities directorates, and two vehicles bridges of Constructing Urban Societies. On the other hand, through the data achieved from Egypt's Project Map Web site between the year 2018 and 2022, 104 bridges were constructed in Cairo only (Fig. 1). Moreover, other infrastructure projects were constructed at the expense of public spaces. Heliopolis witnessed numerous urban developments aiming to keep the city up with the continuous growth of the population (Gouda, 2015). Recently, Heliopolis is witnessing a new urban development that changed and affect the city morphology and the ability of urban spaces to serve different activities. The new urban development taking place in Heliopolis added a new layer of modification to Heliopolis' urban nodes. This layer created a new mobility network in the city. The new modification added to Heliopolis' urban nodes affected the city street pattern and the movement organization.

Moreover, it introduced new elements to public spaces and eliminated other elements considered as one of the city features. The added new elements aimed mainly to solve traffic problems and enhance the mobility network. However, this new development addressed only vehicular movement and neglected sustainable transportation modes like walking and cycling.

This study aims to investigate the changes that happened in Heliopolis' urban nodes and review them to contribute to understanding the relationship between the city's urban mobility and urban form. More specifically, the aim is to establish an understanding of the relationship between street patterns and transportation modes, furthermore, explaining that the changes in the urban form can encourage and diminish the city transportation modes.

The research starts by reviewing the definition of urban morphology and explaining the principles of different schools that addressed urban morphology over the years. Consequently, it discusses the principle of Robb Krier in approaching urban morphology. The research then reviews the definition of urban mobility and focuses on connecting urban morphology, urban mobility, and transportation modes. Finally, it identifies the morphological changes in the selected case studies and how they affected urban mobility and transportation modes.

This research discusses Heliopolis urban nodes' characteristics before and after the new urban development in a comparative method to explain how these changes affected the existing activities, whether positively or negatively. The research then focuses on analyzing the typo-morphology of the urban nodes to identify the city's changes in street patterns and urban mobility patterns. This study contributes to the literature to document the current situation of Heliopolis urban nodes to create a better understanding of the relationship between urban forms and urban mobility.

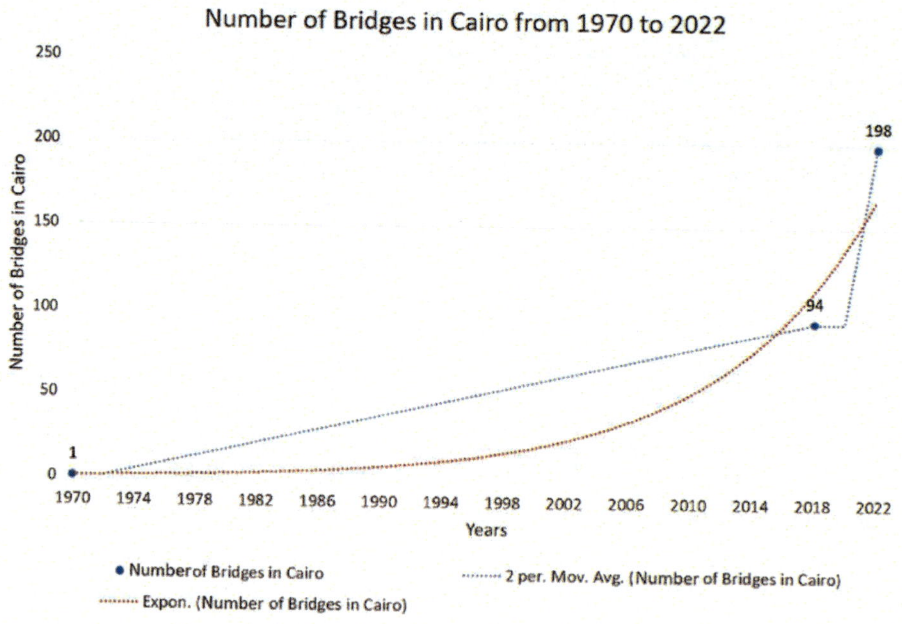

Fig. 1 Number of bridges in Cairo from 1970 to 2022 (by the author)

2 Literature Review

2.1 Identity

According to the Cambridge dictionary, the word character refers to a set of particular common qualities in a place or a person that makes them different and unique (Cambridge Dictionary, n.d.). Cullen (1961) addresses the skill of tying together all the design components to produce a single, cohesive concept that gives the city a distinctive appearance. Lynch (1960) talked about how the perception of a city is a two-way conversation between the observer and the observed. Moreover, he says that the identity, structure, and meaning of the city's image may be divided into three primary categories (Lynch, 1960). According to Kropf (1996), two underlying assumptions may be used to examine identity: The first is that identity is something that people assign to the town, and the second is that identity depends on the distinctions that set the town apart. Mansy (1991) argues that after reviewing the previous literature and research, the character can be identified as the common visual features that provide distinctiveness and uniqueness. Moreover, he emphasizes that character is a result of continuity (Mansy, 1991). Shamsuddin et al. (2012) explain that character is a recognizable group of design elements that differentiate one design from the other. Moreover, they argue that the townscape character has a dual nature: individual and social (Shamsuddin et al., 2012). The spatial characteristics and the quality of the built environment in cities define the cities' look and character (Larkham & Conzen, The Making of Urban Landscape, 2014).

2.2 Urban Morphology

2.2.1 Overview

Castells (1983) explains that urban forms are dynamic systems affected by the interaction between the different layers of space and society, moreover, the continuous economic growth, and political pressures. The impact of these factors on urban spaces has been remarkable as they profoundly affected the development of urban forms. Furthermore, settlements have always been an indicator of human activities and culture throughout the years. Urban morphology is the study of transformations in the physical form of settlements over time, focusing on the city's urban pattern and continuous growth process (Carmona et al., 2010). Moudon (1997) argues that urban morphology is the study of the city as a human settlement; moreover, urban morphologists' role is to analyze the continuous evolution of city patterns and networks to clarify and expound the city components.

Conzen (1960) explains urban morphology by breaking it down into main vital elements: street pattern, plot pattern, and buildings with their structure and land use. Conzen (1960) further emphasized that not all the elements have the same level of resilience, as some directly respond to dynamic actions in the city. Whitehand (2001) argues that the term 'urban morphology' in the United Kingdom is used in different types of analysis and research; however, nearly all these studies target the physical form of urban spaces. Kropf (2013) argues that urban morphology is not about the physical built environment of cities but about studying the transformation process and the factors that generated these physical forms. In 1996, the formalization of the International Seminar on Urban Form (ISUF) provided a better opportunity for the urban morphology field to expand beyond its borders; moreover, this seminar created a common ground suggesting that the city can be analyzed through its physical environment (Moudon, Urban morphology as an emerging interdisciplinary field, 1997). Moudon (1997) further explains that the morphological analysis is based mainly on three levels:

- Buildings and spaces attached to them, streets, and plot patterns are the main physical elements that define the urban form.
- The understanding of urban form can be understood through different levels; buildings, streets network, the city, and the region.
- The dimension of time is a primary element in understanding the urban form as its elements undergo continuous transformations and changes.

2.2.2 Three Schools of Urban Morphology

Over the years, numerous studies were conducted on urban morphology. Moudon (1997) argues that the ISUF meeting in 1996 concluded three main classifications for urban morphology: Conzenian (British School), Muratorian (Italian School), and Versailles (French School). The primary purpose of the three schools is to provide a proper process for studying urban morphology (Sadeghi & Li, 2019). Moudon (1997) further argues that the German geographer M.R.G Conzen and the Italian architect Saverio Muratori were two individual figures who affected the urban morphology field profoundly moreover, established a ground for two early urban morphology schools—then followed by the French school in the late 1960s when the architects Philipe Panerai and Jean Castex with the sociologist Jean-Charles Depaule established the school of architecture in Versailles.

The raising and education of M.R.G Conzen in Berlin provided him an opportunity to reach new perspectives in art, humanities, and natural sciences that encouraged and developed his thoughts and ideas (Larkham, Urban Morphology and Typology in the United Kingdom, 1998).

Whitehand (2001) argues that the German geographer Otto Schlüter had a profound influence on the development of Conzen's thoughts over time. Whitehand (2001) further explains that Conzen concluded a ternary approach to understanding the townscape or urban landscape. Conzen (1960) explains that it is essential to establish what is meant by a 'town plan'; moreover, it is necessary to take into consideration the relevant geographical factors as the geographical arrangement and organization of the built-up area enhance its relation to the other morphological aspects; buildings fabric and land use. Conzen (1960) further explains that a town plan can be defined through the organization of human-made elements of the built-up area: street system, street blocks, and block plans.

Studying the typology of houses and their location in the city was the first step for Muratori toward understanding urban morphology (Sadeghi & Li, 2019). Moudon (1997) explains that Muratori depended on his studies on history as a theoretical basis rejecting modernist architecture's traditions, and neglecting history. Cataldi et al. (2002) argue that Muratori's experience in Venice in 1952 was the main inspiration behind his idea of operative history that aims to understand the inherited values in urban tissue resulting from the history of the city. Consequently, the ideas of the Italian school focused on studying building typology as buildings are the cells that create the urban tissue (Sadeghi & Li, 2019). However, Moudon (1998) argues that Italian urbanists believe that studying morphology is only applicable to historic cities and limited to contemporary fabrics because of the fragmented patterns. Marzot (2002) argues that the Italians had a unique perspective in understanding urban forms that consider the city as a contextualized system of knowledge. Moudon (1997) explains that the strength of Muratori's teaching inspired many followers to see the importance of the city's mnemonic powers. Moudon (1997) further explains that the research and studies of Muratori's followers, like Gianfranco Caniggia, developed the Italian approach to understanding urban forms and generated more comprehensive theories.

The French school that emerged in the late 1960s shared the rejection of modernist architecture and neglecting history with the Italian school; however, it benefited from the vibrant ideologies and discussions on urban life in this era (Moudon, Urban morphology as an emerging interdisciplinary field, 1997). Sadeghi and Li (2019) argue that the French school's thoughts and ideologies were a resumption for the Italian school with the addition of socialists' critical thoughts like Henry Lefebvre and historians like Francoise Boudon and Andre Chastel. Gauthiez (2004) explains that the French school of urban morphology discussed numerous approaches. Moreover, the French school works focused on examining urban theories and relating them to urban forms as they believed a new approach is needed to understand modernist urban planning (Sadeghi & Li, 2019). However, Darin (1998) questions the validity of considering French ideologies and thoughts, as a school in urban morphology. Darin (1998) explains that studying urban forms in France is a fragmented field due to the lack of uniformity, and connectivity between the different studies and hierarchy.

In conclusion, firstly, the Conzenian or the British school founded by M.R.G. Conzen and then developed by J.W.R. Whitehand mainly depended on breaking down settlements into four main elements: land use, building structure, plot pattern, and street pattern. Thus, it can be concluded that it depended on the map unit in their process. Secondly, the Muratorian or the Italian school; the Italian School was founded by Saverio Muratori, and developed by Gianfranco Caniggia, and it depended on studying buildings as a single element generating the urban tissue with the consideration of history. Thirdly, the Versailles, or the French school founded by Philipe Panerai and Jean Castex, and Jean-Charles Depaule, focused on studying urban theories and how they affected the urban forms and morphology of cities. However, the three schools shared universal principles and ideas: the time dimension and how it impacts urban forms profoundly, studying urban form should be conducted through different levels and layers, and part of studying urban form is studying the open spaces related to it.

2.2.3 Robb Krier Principles

Carmona et al. (2010) argues that in the mid-1960s, a new morphological approach was generated by the Italian Rationalist School and initially Aldo Rossi then followed by many others, including the brothers Rob and Leon Krier. Carmona et al. (2010) further explains that the typology developed by Rob Krier in his book '*Urban Space*' is in contrast to Camillo Sitte and Paul Zucker's principles as Krier studied the basic geometry of the space abstracting the esthetic effects.

Krier (1979) argues that urban squares can be described as the first way humans discovered as an urban space. He further explains that urban squares are generated by the grouping of buildings around a space to produce an articulated controlled space. Krier then discusses that the typology of urban space can be classified into three groups according to its spatial quality and geometrical pattern in layout; these groups are obtained from the square, the circle, and the triangle. Moreover, Krier emphasizes the vital influence of elevations and sections of houses on the space. Rob Krier's principles show that urban squares can be classified into three types of layouts, and each type can be modified or developed in many methods like distortion, adding, subtracting, and merging with other shapes. Moreover, the elevations and sections of the building's surroundings also affect its degree of enclosure and articulate specific activities in the space.

2.3 Urban Mobility

2.3.1 Overview

The capability to transfer from one point to the other can be described as 'mobility' (Costa et al., 2016). Jiron (2008) argues that daily urban mobility discusses the relationship between people and the change of place daily. The term urban mobility can be explained as how people move in the city (Hossam, 2019). People's movement in cities has always been a field of study as people do not move randomly in space (Noulas et al., 2012). Carmona et al. (2010) clarifies that movement is a primary aspect of understanding how spaces function; moreover, the heart of the urban experience can be observed through pedestrian movement in space. Carmona et al. (2010) further explains that mobility can be discussed in terms of car-based and non-car-based accessibility; however, spaces are inaccessible if they depend mainly on users reaching them with private modes of transportation. Sheller and Urry (2000) argue that automobility defines cities as a layer affecting the other city aspects; moreover, mobility is an indispensable feature in the continuous development and transformation of cities.

2.3.2 Urban Mobility and Transportation Modes

The evolution and transformation of travel forms throughout history profoundly affected the development of urban mobility and the city's public space network (Carmona et al., 2010). Carmona et al. (2010) further explains that when the principles of transportation modes depended on foot and horses, the realm of movement and public spaces overlapped but with the development of new transportation modes depending mainly on private cars this relation between the spaces was diminished which resulted in a separation between the space different movements. Jones (2014) argues that the development of transportation modes and networks in the late nineteenth century profoundly affected urban forms. However, the considerable development of motor vehicles in the mid of the twentieth century had the most significant influence on our cities and towns.

Norman (the minister of State for the Department for Transport) (2019) argues that the evolution of transportation technology and networks will be even faster in the next two decades, and he further emphasizes the importance of considering mobility and transportation modes as primary elements in the urban design process. Hossam (2019) argues that both transportation modes and urban forms profoundly affect each other; moreover, city mobility and accessibility are articulated by the relation between transportation patterns and urban forms. Rodrigue (2018) explains that urban transportation can be studied and discussed in three aspects: public transportation, individual transportation, and freight transportation. Rodrigue (2018) further explains the three aspects; firstly, public transportation's primary purpose is to provide easy accessibility for people to move in the different parts of the city and transfer large numbers of people at the same time, buses and tramways, trains, and ferryboats. Secondly, individual transportation is a result of personal choices like walking, cycling, or using personal cars. Thirdly, freight transportation and its related to the industrial and productive nature of cities and can be observed through the network connecting distribution centers or major transportation nodes like airports and train stations.

2.4 Urban Morphology and Urban Mobility

The heavily depending on mechanical transportation modes raised many concerns that shaped the new interest in relating cities' urban forms and mobility (Hossam, 2019). The design of neighborhood elements can reduce this depending on mechanical modes and encourage sustainable transportation modes. Cervero and Cockelman (1997) argue that throughout history, many urban design approaches like new urbanism, transit-oriented development, and traditional town planning discussed city mobility and travel demand. They further explain that these approaches shared common objectives which are: (1) decreasing the dependency on motorized modes, (2) encouraging and increasing sustainable modes like walking and cycling, (3) decreasing traveling distance and increasing vehicle occupancy levels. Ewing and Cervero (2010) explain that understanding the effect of the built environment on urban mobility can be discussed through the main five elements. They further explain that the first three elements originated in 1997 by Robert Corvero and Kara Cockelman were density, diversity, and design, followed by destination accessibility and distance to transit by Reid Ewing and Robert Cervero in 2001.

Density: the variable of interest per area can identify density—for example, the number of people occupying a specific land area—Fig. 2a.

Diversity: diversity deals with the varieties of land use in a given land and the variety of mobility options—Fig. 2b.

Design: the design aspect includes the public space network: the street patterns and plot patterns. It includes the urban network layout and its characteristics—Fig. 2c.

Destination accessibility: it is evaluating the degree of accessibility to reach a trip destination—Fig. 2d.

Distance transit is the average of routes connecting people from their residence or workplace to the nearest transportation nodes—Fig. 2e (Ewing & Cervero, 2010).

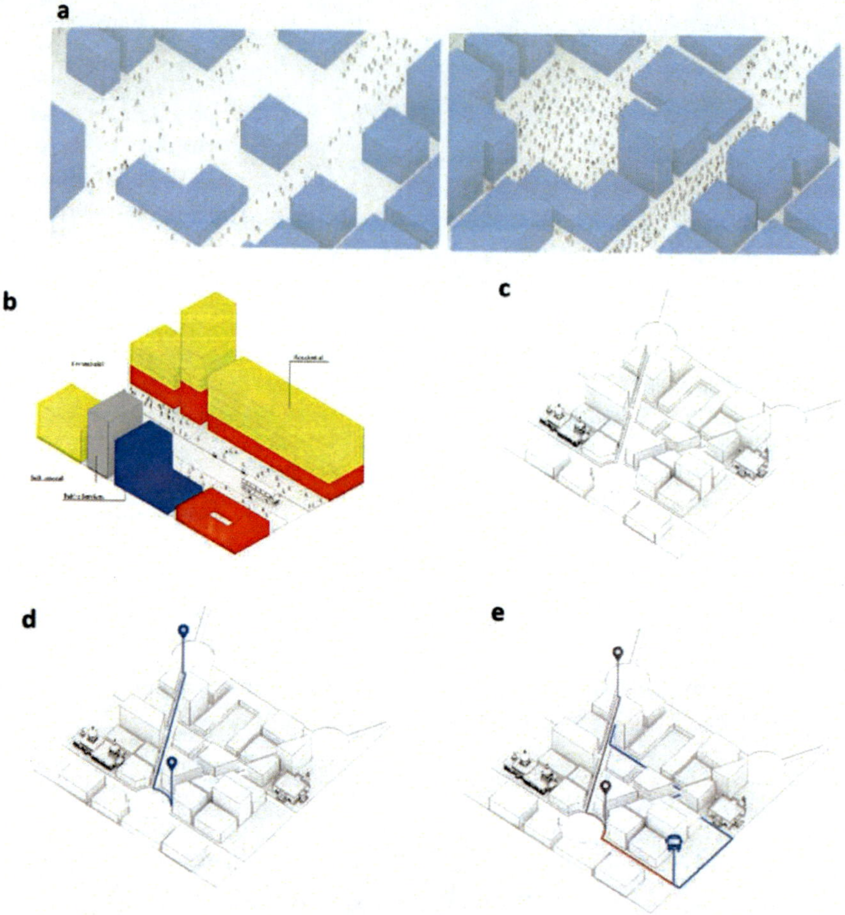

Fig. 2 **a** Density as an element connecting urban morphology and urban mobility. (by the author). **b** Diversity as an element connecting urban morphology and urban mobility. (by the author). **c** Design as an element connecting urban morphology and urban mobility. (by the author). **d** Destination accessibility as an element connecting urban morphology and urban mobility. (by the author). **e** Distance transit as an element connecting urban morphology and urban mobility. (by the author)

2.5 Heliopolis Planning

2.5.1 Background

Heliopolis is one of the cities located in the northern part of Cairo. The city consists mainly of four sub-districts: al Bostan, Almaza, al Montazah, and Mansheyet el-Bakry. Built in 1906, Heliopolis was not a result of private enterprise development; however, it was a creation of a city with a solid identity, unique characteristics, and multiple services (Ilbert, 1985). Baron Empain and Boghos Nubar had a vision and believed that Cairo's development should be oriented toward the desert instead of the Nile banks; their dream and ambition came true in 1906 when the Egyptian government provided them with a piece of land in the desert (Gouda, 2015). Ilbert (1985) explains that the Heliopolis project had three main features:

Firstly, it was an independent project without the need for assistance from the government.
Secondly, the project started from scratch in the desert.
Thirdly, it was a vision of one person not a result of a planning project; however, it followed the 'garden city' theory.

Heliopolis' structure followed the garden city and satellite towns principles as it was designed like one remote unit in the desert (Ilbert, 1985). Ilbert (1985) further explains that the general design of Heliopolis was an implementation of Ebenezer Howard's planning theories. Nowier (2014) explains that the original planning of Heliopolis can be described as establishing two connected oases: a residential oasis near Cairo and the other for industrial and low-income housing. According to the original plan, Heliopolis was supposed to cover 2500 hectares, including various land uses (Ilbert, 1985). Ilbert (1985) further explains that there were differences between the original plan in 1906 and the result in 1908.

The map of Heliopolis in 1920 shows its original planning of it. The central node (Basilique Square) radiates five streets to five main elements of the city which are as follows: Baron Empain Palace, Heliopolis Palace Hotel, Heliopolis racing course (Hippodrome), the primary market, and Gamei-square (mosque-square) (Gouda, 2015). The communication routes were the main factors affecting the survival of the city (Ilbert, 1985). Gouda (2015) explains that the two streets leading to the market and mosque-square led to the extensions in the '30s and 40s.' Ilbert (1985) argues that Heliopolis gained its originality and reputation through its form, similar to the theories and planning approaches carried out at the same time in Belgium, Britain, and France. Heliopolis' urban pattern can be explained as a radial-concentric grid with green spaces surrounding buildings from all sides (Gouda, 2015). Haussmann's Paris influence can also be observed in the planning, and the city's spatial organization, such as the wide streets and avenues connected to the city, squares together adequate placing of landmarks (Nowier, 2014). Nowier (2014) further explains that the strong urban and architectural identity, along with the strict obligations applied by Heliopolis company, sustained its public spaces layout over the years.

2.5.2 Heliopolis Urban Growth

The years following 1937 can be considered a transitional time in Egypt's history; moreover, they include Cairo's massive demographic evolution (Nowier, 2014). Nowier (2014) further explains that private investment took over after WWI profoundly affected the city's growth. All these aspects considered changed the urban characteristics of Heliopolis over the years. Gouda (2015) explains that neglecting Heliopolis' original urban pattern and its architectural style, along with the decisions aimed only at the economic aspect, ruined the city's identity. Consequently, the continuous urban growth to satisfy the increasing of population Heliopolis urban nodes witnessed many changes. New squares were created by adopting new activities. Moreover, the significance of some old urban nodes decreased due to the development of street networks.

2.5.3 Heliopolis' New Urban Development

Recently, Heliopolis is witnessing a new urban development in its nodes and streets. This urban development aims mainly to enhance the vehicles' movement by constructing new bridges and widening streets. However, these intrusions neglected the original morphology of Heliopolis squares. Depending on Rob Krier's principle in studying the morphology of urban squares, the geometrical form of Heliopolis squares was derived from the circle to serve the radial-concentric grid of the city. Most of the time, the squares' circular shape included a central element like a building with great value, public space, or a tramway, which was one of Heliopolis' important features. Moreover, these squares were opened and radiated a street network connecting the city. Consequently, their form was modified by the subtraction of the city plot pattern from them.

The first case study is Heliopolis square. Figure 3 shows a figure–ground analysis for Heliopolis square before and after the transformation process. Figure 3a shows the square geometrical shape and its relation with the street networks before the new urban development. It shows the radial-concentric nature of the city. On the other hand, Fig. 3b shows the geometrical form of Heliopolis square after the new development and how streets penetrated and disturbed its form instead of radiating from it.

Furthermore, these morphological transformations also affected the different kinds of movement in the square. As a result, the city's urban mobility was influenced. Figure 4 shows Heliopolis square before the urban development, where the central element is serving the tramways as one of the features and transportation modes of the city. It also shows the green areas, vehicle movement areas, and plot areas where pedestrian movement can occur. Figure 5 shows the same elements after the urban development. This transformation affected profoundly public transportation and individual transportation in the city. Firstly, after the

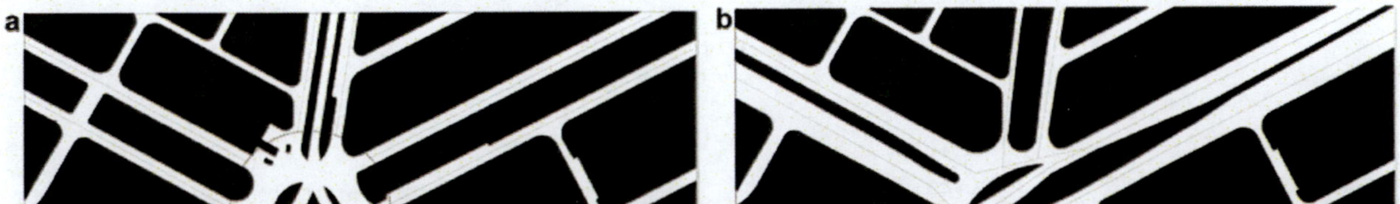

Fig. 3 **a** Figure ground analysis for Heliopolis square in 2014 (by the author). **b** Figure ground analysis for Heliopolis square in 2020 (by the author)

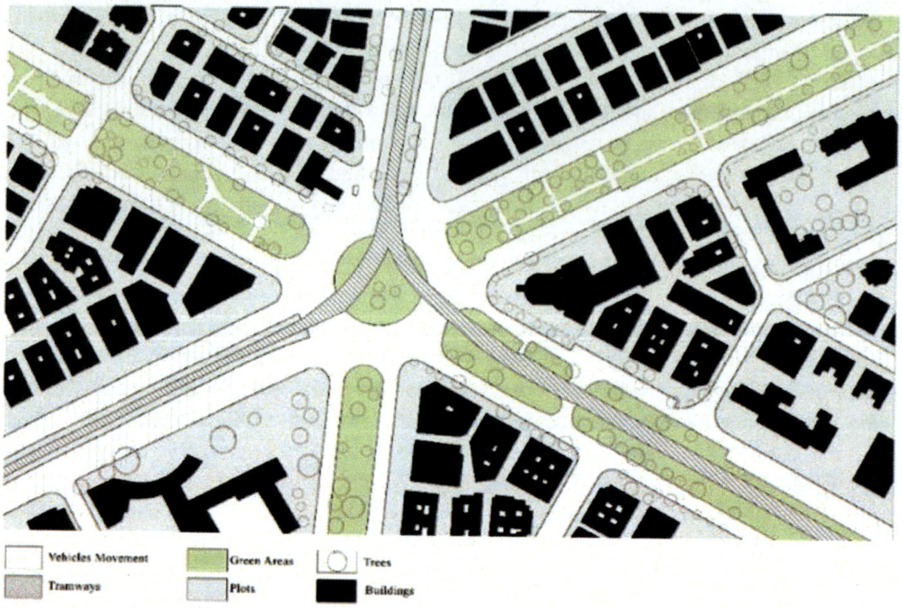

Fig. 4 Different movement in Heliopolis square in 2014 before the new urban development (by the author)

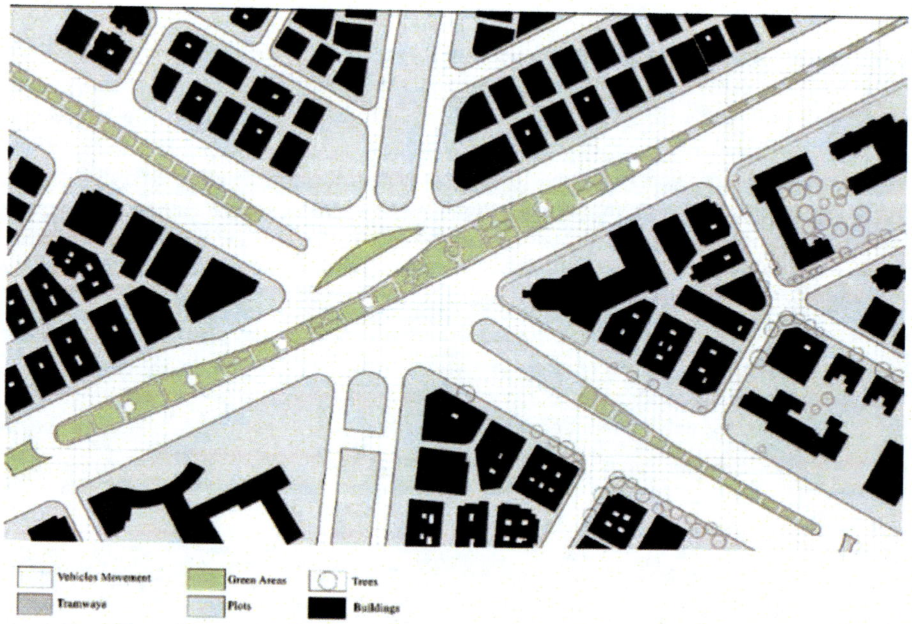

Fig. 5 Different movement in Heliopolis square in 2020 after the new urban development (by the author)

new development tramway was removed from the space, eliminating the opportunity to develop this public transportation mode in future moreover, it neglected that the tram is considered one of the city features over the years. Secondly, despite that widening the streets at the expense of pedestrian paths, green areas and trees served vehicle movement. They increased the probability of using personal cars; it decreased the probability of selecting sustainable transportation modes like walking, running, and cycling. Despite the new urban development aimed to deal with traffic congestion and increase city accessibility, it encouraged the usage of personal cars instead of sustainable transportation modes, which will lead to more congestion and traffic problems. The wide streets will create an inner edge between the city neighborhoods.

The second case study is El-Hegaz square. Figure 6 show a figure–ground analysis of El-Hegaz square before and after the transformation process Fig. 6a, respectively.

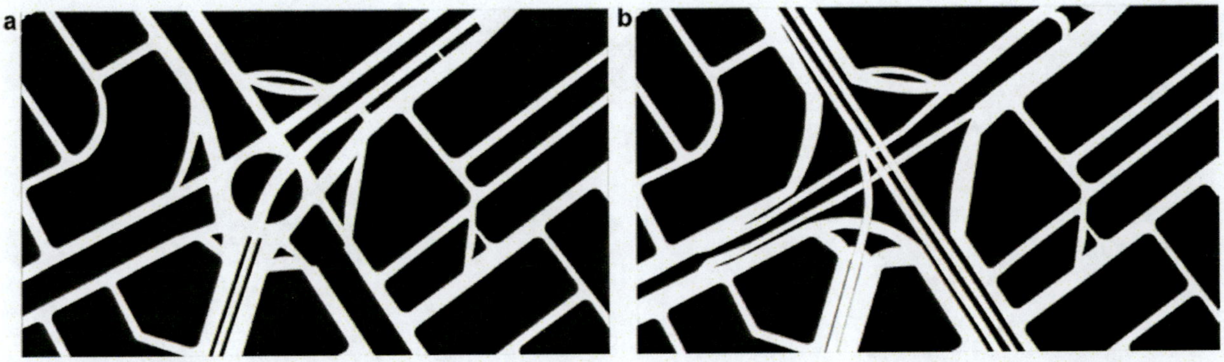

Fig. 6 **a** Figure–ground analysis for El-Hegaz square in 2014 (by the author). **b** Figure–ground analysis for El-Hegaz square in 2020 (by the author)

Before the new urban development, the square geometrical form is derived from the circle and modified by the subtraction of the plot pattern to organize the public space network.

The new urban development created a new element in the square that deformed its geometrical form. As shown in Fig. 6b, the new bridges added a new layer of vehicle movement. This new layer created new spaces in the square that encourage undesired activities. Similar to Heliopolis square, the movement in El-Hegaz square was affected profoundly. The tramway was removed, the pedestrian pathways were narrowed, and the vehicle spaces were widened. The new bridges developed vehicle movement in the city as they provided a new layer of movement for vehicles. This modification can be used in enhancing the public and freight transportation network. However, new spaces were generated under the bridges. These spaces without control can encourage undesired activities and threaten pedestrian movement in the square. Moreover, introducing the bridges to the square created a barrier to the movement between neighborhoods. Thus, the urban fabric of the city became fragmented and separated. This difficulty of movement and the dominance of vehicles network increased the risk of causing accidents. Figures 7 and 8 show the movement in the square before and after the new urban, respectively.

3 Methodology

The nature of this study is comparative; it aims to compare the ability of Heliopolis urban nodes to serve the city's urban mobility before and after the new urban development. The first part of the study adopts a content analysis to review the available literature to identify the relation between urban morphology and urban mobility to identify the comparison factors. The second part is selecting the case studies: Heliopolis square and El-Hegaz square; the two squares were selected as they are two connected urban nodes resulting from an old urban development that respected the city's urban pattern and identity. Furthermore, recently witnessed a significant change in their form and planning due to the new urban development taking place in Heliopolis. The third part is documenting the current situation by using satellite Google Earth maps to draw and generate the analysis maps. The fourth part applies a morphological analysis of the geometrical form, street pattern, and movement in both squares before and after the new urban development to trace the morphological changes and identify their impact on city mobility. The study will depend on reviewing available documents and publications such as the publications of Heliopolis Company as well as online publications and printed documents gathered from libraries. Historical maps are primary elements in the analysis process; two main maps are used (2014) and (2020). Figure 9 shows the methodological framework adopted in this study.

4 Results and Discussion

The paper mainly focuses on the changes that occurred in the street pattern, geometrical form, and different movement patterns in Heliopolis urban nodes following the new urban developments. The changes are analyzed and comparatively investigated the effects of the new urban developments on the capability of the suggested nodes to serve the area mobility and transformation.

The different patterns of the city profoundly affect urban mobility. The design of the city's urban form also can produce an efficient public space to enhance urban mobility. Moreover, there is a direct relationship between transportation modes and city mobility. Mobility can be divided into car-based and non-car-based movements (Carmona et al., 2010). The main drive behind most of the new urban development taking place in Cairo is traffic congestion. Thus,

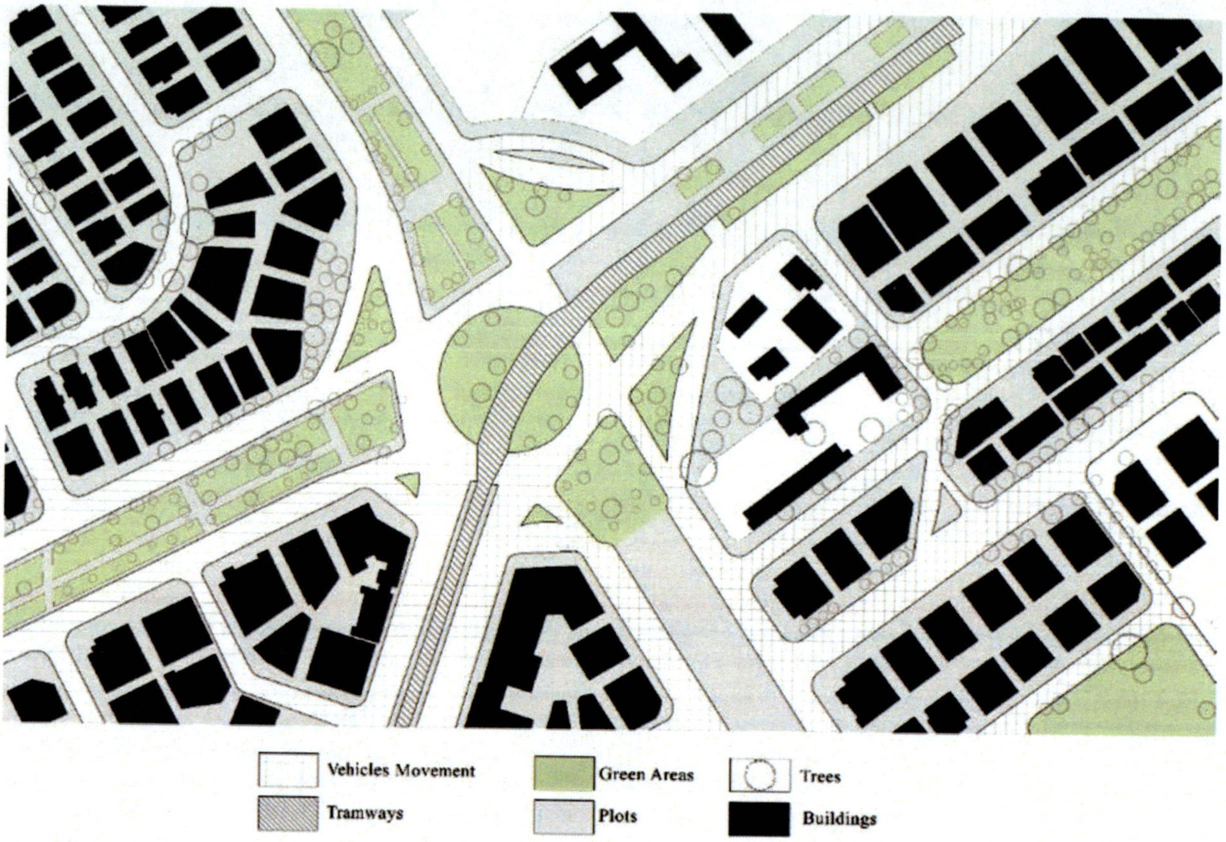

Fig. 7 Different movement in El-Hegaz square in 2014 before the new urban development (by the author)

widening streets and constructing new elements like bridges to facilitate vehicle movement is the main theme and approach of the new urban development. The new urban development in Heliopolis is following the same approach and theme. However, this approach cares about the car-based movement and neglects the other non-car-based movement.

The comparison between the design of Heliopolis and El-Hegaz squares before and after the new urban development resulted in an apparent change in the movement between the two cases. On the first hand, the car-based movement was enhanced by providing wider streets and increasing accessibility in the city. However, this change will lead to more congestion problems as it encourages depending on vehicles by providing a suitable environment. On the other hand, most of the elements used to enhance and encourage the non-car-based movement were neglected. For example, the islands between streets were reduced to widen the streets by removing most of the green areas and the trees providing shade for pedestrians.

Moreover, widening the streets created an edge or barrier between neighborhoods, which decreased the connectivity between the neighborhoods. Moreover, the fragmentation of the space by using bridges created new undesired spaces that decreased the sense of safety. All these aspects decreased the city's degree of mobility as the city is designed for people, not vehicles, and the central heart of the city experience is the pedestrian and people movement. The literature review shows that there are five main elements to be discussed when studying the relationship between urban mobility and urban morphology which are as follows: density, diversity of land use, design, destination accessibility, and distance transit. This study focuses on the design element as the other elements did not witness a fierce change during the transformation process. Table 1 shows the changes that happened to the squares' design, movements, trees, and green areas after the new urban development and their effects on mobility modes.

Additionally, one of the findings of this paper is that there is a lack of literature on studies that investigate

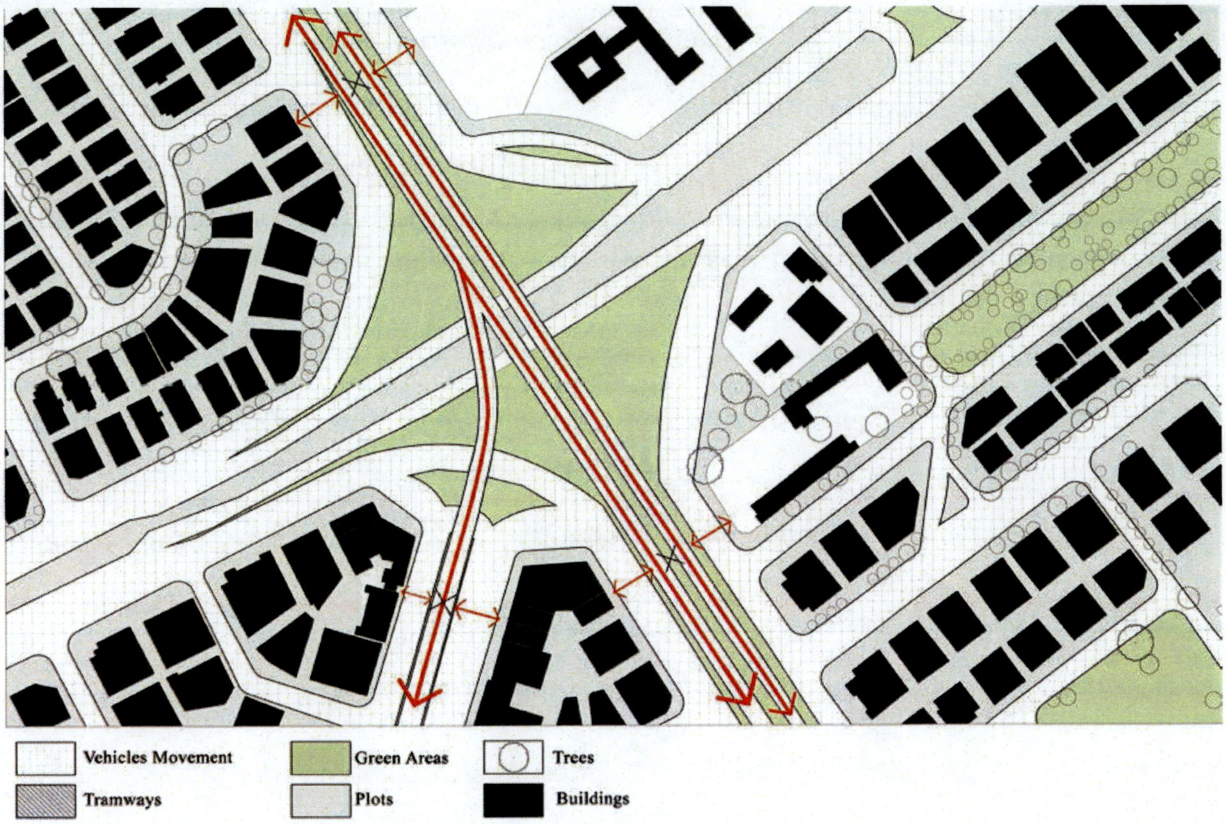

Fig. 8 Different movement in El-Hegaz square in 2020 after the new urban development (by the author)

the relationship between identity and modes of mobility in Egyptian neighborhoods. Mansy (1991) discusses Egyptian cities' architectural and urban character and their relation with regulations and law. While StampsIII (1994) assessed the relationship between character, esthetics control, and environmental preferences, Nasar (1994) discussed character by studying the exterior qualities of buildings to evaluate urban esthetics. Sisiopiku and Akin (2003) studied the behavior of pedestrians toward walkability in the neighborhood and the effect of the crosswalks and mobility network on their perception. In the SOAC 2005 (State of Australian Cities Conference), Dovey et al. (2005) discuss the relationship between new urban development, the future planning framework, and the construction of character and place identity. Shamsuddin et al. (2012) study the factors that affect the historical character of urban landscapes. Wanas (2013) discusses the relationship between identity and character of urban spaces and places with political factors. Yu (2013) studies the relationship between regulations and forming the character of new Chinese cities. Mohamed (2021) explores the deterioration of character in historical Egyptian cities due to globalization and the blind copying of western architecture and urban planning. Shafik et al. (2021) discussed the impact of the urban development process in Cairo regarding Heliopolis on the mobility of children in neighborhoods. Hefnawy et al. (2022) analyze the relationship between the new urban development in Heliopolis and the urban heat islands effect.

Despite the aforementioned investigations, limited studies addressed the mutual relationship between rapid urban transformation and identity degradation on the one hand and mobility on the other. Therefore, the study expounds that there are future research opportunities to explore the

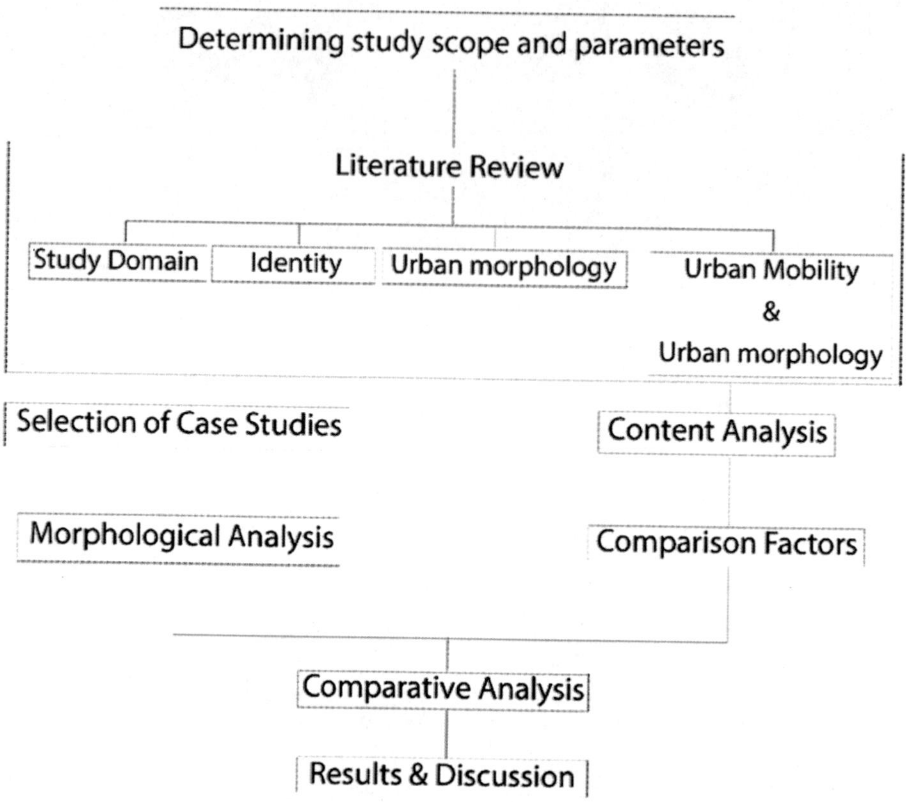

Fig. 9 The methodological framework for the study (by the author)

direct effect of these rapid transformations in the urban environment and its character on mobility modes in Cairene neighborhoods.

5 Conclusion

In Cairo, new urban development projects have been taking place in cities. This research discussed the recent urban development taking place in Heliopolis. This development resulted in a significant morphological change to the city's urban nodes and streets. These intrusions vary between adding new elements to the street network or removing already existing features that contribute to the city's identity. Thus, the city's urban mobility was also affected profoundly.

Considering the literature on the relation between urban form and urban mobility, the definition and factors of both urban form and mobility were discussed. Consequently, the Heliopolis squares' morphing process was investigated to identify its positive and negative impact on the city's urban mobility. The morphing process enhances city mobility, which depends on car-based movement by widening the streets at the expense of pedestrian paths and adding new elements to facilitate vehicle movement like bridges. However, it discouraged other methods of mobility, like walking, running, and tram. Moreover, it created new spaces that encourage undesired activities if left uncontrolled.

In conclusion, it is essential to study cities' morphing process, especially cities with a unique identity and character like Heliopolis. This study can provide more opportunities to understand the relationship between urban forms and other city patterns and how to intervene in cities with distinctive identity and culture.

Table 1 Explains the changes that resulted from the new urban development and their effect on mobility modes (by the author)

		Heliopolis square	El-Hegaz square	Mobility mode		
				Vehicles	Walking	Tram
Design layout	Before		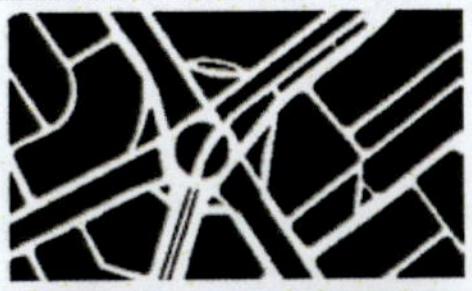	✓	✓	✓
	After		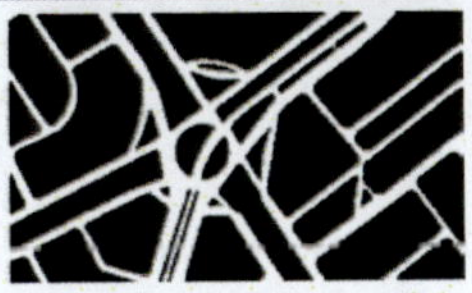	✓	X	X
Pedestrian area	Before	The islands between streets were wide with articulated pathways that encouraged the pedestrian movement			✓	
	After	The street sections were changed, and the pedestrian movement paths were removed to create wider streets for cars which discourage the pedestrian movement			X	
Vehicles area	Before	The organization of movement depended on the central space created by the geometrical form of the square and the radial pattern of street networks		✓		
	After	The new urban development enhanced the movement of the vehicle by adding new layers of circulation and widening the streets to can bear more cars		✓		
Tramway	Before	The tramways existed in the middle of the square connecting the square and the urban fabric				✓
	After	The tramways were removed from the square wasting the opportunity to revive this transportation mode				X
Green areas and trees	Before	Green areas and trees were one of the features of Heliopolis squares and streets. They existed heavily in the pedestrian area to enhance the pedestrian movement			✓	
	After	Green areas decreased significantly with decreasing the pedestrian area in the main streets where the urban development took place			X	

References

Cambridge Dictionary. (n.d.). *Meaning of character in English*. Retrieved June 26, 2021, from https://dictionary.cambridge.org/dictionary/english/character

Carmona, M., Tiesdell, S., Heath, T., & Oc, T. (2010). *Public places urban spaces; The dimensions of urban design* (2nd ed.). Routledge.

Castells, M. (1983). *The city and the grassroots: A cross-cultural theory of urban social movements* (1st ed.). University of California Press.

Cataldi, G., Maffei, G. L., & Vaccaro, P. (2002). Saverio Muratori and the Italian school of planning typology. *Urban Morphology, 1*(6), 3–14.

Cervero, R., & Cockelman, K. (1997). Travel demand and the 3Ds: Denisty, diversity and design. *Transportation Research,* 2(3), 199–219.

Conzen, M. R. (1960). Alnwick, Northumberland: A study in town-plan analysis. *Transactions and Papers (institute of British Geographers), 1*(27), 1–140.

Costa, P. B., Neto, G. M., & Bertolde, A. I. (2016). *Urban mobility indexes: A brief review of the literature*. Elsevier.

Cullen, G. (1961). *The concise townscape*. The Architectural Press.

Darin, M. (1998). The study of urban form in France. *Urban Morphology,* 2(2), 63–76.

Dovey, K., Woodcock, I., & Wood, S. (2005). What is urban character?: The case of Camberwell. *SOAC [State of Australian cities national conference]* (pp. 1–14). Griffith University.

Ewing, R., & Cervero, R. (2010). Travel and the built environment: A meta-analysis. *Journal of the American Planning Association, 76*(3), 265–294.

Gauthiez, B. (2004). The history of urban morphology. *Urban Morphology,* 2(8), 71–89.

Gouda, S. M. (2015). *Towards a sustainable preservation approach to Egyptian heritage neighborhoods*. Technical University of Dortmund.

Hefnawy, H., Fayoumi, M. E., Elshater, A., & Gulsrud, N. M. (2022). Urban heat Islands in degenerative and regenerative cities: Heliopolis, Cairo, Egypt. In H. Abusaada, A. Elshater, & M. Khalifa, *Remapping urban heat Island atlases in regenerative cities* (pp. 145–168). IGI Global. https://doi.org/10.4018/978-1-6684-2462-9.ch007

Hossam, M. A. (2019). *Urban mobility and urban form towards a sustainable neighborhood*. Arab Academy for Science, Technology and Maritime Transport.

Ilbert, R. (1985). Heliopolis: Colonial enterprise and town planning success? In A. Evin (Ed.), *The expanding metropolis: Coping with the urban growth of Cairo* (pp. 36–42). Concept Media/Aga Khan Award for Architecture.

Jiron, P. (2008). *Mobility on the move: Examining urban daily mobility practices in Santiago de Chile*. London School of Economics and Political Science.

Jones, P. (2014). The evolution of urban mobility: The interplay of academic and policy perspectives. *IATSS Research, 38*(1), 7–13.

Krier, R. (1979). *Urban space* (1st ed.). Academy Editions.

Kropf, K. (1996). Urban tissue and the character of towns. *Urban Design International, 1*(3), 247–263.

Kropf, K. (2013). What is urban morphology supposed to be about? Specialization and the growth of a discipline. *Urban Morphology, II*(17), 128–131.

Larkham, P. J. (1998). Urban morphology and typology in the United Kingdom. In A. Petruccioli (Ed.), *Typological process and design theory* (pp. 159–177). Aga Khan Program for Islamic Architecture.

Larkham, P. J., & Conzen, M. P. (2014). The making of urban landscape. In P. J. Larkham & M. P. Conzen (Eds.), *Shapers of urban form: Exploration in morphological agency* (pp. 3–24). Routledge.

Lynch, K. (1960). *Image of the city*. The M.I.T. Press.

Mansy, K. (1991). الطابع المعماري و العمراني و نظم التحكم في العمران في مصر *[Architectural and urban character: Urban controlling systems in Egypt]*. Helwan University.

Marzot, N. (2002). The study of urban form in Italy. *Urban Morphology, II*(6), 59–73.

Mohamed, D. H. (2021). Cairo: An Arab city transforming from Islamic urban planning to globalization. *Cities*, 1–10.

Moudon, A. V. (1997). *Urban morphology as an emerging interdisciplinary field*. University of Washington.

Moudon, A. V. (1998). The changing morphology of suburban neighborhoods. In A. Petruccioli (Ed.), *Typological process and design theory* (pp. 141–157). Aga Khan Program for Islamic ArchitectureAttilio Petruccioli.

Nasar, J. L. (1994). Urban design aesthetics: The evaluative qualities of building exteriors. *Environment and Behavior, 26*(3), 377–401.

Norman, J. (2019). A time of unprecedented change in the transport system. The future of mobility. In G. O. Science (Ed.), *A time of unprecedented change in the transport system. The future of mobility* (pp. 3–4). UK Government.

Noulas, A., Scellato, S., Lambiotte, R., Pontil, M., & Mascolo, C. (2012). A tale of many cities: Universal patterns in human urban mobility. *PLoS ONE, 7*(5), 1–10.

Nowier, M. M. (2014). *Urban space and politics of transition in contemporary Cairo*. Ain Shams University; University of Stuttgart.

Rodrigue, J.-P. (2018). *The geography of transport systems* (3rd ed.). Routledge.

Sadeghi, G., & Li, B. (2019). Urban morphology: comparative study of different schools of thought. *Current Urban Studies, I*(7), 562–572.

Shafik, N., Mansour, Y., Kamel, S., & Morcos, R. (2021). The impact of the Cairo streets development project on the independent mobility of children: A field study on the streets of Heliopolis, Egypt. *Infrastructures, 6*(98), 1–20. https://doi.org/10.3390/infrastructures6070098

Shamsuddin, S., Sulaiman, A. B., & Amat, R. C. (2012). Urban landscape factors that influenced the character of George Town, Penang Unesco world heritage site. *Procedia—Social and Behavioral Sciences, 50*, 238–253.

Sheller, M., & Urry, J. (2000). The city and the car. *International Journal of Urban and Regional Research, 24*(4), 737–757.

Sisiopiku, V., & Akin, D. (2003). Pedestrian behaviors at and perceptions towards various pedestrian facilities: an examination based on observation and survey data. *Transportation Research Part F: Traffic Psychology and Behaviour*, 249–274. https://doi.org/10.1016/j.trf.2003.06.001

StampsIII, A. A. (1994). A study in scale and character: Contextual effects on environmental preferences. *Journal of Environmental Management*, 223–245.

Wanas, A. (2013). Political drifts: Implications upon urban space and place with special reference to Cairo, Egypt. *International conference on cities, people and places*. University of Moratuwa.

Whitehand, J. (2001). British urban morphology: The Conzenian tradition. *Urban Morphology, 2*(5), 103–109.

Yu, Z. (2013). *Regulation of urban character: Style, colour and historic character in a modern Chinese City—The case of Harbin*. University of Manchester.

Enhancing the Sense of Identity in the Public Places of Administrative Areas in Greater Cairo

Omnia Ali, Yasser Mansour, Abeer Elshater, and Ayman Fareed

Abstract

Contemporary urban design, as well as sustainable development calls, pays serious attention to the notion of identity in the urban context. Furthermore, there is an increase in scientific research in this field. This paper revisits identifying place identity components. Therefore, reliable literature and research articles were reviewed to identify the components that can create place identity. The paper empirically analyzes the checklist of place identity components by comparing two case studies. As identity is affected by the place's culture and the nature of the people in it, this paper focuses on public places in administrative areas in Greater Cairo. This comparison was made using the collected data from documents and archives and the observation during the site visits. An interview with experts was also done. Based on the data, the study assessed the sense of identity in the two cases. According to the results, this study identified the strengths and weaknesses in each assessment category, then presented identity enhancing strategies (IES) to enhance identity. The findings came in three categories, environment-related strategies, people-related strategies, and strategies for the relationship between the place and its people. This approach adopted by the study can be used to derive other strategies for different places.

Keywords

Place identity · Enhance identity · Measuring identity · Identity components · Public place · Greater Cairo

1 Introduction

Identity has consistently been a popular research subject in various disciplines, such as psychology, philosophy, sociology, and human geography. In the last few decades, urbanization and globalization processes caused rapid environmental changes, and their strategies created cities with a similar, repetitive, and monotonous image. This influence included the existing and newly established cities. The old cities have changed to accommodate the new needs, whereas the new cities, in most cases, are indifferent to projecting their own identity. This situation led to the notion of identity becoming an essential topic on the agendas of planners and designers; several UN-Habitat reports confirm this (UN-HABITAT, 2012, 2013a, 2013b, 2018). As public places life is a major significant aspect in contemporary urban design, the interest in the quality of public places is one of the focuses of attention in sustainable development and the New Urban Agenda (NUA), which clearly states that urbanization needs to focus on the three elements of urban planning policies: standard of living, quality of life, and well-being (UN-HABITAT, 2020). The latter is a complex but comprehensive notion, where identity, or Eigenart, is one of the various attributes of well-being (Shekhar, 2017). For scientific research, the literature shows a growing interest in urban identity. A recent study (Peng et al., 2020) through 1011 bibliographic records from the Web of Science core database from 1985 (the first year for data to be available in the database core) and 2019 is surveyed. Using CiteSpace (a scientometric tool for analyzing and visualizing patterns and trends in scientific publications), this study shows that academic journals have published

O. Ali (✉) · Y. Mansour · A. Elshater · A. Fareed
Engineering Faculty, Ain Shams University, Cairo, Egypt
e-mail: Omnia.monier@hti.edu.eg; omnia.monir@gmail.com

Y. Mansour
e-mail: yasser_mansour@eng.asu.edu.eg

A. Elshater
e-mail: abeer.elshater@eng.asu.edu.eg

A. Fareed
e-mail: aymanfaridafh@eng.asu.edu.eg

O. Ali
Higher Technology Institute, 10th of Ramadan City, Ramadan, Egypt

M. F. Shahidan et al. (eds.), *Urban Narratives: Exploring Identity, Heritage, and Sustainable Development in Cities*, Advances in Science, Technology & Innovation, https://doi.org/10.1007/978-3-031-48517-6_13

more articles about place identity over the past 40 years, particularly since 2006. Despite this momentum in research, a theoretical contribution is needed to connect the different perspectives of urban identity to define specific elements that create and affect the place identity.

At the regional level, the recent monumental urbanization changes in Egyptian cities raise concerns about the quality of the identity of its public places. Although there are many masterful buildings and brilliant architecture in Egyptian cities, particularly Cairo, the capital city of Egypt, these changes raise a concern about identity and summon the need to intensify the studies to determine their impact on the quality of identity in its places. The motivation behind choosing Cairo for studying was that it is the largest urban area in Egypt, Africa, the Middle East, and the Arab world, and the thirteenth in the world (Demographia, 2020). Since identity is affected by the place's culture and the nature of the people in it, this paper focuses on public places in administrative areas in Greater Cairo. Where this study adopts the opinion that the most important of urban identities is the experience within urban places (Cheshmehzangi, 2020), it is necessary to consolidate and enrich this experience. Therefore, this study examines the enrichment of this experience through enhancing the identity. Hence, this paper is intended to present specific strategies to enhance the sense of place identity in Cairo case studies by answering three questions:

What are the place identity components that the literature shows?
How can these components be measured in case studies and identify strengths and weaknesses?
What is identity status in case studies, and how can it be enhanced?

To this end, the present paper is structured into three stages to reach the research objective. The first stage is theoretical, which starts with this introduction, then includes a background of identity in the relevant literature to identify place identity components, and ends with a reliable tool (checklist) to assess the identity of public places based on the literature review. In stage two (Field study), the methodology of this study is described in three steps: the criteria for selecting a case to study, types of data resources, and finally, set up the evaluation methods of case studies. Then, the paper describes and assesses the two selected cases for studying to evaluate the identity of those places. In the third stage, positive and negative aspects of identity in each case are explored based on the matrix assessment. Then, the evaluation results are discussed through a comparative analysis to present specific strategies for enhancing identity in similar places. Finally, the conclusion states the whole process of establishing identity enhancement strategies (IES) for enhancing the sense of identity of public places in the administrative areas of Greater Cairo, based on the previous discussion.

2 Research Background

2.1 Place Identity Overview

Humans use the term identity to define their origin, race, language, belief, class, esthetics, behavior…, and most importantly, their selves. Thus, the notion of identity is a common term in various sciences and defined from different perspectives. Likewise, place identity is a versatile concept studied in multiple disciplines, such as geography, sociology, psychology, environmental sciences and ecology, public administration, spatial planning, and more (Peng et al., 2020). In the 1960s, place identity studies appeared as a scientific subject. It was introduced by Lynch (Chapman & Lynch, 1962) when he said that the identity of a place is a response to the identification and recognition of objects in space, primarily to distinguish one from another. He also asserts that the meaning of the objects can be both practical or/and emotional. Place identity was also introduced by Proshansky (1976) as "those dimensions of self that define the individual's personal identity concerning the physical environment through a complex pattern of conscious and unconscious ideas, feelings, values, goals, preferences, skills, and behavioral tendencies relevant to a specific environment." For Relph (1976), however, the place is a humane environment with a mixture of meaningful significance and social concerns; thus, we can elucidate that place is more than a location. Successively, the researchers enriched the studies related to the identity of the place. According to a survey about the active topics and new developments of publications that study place identity, studies in this field experienced an active exploration in plural disciplines after 2000 (Peng et al., 2020).

In the urban context, some literature distinguished place identity into "insiders" and "outsiders" as types of place identity, where the "essence of place" is divided into the occasionally unconscious experience of an "inside" as opposed to an "outside." (Norberg-Schulz, 1971). Consequently, it comes in numerous purposes and forms. Place identity comprises three components essential for an intersectional understanding: activities and functions, physical features, and meanings and symbols. These elements are intertwined, showing an interconnected complexity. (Relph, 1976). Some other literature divided the identity of place into layers, extending from a person to the entire group in four layers; the personal layer: self-image; the enactment layer: performance; the relational layer: social interaction; finally, the communal layer: existing within a group (Hecht et al., 2005). This differs from the literature that divides urban identity into levels: the global outlook or "global level," the urban setting or "macro level," the environmental framework or "medium level," and the personal perspective or "micro-level" (Cheshmehzangi, 2020). The place identity incarnates in the third level (the medium level). The experience and events within this level are the most critical elements of urban identities, as the image remains in mind, and the experience is in the heart.

The dimensions of identity have been studied from different perspectives as well. In sociology, identity has four dimensions; personal: related to a humane perspective; social: related to human-society relations; cultural: related to human and everyday life; and locational: related to human–environment relations. Thus, one may embrace multiple identities. He can have a specific personality with a set of esthetics and characteristics while at the same time being associated with the social world, defined in groups, classes, and cultures, and inter-relate with various contexts (Cheshmehzangi, 2020). The literature discussed the dimensions of place identity in different ways. Lefebvre's three key identity factors are self, reality, and relations (Lefebvre, 1991). Consequently, identity was defined in three intertwined dimensions that refer to the essence of human–environment relations, perceptual or psychological, behavioral, and social (Cheshmehzangi, 2020). A recent study summarized place identity in four primary dimensions: place, people, external look, and internal thoughts (Peng et al., 2020).

In sum, the identity of place is a set of various and reciprocal relationships between the context and the contents. It is a set of relationships, a social relationship between people and their surroundings, place and their constituent elements, and between elements and each other. See Fig. 1 for the summary of the main topics in this section.

2.2 Place Identity Components

This section will list identity components from many scholars' perspectives. Reliable studies demonstrate that scholars define the components of place identity in diverse categories. Relph (1976) comprehensively analyzed place identity and outlined three main elements that contribute to its formation: observable activities and functions, physical features or appearance, and meanings or symbols. He posits that a sense of place is created through experiencing an inside distinct from an outside. For Lefebvre (1991), it was self, reality, and relations. Shekhar (2017) listed elements enhancing urban identity based on Lalli (1992). These elements were land ownership that impacts positively on the identity of place for a citizen and in like manner, the place of birth, social relationships, and quality of place (amenities and facilities). Lalli also highlighted the main elements that Lynch mentioned in the image of city (Chapman & Lynch, 1962), such as a landmark, the history of a place, and its esthetics (visual form and appearance). All of the previous elements positively impact the identity of the place; in addition, the duration of stay impacts positively but not very significantly. In contrast, pollution and noise negatively impact the identity of the place.

The literature also shows other approaches to identifying the elements of urban identity. Rapoport (1990) argued that symbols and signs play the leading roles in creating the perceptional meaning of the places users. Thus, environments can be studied through signs that guide behaviors, affective signs that elicit feelings, and symbols that influence thoughts. In contrast, Valera (1998) discusses urban identity from a social perspective. For him, the character of the place and its meaning for users can be shaped through environmental and social imageability attitudes.

Moreover, available studies assessed identity through the cultural landscape elements. Ziyaee (2018) suggests three descriptive categories—materials, immaterials, and links—to summarize the key components of the cultural environment; see Table 1. Surchi and Nafa (2021) explored the interplay between place identity elements and cultural landscape elements, examining both tangible components of identity and culture. Their model was used to assess this connection. (See Fig. 2).

These different approaches and more have been discussed in a previous recent study by the same authors of this paper (Ali et al., 2022), who have discussed these previous approaches and more. Elements and components of place identity mentioned in the literature have been organized into three classifications: environment, people, and interaction between people and place, under two forms: tangible and intangible (See Fig. 3). These components will be used as a checklist to assess the identity of place in two Egyptian case studies to re-acquaint them through the field study.

3 Materials and Methods

3.1 Case Selection

The research is interested in examining the identity of two public places in administrative areas in Greater Cairo. The study sets three conditions for selecting the study case, classified as a public place, located in an administrative area, and existing in the Greater Cairo region (the capital). These conditions are defined as follows:

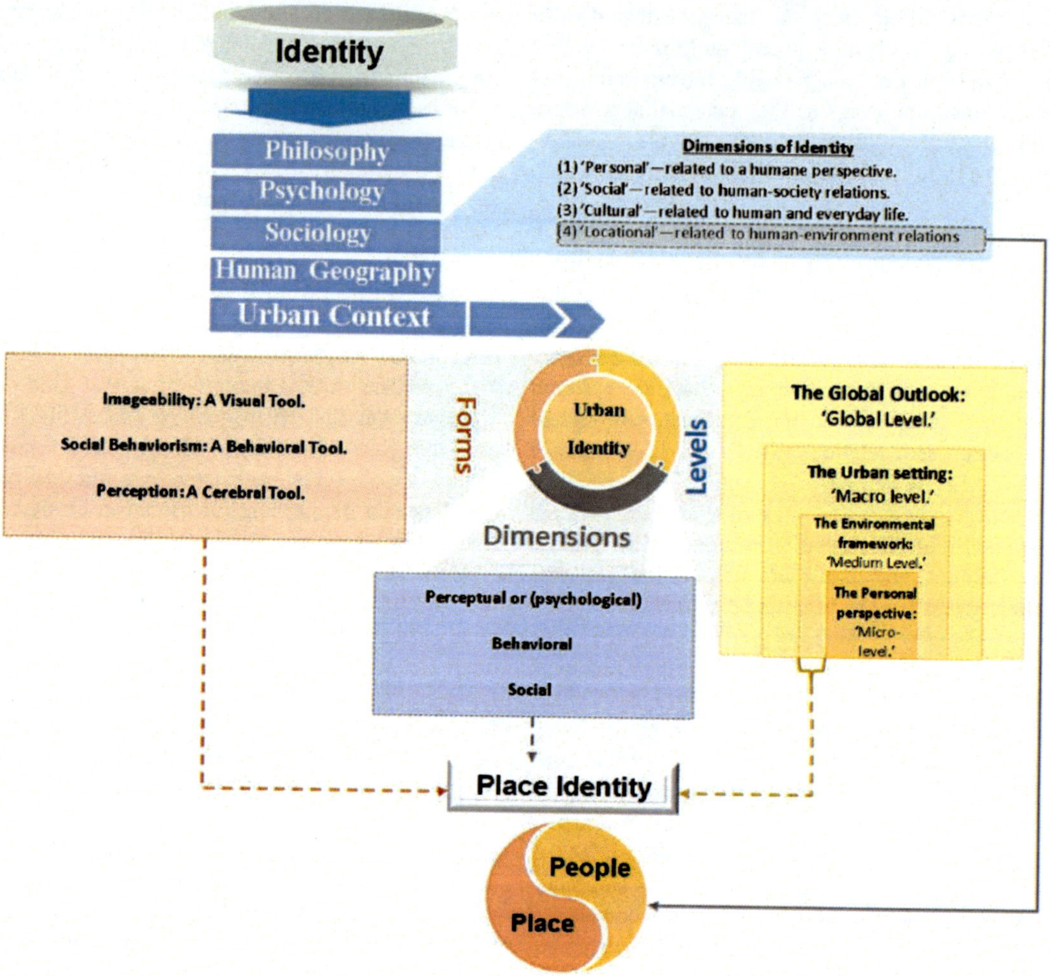

Fig. 1 Place identity overview. *Source* The Authors

Table 1 Components of cultural landscapes

Co.	Main elements	Sub-elements
Materials	Natural forms	Topology, geography, hydrography. Texture and surface materials, vegetation. The climatic situation, deserts, seaside
	Manmade forms	Buildings, roads, bridges, monuments, urban artifacts, gardens, agricultures
Immaterials	Beliefs	Values, visions, religions, ideologies
	Rules	Policy, economy, power
	Behaviors	Social practices, place name, patterns, symbols, lifestyle, icons, meanings, activities, myths, stories
Links	Time/process	History, memory
	Method and technique	Style, dance, paintings, clothes, foods, singing

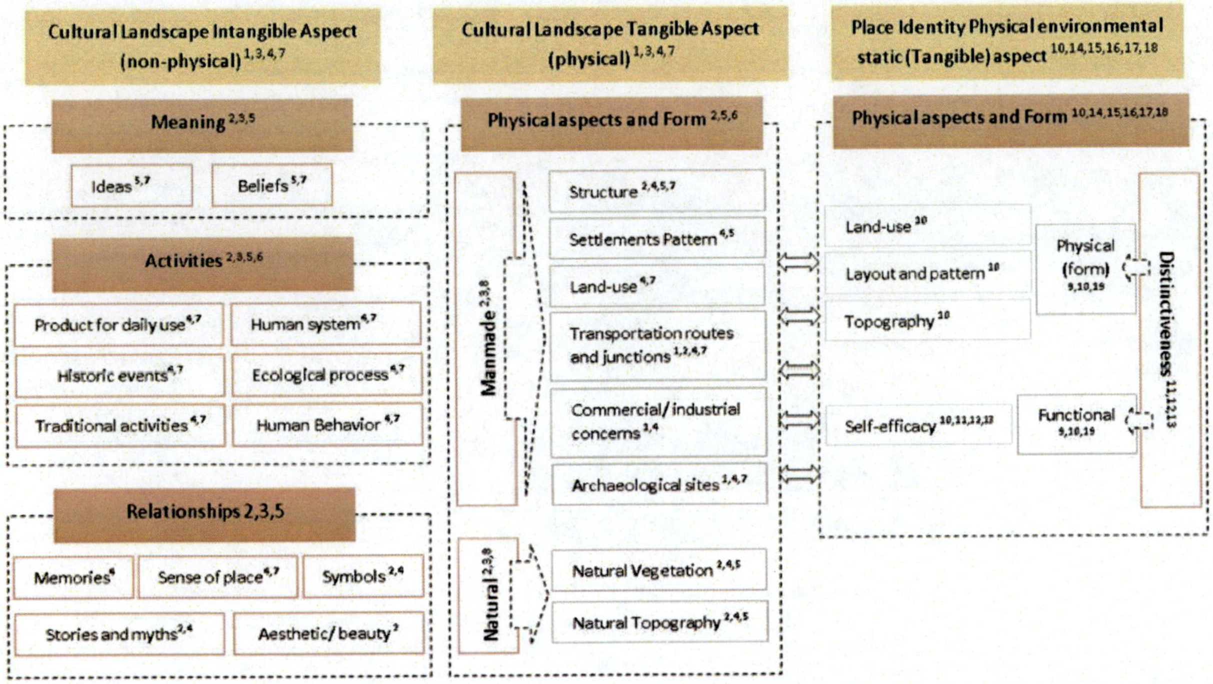

Fig. 2 Suggested model for the relationship between culture landscape and place identity. *Source* The Authors based on (Surchi & Nafa, 2021). (1) Brown (2001), (2) Stephenson (2008), (3) UNESCO and World Heritage (2008), (4) O'Donnell and Fasla (2008), (5) Vogeler (2010), (6) Jones (2003), (7) Sonkoly (2017), (8) Rees and Cosgrove (1986), (9) Kermani et al. (2016), (10) Ibrahim et al. (2014), (11) Stokols and Shumaker (1981), (12) Breakwell (2015), (13) Twigger-Ross and Uzzell (1996), (14) Shawesh (2000), (15) Greene (1992), (16) Al-naim (2008), (17) Lynch (1981), (18) Tomlinson (2003), (19) Ziyaee (2018)

Public Places: These can be defined as physical spaces that, in the ideal, are open to all people to exercise their rights (Childs, 2006). In urban planning, public space has generally been characterized as an open space and ranges from informal alleys to the main arenas of the urban environment. Widely, formal public spaces play a significant role as presumed meeting centers and venues for public life, programs, and events. On a smaller scale, they might be just a place of rest, hanging out, or playing while providing a visual pause in the street flow through the urban area (Carmona, 2019). From another perspective, public areas must be open to everyone regardless of their culture, religion, or social status. In this context, Francis (1989) considered three types of access necessary to name any area as a public place. The first is physical access, where doors, walls, and locked passages block the accessibility of the space physically. The second type is social access, which means the place is open to different kinds and groups of users. The third type is visual access. If citizens can see inside any public area, it is visually accessible.

Administrative Area: According to the General Authority for Urban Planning of Egypt (General Authority for Urban Planning, n.d.), the government constructed the administrative area. It has different buildings, such as the House of Representatives, senate, ministries, and governmental agencies. Besides, it may have financial and business companies, provincial city hall, local councils and units, popular councils for governorates, cities, villages, and neighborhood headquarters. Public services directorates (education, health, social, security, supplying, etc.) and the tax authority also might be there.

Greater Cairo (GC): Egypt's political and economic capital. It comprises three governorates: Cairo governorate, Giza governorate, and Qalyubia governorate. It is one of the fastest-growing megacities worldwide, the sixth-largest city in 2018 with a 21.581 million population and the fifth in 2030 with a 25.517 million population (United Nations, Department of Economic and Social Affairs, Population Division, 2018). It has a historical administrative center, Downtown (DTC), surrounded by old suburbs. These suburbs are bordered by the ring road and satellite cities, Sheikh Zayed and the Sixth of October in the west, and New Cairo, Obour, El-Shorouk, and Badr in the east, in addition to the new administrative capital (NAC) that has currently constructed with a central administrative district as well.

Considering the three conditions, the study selected two cases to study (Fig. 4). The first case is Ministries Square

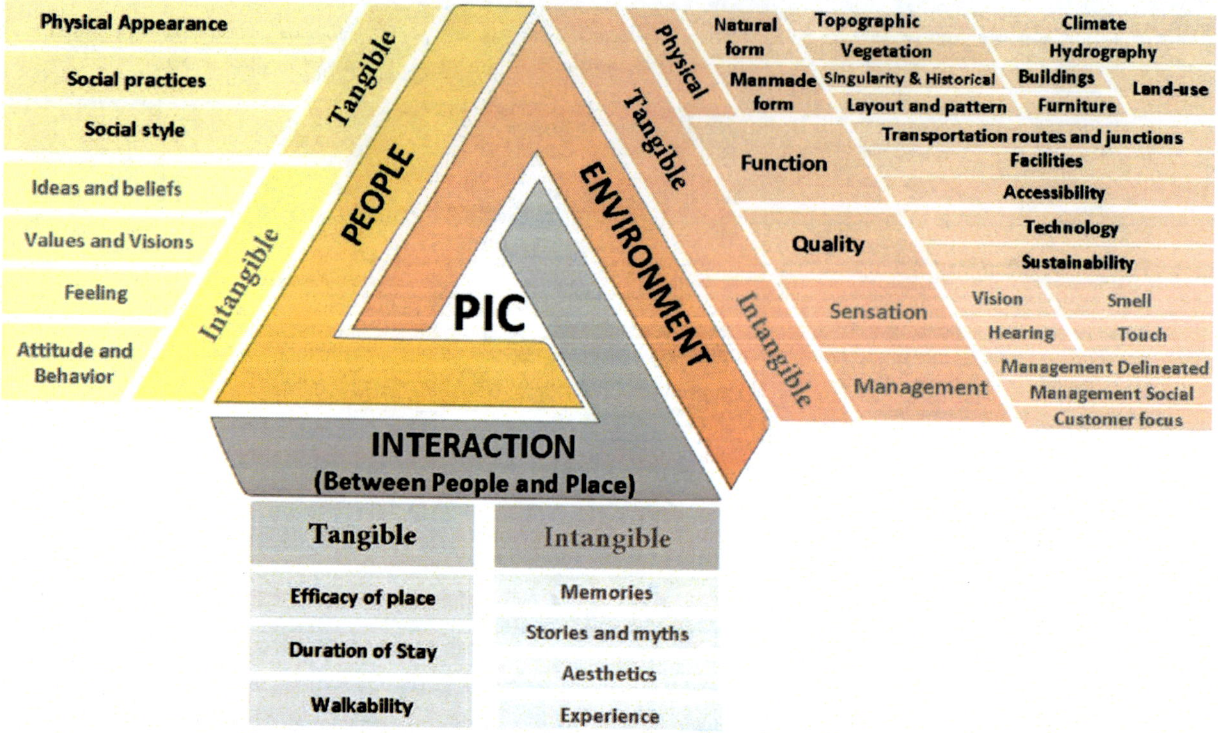

Fig. 3 Place identity components (PIC). *Source* The Authors based on (Ali et al., 2022)

Fig. 4 Location of the two case studies: The Ministries District at Downtown Cairo (DTC) and new administrative capital (NAC). *Source* The Authors

in the downtown area of Khedivial Cairo (DTC), while the second is the governmental district in the new administrative capital (NAC). These two places have the same function as the central ministries' buildings will be moved from the Ministries Square Downtown to the Ministries District in the new administrative capital. Ministries Square contains heritage buildings and buildings of a distinguished nature (palaces converted into public buildings) (Hawwas, 2002; UNESCO-WHC, 2012). Thus, Downtown Cairo was classified as "a protection range," according to National Organization for Urban Harmony (NOUH) (2022). It has also buildings built as governmental buildings. Opposing the ministerial district was built in the new administrative capital as a governmental district from the beginning. So, every case reflects its identity differently.

3.2 Data Sources

This study aims to explore practices for promoting a stronger sense of identity in public places of the administrative districts in Cairo and evaluate the current levels of identity present in two chosen case studies from Cairo. In light of the literature results, identity composition elements (environment, people, and interaction between them, in tangible and intangible forms) were used to evaluate the sense of identity in the two case studies. The data of study cases were collected from three primary sources:

- **Documents and Archives**: Due to the sensitivity of the study cases as they are central government places, the study will depend on all available data about the area and its buildings from previous studies, government reports, and the competent authorities, in addition to the government's websites.
- **Site Inventory**: It involves collecting scenes and impressions through four visits to each site on working days, morning and afternoon. The scenes' choices were based on atmospheres in the public place, facial details of users, and their attitudes, in addition to physical traces. Recording notes and cameras were used to take images (whenever possible) documenting the scenes, outdoor activities, and life in public places.
- **Interviews with Experts**: It was challenging to survey users, given the few visitors to the new administrative capital as it is nascent. So, the study relied on expert interviews (focus groups) to explore the users' opinions of the two study cases. The first interview was with four members of the Higher Scientific Committee of the National Organization for Urban Harmony to discuss the identity of public places in case studies and the factors that influence and constitute it. The second interview was with three real estate brokers, specifically those who market the administrative units in the new administrative capital, to know the elements of attraction and interest of clients (future users) and the future life the place will give. In addition, their personal opinions as visitors to the place.

3.3 Methods Setup and Analysis

Since the research aims to enhance a sense of identity in study cases, the field study was conducted in three steps:

- **The First Step (Description)**: The collected data from documents and archives, the observation during the site visits, and interviews were used to describe each case study's main feature for outlining the place's public image.
- **The Second Step (Evaluation)**: The collected data from documents and archives, the observation during the site visits, and interviews were used to assess the sense of identity by evaluating all elements of place identity components (checklist) in each case study. The components are evaluated based on their contributions and are divided into three scours: "1" for element contribution positively and "−1" for negatively, while "0" is for the neutral or non-contribution element.
- **The Third Step (Analysis)**: Based on the evaluation in the second step, a SWOT analysis was done to identify strengths and weaknesses that affected the quality of identity in the two study cases to design the identity enhancement strategies (IES).

4 Case Study Setting

4.1 Case 1: Ministries Square in Downtown Cairo (DTC)

The Downtown Cairo, or Khedivial Cairo, is to its founder Khedive Ismail. It has a rich historical value. According to (NOUH), this area has 633 heritage buildings with distinctive values (National Organization for Urban Harmony, 2022) on about (6 km^2) on the eastern bank of the Nile (UNESCO-WHC, 2012). This area combines European styles with local materials and features designed by famous national and international architects; thus, it was considered a turning point in urban design in Egypt in the nineteenth century (Hawwas, 2002). This area was occupied by the wealthy class, making it an area for the most profitable and international activities, and through time, it became the central governmental area.

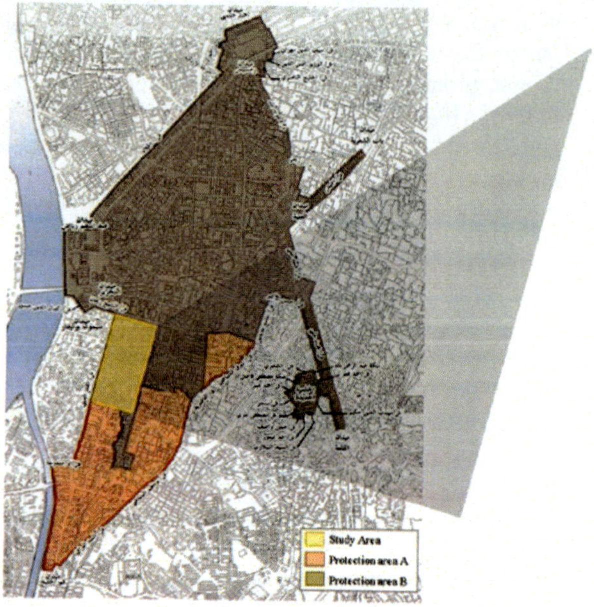

Fig. 5 Protection zones map in Khedive Cairo. *Source* Adapted by the Authors based on (National Organization for Urban Harmony, 2022)

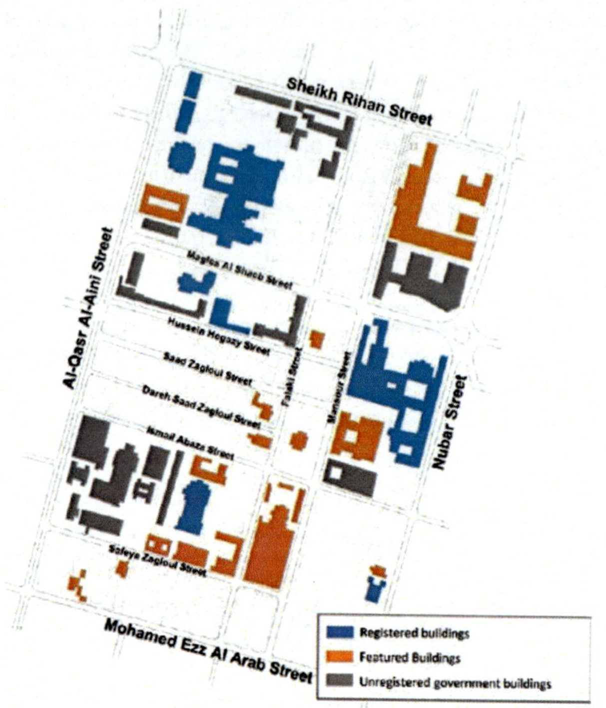

Fig. 6 Types of governmental buildings in the study area. *Source* The Authors based on (The National Committee for the Development and Protection of Heritage Cairo, 2019)

4.1.1 Description

The study area is in protection zone B in Khedivial Cairo; see Fig. 5. The area is named the "Ministries Square" by The National Committee for the Development and Protection of Heritage Cairo. It is about 900 × 500 m, occupied by ministerial and administrative buildings scheduled to be moved to the new administrative capital. The region's borders are Sheikh Rihan Street from the north, Nubar Street from the east, Muhammad Ezz Al-Arab or Al-Mubtadayan Street from the south, and Al-Qasr Al-Aini Street from the west. The area is also crossed by six horizontal streets: Magles Al Shaeb, Saad Zaghloul, Dareh Saad Zaghloul, Ismail Abaza, and Safia Zaghloul, and vertically, Falaki Street and Mansour Street. The technical committee divided the buildings in this sector into three types: buildings registered as antiquities, others as distinctive architectural styles, and third as unregistered government buildings, see Fig. 6.

"The Ministries Square" contains many ancient and historical buildings, such as the Scientific Complex, the Egyptian Geographical Society, the Council of Ministers, the House of Representatives, the Ministry of Health, the Ministry of Housing, the Ministry of Education, the House of the Nation, the mausoleum of Saad Zagloul, and the Ministry of Justice. These buildings are lofty with their unique designs (Fig. 7).

The interviewees from the Higher Scientific Committee of the National Organization for Urban Harmony maintained that the main problem in the area is that it was fundamentally built as an upscale neighborhood for the wealthy class. Therefore, the buildings of most of the ministries are residential palaces that have been converted into ministries. So, governmental administrative buildings were built in the gardens and courtyards of those palaces to adapt to new functions' needs, which led to an increase in the building density of the area, thus reducing the visual perspective and the open space scenes and, unfortunately, limiting the enjoyment of the region's unique architecture. The interviewees emphasized that the essential element of identity is the building, as buildings translate the nature of the community in which the building is built. The difference in the urban fabric in the region indicates the different patterns of society during those eras. The interviewees added that the identity without people occupying the place is not accurate and incomplete.

From the site visits, the authors found that most of the area's buildings were mixed-use, with administrative, commercial, and residential overlapping activities in them. There are street vendors, especially near the Saad Zaghloul metro station. The health and judicial services areas were overcrowded by citizens seeking these services. The possibility of walking in the place is good, as the sidewalks and traffic lights intersect, and the nature of the urban fabric

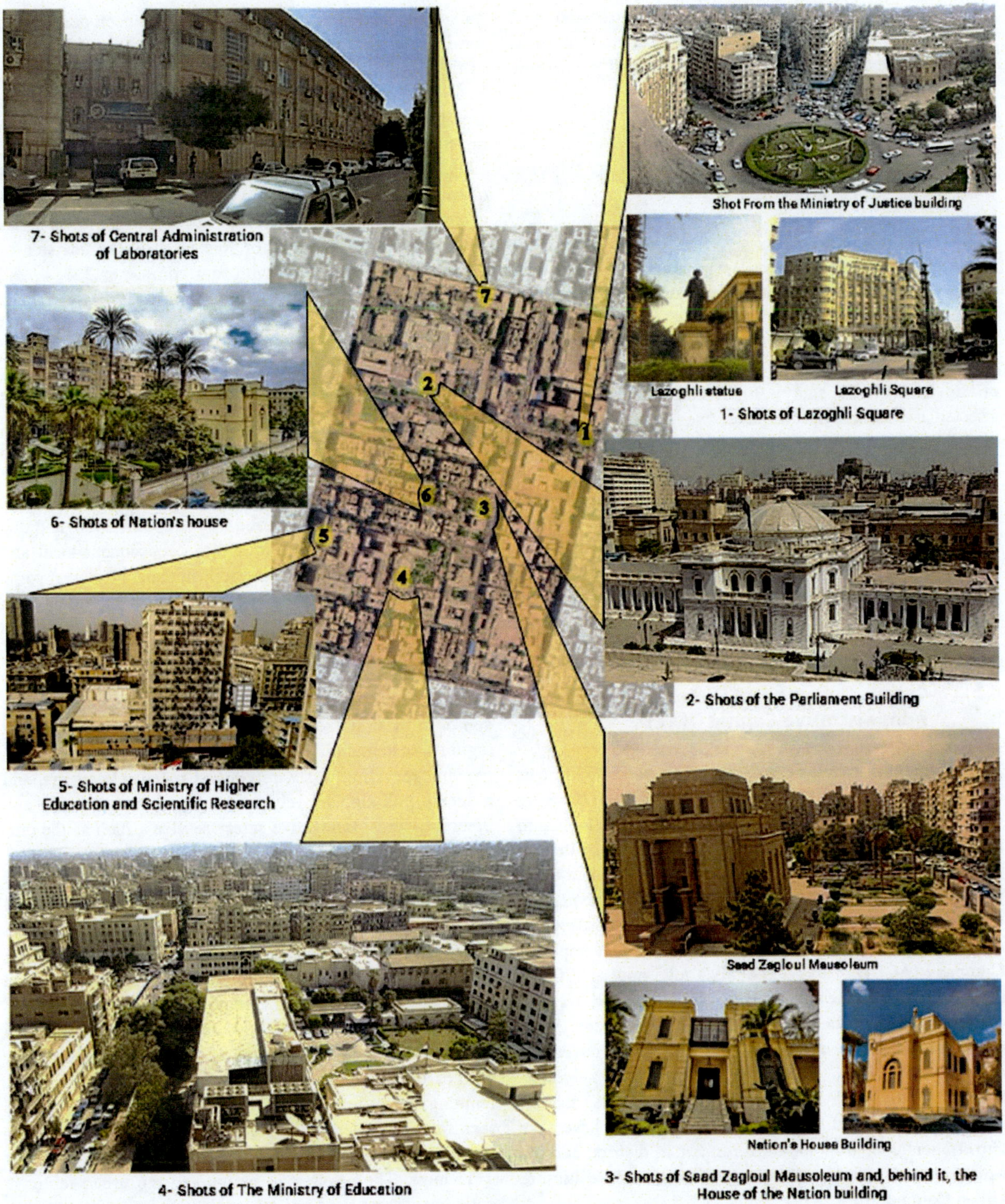

Fig. 7 Panoramic shots of the outdoor areas. *Source* The photograph was taken in the last two years, from 2020 to 2022, The shot from (https://www.google.com/maps), and it represents the current situation, according to the author's observation of the site. Edited by The Authors

provides shade for the sidewalks. As for transportation, it varied between bus and metro stations and, more recently, bicycle stations (Cairo bike), as the region had two nearby bicycle stations.

4.1.2 Evaluation

Based on the place identity components checklist, this case study is evaluated as given in Table 2.

4.1.3 Analysis

According to the two previous steps (description and evaluation), strengths, weaknesses, opportunities, and threats can be determined as follows, see Fig. 8. The results of the assessment came to show that 22 components had a positive impact on enriching the identity of the place (strengths of the case of the study). Eight components negatively affected the identity in the place (weaknesses of the case of the study). And, five components did not negatively or positively affect identity (potential opportunities to improve identity in the case of the study). Perhaps, the recurring impediment as a cause of weakness in several identity components was the overlapping activities in the area's buildings, overcrowding, and the new buildings in the yards of heritage buildings that led to the distortion of the area's architecture (threats in the case of the study).

4.2 Case 2: Government District in the New Administrative Capital (NAC)

The new administrative capital was established 35 km east of Cairo to distinguish the location as it is close to the Suez Canal area, regional roads, and main axes to develop Cairo into a leading political, cultural, and economic center for the Middle East and North Africa region through a prosperous economic environment supported by diversified economic activities and to achieve sustainable development (The Administrative Capital for Urban Development Company (ACUD), 2017). The city's total area is 170,000 acres, the target population is 6.5 million, and the job opportunities generated are about 2 million (New Urban Communities Authority, 2022). The first phase covers an area of 40,000 acres. It contains six residential neighborhoods and the most important ongoing projects, including the Business and Money District, the Green River, the Governmental District, the third residential district, and the fifth residential district (Ministry of Housing Utilities & Urban Communities, 2022).

4.2.1 Description

The government district, "The Ministries District," in the new administrative capital, consists of 34 buildings. It includes ten complexes, each containing a group of ministries. Some of these complexes have been designed in the Islamic style (such as complexes no. 5 and10), others in the Egyptian Pharaonic style (such as complexes no. 2, 3, 4, 7, 8, and 9) as shown in Fig. 9, in addition to the sovereign nature of the Ministry of Foreign Affairs (complexes no. 6), the Ministry of Justice (complexes no. 1), and basically, the Council of Ministers. In addition, the government district contains a place called "Plaza," which is a yard to entertain employees, and the government district includes many parks. The Ministry of Justice is on the left side of the Council of Ministers, with four complexes 2, 3, 4, and 5. The Ministry of Foreign Affairs building is on the right of the Council of Ministers, with four complexes 7, 8, 9, and 10 (see Fig. 10).

The area of one complex within the government district is 170 thousand square meters, with a total of 1.5 million square meters for the ten complexes, which is equivalent to 360 acres out of 40 thousand acres, which is the area of the first stage of the new administrative capital (Ministry of Housing Utilities & Urban Communities, 2022). The ministries were due to be transferred by the end of 2020, and due to the COVID-19 pandemic and the consequent local and global repercussions, it has been postponed. However, some ministries are partly operating there during this research in 2022.

The interviewees, from real estate marketing experts, emphasized the importance of the view of the iconic tower and the central area as the prices of projects rise as we approach these influential areas. They also emphasized the region's sustainability by recycling water and using it in gardens, recycling waste, relying on renewable energy, and reducing pollution by relying on green transportation and other measures. Innovative safety services, such as the citizen application, are also available in the capital, making it easier for citizens to obtain intelligent and fast security services. They also confirmed that most of what was promised when establishing the new capital has become a tangible reality and that the mental image of the capital is strong enough to become the latest icon for Egypt that will be used as branding for Egypt soon abroad.

From the site visits, the authors found that the scene is open and highlights the important surrounding projects, such as the Masr mosque, People's Square, the parliament building, the opera, and the iconic tower. All of this can be seen from the plaza of the place. Walking around and moving between buildings was very difficult as the distances were huge. The weather is sunny, and the urban fabric is not compact and does not provide shade for passers-by, see Fig. 11.

4.2.2 Evaluation

Based on the Place identity components checklist, this case study is evaluated as given in Table 3.

Table 2 Sense of identity evaluation—Case study 1

Classification			Element		Description	Mark
Environment	Tangible	Physical	Natural forms	Topography	No natural or open views, so this is non-contribution element	0
				Vegetation	Although there are gardens and plants, it is few and limited to its buildings. And because vegetation is essential in public places, the loss of this element negatively affects the evaluation	−1
				Climate	The streets are narrow and shaded, and the weather is fresh, so this element's effect is positive	1
				Hydrography	There are no hydrography elements in the place despite the proximity of the Nile River, so this is non-contribution element	0
			Manmade forms	Singularity and historical	The place has landmarks, individual semantics, and symbols. The historical image is present, so this element's effect is positive	1
				Layout and pattern	The place is designed in a radial style and respects the human scale. The urban fabric is compact and integrated, so this element's effect is positive	1
				Buildings	Unique cultural style, heritage, and historical buildings. Buildings are one of the most important elements of strength in this place, so this element's effect is positive	1
				Furniture	Rich urban artifacts: sculptures, lights, paving, etc., so this element's effect is positive	1
		Function	Land-use		There are archeological and unique places, so this element's effect is positive	1
			Transportation routes and junctions		Various transportation declared and clear conductors; subway, bikes, buses, and taxis, so this element's effect is positive	1
			Facilities		Infrastructure is adaptable with time and history, so this element's effect is positive	1
			Accessibility		Vividness and openness, free, secure, and pedestrian friendly, so this element's effect is positive	1
		Quality	Technology		There are no elements of intelligent technology in the place. While the reliance on modern technology services is essential in enhancing the experience in public places today, the loss of this element negatively affects the evaluation	−1
			Sustainability		There is a subway, and stations and roads for bikes. Nevertheless, sustainability applications were insufficient, so that this element would be neutralized	0
	Intangible	Sensation	Vision		The mental image is singular and unforgettable, so this element's effect is positive	1
			Hearing		Most of the time noisy, because of the public services there, traffic, and mixed-use building. As it is known, noise has a negative effect	−1
			Smell		Pollution in the area is high due to crowding and various activities in the place. Pollution has a negative effect	−1
			Touch		The texture of the place is diverse and valuable, so this element's effect is positive	1
		Management	Delineated		Ownership of places varies between private, public, and endowments. So, this is a non-contribution element	0
			Social		The place encourages social engagement. It is a positive element	1
			Customer focus		The human spatial experience is dispersed, and the convenience of users and recipients of services is not focused on them. The loss of this element negatively affects the evaluation	−1

(continued)

Table 2 (continued)

Classification			Element	Description	Mark
People	Tangible		Physical appearance	Uncomfortable. The residents and owners of shops and non-administrative activities are present, negatively affecting the evaluation	−1
			Social practices	Although there are cafes and public spaces in the place, they are limited and do not serve to activate social activities, it is a non-contribution element	0
			Social style	The place is an affinity of different social styles and supports localism. It is a positive element	1
	Intangible		Ideas and beliefs	Ideas of originality and richness influence the people in the place. It is a positive element	1
			Values and visions	Cultural and historical heritage are reflecting on their values. It is a positive element	1
			Feeling	Feeling safe and relaxed, as people in the place represents segments of society without discrimination. It is a positive element	1
			Attitude and behavior	The overlap of activities, thus, the users' goals affect understanding and acceptance of each other. So, the attitude of users has a negative effect	−1
Interaction	Tangible		Efficacy of place	Various activities negatively affect the efficiency of the place. Also, reusing buildings for non-essential functions. It is a negative element	−1
			Duration of stay	The place is a Humanscale and has different ways to access it. It is a positive element	1
			Walkability	There are unique places and architecture that affect passers-by. The possibility of walking in the place is good. It is a positive element	1
	Intangible		Memories	The memories of the place could be imageable and memorable. It is a positive element	1
			Stories and myths	The place has stories that reflect warmth and richness. Positive element	1
			Esthetics	The designs of buildings and squares give the impression of pride and glory. It is a positive element	1
			Experience	This site is familiar with being an administrative center for many centuries. The experience in general is good. It is a positive element	1

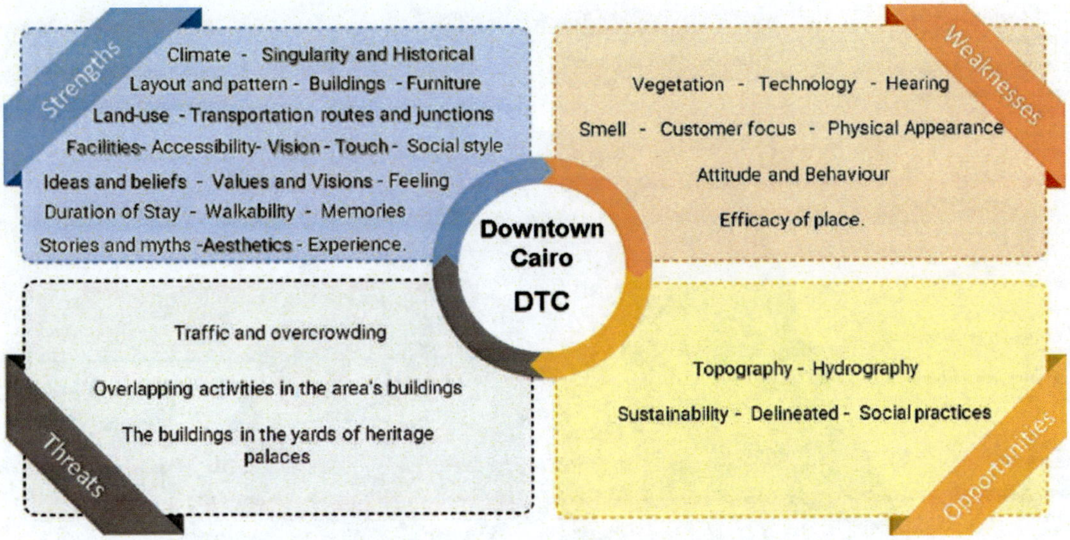

Fig. 8 SWOT of Ministries Square in Downtown Cairo (DTC). *Source* The Authors

Fig. 9 Examples of Islamic and Egyptian Pharaonic style of ministries. *Source* The Authors

4.2.3 Analysis

According to the two previous steps (description and evaluation), strengths, weaknesses, opportunities, and threats can be determined as follows, see Fig. 12. The results of the assessment came to show that 22 components had a positive impact on enriching the identity of the place (strengths of the case of the study). Six components negatively affected the identity in the place (weaknesses of the case of the study). And, seven components did not negatively or positively affect identity (potential opportunities to improve identity in the case of the study). Perhaps, the recurring impediment as a cause of weakness in several identity components was the lack of respect for the human scale in the distances between buildings, the modernity of the place,

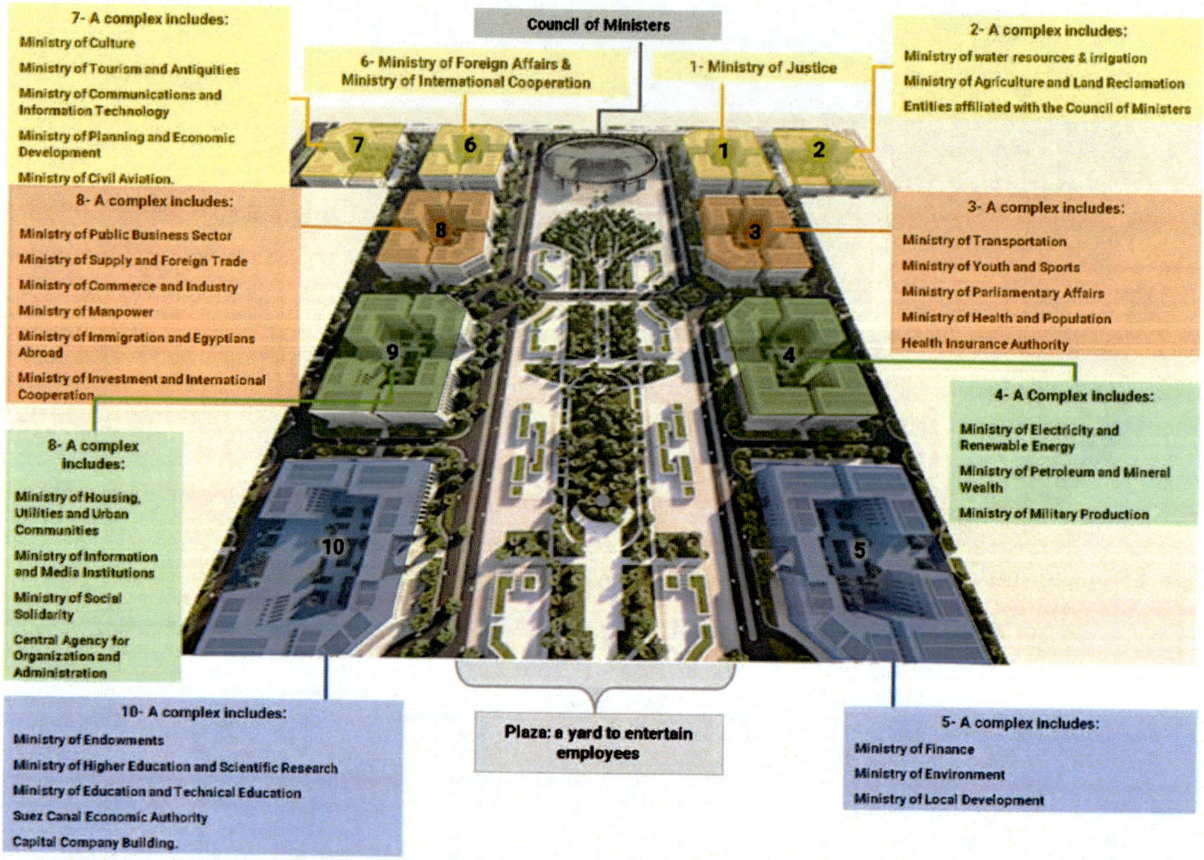

Fig. 10 Distribution of ministries in the government district in the new administrative capital. *Source* The shot from (www.egy-maps.com), Edited by The Authors based on (Ministry of Housing Utilities & Urban Communities, 2022)

which made it difficult to evaluate some components, and the non-overlapping urban fabric, thus no shaded places, which negatively affected the climate (threats in the case of the study).

5 Discussion and Results

The analysis shows that each case study has 22 components that build place identity. Although the number of components is the same in both cases, the components differ. Consequently, this proves the hypothesis assumed by the study at the beginning, which is that each case reflects its identity differently. Of the 22 components, there were 11 components commons, and they were:

(1) Singularity and historical,
(2) Buildings,
(3) Furniture,
(4) Land-use,
(5) Transportation routes and junctions,
(6) Facilities,
(7) Vision,
(8) Touch,
(9) Ideas and beliefs,
(10) Values and visions,
(11) Esthetics.

In contrast, there were no common components in the negative or neutral effects. Thus, this increases the opportunity to take advantage of each case to extract ways to enhance the deficiency in the other. In the following, we will discuss the strengths and weaknesses extracted from the two cases in preparation for developing strategies to avoid weaknesses and enhance strengths in the next step.

5.1 First Results: Exploring the Positive and Negative Aspects

Based on the components of place identity, Table 4 presets the positive and negative aspects from the two cases.

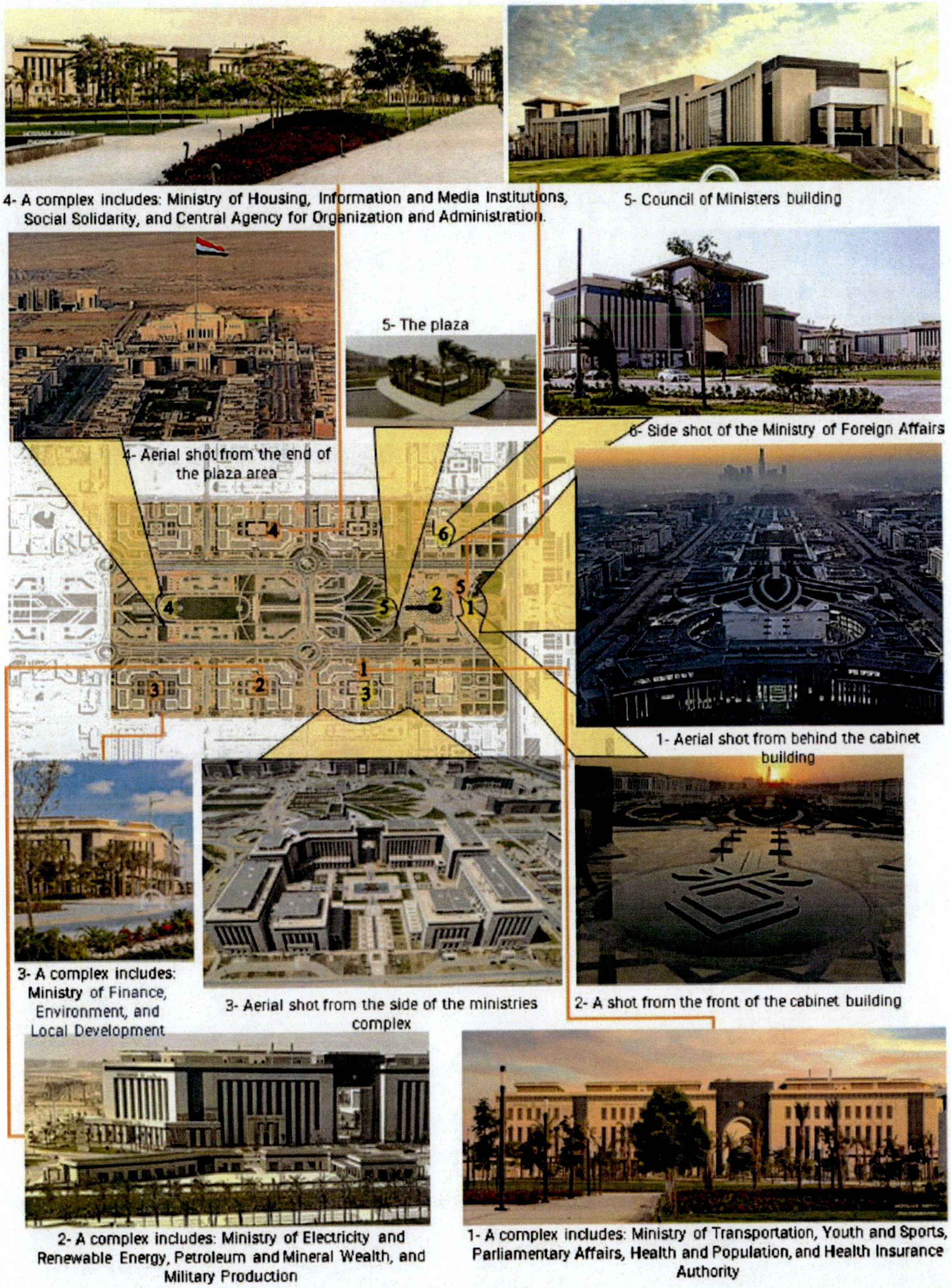

Fig. 11 Panoramic shots of the outdoor areas and shots for main buildings. *Source* The shot from (New Urban Communities Authority, 2022) in 2022, and it represents the current situation, according to the author's observation of the site. Edited by The Authors

Table 3 Sense of identity evaluation—Case study 2

Classification			Element		Description	Mark
Environment	Tangible	Physical	Natural forms	Topography	The scene is open and highlights the most important surrounding projects, so this element's effect is positive	1
				Vegetation	A variety of forms and colors of vegetation, so this element's effect is positive	1
				Climate	The climate is sunny and lacks shade, and the loss of this element negatively affects the evaluation	−1
				Hydrography	There are lakes and fountains for esthetic purposes and to moisten the atmosphere, so this element's effect is positive	1
			Manmade forms	Singularity and historical	The place views the tallest tower, the largest mosque in Africa, and the tallest flagpole in the world, so this element's effect is positive	1
				Layout and pattern	The place is designed in a pattern but does not respect the human scale, So, this element negatively affects the evaluation	−1
				Buildings	Iconic buildings. Buildings are one of the most important elements of strength in this place, so this element's effect is positive	1
				Furniture	Rich urban furniture: fountains, stairs, lights, paving, etc., so this element's effect is positive	1
		Function	Land-use		There are iconic places, so this element's effect is positive	1
			Transportation routes and junctions		Clearly public transportation such as monorail and bus stations, so this element's effect is positive	1
			Facilities		Infrastructure is adaptable with time and history, so this element's effect is positive	1
			Accessibility		Fortified and not pedestrian friendly, the loss of this element negatively affects the evaluation	−1
		Quality	Technology		Although many technological elements are included in the city's design, it is difficult to evaluate it. So, it is a non-contribution element	0
			Sustainability		The buildings and lighting are powered by solar energy, and there is a monorail nearby, and there is water recycling to use to irrigate the gardens. It is a positive element	1

(continued)

Table 3 (continued)

Classification			Element	Description	Mark
People	Intangible	Sensation	Vision	The mental image has a pattern, so it is imageable and coherent. It is a positive element	1
			Hearing	Tranquility. It is a positive element	1
			Smell	The plaza and the multi-gardens make the smell likable. It is a positive element	1
			Touch	The texture of the place is diverse and valuable, so this element's effect is positive	1
		Management	Delineated	Ownership is clearly defined. It is a positive element	1
			Social	The place is designed to be encouraging social interaction (Plaza). It is a positive element	1
			Customer focus	The human spatial experience is clear. It is a positive element	1
	Tangible		Physical appearance	Everyone in the place is there for matters related to the function of the place, so their appearance is comfortable and acceptable. It is a positive element	1
			Social practices	The place was designed to activate social activities, but in the current situation, it is not possible to measure the extent of its success	0
			Social style	The place was planned to be an affinity of different social styles, but it is not possible to admit it now	0
	Intangible		Ideas and beliefs	The people in the place are influenced by ideas of progress and prosperity. It is a positive element	1
			Values and visions	Smart, sustainable societies are reflecting on their values. It is a positive element	1
			Feeling	Feeling of isolation and insignificance as access to the place is limited to specific groups. Negative element	−1
			Attitude and behavior	The place was designed to contain future activities that reflect the attitudes and behaviors of users, but it is not realized yet	0

(continued)

Table 3 (continued)

Classification			Element	Description	Mark
Interaction	Tangible		Efficacy of place	The place has been designed to its function, so it is flexible and caters to its different needs	1
			Duration of stay	The place is vast (monumental scale), and it takes a lot of time to access or cross. It was hard to evaluate this element in the current situation	0
			Walkability	The place has a monumental plaza and unique views. But because of the distances, it is difficult to walk. Negative element	−1
	Intangible		Memories	The place is new, so it has no memories. The loss of this element negatively affects the evaluation	−1
			Stories and myths	The place is new, so it lacks stories, but there is a legendary story, but it cannot be evaluated currently	0
			Esthetics	The buildings and landscape designs of the place give the impression of elegance and pride. It is a positive element	1
			Experience	This site is unfamiliar to users, and therefore, there is no depth in-place experience and at the same time is not objectionable. Neutral element	0

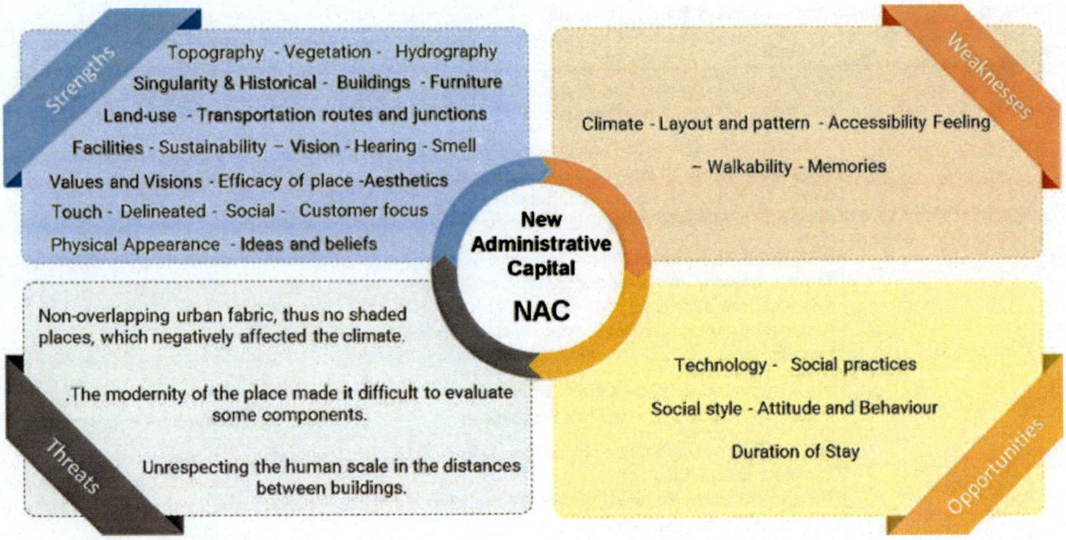

Fig. 12 SWOT of government district in the new administrative capital (NAC). *Source* The Authors

5.2 Second Results: Informed Decision-making

Based on previous analysis, the strategies for enhancing the sense of identity of public places in the administrative areas of Greater Cairo, Fig. 13 (identity enhancement strategies—IES) can be designed in three categories as following: enhancing identity through the environment's elements, enhancing identity by impressing people, and finally, enhancing identity by strengthening the interaction between the environment and people positively.

6 Conclusion

This paper concluded by presenting strategies that enhance the sense of identity in public places. The strategy came in three categories: enhancing identity through the environment's elements, enhancing identity by impressing people, and finally, enhancing identity by strengthening the interaction between the environment and people positively. To achieve those results, the study was carried out in three stages. The theoretical study included a review of the concept of place identity in the literature and ended with identifying place identity components emanating from the reliable literature that could be used for assessing identity in public places. Then, the empirical study evaluated two cases study of public places in administrative areas in Greater Cairo. The evaluation was based on three types of data, archives and available documents of those places, site visits (observation), and interviews with experts by authors as well. Based on the evaluation and analysis of the results, the strategies were set to enhance identity in public places in Greater Cairo. However, the results of this study need to be supported through a survey of users' opinions of those places and compared with the study's outputs. These strategies require multi-stakeholders serious work translating them into policies that can help government officials, urban planners, and academics alike comprehend, gauge, and provide practical solutions to enhance identity. This paper gives a theoretical foundation that can be starting point to design projects and city-wide programs in the future. The added value of this study is that its methodology can be followed to evaluate and develop other places with different functions and characteristics.

Table 4 Points of strength and weakness of identity from case studies

Component	Strength	Weakness
Physical	Narrow-shaded streets. Lagoons and fountains. A variety of forms and colors of vegetation. Open scene	No natural views and no hydrography elements. The lack of shading
	The landmarks, individual semantics, and symbols. The place is designed in a pattern and respects the human scale Iconic, heritage, and unique buildings. And rich urban furniture: fountains, sculptures, stairs, lights, paving, etc	Place needs to respect the human scale
Function	Archeological, iconic, and unique places Transportation is, evolving and diverse, and infrastructure is adaptable with time and history Accessibility is vividness and openness, free, secure, and pedestrian friendly	Accessibility is fortified and not pedestrian friendly
Quality	Subway, monorail, bikes stations and road, solar energy, and there is water recycling to use to irrigate the gardens	There are no elements of intelligent technology in the public space
Sensation	The mental image is imageable, coherent, singular, and unforgettable. The place is tranquil The gardens make the smell likable. The touch of the place materials in place is diverse and valuable	The place could be quieter The multi-use buildings and the traffic density often make the smell unlikable
Management	The place encourages social engagement, and the human spatial experience is clear. Ownership is clearly defined	Ownership is vague. The human spatial experience is dispersed
Physical appearance	Physical appearance is comfortable and amusing	Physical appearance is surprising and shocking
Social practices	The design supports community participation and civic initiatives	Lack of community participation and civil initiatives
Social style	The affinity of different social styles supports localism	–
Ideas and beliefs	Progress, prosperity, originality, and richness	–
Values and visions	Cultural and historical Heritage, and Smart and sustainable societies	–
Feeling	Feeling safe and relaxed	Feeling of isolation and insignificance
Attitude and behavior	Social practices, activities, understanding, and acceptance	The overlap of activities and the users' goals
Efficacy of place	Design the place to its function, flexible and caters to its different needs	Reusing buildings and various activities
Duration of stay	Humanscale, and rich experience	Monumental scale, and discomfort
Walkability	Plaza and unique views	The congestion
Memories	Old places have memories that could be imageable and memorable	New places have no memories
Stories and myths	Stories reflect warmth and richness	New places need more depth of in-place experience
Esthetics	The design gives the impression of elegance, pride, and Glory	–
Experience	Old places are comfortable and familiar	New places need more depth of in-place experience

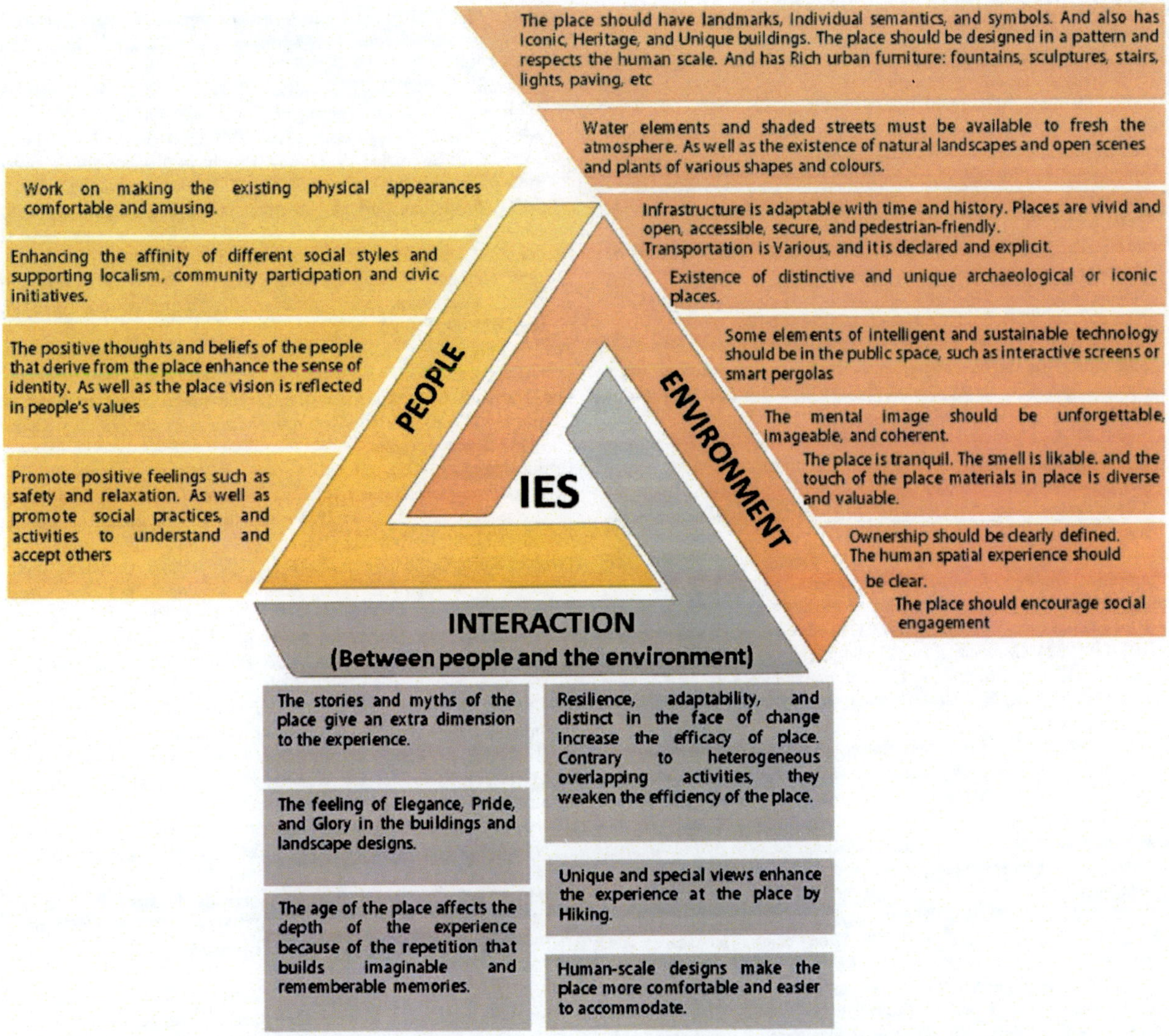

Fig. 13 Identity enhancement strategies. *Source* The Authors

References

Al-naim, M. A. (2008). Identity in transitional context: Open-ended local architecture in Saudi Arabia. *Archnet-IJAR, 2*(2), 125–146.

Ali, O., Mansour, Y., Elshater, A., & Fareed, A. (2022). Assessing the identity of place through its measurable components to achieve sustainable development. *Civil Engineering and Architecture, 10*(5), 137–157. https://doi.org/10.13189/cea.2022.101407

Breakwell, G. M. (2015). Coping with threatened identities. In *Gastronomía ecuatoriana y turismo local.* (Vol. 1, no. 69). https://doi.org/10.4324/9781315733913

Brown, D. G. (2001). Characterizing the human imprint on landscapes for ecological assessment. In *A guidebook for integrated ecological assessments* (pp. 404–415). Springer-Verlag. https://doi.org/10.1007/978-1-4419-8620-7

Carmona, M. (2019). Principles for public space design, planning to do better. *Urban Design International, 24*(1), 47–59. https://doi.org/10.1057/s41289-018-0070-3

Chapman, E. H., & Lynch, K. (1962). The image of the city. *The Journal of Aesthetics and Art Criticism, 21*(1), 91. https://doi.org/10.2307/427643

Cheshmehzangi, A. (2020). Identity of cities and city of identities. *Springer Nature Singapore*. https://doi.org/10.1007/978-981-15-3963-3

Childs, M. (2006). Squares: A public place design guide for urbanists. *University of New Mexico Press*. https://doi.org/10.5860/choice.42-4460

Demographia. (2020). Demographia world urban areas. In *16th Annual Edition* (Vol. 92). http://www.demographia.com/db-worldua.pdf

Francis, M. (1989). Control as a dimension of public space quality. In I. Altman, & E. Zube, (Eds.), *Public places and spaces* (pp. 147–172). Plenum Press. https://doi.org/10.1007/978-1-4684-5601-1

General Authority for Urban Planning. (n.d.). *Part 2: Concepts and terms used in land use*. chrome-extension://efaidnbmnnnibpcajpc

glclefindmkaj/http://gopp.gov.eg/wp-content/uploads/2015/11/2-land-uses-.pdf

Greene, S. (1992). City shape: Communicator and evaluating community design. *Journal of the American Planning Association, 58*(2), 177–189. https://doi.org/10.1080/01944369208975792

Hawwas, S. Z. (2002). *Khedivian Cairo*. Architectural Design Center.

Hecht, M., Jackson, R., & Pitts, M. J. (2005). Culture: Intersections of intergroup and identity theories. In J. Harwood, & H. Giles (Eds.), *Intergroup communication: Multiple perspectives* (pp. 21–42). peter Lang.

Ibrahim, R., Mushatat, S., & Abdelmonem, M. G. (2014). Authenticity, identity and sustainability in post-war Iraq: Reshaping the urban form of Erbil City. *Journal of Islamic Architecture, 3*(2), 58. https://doi.org/10.18860/jia.v3i2.2533

Jones, M. (2003). The concept of cultural landscape: Discourse and narratives. In H. Palang, & G. Fry (Eds.), *Landscape interfaces. Landscape series,* (Vol. 1, pp. 21–51). Springer. https://doi.org/10.1007/978-94-017-0189-1_3

Kermani, A. A., Charbgoo, N., & Alalhesabi, M. (2016). Developing a model for the relation between heritage and place identity. *International Journal of Civil, Environmental, Structural, Construction and Architectural Engineering, 10*(3), 391–396. https://doi.org/10.5281/zenodo.1123697

Lalli, M. (1992). Urban-related identity: theory, measurement, and empirical findings. *Journal of Environmental Psychology, 12*.

Lefebvre, H. (1991). *The production of space* (Vol. 53, no. 3). Basil Blackwell. https://doi.org/10.2307/378107

Lynch, K. (1981). *Good city form*. The MIT Press.

Ministry of Housing Utilities and Urban Communities. (2022). *The new administrative capital*. https://mhuc.gov.eg/العاصمة_الادارية_الجديدة/المدن_الجديدة

National Organization for Urban Harmony. (2022). *Boundaries and foundations of preserving the Khedivial Cairo area of outstanding value*. http://urbanharmony.org/download/pdf/khdweic_cairo_boundries2022.pdf

New Urban Communities Authority. (2022). *The new administrative capital*. http://newcities.gov.eg/know_cities/NewCapital/default.aspx

Norberg-Schulz, C. (1971). *Existence, space and architecture, 120*.

O'Donnell, P. M., & FASLA, A. (2008). Urban cultural landscapes and the spirit of place. *ICOMOS QUEBEC, 8*. http://www.academia.edu/5033361/Augmented_Reality_AR_in_Urban_Heritage_Tourism

Peng, J., Strijker, D., & Wu, Q. (2020). Place identity: How far have we come in exploring its meanings? *Frontiers in Psychology, 11*, 294. https://doi.org/10.3389/fpsyg.2020.00294

Proshansky, H., Ittelson, W., & Rivlin, L. (1976). *Environmental psychology: People and their physical settings*. Holt.

Rapoport, A. (1990). *The meaning of the built environment: A nonverbal communication approach*. University of Arizona Press.

Rees, R., & Cosgrove, D. E. (1986). Social formation and symbolic landscape. *Geographical Review, 76*(1), 119. https://doi.org/10.2307/214798

Relph, E. (1976). *Place and Placelessness*. Pion Limited. https://doi.org/10.4135/9781446213742.n6

Shawesh, E. M. (2000). The changing identity of the built environment in Tripoli City Libya. In *School of architecture, planning and landscape* (Vol. PhD, no. L6675). University Of Newcastle Upon Tyne.

Shekhar, H. (2017). *Urban identity as a complementary tool to urban wellbeing*. University of Duisburg-Essen.

Sonkoly, G. (2017). Historical urban landscape. In *Historical Urban Landscape*. Springer Nature. https://doi.org/10.1007/978-3-319-49166-0

Stephenson, J. (2008). The cultural values model: An integrated approach to values in landscapes. *Landscape and Urban Planning, 84*(2), 127–139. https://doi.org/10.1016/j.landurbplan.2007.07.003

Stokols, D., & Shumaker, S. (1981). People in places a transactional view of settings book chapter.pdf. In J. H. Harvey, (Ed.), *Cognition social behaviour and the environment Erlbaum* (pp. 441–488), New Jersey.

Surchi, Z., & Nafa, H. (2021). Appraisal of place identity through tangible elements of cultural features in Erbil city. *Civil Engineering and Architecture, 9*(7), 2528–2541. https://doi.org/10.13189/cea.2021.090736

The Administrative Capital for Urban Development Company (ACUD). (2017). *The administrative capital*. http://www.acud.eg/

The National Committee for the Development and Protection of Heritage Cairo. (2019). *Development plans for the Ministry Square in the downtown area*.

Tomlinson, J. (2003). Globalization and cultural identity. In *The global transformations reader: An introduction to the globalization debate* (pp. 269–277). Polity Press. http://books.google.com/books?id=xYLdRrJSxW8C&pgis=1

Twigger-Ross, C. L., & Uzzell, D. L. (1996). Place and identity processes. *Journal of Environmental Psychology, 16*(3). https://doi.org/10.1006/jevp.1996.0017

UN-HABITAT. (2012). *The state of Arab Cities 2012 challenges of urban transition*. (UN-Habitat). https://unhabitat.org/the-state-of-arab-cities-2012-challenges-of-urban-transition

UN-HABITAT. (2013a). *Time to think urban*. (UN-Habitat). https://unhabitat.org/time-to-think-urban-un-habitat-vision-on-urbanisation

UN-HABITAT. (2013b). *Turning spaces into places*. (UN-Habitat). https://unhabitat-kosovo.org/un_habitat_documents/turning-spaces-into-places-handbook/

UN-HABITAT. (2018). Training module: Public space. In *Continental philosophy review*. (UN-Habitat). https://doi.org/10.1007/s11007-006-9038-x

UN-HABITAT. (2020). The new urban agenda. In *Australian planner*. UN-HABITAT. https://doi.org/10.1080/07293682.2003.9995262

UNESCO-WHC. (2012). *Urban regeneration project for historic Cairo*. http://www.urhcproject.org/

UNESCO, & World Heritage, C. (2008). Operational guidelines for the implementation of the world heritage convention. In *Chirurgische Praxis* (Vol. 74, no. 4). http://whc.unesco.org/en/guidelines

United Nations, Department of Economic and Social Affairs, Population Division. (2018). The world 's cities in 2018. In *Data booklet (ST/ESA/ SER.A/417)*.

Valera, S. (1998). Public space and social identity. *On the W@terfront, 0*, 1–20–20.

Vogeler, I. (2010). *Critical cultural landscapes of North America*. https://docplayer.net/40883365-Critical-cultural-landscapes-copyright-2010-ingolf-vogeler-chapter-4-i-racist-landscapes-indian-reservations-1.html

Ziyaee, M. (2018). Assessment of urban identity through a matrix of cultural landscapes. *Cities, 74*, 21–31. https://doi.org/10.1016/j.cities.2017.10.021

Historical and Cultural Impact on Banja Luka City (Bosnia and Herzegovina)

Nataša Grgić, Dragana Popović, Maja Odobašić, and Dragana Kuzmanović

Abstract

A city is a densely populated, compactly built and functionally organised spatial unit within which specific relationships between people and the environment are established. The functional and physiognomic development of a city is a reflection of various civilisational, cultural, socio-political, socio-economic and natural influences. These are, to a specific extent, determined by historical heritage. As a consequence, no settlement, Banja Luka included, can be viewed isolated from space and time. Historical and cultural influences on Banja Luka city in the process of its urban development are the subject and the thematic framework of this research. The development of Banja Luka was influenced not only by its geographical and spatial position but also by the roles it played and the significance it held throughout history. Its life and development were conditioned by turbulent socio-historical factors, the clash of civilisations, earthquakes, wars, socio-administrative status, etc. The cultural and historical influence will be analysed based on the literature, written and graphic archival material, the facts determined on site and the development trends of the city through specific historical periods: the Ottoman Empire, the Austro-Hungarian period, the Kingdom of Yugoslavia, SFR Yugoslavia and recent history (period from 1992 to the present). Each of these epochs, with its specific development characteristics, caused migrations of population and unplanned expansion of the city. This made planning and urbanism a neglected social discipline. The canvas of the city was painted with eclectic architectural and construction strokes all fatefully reflecting the period and the culture of their making. This paper will talk about shaping urban space through historical and cultural influence. It will also discuss the true state of urban space and the practical application (non-application) of ethical rules and norms (not) built into the city of Banja Luka.

Keywords

Banja Luka · Urban space · Historical significance · Cultural influence · Urban space ethics · Urban development

N. Grgić (✉)
Civil Engineering Institute "IG", Banja Luka, Bosnia and Herzegovina
e-mail: natasa.grgic@institutig.ba

D. Popović
Institute for Scientific Research On the Independent University Banja Luka, Independent University Banja Luka, Banja Luka, Bosnia and Herzegovina
e-mail: draganapavlovic78@hotmail.com
dragana.popovic@nubl.org

M. Odobašić
Civil Engineering Institute "IG", Banja Luka, Bosnia and Herzegovina
e-mail: majci98@gmail.com
maja.odobasic@institutig.ba

D. Kuzmanović
Spatial Planning, Civil Engineering Institute "IG", Banja Luka, Bosnia and Herzegovina
e-mail: dkuzmanovic.dragana@gmail.com
dragana.kuzmanovic@institutig.ba

1 Introduction

The balance of the state, in which parts of a system and their mutual interaction are established as a set of spatial connections (i.e. traffic connections) and spatial structures (i.e. the way land is used), represents a successful adaptation to new conditions and an adequate perception of the environment within which a city grows and develops. According to IEREK, urban planning is a large-scale concept concerned with planning and development at all levels (architecture, infrastructural, ecological, economic and even political) (IEREK, 2022). "It attempts to organize socio-spatial relations across different scales of government and governance" (Huxley & Inch, 2022, p. 87).

M. F. Shahidan et al. (eds.), *Urban Narratives: Exploring Identity, Heritage, and Sustainable Development in Cities*, Advances in Science, Technology & Innovation, https://doi.org/10.1007/978-3-031-48517-6_14

Urban growth and changes in its structure do not occur independently of external influences. It is a well-known fact that urban agglomeration is only one living part of a wider spatial system constantly interacting with the environment. For this reason, urban planning must be accompanied by sustainability concept which enables us to carry out the urban development process in the right way to keep up the sustainable environment for future generation (IEREK, 2022). This serves to prove that no settlement, Banja Luka being no exception, can be perceived as isolated from time and space. Models of urban renewal that are successful for small historic towns must have the capability to strike a balance between preserving their cultural heritage and meeting the needs of development, thus increasing their competitiveness whilst still preserving their unique qualities (Božić et al., 2019).

"Urban planning is concerned with the social, economic, and environmental consequences of delineating spatial boundaries and influencing spatial distributions of resources" (Huxley & Inch, 2022, p. 87). In the same way, the development of Banja Luka is a reflection not only of its geographical and spatial position but also of the city's importance and functions throughout history. Its life and development were shaped by tumultuous socio-historical factors, change of civilisations, earthquakes, wars, administrative-social status, etc. Due to its geostrategic position, Banja Luka was under the influence of different civilisations, all of which left a mark on its development. Each civilisation carried its own specific development characteristics and thus influenced the migration of the population and often caused an unplanned expansion of the city, making planning and urbanism a neglected social discipline. The canvas of the city got painted with eclectic architectural and construction strokes all fatefully reflecting the period and the culture of their making.

Different civilisations present in these areas imposed their identity on urban development. This had a heavy impact on how Banja Luka looks today. Analysis of professional and scientific literature indicates one problem—a lack of synthetic presentation of historical and cultural impacts on the development of Banja Luka through various epochs. This particularly refers to presenting positive and negative impacts. This study aims to highlight the importance of perceiving urban development through the perspective of historical events and the impacts of various cultures in order to set the course of future urban development. At specific points in this paper, a comparison will be made between Banja Luka and Graz (Austria). The reason for such a comparison is the fact that both cities experienced continuous development under different social and economic circumstances and similar natural and geographical conditions. This resulted in the formation of similar urban identities.

The main objectives of the paper are a synthesis of the historical and cultural influences on Banja Luka in the process of its urban development and a concise presentation of the problems regarding its current urban development. Finally, the paper also aims to show the true state of the urban space and including possible directions for its future urban development.

2 Theoretical and Methodological Approach

Establishing a unique methodology for studying the spatial-functional structure of a city is almost impossible. This is because each urban system is extremely specific, as it is a unique result of a particular geographical position, physical and geographical characteristics of the area and socio-economic conditions governing its development. All of the above is further complicated by the influence of various cultures and civilisations that have intertwined in these areas. So, in order to answer the questions raised, and to achieve the objectives defined above, it was necessary to apply various techniques (cabinet, field), work methods (theoretical, cartographic, archival, documentation, media) as well as synthesis and generalisation. It is important to note that the literary-documentary research method, along with the cartographic and statistical-mathematical method, is the base of this specific methodological approach.

The paper also implies spatial research, which entails the need to use the cartographic method, i.e.to apply modern cartographic tools (geographical information system—GIS), as an indispensable asset in spatial/urban planning.

3 The Most Important Natural and Socio-Geographical Processes in Shaping Banja Luka

The foundation, development and expansion of Banja Luka and the development of its material and spiritual culture were determined by natural (morphometric features of the landscape, hydrography, climate, floods, fires, earthquakes, etc.) and socio-geographical processes (civilisations with their cultures: Roman, Slavic and Hungarian, Ottoman, Austro-Hungarian, Yugoslav, Dayton):

- Settlement construction began in the southern part of the Banja Luka basin in all historical periods (the high ground at the entrance to the gorge was strategically extremely favourable for the construction of fortifications and control of the traffic direction towards Jajce);
- A fortress was built, and a civilian settlement has been developing in this location since the period of ancient Rome, the Middle Ages and the Ottoman period (Mandić, 2013);

- The Ottoman period entailed changes in a physiognomic, functional and architectural sense—a differentiation occurred into the business part of the settlement (bazaar) and the residential part (mahalla). The construction of the Gornji Šeher settlement began. The settlement gradually expands along the right bank of the River Vrbas. Settlement Donji Šeher began developing and construction is directed towards the north (Bejtić, 2007). During this period, significant cultural, religious and economic facilities were built, such as the Kastel fortress (it got its present form during the Ottoman period) and bridges over the River Crkvena and the River Vrbas (these enabled west-east development);
- At the end of the seventeenth and the beginning of the eighteenth century, the image of Banja Luka changed, as a result of the Austro-Turkish wars (Skarić, 1924);
- "The oriental concept of the city was conditioned by the direction of communication flow along the river; it kept its basic longitudinal form which was the basis for the linear organisation of the city, inherited from ancient times, still living in the modern age" (Mandić, 2013, p. 63);
- The Austro-Hungarian occupation led to a change in the oriental character of Banja Luka;
- The fire that happened between 14 August 1878 and 16 August 1878, "turned a third of Banja Luka into ashes and destroyed wealth and trade" (Mikić, 2004, p. 82);
- Construction of Banja Luka (end of the nineteenth and beginning of the twentieth century) began without an urban plan. However, blocks, squares and streets were formed, shaping a silhouette of a European city, with architectural and stylistic forms and artistic characteristics (Mikić, 2004); green avenues are constructed; the city spreads towards the north; road communications are built, and industry and trade are developing, and cultural and educational activities are progressing; (Mandić, 2013);
- The city structure and morphology that was formed during the Austro-Hungarian era was the basis for the development of Banja Luka in the period of the Vrbas Banate;
- After 1929, the city became the centre of the Vrbas Banate. This is when the urban foundations of its future development were laid. The Banski Dvor Banate Administration, Mortgage Bank, Palace Hotel and theatre were built; some representative buildings of this period were destroyed during the Second World War (Main Orthodox Church, General Hospital); substantial investments were made in building physical and social infrastructure during the time of Ban Svetislav Milosavljević;
- As the centre of the Vrbas Banate, from 1929 to 1941, Banja Luka became an administrative centre with a gravity zone (19.917 km^2) of about one million inhabitants;
- Between the two wars, the administrative role of the city became the foundation of its development; in the physiognomic sense, the city adopted new architectural styles with elements ranging from Serbian medieval architecture to the spirit of functionalism; the city plan kept the basic longitudinal concept, whilst also developing the transversal ones (Mandić, 2013);
- From 1945 to 1948—reconstruction of buildings and infrastructure;
- From 1948 to 1954—first industrial development;
- From 1954 to 1969—orientation towards economic development, infrastructure and housing construction;
- From 1969 to 1992—planned reconstruction and construction after the earthquake, industrial development;
- From 1992 to 1996—war, stagnation of economic development, no changes in the physiognomy and spatial structure of the city (Mandić, 2013);
- In the period after 1996, Banja Luka becomes the administrative, economic, political and cultural centre of the Republic of Srpska, characterised by a (un)planned construction, formation of peri-urban settlements, strengthening of the private sector, etc.
- Current development is directed towards the north, towards the city of Laktaši (two populated areas between these cities, Zalužani in Banja Luka and Trn in Laktaši, have spatially "fused together").

Looking from the perspective of natural-geographical and social-geographical characteristics, Graz and Banja Luka have the following in common (Arandelovic, 2015; Simonović, 2010):

- Favourable natural conditions for living and development of settlements (the Alps embrace the Styria region, opening towards the valleys of two rivers, Drava and Mura. Similar to the Banja Luka region, the area combines spacious and fertile river valleys at the southern edge of the Panonian Basin and a mountainous heavily forested hinterland, intersected by numerous rivers);
- Natural protection by the river valley (both settlements were formed in the river basins, sheltered by forested mountain hinterland, at the location where the mountains transform into plains; rivers Vrbas and Mura with their tributes organically penetrate the urban fabric);
- The settlements are formed around the fortress, as a core, and along the river, as the axis;
- Initial fortifications, the Gradec (in Graz) and the medieval Kastel (in Banja Luka) were built by the Slavic people;
- Both belonged to border regions, with Graz always having the status of the capital of the region;
- The areas surrounding Graz and Banja Luka as their capitals, throughout a great part of their history, formed a part of border regions of the countries they belonged to at the time (Donji Kraji, Bosanska Krajina and Republic of Srpska on one and Ostmark, Narrow Austria and Styria);

- "The cultural revival in the second half of the twentieth century brought innovative architecture and urban modernization to Graz" (Arandelovic, 2015, p. 78); Banja Luka experienced a similar development trend.

Emigration from Bosnia and Herzegovina during the civil war (1991–1995) had a major impact on ethnic and social structure of Graz.

4 Evaluation of Morphometric Characteristics from the Aspect of the Development of Banja Luka

The city of Banja Luka, as a unit of local self-government within its current borders, is located in the north-western part of the Republic of Srpska entity. Orographically, the entire area is characterised by a gradual elevation of the relief, from north to south (Fig. 1).

The evaluation of natural potentials aims to justify the existing population. It also outlines the most optimal areas for future construction. The same is done by analysing hypsometry, the height of groundwater, the degree of soil erosion, the possibility of water supply, the stability of the terrain, the slope of the terrain, the exposure of the terrain, etc. This paper will focus on the morphometric characteristics of the relief—elevation and slope. Exposure was not considered. It was rather taken as a secondary factor to determine the quality of the locality (residence conditions) (Đorđević, 2006). The most favourable terrains for settlement are those with a slope of 0–5° and hypsometry of up to 500 m. Favourable ones have a slope of 5–12° and an elevation of 500-600 m. Terrains with a slope of 12–16° and elevation of 600-800 m are considered conditionally favourable whilst unfavourable ones have a slope higher than 16° and an elevation over 800 m (Đorđević, 1996). The terrain was classified into these categories by the use of a digital elevation model and GIS and functions surface slope, reclass–reclassify and intersect. It is very clearly observed, as shown in Fig. 2, that the (historical) development of Banja Luka (a vectorised graphical presentation, author Mandić (2013), overlapped with natural conditions) is closely related to the most favourable natural conditions, considering morphometric characteristics of

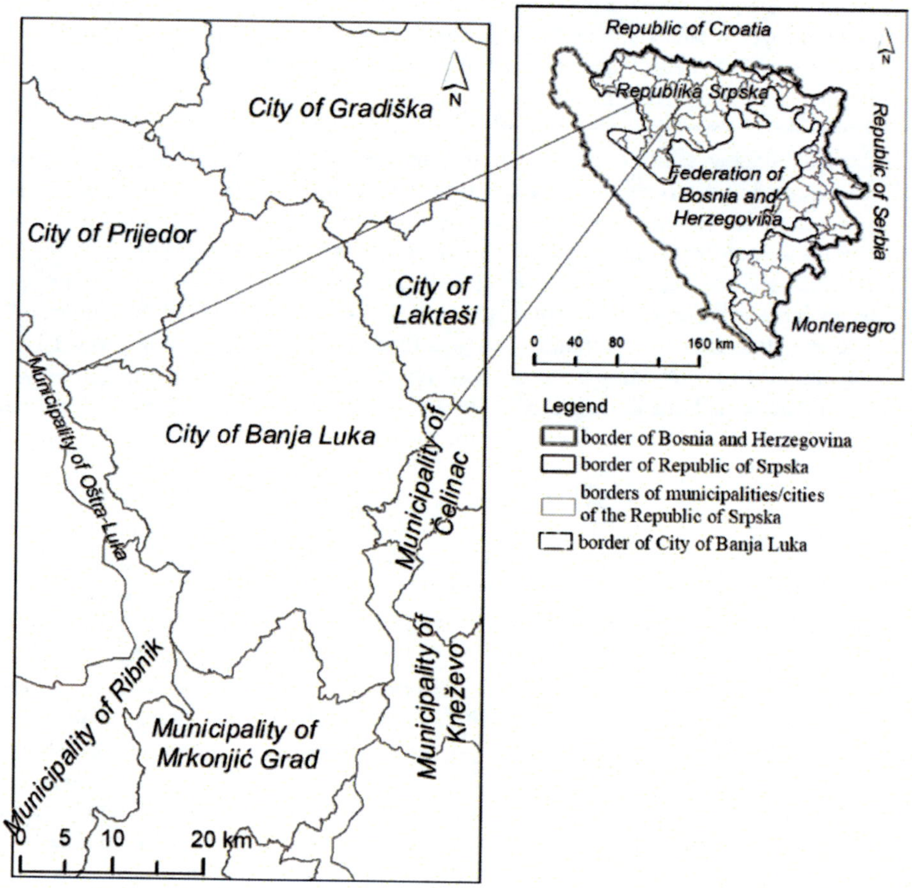

Fig. 1 Geographical location of the city of Banja Luka (vector data of borders from reference spatial plan of the city of Banja Luka (Projekt a.d., Institut za građevinarstvo "IG" d.o.o. Banja Luka. 2014)

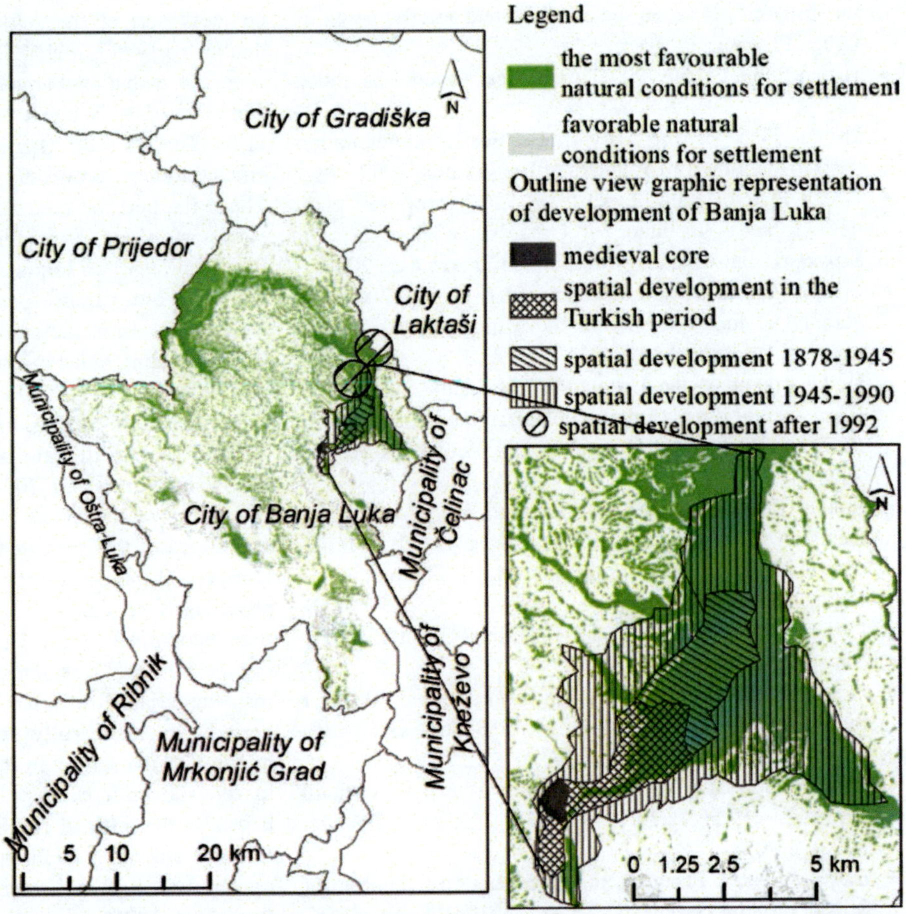

Fig. 2 The most favourable and favourable natural conditions (slope and hypsometry) for settlement and outline graphic representation of development of Banja Luka—interpretation of the author based on the following sources: vectorisation from Mandić (2013), vector data of borders and DEM from reference spatial plan of the city of Banja Luka (Projekt a.d., Institut za građevinarstvo "IG" d.o.o. Banja Luka. 2014)

the terrain—hypsometry and the slope. At the same time, the city is currently mostly developing in the direction of the northern lowlands, towards Laktaši. The assumption is that this trend of expansion towards the north will continue precisely because of the favourable terrain in the area. The north-west direction towards the city of Prijedor can also be singled out as an area of future development of Banja Luka. This is because the region has great natural potential and developed infrastructure, both of which are prerequisites for intensified urban development.

5 Urban Development and History of Planning of Banja Luka

The cultural and historical influence will be analysed based on the literature, written and graphic archival material, facts established on-site, and development trends of the city through specific historical periods: the Ottoman Empire, the Austro-Hungarian period, the Kingdom of Yugoslavia, SFR Yugoslavia and recent history (period from 1992 to the present).

Revisions, modifications and amendments to the planning documentation take place under the influence of various factors, natural, political and economic. It was only in the twentieth century that a more serious planning and legal framework for the development of the urban fabric of Banja Luka was created. The chronology of preparation/amendments of respective planning documents and legal acts is given below:

- Urban documentation of the city starts with the "Conceptual Design of the General Urban Plan", 1952;
- "Urban Planning Program", 1967;
- Revision of the 1967 "Urban Planning Program", caused by general development trends in the wider area of Banja Luka, the disproportion between population increase and the planned growth—the development was much faster than anticipated and the catastrophic earthquakes of 26 October and 27 October and 3 December 1969; after the earthquake, it was concluded that the destruction of the basic housing stock differed amongst various areas of the city. There were indications that specific urban zones required a special regime of use, the problem of

inconsistency between the existing state and the previous Urban Planning Program;

- Urban Plan, 1975–1990, (Urbanistički zavod Banjaluke, 1975);
- Urban Plan of the City—Analytical documentation basis—Institute for Urbanism of Banja Luka, 1992;
- Urban Plan of Banja Luka 2008–2020 (proposed plan, not adopted)—Institute for Urbanism of Republic of Srpska (Urbaistički zavod Republike Srpske a.d., 2009);
- Spatial Plan of Banja Luka Municipality 1986–2005 was adopted at the Assembly of the Municipality of Banja Luka on 7 February 1990;
- Development of the Urban Plan of Banja Luka, 2021 (Urbis centar d.o.o, Centar za planiranje urbanog razvoja (CEP), Institut za zaštitu i ekologiju Republike Srpske, 2021).
- Spatial Plan of Banja Luka (adopted on 3 April 2014, Decision of the Assembly of Banja Luka City, 11/14), (Projekt a.d., Institut za građevinarstvo "IG" d.o.o. Banja Luka, 2014)
- Law on Urban Planning (SRBIH Official Gazette, 4/65);
- Law on Spatial Planning, 1987;
- "Instructions on Mandatory Uniform Methodology for the Preparation and Development of Spatial and Urban Plans and Zoning Orders", 1987;
- Law on Spatial Planning (Republic of Srpska Official Gazette, 19/96, 84/02, 14/03, 112/06 (Pajović, 2006), 53/07),
- Law on Spatial Planning and Construction (Republic of Srpska Official Gazette, 55/10, 40/13, 106/15 and 3/16)
- Rulebook on the Manner of Preparation, Content and Formation of Spatial Planning Documents (Republic of Srpska Official Gazette, 69/13) also denoting general rules of professional conduct.
- Rulebook on General Rules of Urban Regulation and Land Subdivision (Republic of Srpska Official Gazette, 115/13, 52/20)

5.1 Period of Ottoman Empire

After 1528, when the Turkish army occupied this area, the city started developing in significantly changed conditions. With the establishment of the Turkish administration, Banja Luka and other settlements in Bosnia and Herzegovina experienced a strong physiognomic and functional transformation. The transformation was a result of new socio-economic relations and the introduction of numerous elements of Turkish-oriental culture. These new relations and elements left their mark on the demographic composition, economy, exterior appearance of the houses and their interior furnishing. A change occurred in the spatial and functional structure of the settlement. The division into the business part of the settlement—the bazaar and the residential part—the mahalla became prominent. At the same time, the mahallas themselves got divided into Muslim and Christian ones. Being located for a long time on the border of the Turkish state, Banja Luka developed under very specific conditions prevailing in the border areas of the empire. These implied the constant presence of military garrisons. However, along with this military-strategic importance, Banja Luka got an important role in the sphere of trade and craft. The city's transit position in trade between Italian cities and Bosnia in the sixteenth century made it one of the leading trading cities in BiH.

Looking through historical facts, the beginning of the development of Banja Luka as a city is related to the southernmost part of the Banja Luka basin. Here, the width of the basin narrows to 200 m and then transits into a narrow gorge of the Vrbas River. From the sixteenth century and forward, this part of the city was called Gornji Šeher (Bejtić, 2007). Its current name is Srpske Toplice. The advantages of this location for urban development were first recognized by the Romans. The trend continued during the medieval Bosnian state as well as during the reign of the Hungarian–Croatian kings and throughout the Turkish period. So, historically speaking, the urban development of Banja Luka started in this southern part of the city, back in the sixteenth century, i.e. in 1528. The development began on the left and continued on the right bank of Vrbas. Housing and other facilities were being built, and the development of crafts and trade had begun (Urbanistički zavod Banjaluke, 1975).

Some of the most prominent examples of construction activities of this period include the Mosque (1579), the Mosque (1595), the Kastel fortress, the clock tower, etc. Other buildings were also built during this period, all in the spirit of oriental architecture. These include numerous mosques, bridges, a caravanserai, mills and hamlets with residential buildings and a large number of craft and trade shops.

For comparison purposes, during the same period, Graz developed as a fortified city with compact suburbs whilst Banja Luka, mainly stretching in the longitudinal direction (sequencing and branching of mahallas), also developed as a Christian plains town (Simonović, 2010).

5.2 Austro-Hungarian Period

In the second half of the nineteenth century, elements of various European construction styles were introduced into the oriental architecture of Banja Luka (Kumarić, 1932). These were first adopted by richer Serbian merchants. Šeher was the oriental part of the city, whilst Varoš and the buildings on the left bank of Crkvena were European in their individual architectural solutions. However, it was only after the Austro-Hungarian occupation and the introduction of the cadastre and construction law that Banja Luka acquired the appearance of a European city. This was reflected in the general and urban concepts, as well as in the architectural and stylistic forms of individual buildings. So, a true physiognomic and functional transformation of Banja Luka started with the establishment of the Austrian administration in 1878. The transformation came as a result of changes in social relations, legislation, economy, valorisation and exploitation of space. The changes in the demographic structure also contributed to the transformation, since they brought new socio-economic and cultural opportunities. The Trappists, a Catholic religious order of monks, also settled in Banja Luka in this period. Their arrival started modernisation in agricultural production and processing of various products. It is safe to say that it was precisely the Trappists who laid the foundation of industrial production in the area. The structure of the city and the urban morphology formed during the Austro-Hungarian occupation was the basis for the development of Banja Luka in the period of the Vrbas Banate (Živković, 2005).

In the period from 1878 to 1918, Banja Luka was under the occupation of Austria-Hungary. Although it was the administrative centre of the region during Turkish rule, Banja Luka met the Austro-Hungarian occupation as a backward oriental settlement with less than 10,000 inhabitants. Under the new government, the city progressed significantly. Roads, water supply, sewerage and electric lighting were being provided. The economy, industry and trade were growing. Timber was extracted by railway and further distributed to all corners of the "dual monarchy". Banja Luka was connected by traffic with Prijedor, Prnjavor, Kotor-Varoš and Jajce, whilst several forest railway lines were used for the extraction of wood resources. Vibrant trade between Banja Luka and Vienna, Trieste, Prague and Pest empowered the carriers of its progress and continuously opened new perspectives.

Changes occurred in the entire social system. These included a transformation of the economy, traffic, education and other areas of life and had a direct impact on the territorial expansion and urban physiognomy of Banja Luka. The territorial expansion of the city continues north of the River Crkvena, on the left bank of the River Vrbas and along the main road towards Gradiška. With the expansion of the city towards the north, three new settlements were formed in the area: Kumsale, Predgrađe and Nova Varoš. As new settlements emerged, the very centre of the city moved to a new business area known as Gospodska Street.

This period is characterised by the construction of many public, educational and cultural institutions. Another typical trait of the period was the rows of trees along the main street (the so-called Carski drum (*Imperial Road*)), planting of which started in 1882. Planned construction was reflected in the introduction of order into Gospodska Street and the construction of prominent buildings. This conditioned a planned development of the urban core of Banja Luka (Romić-Perić, 2018). The modernisation of Banja Luka in the decades that followed made it become an important industrial, economic and cultural centre of Bosanska Krajina. By 1895, the city already had a hospital, a tobacco factory, a railway connection and a grammar school. Banja Luka was being built expansively and was adopting a modern urban structure.

During the period of Austro-Hungarian rule, numerous important buildings were built. Some of these were destroyed during the Second World War and later in the devastating earthquake of 1969. Recognisable buildings of that period are the Velika Realka grammar school (destroyed in the 1969 earthquake), many family houses, which stylistically supported the Neo-Renaissance concept, the hospital, etc. One of the more significant buildings, still standing today as a part of cultural heritage, is the old Railway Station built at the end of the nineteenth century in the Neo-Renaissance style. After the railway station got relocated from the city centre to the suburbs in the second half of the twentieth century, the building adopted a new purpose. Today, it houses the Museum of Contemporary Art of the Republic of Srpska and has been declared a national monument of Bosnia and Herzegovina.

Graz and Banja Luka were developing in a similar direction during this period. However, the similarity is not complete as it reflects the importance the cities had within the monarchy. Graz was far more prominent and closer to the source of influence, making the intensity of the influence much stronger (Simonović, 2010). The greatest architects and artists of different regions (Germanic region, the Balkans and the Mediterranean) expressed themselves forcefully in Graz and thus created a brilliant syntheses (UNESCO. World Heritage Convention, 2023).

5.3 Period of the Kingdom of Yugoslavia

The structure of the city and its morphology created during the period of Austro-Hungarian occupation were the basis for the entire further development of Banja Luka, especially in the period between the two world wars (Urbanistički zavod Banjaluke, 1996). Before the formation of the Banate (1929), Banja Luka was characterised by a great economic depression and the absence of construction activity, with travel writers describing it "as a desolate, orientally neglected village" (Kumarić, 1932).

Banja Luka met the end of the First World War as an agricultural trading centre whose poorly developed industry was mainly focused on the processing of agricultural products. It was only after 1929, when it became the centre of the Vrbas Banate, that the city experienced a cultural and economic awakening. This was the period of one of the most prominent figures in the history of Banja Luka-Ban Svetislav Tisa Milosavljević. His governance was characterised by significant investments in physical and social infrastructure. Many roads, bridges, schools, hospitals and representative buildings were built during this period. Gospodska Street also got its famous shape during this period. Despite the fact that the industry had not made significant progress from 1929 to 1941, as the centre of the Vrbas Banate, Banja Luka grew from a second-rate provincial town to an administrative centre of a higher rank, with a gravitational zone of 24 districts, a total area of 18,917 km^2, and one million inhabitants according to the 1931 census (Živković, 2005).

During the Kingdom of Yugoslavia, the city developed rapidly, especially under the administration of the Ban Svetislav Tisa Milosavljević. Many prominent buildings were built during this period, such as the Mortgage Bank, the Banate Administration, the Banski Dvor, the Sokolski Dom, the Institute of Hygiene, the National Theatre of the Vrbas Banate and the Church of the Holy Trinity, which was destroyed in the German bombing (today the Temple of Christ the Saviour). A great number of new schools were built, along with the Ethnographic Museum of the Vrbas Banate, the City Bridge and the City Park. This made the cultural life of Banja Luka flourish.

The First World War impoverished not only Banja Luka, but Graz also, jeopardising the urban development of the city and the economy of the state. When the Kingdom of Yugoslavia was declared in 1929 and administratively divided into banates, Banja Luka became the capital of the Vrbas Banate and thus got the status of a centre of a wider region. Graz was also a centre of the Styria region (Simonović, 2010).

Between the two wars, urban development still gravitated north. In terms of physiognomy, between the two world wars, the city acquired new architectural expressions combining elements of Serbian medieval architecture and the spirit of functionalism. These expressions were embedded in many general-purpose buildings and private villas of wealthier residents, which still stand today (the famous Streets of Kralja Petra I Karađorđevića and Mladena Stojanovića). During this period, the city centre moved to its current location. The City Plan kept the basic longitudinal concept, also developing the transversal ones, particularly in the city centre. Construction was concentrated on the left bank of the River Crkvena, whilst the settlements on the right bank of the river, the area around the Kastel fortress and the oriental part of the city were neglected.

After the Second World War, the country experienced accelerated development. The new conditions prevailing in the country created a need to intensify planning activity and develop planning institutions. Before the breakup of the former Yugoslavia, Banja Luka was the second largest city in the Socialist Republic of Bosnia and Herzegovina and the tenth largest city in Yugoslavia. Intensive industrialisation and modernisation of the transport infrastructure in the 1950s accelerated the development of this city. This caused significant changes in the socio-economic structure of the population, functional transformation and stronger cohesion with the immediate environment. According to the 1991 census, Banja Luka had 195,000 inhabitants.

After the Second World War, Banja Luka and Graz develop in different social-economic conditions. Different urban development approaches were applied. The socialist system of former Yugoslavia adopted a new set of urban values (through interpretation of the Athens Charter of 1933). For Banja Luka, this implied a change in the urban identity and character of the city. Graz, on the other hand, managed to adapt the implementation of the proclaimed system of values to its local urban identity. The identity of the city did not suffer significant changes, particularly not those implying negative consequences. This concept of city construction was practised on an almost experimental level. Afterwards, it did not become generally accepted. Urban archetypes of a European city were preserved in Graz whilst inherited urban structure became disintegrated in Banja Luka (Simonović, 2010). Urban planning, housing and transport are important issues from 1945 to this day in Graz (Graz Museum, 2023).

5.4 Recent History

During socialism, spatial planning experts were guided by the common good and common interests of all parties involved in the planning process. This concept, combined with a one-party centralised system and improved use of public land, annihilated aspirations of political and economic currents (Milojević, 2015). In this period, at the beginning of the 60 s of the last century, economic entities with state capital were being established and industrial production developed rapidly. The two significantly increased immigration and consequently created a need for the construction of new settlements.

Banja Luka suffered the worst blow in the catastrophic earthquake of 1969 (the foreshock with the intensity of 7.5 degrees Mercalli scale occurred on 26 October 1969; the 8.5 degrees Mercalli scale mainshock followed the next day). Human lives were lost, and enormous material damage was incurred.

Accelerated reconstruction and rehabilitation of devastated buildings followed. The city infrastructure was restored, and rapid urbanisation began. Banja Luka got its distinctive look.

On the other hand, this experience with the destructive power of a natural disaster, and the rehabilitation and reconstruction of a city after an earthquake, got embedded in the Planning Documentation in the form of requirements defining short-term and long-term protection measures. These measures had a significant impact on the concept of the spatial organisation of the city. They were reflected in the arrangement of city functions, the construction system, the traffic concept and in the system of important infrastructure installations and devices. They were also considered in the dimensioning of green areas which are an important planning factor when seismic activity is prominent and the degree of intensity is high. It should be emphasised that rules and laws established at the time were abided by in the period from 1975 until the beginning of the nineties.

Due to specific circumstances, the 1975 Urban Plan has been applied until this day, whilst the amendments adopted in 1993 have been used as planning guidelines. It is a fortunate circumstance that the 1975 Plan, time-oriented to 1990, also contains basic provisions for the so-called post-plan period. This enabled adoption of regulatory plans in period from 1990 until present day, for a significant part of the urban area of the city, especially the so-called core urban zone. These regulatory plans are based on and comply with the 1975 provisions as well as with the provisions of the 1993 Draft.

During the civil war, which was fought on the territory of the former Yugoslavia from 1992 to 1995, there was a "transition from state centralised planning in the period of socialism, focused on the public interest (jobs, housing stock, public services, physical structure) to market-oriented disposal of urban space, resulting in social, functional and ecological conflicts" (Mandić, 2019, p. 33). Unfortunately, the events of the period interrupted the spatial planning momentum, and a major planning crisis occurred in BiH. Not only physical structures were destroyed. Demographic and cultural destruction occurred, with an aspiration to completely annihilate the cultural values and the identity distinguishing BiH as a specific entity, not only in the Balkans but also in Europe. All of the above left a mark on the urban development of Banja Luka.

At the end of the last, and throughout this century, during the post-war period and the period of transition, Banja Luka and a great majority of other cities in the former Yugoslavia faced the impact of private initiatives, working in the interest of private capital. These initiatives became a significant generator of changes in the area. For this reason, for more than two decades, Banja Luka has been experiencing a quantitative development of the urban structure, not accompanied by a qualitative one, expressed through public content and functions that contribute to the general increase of the urban standard (Urbaistički zavod Republike Srpske a.d., 2009). Accordingly, the construction is directed towards the privately owned residential sector, whilst public facilities are left to the city authorities. Graphically, current urban development is characterised by a lack of green areas and the presence of large residential and commercial buildings, which not only impede traffic but also diminish the architectural and urban significance of the space. Construction of public spaces, public facilities and public services is left to the city authorities who lack the financial resources to perform such activities. New housing blocks

Fig. 3 Contrast in the planning of residential blocks. *Source* Authors

are left without supporting facilities, thus creating inhumane settlements (Fig. 3a, b as examples of blocks from the 1970s, and Fig. 3c, d as examples of blocks from the 2020s). Amendments are also frequently made to the implementing planning documentation, which directly "dictates" construction conditions. There is a lack of respect for the professional view and opinion.

A unique, purposeful and integral approach to planning the urban matrix as a whole was replaced by political and economic mechanisms and an approach to planning known as "investor-led" urbanism. Due to the decrease in public ownership and the fact that investments in public content and projects of general interest are limited by the city administration budget, the participation/impact of the city administration in shaping the physiognomy and urban structure of Banja Luka has decreased. Positive developments do exist, but they are still not dominant and clearly recognized.

Today, Graz has an excellent architectural reputation on the international level, whilst at the same time, the city is characterised by high fragmentation. The contemporary urban design supports the new model of city disintegration, acknowledging the new, fragmented state (Holub, 2020). Reconciliation with historical trends, social and spatial context or spatial morphology of the current fabric of the city does not occur. For this reason, Banja Luka, Graz and many other European cities become/have become a part of in situ fragmentation through modern urban planning and design.

6 Conclusion

Due to its geostrategic position, Banja Luka was under the influence of various civilisations, all of which left their mark on the development of both, the city itself and the area that gravitates to it. Through a comprehensive analysis, it was determined that the respective area was and still remains in the vortex of various influences.

The state of the urban space and the practical application (non-application) of ethical rules and norms (not incorporated) incorporated into the city of Banja Luka are shown from the perspective of historical and cultural influence. Each historical period lefts its mark on the urban matrix and development of Banja Luka as a city. It appears that ethics was present throughout all periods. Nevertheless, it also appears that with urbanisation and modernisation, ethics is slowly dying away and disappearing, more progressively so since the 1990s.

The synthesis approach was used to present the development path of Banja Luka through different eras. It is perceived that the material and spiritual cultures of the nations who occupied this area left their mark on its spatial-urban development. However, throughout its history, the city has maintained a longitudinal direction of expansion. The city's structure and morphology, formed during the Austro-Hungarian era, was the basis for the development of Banja Luka in the period of the Vrbas Banate. Urban development of the city began at that time. Between the two world wars, the city became the centre of the Vrbas Banate. This is when the urban foundations of its future development were laid. The time between the 1950s and the 1990s represents a period of accelerated construction and changes in the physiognomy of the city. These changes were mostly caused by economic development (the establishment of economic entities with state capital and the development of industrial production), which increased immigration and created a need to build new settlements. The events of the 1990s interrupted the spatial planning momentum. A planning crisis occurred, and a period of spontaneous construction and development followed.

More significant deviations from the planned development, visible in the spatial planning documentation valid at the time, are reflected in the following: devastation and erosion of cultural heritage buildings, insufficient compliance with water protection regulations, damage to green areas, the unsolved problems regarding temporary facilities, arrangement of specific functions and similar. If Banja Luka intends to keep its identity as a "city of greenery", the course of action concerning the arrangement and management of space must change. Good example is Graz, which was adopted in 2007 a greening policy (Grünes Netz Graz) aimed at interconnecting green spaces across the city (Hoeben & Posch, 2021).

An integral and multidisciplinary approach to planning, being the most important segment in shaping the space, the city and the entire society, started in the middle of the twentieth century. It was abruptly interrupted in the 1990s, but it must continue. It will enable the restoration of the currently disturbed balance between private and public interests. At the same time, the creation or modification of mechanisms (legislation) will direct future decisions towards strengthening the professionals in the planning process, creating as a result a space made by a man and to the man's measure.

Following Austria's accession to the European Union, Graz experiences many changes—economic, social, demographic, etc. The urban identity of the city changed too, as a consequence. We have shown how Banja Luka developed as a city through historical and cultural impact. However, changes are yet to come. This leaves a possibility to further develop this paper, hoping that respective changes will be based on proper planning foundations and readiness for better urban development.

References

Arandelovic, B. (2015). Graz, UNESCO city of design and historical heritage. *Cities*, *43*, 78–91. https://doi.o.1016/j.cities.2014.11.014

Bejtić, A. (2007). *Banja Luka pod osmanskom vladavinom: arhitektura i teritorijalni razvoj grada.* Sarajevo.

Božić, N., Dumbović Bilušić, B., & Kranjčević, J. (2019). Urban transformation and sustainable development of small historic towns. In M. Obad Šćitaroci, B. Bojanić Obad Šćitaroci, & A. Mrđa (Eds.), *Cultural urban heritage. The Urban book series* (pp. 113–115). Springer. https://doi.org/10.1007/978-3-030-10612-6_10

Đorđević, J. (1996). *Evaluacija prirodnih potencijala na primeru slivova Jablanice i Veternice.* Srpska akademija nauka i umetnosti.

Đorđević, J. (2006). *Evaluacija prirodnog kompleksa na primeru područja opštine Ljig.* Geografski institut Jovan Cvijić.

Graz Museum. (n.d.). *The Story of Graz.* Retrieved April 1, 2023 from https://360.grazmuseum.at/en/

Hoeben, D. A., & Posch, A. (2021). Green roof ecosystem services in various urban development types: A case study in Graz, Austria. *Urban Forestry & Urban Greening.*, *62*, 127167. https://doi.org/10.1016/j.ufug.2021.127167

Holub, B. (2020). *NORMAL x 4 - Direct Urbanism in Graz: catalog/Culture Year GRAZ 2020.* https://www.linkedin.com/pulse/normal-x-4-direct-urbanism-graz-catalog-culture-year-barbara-holub/

Huxley, M., & Inch, A., (2022). Urban planning. In *International encyclopedia of human geography* (2nd Edn, pp. 87–92). Elsevier. https://doi.org/10.1016/B978-0-08-102295-5.10228-8

IEREK (n.d). *Urban planning: Definition, problems and solutions.* Retrieved September 22, 2022, from https://www.ierek.com/news/urban-planning-definition-problems-and-solutions/

Kumarić, M. (1932). *Monografija Vrbaske banovine.* Arhiv Republike Srpske. https://arhivrs.org/?blog_post=monografija-vrbaske-banovine-1932

Mandić, M. (2013). *Karakteristike urbanog razvoja Banja Luke – savremena prostornofunkcionalna transformacija grada.* Geografsko društvo Republike Srpske.

Mandić, M. (2019). *Postsocijalistička urbana transformacija Banja Luka. SKUP*, *10*(2), 31–40. https://doi.org/10.7251/SKPSR191002031M

Mikić, Đ. (2004). *Banja Luka kultura građanskog društva.* Institut za istoriju.

Milojević, B. (2015). *Gradovi u promjenama: Urbane forme, transformacije, uticaji.* Univerzitet u Banjoj Luci.

Pajović, S. D. (2006). *Urbanistički zakoni južnoslovenskih zemalja, Bosna i Hercegovina, Crna Gora, Hrvatska, Makedonija, Slovenija, Srbija, sa objašnjenjima.* Javno preduzeće "Urbanizam", Zavod za urbanizam.

Projekt a.d., Institut za građevinarstvo "IG" d.o.o. Banja Luka. (2014). *Prostorni plan Grada Banjaluka.* Administrativna služba Grada Banja Luka – Odjeljenje za prostorno uređenje. https://www.banjaluka.rs.ba/wp-content/uploads/pdf/prostorni_plan_grada_bl/0_Prostorni_plan_BL-tekst.pdf

Romić-Perić, R. (2018). *Banja Luka u urbanoj transformaciji.* Fakultet političkih nauka Univerziteta u Banjoj Luka.

Simonović, D. (2010). *Pejzažni gradovi, poređenje razvoja urbanog identiteta Banja Luke i Graca.* Arhitektonsko-građevinski fakultet Univerziteta u Banjoj Luka.

Skarić, V. (1924). *Banja Luka i njena okolica u davnini.* Otadžbina.

UNESCO. (n.d.). World heritage convention. *City of Graz-Historical Centre and Schloss Eggenberg.* Retrieved March 29, 2023 from https://whc.unesco.org/en/list/931

Urbanistički zavod Banjaluke. (1975). *Urbanistički plan Banja Luke* – sinteza. Urbanistički zavod Banja Luka.

Urbanistički zavod Banjaluke. (1996). *Urbanistički plan Banja Luke.* Urbanistički zavod Banja Luka.

Urbanistički zavod Republike Srpske a.d. (2009). *Urbanistički plan Banja Luka od 2008 do 2020. godine.* Urbanistički zavod Republike Srpske.

Urbis centar d.o.o., Centar za planiranje urbanog razvoja (CEP), Institut za zaštitu i ekologiju Republike Srpske. (2021). *Nacrt Urbanističkog plana grada Banjaluke (2020–2040.godina).* Administrativna služba Grada Banja Luka – Odjeljenje za prostorno uređenje. https://www.banjaluka.rs.ba/wp-content/uploads/2021/12/UPBL_NACRT-PLANA_javni-uvid.pdf

Živković, M. (2005). *Banja Luka – savremena prostorna i funkcionalna transformacija grada*, [Doktorska Disertacija, Univerzitet Banja Luka]. Prirodno-matematički fakultet Banja Luka

Heritage Arts: Sustainability, Identities and Public Regulations

Architectural Heritage and Site Management of Shuar Culture in the Amazon Region—Ecuador

David Eduardo Morocho Jaramillo, Camilla Mileto, and Fernando Vegas

Abstract

In recent decades, modern architecture has focused excessively on the search for new construction techniques to improve the city construction processes. This has led to a reduction in local construction techniques and materials, deteriorating cultural heritage and architectural landscape. Vernacular architecture is the result of a symbiosis of social, economic, and environmental aspects. It is therefore essential to understand the influence of architecture on lifestyle, as material, aesthetic, and functional aspects are also influential. Anthropological factors are also key, as the phenomena of change are shown in order to create guidelines for the analysis of safeguarding strategies. The Shuar are one of the 14 indigenous nationalities of Ecuador and are located in the south-east of the country, in the Amazon region. Their specific way of viewing the world involves symbolism in their spaces. During the last century, after colonization by religious expeditions, their lifestyle and construction techniques have been distorted, leading to a loss of their traditional heritage values, a situation which has worsened in recent years due to the arrival of new materials. This article aims to link the most important aspects of the Shuar settlements and their architecture to examine the problem which puts their vernacular architecture and culture at risk. Hence, strategies will be proposed to cover aspects of the indigenous territories, mitigate the threats, and identify a sustainable approach to their opportunities. The starting point of the conservation and site management proposal will be restoring value to the heritage aspects of its architecture through anthropological features.

Keywords

Identity · Shuar · Vernacular architecture · Sustainability · Cosmovision · Heritage conservation

1 Introduction

Ecuador has a vast array of cultures, with 14 recognized indigenous nationalities (Tsa'chila, Chachi, Épera, Awá, Kichwa, Shuar, Achuar, Shiwiar, Cofán, Siona, Secoya, Zápara, Andoa, and Huaorani). One of the most important of these is the Shuar nationality which is found in the south-east of the country in the Amazon region and shares the "Aénts Chicham" (Deshoullière & Utitiaj, 2019) language with the Achuar nationality, and several ethnic groups found in the north of Peru (Awajún, Wampis, Kandozi, and Andoa) (Morocho Jaramillo, 2022).

Nonetheless, the main concentration of Shuar communities can be found in the areas of Morona Santiago province as shown in Fig. 1, followed by Zamora Chinchipe and Pastaza. There are currently 69 centres and 11 associations within the Shuar Federation, covering an area of 887,615 ha, of which 727,615 have deeds, whilst the rest are currently subject to legal proceedings (Juncosa, 2020). In the twentieth century, the Ecuadorean Amazon region underwent a series of major social changes which directly influenced its ways of lifestyle of Shuar people. In the late nineteenth century, the religious missions that had colonized the territory copied the models already found in colonial cities (Bottasso, 2019). The sole difference is that it was important to change the lifestyle and ways of thinking of the local residents through a sedentary way of life to provide work and

D. E. Morocho Jaramillo (✉) · C. Mileto · F. Vegas
Research Centre for Architecture, Heritage and Management for Sustainable Development (PEGASO), Universitat Politècnica de València, Valencia, Spain
e-mail: damoja@upv.es

C. Mileto
e-mail: cami2@cpa.upv.es

F. Vegas
e-mail: fvegas@cpa.upv.es

M. F. Shahidan et al. (eds.), *Urban Narratives: Exploring Identity, Heritage, and Sustainable Development in Cities*, Advances in Science, Technology & Innovation, https://doi.org/10.1007/978-3-031-48517-6_15

Fig. 1 Map of Shuar settlements (Edited by the author from Juncosa, 2020)

establishing urban centres linked to the rest of the country. However, this led to the residents moving around, constantly segregated by territory, where their way of life shifted from nomadic to sedentary (Garzón, 2013). Here, families were grouped into centres organized by communal spaces and institutions which allowed the social change which religious colonizers aspired to. After a series of social and political changes in the country, they were awarded the ownership deeds to their land (under family usufruct)(Gnerre, 2014), providing safety and well-being to the Shuar families and preventing settlers from altering the territory. This allowed Shuar people to secure their own social, economic, legal, and political organization within the country.

Shuar's cosmovision of the world has a direct impact on their ways of life and how they conceive space. "*Arútam*", considered the most important spirit, governs nature and is found in natural sites such as waterfalls, rivers, forests, or orchards. It also takes on anthropomorphic or zoomorphic form or that of natural phenomena, so Shuar architecture is where its inhabitants build their own cosmos.

Shuar dwellings are cabins with elipsoid floor plans which are built within enclosures in the jungle. The structure and enclosures are made of palm, whilst the large pitched roof, made of leaves, can vary greatly depending on the location and existing material (Fig. 2). The spatial organization and configuration of its architectural elements reflect how Shuar people inhabit territory, considered a spiritual temple where space is a microcosmos within a macrocosmos which is the jungle (Chiriap et al, 2012). Its components are elements representing a connection with spirituality, where there is an intense spatial difference

Fig. 2 Traditional Shuar dwelling in Shuar museum "*Shuar Num*" in Zamora Chinchipe (*Source* author)

between the interior (sacred) and exterior (profane) (Eliade, 1981). This architecture represents the cosmovision and ways of thinking of those who live there so that the symbolic elements found in this space should be fundamental pillars for conservation. However, the new materials and the influences of modern architecture and large cities have had a direct impact on the region, where urban growth has only just embarked on a path that will lead to cultural and natural identities being forgotten.

Colonization brought materials from the most influential colonial cities and villages to the Shuar centres in the mountains. These in turn became points of use and distribution serving the most remote centres in the jungle. Materials such as zinc sheets, glass, plywood, and timber accessories (Bianchi, 1978) were accepted and incorporated into the architecture. The qualities of the new colonial dwelling typologies were successfully transferred and applied to brand new natural and cultural settings. The direct influence of colonization on Shuar dwelling typologies brought about the appearance of several new typologies, with materials adapted to a new tradition. Although in some cases, furniture and spatial organization were preserved, and new functional spaces such as bedrooms and dining rooms were added, albeit following a fragmented organization (Figs. 3a, b).

The poor handling of local resources such as timber extraction, livestock farming, urban infrastructure, and large-scale mining hinders the management of the site and its traditional architecture. Palm wood cabins, for example, have been altered with concrete blocks and corrugated sheet metal roofs which do nothing but repeat a new construction system implemented with new changes, highlighting traditional techniques falling into oblivion and decontextualization of the Shuar centres. Therefore, is it relevant to conserve traditional architecture to maintain its heritage or does this reap benefits guiding a sustainable development to cope current necessities? Is it possible to manage both in the same way?

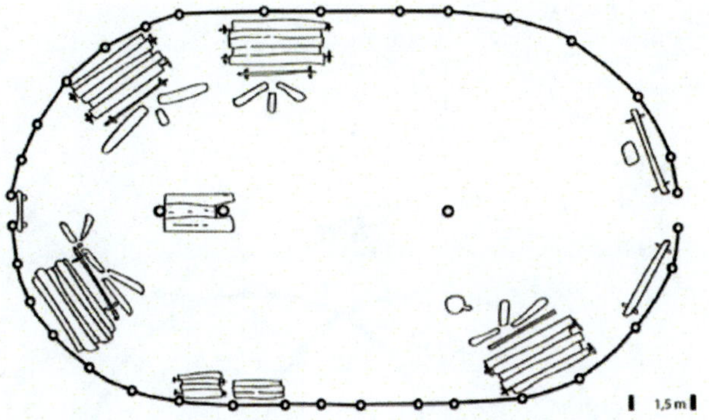

Fig. 4 Traditional Shuar dwelling (*jéa*) with female zone (*ekent*) on the left half and the male part (*tankamash*) on the right half. (edited by the author from Harner, 1978)

2 Methodology

This research aims to establish a correlation between the most important aspects of the Shuar settlements and their architecture in order to examine potential threats to their vernacular architecture and culture. In order to mitigate these and identify sustainable solutions, a series of

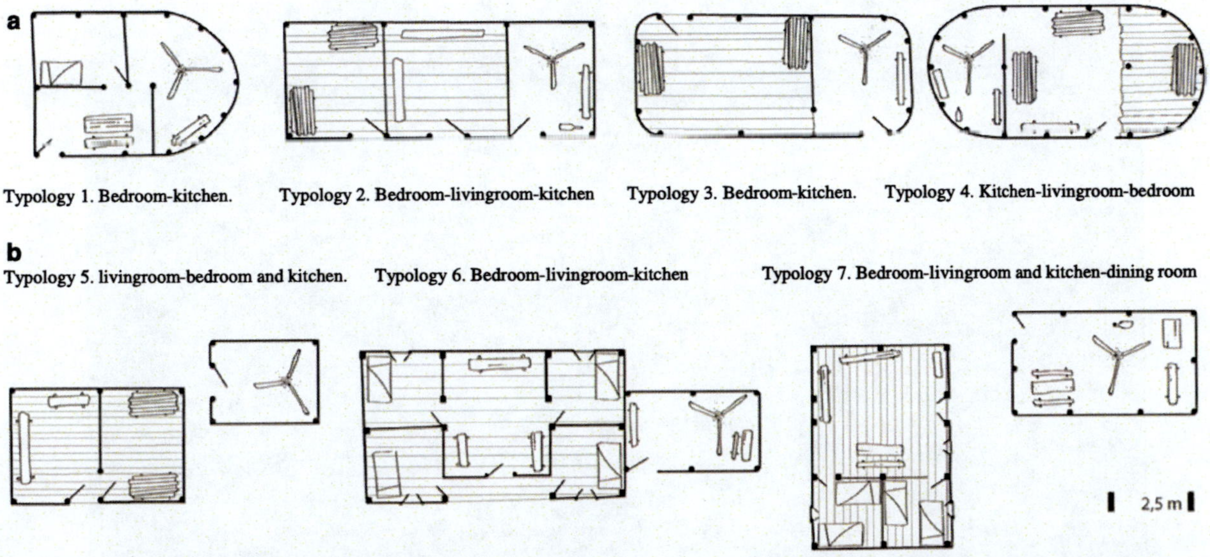

Fig. 3 **a** Colonial dwelling typologies and its internal spatial fragmentation (edited by the author from Pillaceta et al., 1998). **b** Colonialdwelling typologies and its morphological fragmentation (edited by the author from Pillaceta et al., 1998)

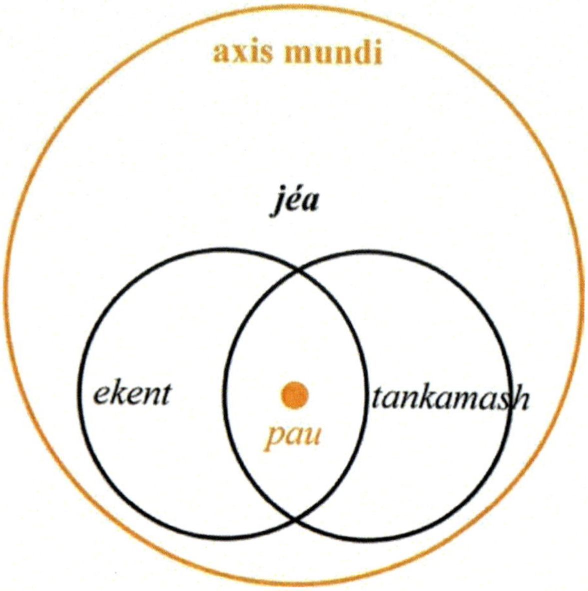

Fig. 5 Anthropological interpretation of the sacralization of Shuar dwelling (*jéa*) *from an understanding of its environment* (*Source* author)

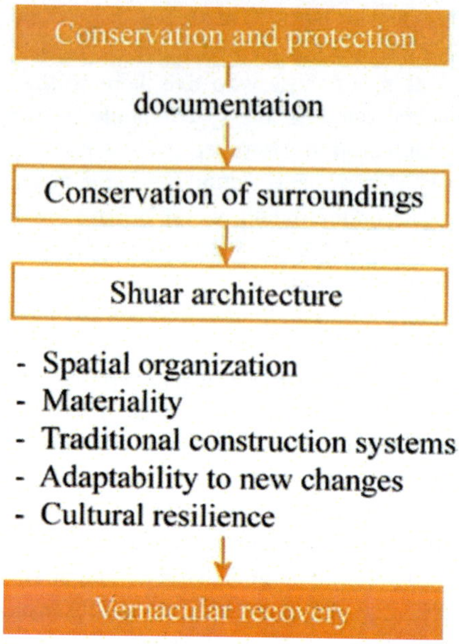

Fig. 7 Main aspects of conservation and protection strategies (Morocho et al., 2020)

strategies are proposed to addressing different aspects of conservation and the promotion of management in indigenous contexts.

The methodology examined takes as its starting point a review of the bibliography, providing an overview of the context for these changes. A historical examination is carried out from an anthropological and theoretical perspective to understand the territory and how it is inhabited, identifying symbolic relationships as values important to the conservation of their architecture. Secondly, information

Fig. 6 New tradition dwelling in "*Tuntiak Nunkeẽ*" natural reserve in Morona Santiago (*Source* author)

Fig. 8 Waterfall as a sacred site where the spirit of the jungle (*Arutam)* is manifested. Transmission of knowledge in Tuntiak Nunkeé natural reserve through lifestyle, (*Source* author)

Fig. 9 Colonial construction detail of the base (*Source* author)

is collected on a sustainable approach to be applied to vernacular heritage solutions. Following the initial analysis of the principles proposed by the "*Versus*" work, this information is contrasted with the current situation of local

Fig. 10 Colonial dwelling façade in Sucúa, Morona Santiago. (*Source* author)

communities. This requires, on the one hand, an assessment of the life plan of Shuar communities, considering the priority needs and objectives in the current ways of life. This methodology therefore validates strengths, weaknesses, and opportunities, analysing threats, and always taking into consideration sustainability principles and lifestyle as major points in the proposal phase. Finally, a series of recovery strategies linked to sustainable principles and local needs are proposed in order to improve actions. This allows the culture to be recovered, carrying out a two-part examination of the tangible heritage and strategies. The first strategy, addressing conservation and protection, follows an interpretation and documentation process to ensure appropriate measures are taken to guarantee enduring heritage values. A second strategy focuses on recovery and promotion, drawing up a proper management plan and using assets to boost sustainable development. It is important to identify input from the stakeholders, incorporating it into the projects and promoting the recovery of heritage values (Table 1).

3 Ontological Relationship Between Dwelling and Nature

Different cultures have their own categorization of nature and its meanings when interpreting the concept of environment. However, nature is a "*mechanical reflection of physical and technical determinations in mind*" (Descola et al., 2001). For the Amazonians, within a single environment, all beings are linked to the same social relationship (Surrallés & García, 2004), presented as a metaphor in rituals. Thus, understanding the variety of cultural positions can represent the opposition to "*epistemological privilege*" and lead us to the decolonization of thought (Descola et al., 2001).

Faced with the duality of nature and culture, Eduardo Viveiros de Castro emphasizes the different relationships between the "*human*" and the "*non-human*", with all beings

Table 1 Summary of methodological proposal

Data collection	Analysis	Strategies
Anthropological theories and arguments to understand the architectural heritage and its relation to culture	Sustainability principles and local needs in order to obtain an overview of the current context (strenghts, weaknesses, opportunities and threats of its settlements)	– Conservation and protection of vernacular dwelling – Recovery and promotion of vernacular dwelling

linked within the same environment although differentiated by their own bodies (Latour et al., 2009). Descola tries to transcend the definition of the implications of nature and society imply to approach another way of understanding autonomy of the parts as a whole (Surrallés & García, 2004). For the Amazonians, these distinctions do not apply to the relationships of beings, and these are viewed in terms of human social kinship where, for example, women see plants in the orchards as their children. In contrast, the Western interpretation distinguishes between nature and culture, embracing the idea promoted by some religions where only humans have a soul and a hierarchy determines relationships between beings: a supreme being, humanity, and nature.

Seeing the Shuar dwelling as a projection of bodies in cosmos helps to understand the symbolic relationships between the "*human*" and the "*non-human*" and define connotations of environmental social dynamics. Amerindian approaches assume the unity of the spirit and the diversity of bodies (Viveiros de Castro, 2010) within a spatial configuration, where materiality becomes sacred. The Shuar people organize the house space into two parts, with specific connotations attached to the axis mundi in either area. The female zone (*ekent*) is an intimate space to be shared by the family for rest and cooking activities and for the storage of utensils or sacred drinks like mandioca *chicha*; it is also linked to the orchard (Descola, 2017). On the other hand, the masculine part (*tankamash*), with its central element in the form of a central column (*pau*) acting as a connector between heaven and the underworld (Sanz & Herrera, 2017), is a space for social activity and for establishing alliances, as well as for celebrating important rituals linked to the jungle. However, both parts were in the same space without any boundary wall (Figs. 4 and 5).

In her analysis of the work of Descola (2017), Sharupi provides clues to understand the Amazonian vision as a native, where body and nature merge into a single perspective, promoting the metaphorical sacralization of space (Sharupi et al., 2022). Sharupi establishes a connection between dwelling structure and the bony rigidity of human body and spatial organization as a projection of the masculine and feminine body linked to sacred symbols.

4 Sustainable Development

In order to propose strategies for the recovery of Shuar vernacular architecture, anthropological aspects such as social organization and its evolution should be considered. It is therefore important to identify the community's social evolution process before proposing lines of action. Prior to religious colonization, Shuar society was traditionally structured around the authority of the head of the family, who had warrior skills and the ability to provide for the family. After colonization and a displacement of certain Shuar centres, a new political and social organization system was set up, defending the territories of the Shuar communities (1960). However, these groups were directed by the religious missionaries who had colonized the Shuar territory and were acting as intermediaries between the mestizo Ecuadorean society and the Shuar indigenous community in order to defend their homes from the colonizing invasion which had already been established in cities. Nowadays, the Shuar lead their own groups through a global organization of the Shuar communities of Ecuador, the "Federación Interprovincial de Centros Shuar" (FICSH) (Rubenstein, 2005). After securing officially recognized ownership deeds, they set out objectives adapted to the Shuar way of life. The government is under the obligation to respect the autonomy of Shuar authorities in the management and administration of their territories; to respect the use of resources for their survival, and to carry out democratic processes whenever a private or public institution plans to carry out activities in Shuar territory. This plan was devised by the Shuar inhabitants and approved in a General Assembly of the authorities of Shuar centres and associations. These objectives, which take agricultural, artisanal, and tourist production as assets for economic and environmental development into consideration, aim to guarantee a better quality of life (Pueblo Shuar Arutam, n.d.). Emphasis is placed on the demand for rights as citizens of a constitutional state and the protection of Shuar culture in keeping with ancestral principles and ways of thinking. This plan is therefore based on the following goals:

- Defence and improvement of traditional orchards and sustainable agricultural and livestock farming which allows families to access the market and be able to provide for themselves,
- Correct use of forests and non-renewable resources,
- Bilingual education and protection of the language,
- Health care which draws from the ancestral knowledge of alternative medicine,
- Improvement of road and services infrastructures.

Table 2 Sustainability principles of vernacular heritage. *Versus* proposal. (Edited by the author from Correia et al., 2014)

Environmental sustainability	Socio-cultural sustainability	Socio-economic sustainability
– Respect for the surrounding environment and landscape – Proper management of natural and climatic resources – Reduced pollution and savings on material expenses – Improved human health and well-being – Reduction in natural hazards effects	– Cultural landscape protection – Transmission of ancestral knowledge – Innovation and improvement of creative solutions – Recognition of intangible values – Improved social cohesion	– Support for autonomy – Promotion of local activities – Optimization of construction processes – Extended useful lifespan for buildings – Savings in resources consumed

Based on the needs identified by the Shuar people, analysis of the current situation focuses on strengths, weaknesses, opportunities, and threats, establishing a connection between cultural and architectural aspects as set out in the methodological proposal "*Versus*" by of Correia, Dipasquale, and Mecca, analysing aspects of sustainability in vernacular heritage. In this study, a framework of 15 sustainability principles (listed in Table 2) drawn up by *Versus* researchers was used as the basis for the interpretation of heritage. The aim is to promote, strengthen, and expand international and transcultural relations; to apply knowledge from lessons and principles of vernacular heritage; and to further disseminate vernacular heritage principles, techniques, and solutions (Bakonirina et al., 2022; Correia et al., 2014).

The analysis of the sustainability principles proposed within *Versus* is contrasted with the objectives of the Shuar life plan in order to understand the current strengths, weaknesses, opportunities, and threats to these settlements in an indigenous context, aiming to assess the potential for preservation of its architecture (Table 3).

This SWOT analysis shows the advantages and opportunities available to Shuar settlements in terms of the opportunities and threats to their architecture. Proposals therefore present solutions for the mitigation of risks and promotion of different possibilities. One of the key points to be revisited is the potential to be viewed globally as the promoter of an identity which self-manages the recovery of its heritage. However, modernity and globalization also pose severe risks, as the acquisition and consumption of material goods are superimposed onto natural and cultural preservation.

5 Conservation and Protection Strategies

Maintaining the historical authenticity and integrity of cultural heritage (Vásquez & Díaz, 2009) is one of the main objectives in the conservation of vernacular heritage, the expression of the identity of a people in its relations with the site, and the cultural expression which makes it unique in the world (ICOMOS, 1994) (Morocho-Jaramillo, 2022).

It is also necessary to understand that this is the way that ancient people have developed for living in their own territory, so that changes will always be present as they must gradually adapt to social and environmental requirements. However, current globalization harms this balance. According to the Charter on the built vernacular heritage (ICOMOS, 1994), this type of architecture will be recognized as: a manner of building of a local nature, with a style and form that are coherent with traditionally established architectural types, and a design dictated by traditional knowledge in keeping with the community's ideologies. The preservation of these dwellings is mostly driven by their occupants, followed by the remaining stakeholders, who must recognize the rights of these people to maintain their traditional ways of life through protection and regulation, so that their legacy may survive as fully as possible. To ensure the conservation of its architecture, it is necessary to establish principles for action and intervention in heritage (Fig. 7), and the charter mentioned above establishes the following points:

Table 3 SWOT analysis of Shuar centres

Strengths	Weaknesses	Opportunities	Threats
High cultural and natural diversity	Lack of management and economic dependence on the cities	Shuar cosmovision as the base of its spatial conception and architecture	Production activities such as livestock farming and mining
Valorization of the natural landscape, revolving around Shuar ways of life	Lack of emerging solutions which adapt Shuar dwellings to current needs	Orchards as a means of subsistence and domestication of surroundings	Natural risks and the vulnerability of settlements
Large young population as a starting point for cultural recovery	Material vulnerability of its architecture	Ecotourism as a possible model for cultural recovery and local development	Loss of cultural identity due to Westernized culture

- The conservation of this heritage must be carried out by multidisciplinary groups of experts in order to better recognize the changes and the identity which represents them,
- Any intervention on vernacular heritage must respect the cultural values of the site,
- The concern for emphasizing the identity values of a territory is reflected in the settlements so that interventions should not merely echo the architecture of single elements, but view it as a whole,
- The collective memory of inhabitants is a basic part of decision-making for conservation,
- Vernacular heritage does obey not only aesthetic, material, or spatial elements but also the value attached to it by the intangible expressions of the people.

5.1 Conservation of Natural Surroundings

Traditionally, the balanced way in which the Shuar people preserved nature ensured that settlements lasted for as long as the fertile soil allowed. This would wear out after about ten years, when the Shuar would move to a new fertile territory, abandoning the land and allowing it to regenerate. They also practised polyculture or crop rotation, planting a variety of products and making sure not to overwork the soil. In the same way, hunting could only be practised in the rainy or fertile seasons, providing abundant food for the wildlife and an optimum environment for them to reproduce. However, hunting is one of the practices which have been practically lost in this area, with local natural species now protected, due to the urban sites have already impinged on their natural environment (Fig. 8).

The environmental imbalance has a negative impact on the natural landscape, as does the indiscriminate felling of trees in their forests, where the road infrastructure has become necessary in the Amazon region to transport timber to other parts of the country and sell it. Furthermore, livestock, one of the main economic assets in the region, also has a negative effect on the virgin forests and the memory of natural heritage, causing erosion and loss of biodiversity in the territories (Chiriap et al., 2012; Magnaghi, 2012). Finally, as one of the most important imbalances causing major changes to the natural landscape is large-scale mining, the following are suggested:

- Promotion of reforestation plans,
- Controlled mining, respecting the environment and ancestral territories,
- Declaration of protected zones which completely conserve the natural landscape and regulation of hunting and fishing, taking advantage of seasons of abundance to encourage reproduction,
- Good handling of the soil using crop rotation to recover fertility,
- Recognition of sacred sites (forests, lakes, waterfalls, rivers, etc.) as part of cultural surroundings.

5.2 Conservation of a New Architectural Tradition

The conservation of this architecture focuses on highlighting values inherent to all its symbolic and anthropological elements, as well as the technical construction ones. However, the changes observed in the culture during the colonization process must also be taken into account, as the colonizing tradition brought about by these has directly affected architecture, prompting serious reflection on what has been lost, what is maintained, what is at risk, and what can be rescued. Some of these changes are as follows:

- Connection with colonization constructions, with the anthropometric measurement system shifting to metric,
- Material form, which proposes to improve the assembly of timber structures, as well as the spatial design currently expected from dwellings,
- Urbanism and the establishment of new facilities such as churches, educational and community institutions, and sports facilities, following religious colonization. Finally, urban growth leads to the provision of a road infrastructure connecting the main centres and promoting an axis for commercial exchange between Shuar centres and other regions.

The new construction methods require a rigorous distinction of the material forms which can help to rescue the memory of the cultural landscape in order to preserve the harmony between traditional methods and new materials when extending the useful life and comfort of the architecture. Self-construction opts for the most viable methods in the setting so that the execution of creative solutions adapted to current needs is a major point of reference in construction. Although the construction solutions appear in response to a socio-cultural setting, it is important to generate emerging forms of adaptation that can accommodate current forms of life and guarantee the direct involvement of design processes. It should be borne in mind that change is ever-present and not always negative to society (Magnaghi, 2012)(Figs. 9 and 10). In fact, these changes help us to understand the new synergies and metabolisms which translate as forms of action and heritage rescue.

5.3 Conservation of Shuar Dwellings

As the monumental architecture of Greek Acropolis, the Egyptian pyramids, and Mayan cities; vernacular architecture owes itself to its tangible and intangible context where the response to place means they can only ever be understood within their own specific singularities. In many cases, the architecture achieved is not controlled by a designer but is rather a spontaneous response to a place, repeated almost intuitively due to either tradition or popular knowledge. Housing is also the faithful reflection of the needs and desires of its occupants, satisfying cultural and physical aspects characteristic of the period (Rapoport, 1969) (Figs. 11 and 12). Furthermore, it should be noted that vernacular architecture is the result of human interaction with nature and its forms of inhabitation. After colonization, Shuar centres began to move away from their traditional ways of life, opting for Westernized ones, maintaining genetic and linguistic characteristics and bringing about major changes in spatial conception and design. A series of aspects can thus be considered for the conservation of the architecture:

- **Material**: It is important to assess traditional materials in terms of their local availability and respect for the environment. A management plan for the natural characteristics of the material form of the Shuar dwellings, mostly palm, should therefore be drawn up.
- **Spatial Organization**: This is an essential aspect of Shuar cohabitation, where surroundings and the dwelling were closely linked. An evaluation will be carried out on aspects which rescue this cultural memory of space, where traditional forms for the inhabitation of space are reinforced in the architecture, not imposed by a nostalgia for the past. In addition, changing dynamics call for an emerging architecture which can be easily adapted to new ways of life.

Fig. 11 Interior of traditional dwelling in *"Shuar Num"* museum located in Zamora Chinchipe (*Source* author)

Fig. 12 Detail of connection column roof (*Source* author)

- **Traditional Construction Systems**: The recovery of traditional construction systems is of importance when showing coherence with the surroundings. It entails a transmission of ancestral knowledge to future generations and is always the best solution. However, it is impossible to guarantee that construction techniques will endure in time, when new materials and forms of construction are accepted by the society. It is understood that the change we are continuously subject to, when adopted in architecture, is what is labelled as "new tradition" (Antoniades, 1971).
- **Adaptability**: When planning new architectural pieces, it is advisable to assess the dynamics between new and traditional in order to propose solutions that fit the site, both in historic and natural terms, without distorting its cultural memory. This architecture should also provide a response to its own symbols, and when new meanings come into play, these should be easily recognized and accepted by local residents to strengthen this cultural memory.
- **Maintenance**: Given the construction conditions of the dwelling, with its palm walls and structure, earth flooring, and plant roof, it is necessary to ensure that these are not jeopardized by climate conditions. It is also necessary to ensure that the more exposed points such as timber in contact with the soil are correctly drained as a protection against rot. This is also the case with the roof, using natural pesticides or drying leaves correctly so that they last longer.

6 Strategies for Recovery and Promotion

These strategies must be adapted to suitable management through a series of negotiations between the numerous government and private stakeholders in the area, along with international organizations which primarily fund projects for the recovery of heritage. However, the Shuar community is also a key player in monitoring this process. Therefore, any plans proposed in these ancestral territories must involve all these stakeholders in the decision-making process. Nevertheless, it is important to note what external agents can and cannot do in Shuar territory, and what can be done, what should be done, and what is desirable should be agreed with the owners of the house.

Within the management process, it is important to consider the ancestral cosmovision of these people, which serves as a basis for social, economic, and environmental aspects. The practice of customs and traditions contributes to the actual conservation of the management of resources and the territory, just as it did in the past and will continue to do in the future. It is therefore important to recover and restore value to this knowledge of the past (Barragán, 2008). When proposing management plans, it is important to first design the organization of processes ensuring the involvement of stakeholders with an interest in the development of indigenous people, and in the search for contact networks for the creation of a communications platform to cover any needs. It is also necessary to guarantee that these plans established are monitored by the main stakeholders, in this case the community and administrators. Education has become a vital asset for the recovery and promotion of cultural and architectural heritage, providing a platform for dissemination to future stakeholders, who will be in control. However, for this dissemination to become real resources and an extensive body of knowledge summarized for education are necessary (Monosalvas & Heras, 2017). Possible ways of dissemination to aid the recovery of heritage awareness in Shuar territory are as follows:

- Educational programmes on traditional heritage,
- Specialization programmes for the community so that they can be in charge of maintenance and trades,
- Information programmes promoting cultural memory and linking the cosmovision and forms of thought for a global recovery and territorial management in keeping with traditions,
- Connection to regional networks for the exchange of experiences and specialists,
- Initiatives for community sustainable development, connecting Shuar centres with each other and cities, generating competences which put inhabitants in charge of promoting their own development.

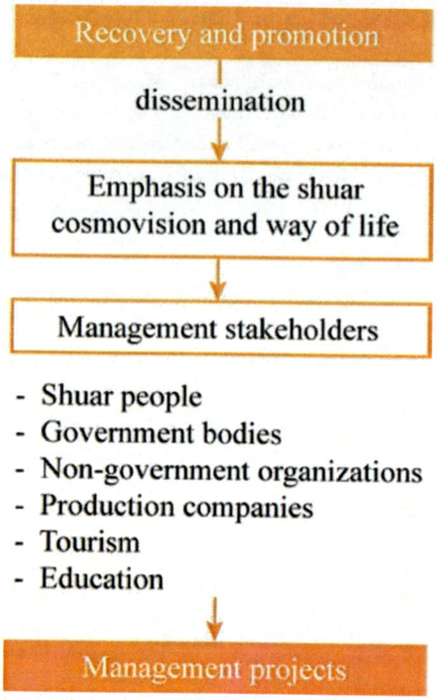

Fig. 13 Main aspects of recovery and promotion strategies (Morocho et al., 2020)

6.1 Conservation of the Cosmovision and Ways of Life

The Shuar cosmovision is an important asset in preserving awareness of local heritage and defending the cultural identity of the people (Morocho-Jaramillo, 2022) (Barriga et al., 2018). As the everyday life of the Shuar revolved around a natural landscape where nature was viewed from a macrocosmos where life unfolds, paying homage to several of these natural sites considered sacred, dwellings become the sacralization of a microcosmos, highlighting these internal–external dynamics. In order to preserve the Shuar ways of thinking, it is important to maintain any symbols inherent to their forms of inhabiting space. The following are important elements to bear in mind:

- Conservation of symbolic elements of dwellings as well as the central columns which represent the connection between earthly and sacred,
- Delimitation of spaces where Shuar life unfolds. In the female zone (*ekent*), household activities such as cooking, sleeping, and everyday family cohabitation take place, and this space even has a door providing direct access to the orchard. The male zone (*tankamash*) is the social space where visitors are welcomed, and major

Fig. 14 Promotion of ancestral ways of thinking through customs and traditions in *"Tuntiak Nunkeē"* natural reserve in Morona Santiago (*Source* author)

agreements with other families are reached; this space is also connected to the jungle,

- The connectivity and dynamism between dwellings and natural sacred sites such as rivers, waterfalls, forests, and lakes. These are places in which the divine figure *"Arútam"* is reflected in different forms telling myths from which Shuar behaviour stems. These are not mere legends but rather accounts of hunting, sowing, and construction and how to behave in different everyday situations,
- The maintenance of the orchard is an important point to be reassessed within the Shuar cosmovision, as this is where knowledge is passed down over generations. The cultivation of medicinal plants, food from the Shuar diet, and forms of cultivation are all experiences which provide sustenance to families. In addition to being a space for the transmission of ancestral knowledge in Shuar families, it allows them to compete in terms of production with other Shuar centres and a regional connection with cities,
- The promotion of traditional activities such as hunting, fishing, agriculture, crafts, collection of raw materials from the forest, and ancestral medicine can be reflected in spatial provisions and territorial structuring, improving the dynamism between dwelling and lifestyle (Fig. 8),
- Traditional construction techniques are an important aspect for the preservation of cultural memory amongst young people, as they help guarantee the presence of specialist labour able to conserve the architecture. Elevated locations are preferred, with a sightline to forests and close to water and natural resources to satisfy any production and spiritual needs,
- The promotion of communal spaces for use as locations for religious celebrations and rites, where the Shuar culture can be expressed. Despite being the product of colonization, these communal spaces and the religious infrastructure encourage social cohesion and strengthen identity, both socio-culturally and in terms of community.

6.2 Management Stakeholders

According to UNESCO, the cultural heritage of a people is the heritage of all individuals. Therefore, the responsibility of rescuing and managing the conservation of tangible and intangible heritage is not only the responsibility of local inhabitants but also the responsibility of stakeholders involved in the cooperation processes for the conservation of cultural expressions and values (ICOMOS, 1994). The management of the Shuar territory must be in keeping with its cosmovision, and this begins with identity being recovered by the direct administrators of the place, its inhabitants. At a later stage, these dynamics with nature and its settlements are the driving force of their own management. Thus, the guidelines set to manage strategies must be the product of a reclaimed awareness of heritage, generating activities that are relevant to this development (Morocho-Jaramillo, 2022).

Furthermore, training is also important for encouraging the participation of communities, guaranteeing the regeneration of values through conservation programmes and plans, ensuring the development of the community and the long-term conservation of tangible and intangible heritage. Training is observed in both the conservation of architecture and the promotion of traditional knowledge through activities for the safeguarding of the natural environment. This training is based on strategic plans incorporating the need for sustainable consumption and the integration on a regional, a national and even international scale, improving the socio-economic conditions of the site. The programmes are also geared towards technical professionals, artisans, conservation, and management experts who can implement and be part of cultural rescue programmes through sustainable projects for mitigating environmental impact and for socio-economic and socio-cultural development (Vásquez & Díaz, 2009) (Fig. 15 and 16).

At present, tourism can generate economic recovery strategies (Carpentier, 2014). However, a more appropriate option is the museum-like reconstruction of the structures within a given landscape, where ethnological and cultural recovery is more important than shallow commerce, interested only in tourist flow rather than the development of the Shuar nation. This entails the recovery of ancestral construction techniques, training specialist workers to maintain continuity in the heritage of construction. The new architecture must accommodate those ancestral paradigms to avoid becoming only a reminder of the past, not a clear reflection of their culture (Núñez, 2012). When proposing management projects, some of the following strategies can be mentioned:

Fig. 15 Community tourism in *"Tuntiak Nunkeẽ"* natural reserve (*Source* author)

Fig. 16 Shuar museum *"Shuar Num"* (*Source* author)

- Interinstitutional coordination between decentralized autonomous governments and the active stakeholders of the Shuar communities, aiming to regulate the use and administration of the ancestral territory;
- Communication with the members of the Shuar organizations who are the main administrations calling for the rights of ancestral people;
- Coordination between the government stakeholders of the environment to regulate and promote the protected areas which border ancestral territories;
- Incorporation of ancestral forests into the National System of Protected Areas;
- Coordination with the "Instituto Nacional de Patrimonio Cultural" (INPC) [National Institute for Cultural Heritage] for the valorization and development of the tangible and intangible heritage of the Shuar culture in the Amazon region;
- Collaboration with other national and international non-government bodies interested in the promotion of ecotourism or the cultural rescue of indigenous people through projects which link the community within a social dynamic that promotes ancestral values.

7 Conclusions and discussion

The issue of the loss of architectural heritage is one that is difficult to resolve as it involves social aspects which must be assessed from the root cause. This in turn makes it easy to comprehend what the situation is and what actions to undertake. However, the influence of a globalized world complicates this task so that the issues examined in this article show the most advisable ways to recover heritage awareness. Firstly, the strategies for conservation and protection of heritage require documentation and research to identify the traditional construction methods and the meanings of the elements aiming to guarantee a habitat suitable for their ways of inhabiting space. Therefore, a starting point should be the conservation of the natural setting and a context which will allow the development of its settlements to answer their architectural needs. The most important aspects of these are spatial design, material, construction system, and finally, adaptability to change and resilience based on cultural legacy. These aspects help identify solutions for an architectural recovery in keeping with their culture. Here, it is important to note that every single aspect of architecture changes. Shuar settlements experienced major changes where the new architecture imposed was not in keeping with the anthropological aspects of its inhabitants. However, these changes were ones which gradually adapted the ways of inhabiting space, unavoidable changes which must be understood in order to guaranteee its conservation.

Secondly, the recovery and correct management of the architectural heritage incorporates dissemination as well as conservation. These processes are observed from the cosmovision and ways of life of the Shuar, and it is advisable to use alternatives for social recovery, with the community managing its own development according to its values and principles. This in turn helps reactivate a cultural identity for the full revalorization of the cultural landscape by strengthening inhabitants' ability to generate employment. This social reactivation requires the involvement of a series of stakeholders in management projects: locals, government, and non-government bodies, production companies, tourism, and education. It is essential to provide training processes involving professionals, specialist artisans, and new generations as they are the ones who will assimilate ancestral knowledge and the sense of heritage recovery which will endure and improve over time.

Finally, understanding different types of environmental relationships and architecture provides us a better understanding of indigenous societies. The examinations carried out in this article outlined the relationship between architecture and culture according to the anthropological discussion by Descola et al., (2001) and Viveiros de Castro, (2010). However, different interpretations can also be found in agreement with Latour's stance (2009) identifying a modern ideological dispute, inherent to a consumerist system which seeks recognition as progress. Although Amazonians make no distinction between *human* and *non-human*, the wide range of discrepancies encountered provides examples of numerous transformation processes and, in turn, different relationships.

References

Antoniades, A. C. (1971). Traditional versus contemporary elements in architecture. *New Mexico Architecture, 13*(N11&12), 9–14.

Bakonirina, R., Moriset, S., Sánchez, N., Mabille, M., & Carbonnelle, A. (2022). *Educational tools to transmit heritage values. Versus Plus heritage for people* (p. 3). ENSAG-CRAterre. https://versus-people.webs.upv.es/booklet/

Barragán, L. (2008). *Manual de Capacitación sobre Pueblos Indígenas y Áreas Protegidas en América Latina. Uso y Manejo de Recursos Naturales*. M. Vergara, & J. Casaza.

Barriga, F., Núñez, J., Guerra, G., & Maldonado, L. (2018). *Historia y antología de la literatura ecuatoriana.* Vol. 1. 317–328.

Bianchi, C. (1978). *La casa shuar. Mundo Shuar*. (Vol. 1). Serie C. Abya Yala.

Bottasso, J. (2019). El proyecto misionero amazónico de los salesianos. Contradicciones y tensiones. Misiones, pueblos indígenas y la conformación de la Región Amazónica: Actores, tensiones y debates actuales. Abya-Yala. 131–136. https://doi.org/10.7476/9789978104941.0008

Carpentier, J. (2014). Los achuar y el ecoturismo: ¿una estrategia sostenible para un desarrollo autónomo?. *Bulletin de l'Institut français d'études andines, 43*(1), 133–158. https://doi.org/10.4000/bifea.4391

Chiriap, N., Jimbiquiti, L., Kayap, O., Kuja, E., Mayak, I., & Sharup, N. (2012). Sabiduría de la Cultura Shuar de la Amazonía Ecuatoriana. Sabiduría Amazónica (vol. 2 (N1), 231). UNICEF, DINEIB, FLACSO. http://8.242.217.84:8080/jspui/handle/123456789/31675

Correia, M., Dipasquale, L., & Mecca, S. (2014). Versus: Heritage for tomorrow: Vernacular knowledge for sustainable architecture. M. Correia & L. Dipascuale. https://www.esg.pt/versus

Descola, P., Pälsson, G., Ingold, T., Hornborg, A., Ellen, R. F., Howell, S., Rival, L., Hviding, E., Ärhem, K., Hell, B., Knight, J., Papagaroufali, E., Nothnagel, D., Richards, P., & Ruivenkamp, G. (2001). Naturaleza y sociedad. *Perspectivas antropológicas. 2*(102), 118.

Descola, P. (2017). La selva culta. Simbolismo y praxis en la ecología de los Achuar. Colección Pueblos del Ecuador. (Vol. 3) 171–215). Abya-Yala. https://doi.org/10.4000/books.ifea.1600

Deshoullière, G. & Utitiaj, S., (2019).Acerca de la Declaración sobre el cambio de nombre del conjunto Jívaro. *Journal de la Société des américanistes,105*(105-2), 167–179. https://doi.org/10.4000/jsa.17370

Eliade, M. (1981). Lo sagrado y lo profano. Guardarrama/Punto Omega (p. 23).

Garzón, V. (2013). Los salesianos en el Oriente Ecuatoriano 1893–1970. El internado de Sevilla Don Bosco y sus repercusiones en la cultura Shuar contemporánea. Universidad Pablo de Olavide de Sevilla. http://hdl.handle.net/10433/801

Gnerre, M. (2014). Los salesianos y los shuar.Construyendo la identidad cultural. *La presencia salesiana en Ecuador: Perspectivas históricas y sociales, 3° ed*, pp567–628. Abya-Yala. https://dspace.ups.edu.ec/handle/123456789/19115

Harner, M. (1978). Shuar. Pueblo de cascadas sagradas. *Mundo shuar*. Abya-Yala. 40–41.

ICOMOS [International Council of Monuments and Sites]. (1994). Documento de Nara. Retrived April 11, 2023, from https://icomos.es/wp-content/uploads/2020/01/21.CONFERENCIADENARASOBREAUTENTICIDAD1994.pdf

Juncosa, E. (2020). Civilizaciones en disputa. Educación y evangelización en el territorio shuar. Serie Investigación decolonial Universidad Andina Simón Bolívar. Abya-Yala. 21–38.

Latour, B., Aspinall, P. J., & Maguire, M. (2009). Perspectivism: Type or bomb? *Anthropology Today, 25*(2), 1–2.

Magnaghi, A. (2012). El proyecto local: Hacia una conciencia del lugar. Arquitectonics. Mind, land & society. Universitat Politècnica de Catalunya. https://upcommons.upc.edu/handle/2099.3/36586

Monosalvas, D. & Heras, V. (2017). Valoración de la arquitectura vernácula de las áreas históricas y patrimoniales. Universidad de Cuenca. http://dspace.ucuenca.edu.ec/handle/123456789/26549

Morocho, D. (2020). Arquitectura Shuar: Estudio de los valores materiales e inmateriales del patrimonio arquitectónico en Yantzaza (Ecuador) y propuesta para unas estrategias de recuperación. Universitat Politècnica de València. http://hdl.handle.net/10251/159781

Morocho-Jaramillo, D. (2022). Shuar architecture as a model of sustainability. edUPV (Eds), *HERITAGE 2022—International conference on vernacular heritage: Culture, people and sustainability.* 499–506. https://doi.org/10.4995/HERITAGE2022.2022.15029

Núñez, P. (2012). Turismo y patrimonio vernáculo: ¿Estrategia de recuperación sustentable?. Paisaje cultural urbano e identidad territorial, 2° Coloquio Internacional RIGPAC. 446–459.

Pillaceta, R., León, S., Martínez, J. (1998). La cultura shuar: La vivienda y su evolución en la Asociación de Bomboiza. Universidad de Cuenca. http://dspace.ucuenca.edu.ec/handle/123456789/15080

Pueblo Shuar Arutam. (n.d.). Plan de vida. Territorio Indígena y Gobernanza. Retrieved April 11, 2023, from http://territorioindigenaygobernanza.com/web/gobernanza/

Rapoport, A. (1969). *House form and culture. Foundations of cultural geography series*. University of Wisconsin-Milwaukee.

Rubenstein, S. (2005). La conversión de los shuar. *Revista de Ciencias Sociales Iconos, 22*, 27–48. https://doi.org/10.17141/iconos.22.2005.99

Sanz, S. & Herrera, I. (2017). La arquitectura shuar: Ordenando el espacio mítico. *Revista Española de Antropología Americana, 47*, 164. https://doi.org/10.5209/REAA.61976

Sharupi, M., Kueva, F., & Durán, A. [Fold]. (2022, October 14). *María Clara Sharupi: An Amazonian thinker #3. Antropología más allá de la traducción [Video]*. Retrived from https://thisisthefold.org/viewpoints11mara-clara-sharupi-jua

Surrallés, A., & García Hierro, P. (2004). Tierra Adentro. Territorio indígena y percepción del entorno. *IWGIA* (vol. 15, p. 34).

Vásquez, V., & Díaz, J. (2009). Optimización de una metodología de análisis para la rehabilitación y protección sostenible de la arquitectura vernácula. Universitat Politècnica de Catalunya. http://hdl.handle.net/2117/93446

Viveiros de Castro, E. (2010). Metafísicas caníbales: Líneas de antropología postestructural (vol. 1, p. 34).

Assessing the Methods of Conservation for Cultural Heritage Values in Cairo: Lessons Learnt or Adoption from the Case Studies

Sally Shafei, Morad Abdelkader, and Akram Farouk Mohamed

Abstract

Scholarly studies have illustrated that the interaction between cities' history and their architecture positively affects the creation of a distinctive architectural style through its cultural heritage. Hence, it is essential to implement accurate conservation methods for cultural heritage in order to raise cities' value. The conservation process of heritage needs to create an oriented evaluation module that can serve as a guide for decision-makers and conservation professionals. Therefore, this study tries to identify the significant preservation methods to design a module that evaluates modern techniques to preserve cultural heritage. To achieve the study objective, the descriptive-analytical approach has been used to examine the conservation process by the comparative analysis between an international case Palazzo D'Accursio in Italy, and a national one, the Technology Innovation and Entrepreneurship Development Center (TIEC) in Cairo. Then, the study involved the application of a strengths, weaknesses, opportunities, and threats (SWOT) analysis in conjunction with a social, technological, economic, environmental, and political (STEEP) analysis to organize the different impacts generated by the adaptive reuse case studies in order to reach the findings of this study which extract the scientific framework in an exact and sequential order, which will be followed by the assessment and then the final report to be adopted in conservation in the future.

S. Shafei (✉) · M. Abdelkader · A. F. Mohamed
Engineering Faculty, Ain Shams University, Cairo, Egypt
e-mail: 2002440@eng.asu.edu.eg

M. Abdelkader
e-mail: morad_abdelkader@eng.asu.edu.eg

A. F. Mohamed
e-mail: akram_mohamed@eng.asu.edu.eg

S. Shafei
Higher Technology Institute, 10th of Ramadan City, Egypt

Keywords

Conservation · Preservation method · Cities' value · Cultural heritage

1 Introduction

Cities' history is considered a witness to their aesthetic. History enriches cities with its character, which in turn builds cities' identity. Since identity is produced through the environment, people, and interaction between them (Ali et al., 2022), the urban form is an expressive characteristic of the cultural dimension of society; it represents a civilizational value. This cultural value of urbanism reaches its highest levels when it is associated with heritage content. Therefore, complacency or neglect regarding the cultural legacies that enrich urbanization is nothing but a lack of national awareness of civilizational values, a waste of national wealth, and a decline in the cultural dimension of society. Therefore, monitoring and documenting the distinctive architectural and heritage value as a memory of cities is essential (Supreme Council for Planning and Urban Development, 2008). Hence, the definition of heritage refers to something that is inherited. Consequently, it incorporates the concept of past and future transmission (Al-sakkaf et al., 2020). A heritage building is generally categorized as an existing value with a high cultural significance to society. It is also defined as a building that offers everyone a sense of admiration and makes the desire to learn more about the inhabitants of this building and their culture in addition to understanding its aesthetic, historical, archaeological, economic, social, and political significance (Hanso, 2016).

The expression of cultural heritage can be represented and distinguished into two kinds: tangible and intangible. Tangible heritage includes materialistic objects, for instance, buildings, monuments, paintings, and statues. In comparison, intangible heritage focuses on immaterial

M. F. Shahidan et al. (eds.), *Urban Narratives: Exploring Identity, Heritage, and Sustainable Development in Cities*, Advances in Science, Technology & Innovation, https://doi.org/10.1007/978-3-031-48517-6_16

cultural heritage, such as literature, music, theater, religious rituals, and traditional performing. UNESCO and the International Council on Monuments and Sites (ICOMOS) agreed that heritage should include cultural and natural heritage. To comprehend global cultural diversity, culture and nature could not be separated. The phrase "cultural heritage" has developed during the last few decades. Initially, it was a masterpiece of historical value. Currently, it is used more broadly and covers everything that has a particular significance for people. Despite the importance of heritage value, it suffers from ongoing problems and threats. This can be attributed to several interrelated factors; the most significant factor is considered the absence of overall control or policy as a result of the multiplicity of overlapping authorities in charge of heritage areas, in addition to the limited resources, and the lack of a complete inventory of heritage buildings (Hesham, 2018), particularly, in Cairo which is distinguished by its historical value. Consequently, the research problem was emphasized.

Heritage conservation is substantial in order to protect and maintain the cultural, spiritual, and architectural significance of heritage value. Conservation refers to any form of place protection that keeps a location's cultural identity (The Burra Charter: The Australia ICOMOS Charter for Places of Cultural Significance, 2013). All activities that extend the existence of cultural and natural heritage are included. The minimum possible action is the most appropriate; if possible, it should be reversible and not have an inverse impact on potential future interventions (Gupta & Prakash, 2018). There are numerous approaches depending on the circumstances and goals of a conservation project; many techniques may be utilized singly or in combination in a conservation plan. Objectively, this study attempts to focus on the conservation method as an approach to protecting heritage value (Aga Khan Trust for Culture (AKTC), 2013). Considering the importance of cultural heritage values, it is necessary to evaluate relevant information to understand the place and its texture. As a result, the proposal has been drawn up, which includes a report with all materials. The report's contents will be organized to fit the site and task constraints.

To accomplish the purpose of the paper is organized into three stages through five sections, as shown in Fig. 1. The first section starts with this introduction. Then, it is followed by the stage which describes the materials and methods in two steps: the case study selection criteria and the case study settings to represent the comparative analytical study for the two case studies following the same hierarchy. Then, in the third section, the paper applies the assessment tool in the two study cases to evaluate the conservation process through the SWOT analysis in the merger with STEEP analysis to organize the many effects produced by an adaptive reuse project. Finally, the framework will be developed in the third stage first section. In the last section, the conclusion summarizes the whole process by establishing a framework to be the guideline for the whole conservation process throughout all its levels.

1.1 Case Selection

In this section, the research will discuss two case studies that applied adaptive sustainable heritage reuse from local and global perceptions, demonstrating how each organization dealt with a heritage building. Analytical hypotheses and analyses will be developed for each case study to determine the appropriate strategies and techniques for restoring heritage buildings. The same hierarchy will be followed in each situation to reach the ultimate results. On the other hand, adaptive reuse is an essential step in the conservation process. Thus, it will be a condition in selecting the study case. This condition is defined as follows:

Adaptive reuse: This term can be defined as an essential idea in heritage conservation policy, and "adaptive reuse" is an efficient approach which also given significant consideration in the heritage conservation efforts of cities around the globe (Amin et al., 2012). Through the action of adaptive reuse, an ancient or ineffective item is transformed into a new one which could serve a different aim for future generations. This long-term, forward-looking approach to heritage protection has numerous positive impacts on the environment, economy, and social well-being (Wangkeot, 2002). Hence, the preservation and reuse of architectural landmarks provide long-term advantages for those who value cultural heritage (Frogé et al., 1999). Based on this definition, the study will analyze some of the economic, social, and environmental benefits of the adaptive reuse of heritage buildings, as illustrated in Fig. 2, and this research selected two cases to examine.

2 Case Study

2.1 Case Study 1: The International Case Study—Palazzo D'Accursio

2.1.1 Building Description

(A) Site context

The Palace is considered the city's headquarters, as shown in Fig. 3. Moreover, it is considered the community's museum which is located in Piazza Maggiore, in the historical center of Bologna, Italy. The building is strengthened by

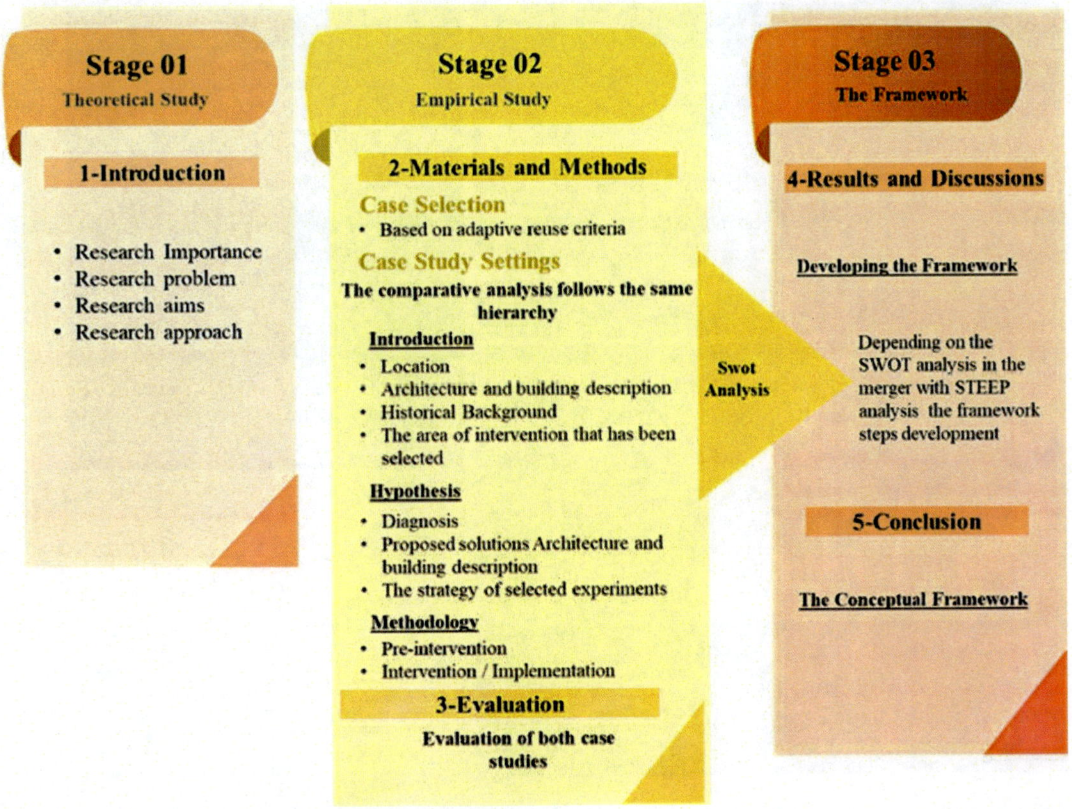

Fig. 1 Research structure (*Source* The Authors) materials and methods

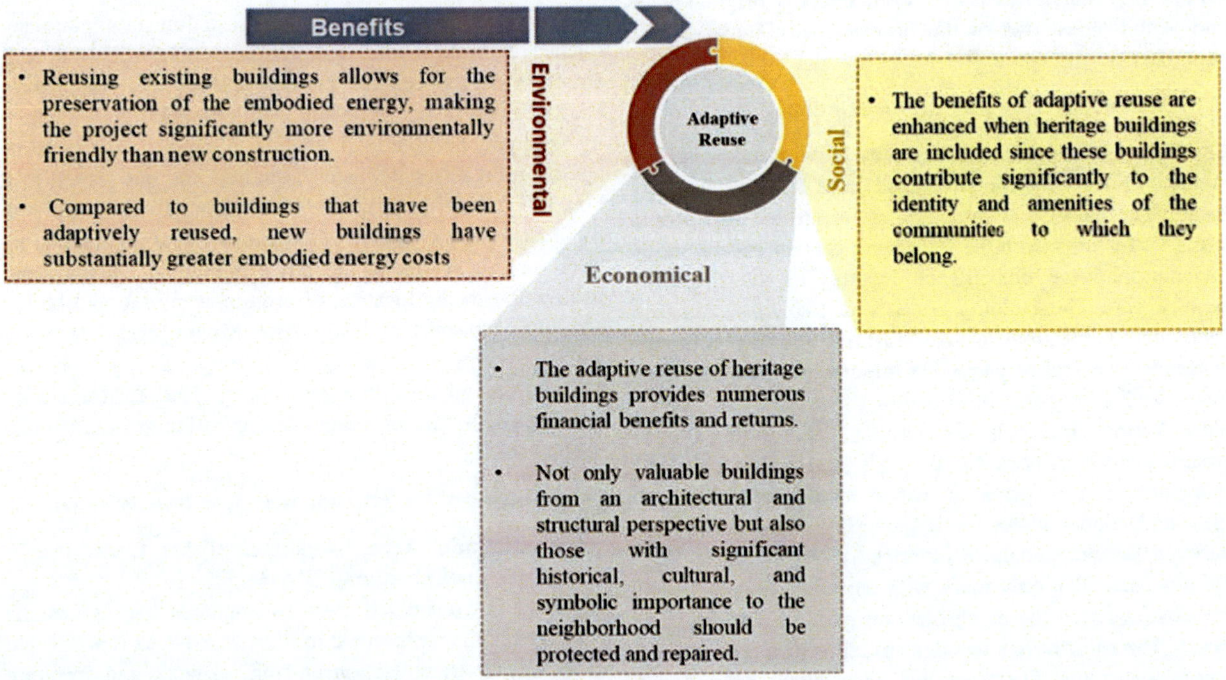

Fig. 2 Adaptive reuse benefits (*Source* (Anna, 2022) Edited by: The Authors)

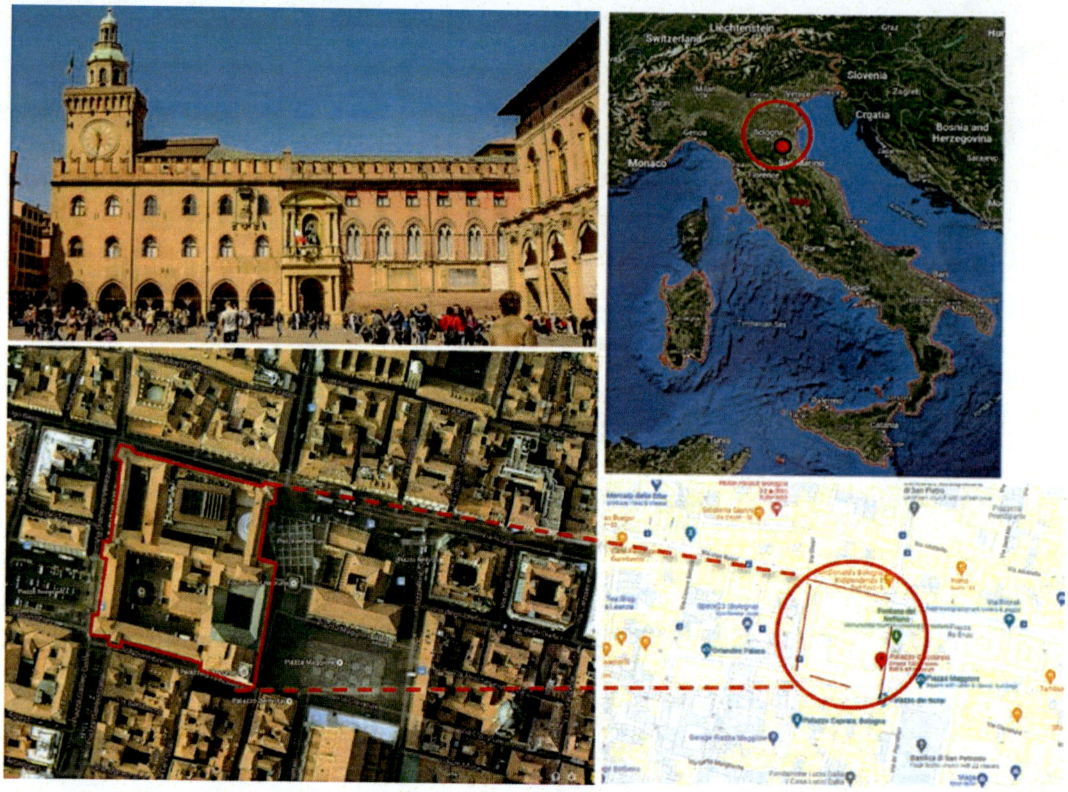

Fig. 3 Location of the first case study—Palazzo D'Accursio (*Source* Google Maps, goo.gl/maps/2A8xb5aDkspGLG8UA, Edited by Author)

its location between Maggiore Square and the historic Via Emilia. It is functioned as the "Roman city's" original heart and is the oldest part of the historical city center (Pesci et al., 2012).

(B) **Architecture and building description**

The Palace is considered a complex building with various architectural styles, Medieval and later Renaissance elements. The palace's architectural styles are heterogeneous, while its features from the fourteenth-century severe architectonic language embellished features of the fifteenth-century part. The building consists of two museums (the Municipal Collections of Ancient Art and the Morandi Museum), the seat of public institutions, a municipal pharmacy, and a citizens' information office (Hall & Maria, n.d.). Various materials are used in the building's construction, such as brick for the wall structure, marble and sandstone for the decoration, wood for the roof structure, tiles, and copper plates for its covering. Additionally, the building facades are made of bricks with decorative marble elements. The only roofs with terracotta tiles and copper coatings face the enormous courtyard, as identified in Fig. 4. The roofs have a wooden structure with primary and secondary beams. There are five floors above ground and two basement floors, as shown in Fig. 5 (Troi, 2014).

(C) **Historical background**

D'Accursio Palace's present construction is the consequence of several interventions: beginning with the construction of the original core of the structure known as Biada Palace in the thirteenth century, which was shielded by a square perimeter of walls and utilized for the storage of grain. Over its history, the building has served as a host to various functions and institutions. Today, it became the Local Authority and Government's official location. Therefore, it serves as the city's municipal headquarters. In addition to the location of prominent museums like the Morandi Museum and the Arts Municipal Collections, which contain artwork and furnishings dating from the Middle Ages to the nineteenth century (Hall & Maria, n.d.; Pesci et al., 2012), the history of the buildings is illustrated in (Fig. 6).

(D) **The area of intervention that have been selected**

- **Functional Area 1—Municipal Art Collections—Collezioni Comunali D'Arte**

This area is considered the most valuable heritage of the Palace which is represented in Figs. 7 and 8, as it is devoted to the city's art collection and has artwork and furniture from many eras. The collections comprise a "furnished

Fig. 4 Building façade facing the monumental courtyard. *Source* (Troi, 2014)

Fig. 5 Building section. *Source* (Troi, 2014)

museum," with the historical decorations still present in the halls where the furniture is the main attraction, as shown in Fig. 9. The Municipal Art Collections is located in the former winter apartment of the Cardinal Legates and consists of fourteen different rooms with 6.51 m height interpolated average net (Bologna, 2022).

- **Functional Area 2—Municipal Police Offices—Pusterla Tower**

The locations of the selected area are the ground level and the first under-floor of the building's Southern Wing as shown in Figs. 10, 11, and 12. The offices and gathering locations for the municipal council staff and the City Council's local political assembly groups have been on the first under-floor. It consists of forty-five rooms with a 3.18 m height interpolated average net (Hubert, 1993).

- **Functional Area 3—Municipal Offices—Pusterla Tower**

The building's Southern Wing is also where the chosen area is situated. The second under-floor and the first floor are included in this functional area specifically AS represented in Fig. 13. There are a total of forty-nine rooms in the building, with twenty rooms on the first floor and twenty-nine rooms on the second as shown in Figs. 14 and 15. Considering that the selected area represents a part of the building's historic core and has unique structural characteristics, the chosen region is given constant heritage value (Hall & Maria, n.d.).

2.1.2 Building Documentation

According to the Urban Building Regulation Code, the building is considered to have historical and architectural value, so it is subject to a high degree of preservation. Hence, only considerate renovation and maintenance interventions are permitted. No single strategy works for all building kinds because the Palace consists of many buildings, and each one is distinct. Hence, the building preservation hypothesis and strategy have been developed. With the analytical approach, the building has no obvious structural problems. On the contrary, some problems have required intervention.

(A) **Diagnosis** (Troi, 2014)

The recommended work involves examining some crucially strategic locations inside the Palace to evaluate its condition. Identifying the type of building's structure and its various implementation details, such as:

- Building envelope characteristics and maintenance conditions.
- Cooling and heating system.
- Adopted lighting style.
- Fixture types.
- Description of any cooling equipment and air-handling units.
- Explanation of the systems used in distribution substations and thermal power stations.
- Technical operational characteristics and control classification.

(B) **Proposed solutions** (Troi, 2014)

- Monitoring of air-handling systems and modern control systems.
- Modern lighting fixtures demand new energy-saving solutions.
- Insulation techniques for floor coverage and walls.
- Windows insulation system.

(C) **The strategy of selected experiments**

- A test that promotes the knowledge of the building materials used and possible building restorations performed (Solla et al., 2012).
- A test that determines the properties of the surface material (Pesci et al., 2012).
- A test measures the entire loss of internal air volume (Dabouh, 2020).
- Estimation of the building's heat losses (Ubertini et al., 2014).
- Monitoring the hygrothermal and environmental status (Al-mukhtar & Hadi, 2019).

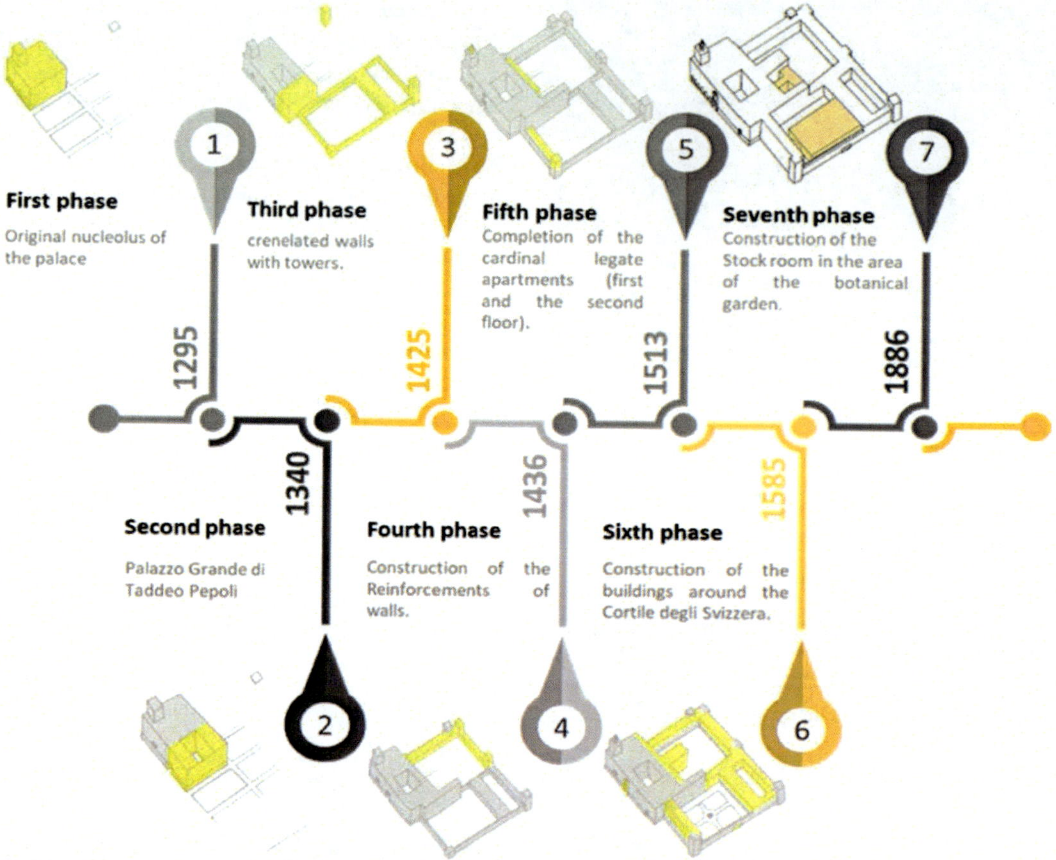

Fig. 6 Construction phases of the building (*Source* (Troi, 2014), Edited by: The Authors)

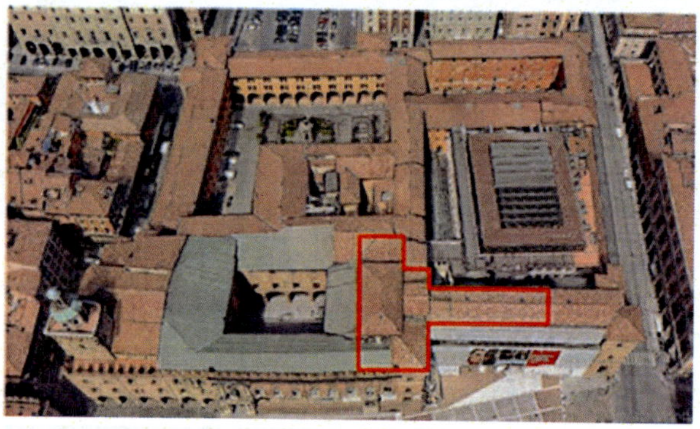

Fig. 7 :Municipal art collections areal views (*Source* https://goo.gl/maps/Md3ShrgWTvuxk38M9 Edited by The Auhors)

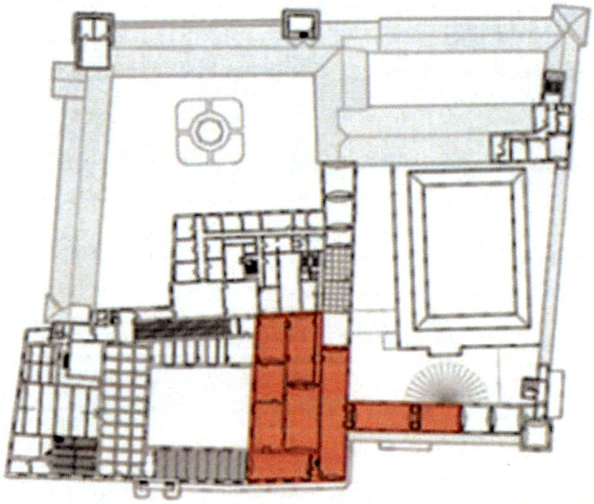

Fig. 8 Municipal art collections plan *Source* (Dabouh, 2020)

2.1.3 Methodology

In order to identify the appropriate modifications to conserve the existing building and enhance its properties to achieve the desired sustainability direction, the analytical

Fig. 9 Municipal art collections interior shot *Source* (Bologna, 2022)

Fig. 10 Municipal police offices façade *Source* (Troi, 2014)

Fig. 11 Municipal police offices areal view (*Source* google maps Edited by Author)

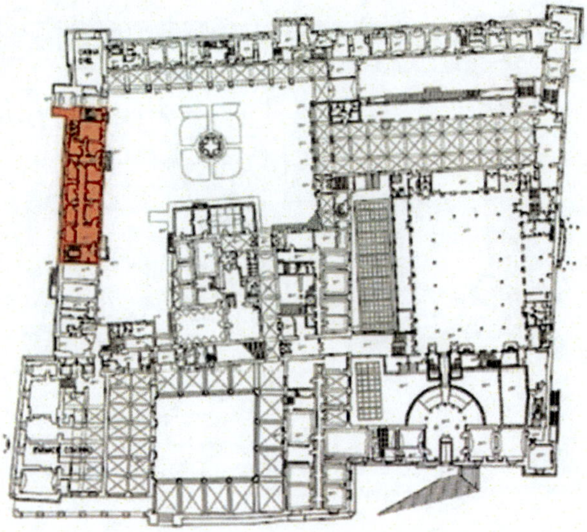

Fig. 12 Municipal police offices plan *Source* (Dabouh, 2020)

Fig. 13 Municipal offices façade. *Source* (Troi, 2014)

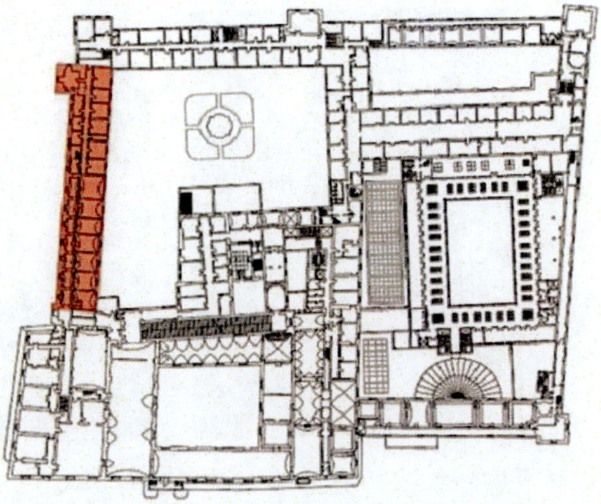

Fig. 14 Municipal offices first floor. *Source* (Dabouh, 2020)

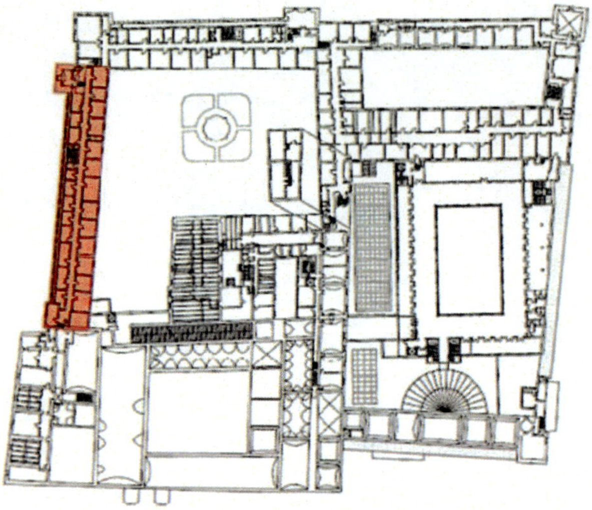

Fig. 15 Municipal offices second under floor. *Source* (Dabouh, 2020)

approach is used. The methodology is divided into two main sections pre-intervention and intervention design status.

(A) **Pre-intervention Status**

- **Architectural Analysis**

The Palace was constructed with locally common materials: With some modifications, marble was utilized to ornament the building, with brick for the structural supports and two or three sandstone heads used for the decorative elements, as shown in Fig. 16. The ceiling of the second level is built of thin plaster arches or, more frequently, a wooden framework covered in a three-cm-thick wooden plank. In addition to the Sala Bosra library, where the ancient windows were substantially replaced by new ones with better air quality, temperature, and penetration of water leakage, all the windows are wooden and single-glazed, while in the southern area under examination, the majority of the wall's thickness varies from 90 to 100 cm (Bergamini, 1976).

The following conclusions have been reached after analyzing the museum's primary exhibition room:

- Considering the heritage aspects, no more openings, additional shade, or adding further layers to the interior surface of the space are allowed. However, in the case of component substitution, which is only authorized if the original components cannot be fixed, the newly inserted elements must have the same partition, material, color, and form as the primary elements (Hubert, 1993).
- It has been studied to modify and replace the current space lighting systems. Moreover, there are serious challenges with the hygrometric control and mixed building features (Troi, 2014).

(B) **Experimental assessment**

At the current stage, full control analysis was used to create the project in order to understand the structure and evaluate its performance in order to select the appropriate interventions through the following techniques, which are demonstrated in Table 1.

(C) **Intervention status**

The intervention concept intends to integrate urgently required non-ordinary maintenance to address the issues of building deterioration and the environment. Following the simulation intervention, the case study was subjected to both the PHPP Protocol and Design Builder Modeling during the as-is state diagnosis phase. The hypothesized solutions are taken into consideration in order to achieve the planned conservation, which will be explained in Table 2.

Fig. 16 Brick walls and marble arches *Source* (Bologna, 2022)

Table 1 Experimental assessment analysis

Technique	Implementation strategy	Findings
Structural analysis and moisture assessment		
Ground-penetrating radar tests for masonry stratigraphy (GPR)	GPR radar tests have been performed throughout six different testing days. Data have been gathered in a reflection model using a GSSI SIR-3000 system and various mediums (900 MHz) to high (2.0 GHz) frequency antennas (Troi, 2014)	The assessment of masonry structures has demonstrated good potential for GPR, which could provide precious information. The proposed approach has the capability to reveal details concerning internal operations Moreover, as illustrated in Fig. 17, evaluating masonry structures has shown GPR to have high ability and the capacity to yield invaluable data. The suggested method can provide information about internal processes (Solla et al., 2012)
Infrared thermography (IRT)	Radiant heat flow is influenced by a variety of properties of the materials and the surroundings of the structural element under examination • The characteristics of the surface • Surface materials • Existence of substrate discontinuities • Substrate materials Several measurement sessions were necessary for the IRT examination of the Palace, as shown in Fig. 18 that indicates the locations where the test was conducted (Pesci et al., 2012)	*Sala Dei Primitivi*: Only a few tiny fractures were found in the roof, that has a well-defined supporting wooden framework. Brick room walls appear to be quite homogeneous. No indication of a water leak or other damage (Ubertini et al., 2014)
		Sala Degli Svizzeri: The IRT inspection's most remarkable finding is visualizing a predictable sequence of hidden structures in the top West wall, whose origin and consistency have been determined from pictures. Other findings include the detection of some water incursion and the identification of a few detachments (Ubertini et al., 2014)
		Sala Urbana: The most concerning findings are the scattered detachments of + the decorative paints in the room as illustrated in Fig. 19. IRT alone cannot evaluate the possibility of the supportive plasters breaking off, so a more full investigation is advised, especially before the anticipated roof rehabilitation work begins. Additionally, some possible water infiltrations have been discovered, and hot spots can be seen in the upper portions of the East and South walls
Blower door test	The test evaluates how long it takes for internal air volume renewals to occur in building enclosing body leaks; this total represents the overall loss in cm2 (leakage area) due to the substantial volume and poor air impermeability	The test revealed exceptionally high findings focused in particular: • Docks that are located close to higher levels in wood made rooms (the naked eye can discern leaks among the beans) • The shutters near the windows all have great dimensions, in addition to wooden frames without seals, are occasionally clearly damaged, and are difficult to close. Both the frame and the window as well as the frame and the walls are where the infiltrations are found • Pipelines next to installations that are under the pavement next to the perimeter walls (Dabouh, 2020)
U-value determination	U-value measurements of the external wall have been performed, in order to compensate for the shortage of documentation on the building's state and to acquire a sufficiently accurate estimate of the structure's heat losses	After the elaboration of the data, the U-value measurements' value in the municipal collection exterior walls were numerically calculated to quantify the heat losses through the floor and the ceiling. Since the floor is already heated on the first and second floors, the heating losses through the ceiling are more significant than those through the floor. However, the direct measure takes longer to implement due to the difficulties in organizing it due to the presence of frescoes and other technical issues (Ubertini et al., 2014)

(continued)

Table 1 (continued)

Technique	Implementation strategy	Findings
Hygrothermal and environmental monitoring		
Wireless sensor network (WSN)	A WSN aims to monitor and understand the physical state (Hygrothermal). The thermometer, humidity transducers, and flow sensor were used to measure the air temperature, relative humidity (RH), and air velocity in various building parts. All were wirelessly connected using the wifi interface protocol to retrieve the data. Then, the data was sent to a laptop so Hygrothermal could be monitored, and climate conditions in the building machine interface (BMS) room, where it is evident that the present situation is not favorable for artwork conservation (Al-mukhtar & Hadi, 2019)	In several areas of the building, it was found that daily air temperature and RH values remained constant during the research periods. The seasonal changes in RH and average air temperature are smoothed out by the room's huge thermal mass. Variations are frequently below the uncertainty of the sensors being utilized, and differences between sensors at various heights are insignificant. In order to set and specify preferable values for environmental quantities from the standpoint of conservation, there are several standards, guidelines, and recommendations (Al-mukhtar & Hadi, 2019) For instance, after monitoring most of the building areas, statistics for the OFFICES area show significant changes, in comparison to the collections' area, primarily as a result of individuals altering the equilibrium of the rooms by opening/closing doors/windows and by the little thermal inertia of the rooms themselves. Additionally, small air conditioning units have a significant impact, and thermal inversion could be seen in the summer. There are no environmental issues in the offices. Therefore, people's comfort needs to be taken into account. Additionally, the temperature distribution inside the space is less uniform, as it is in Sala Urbana, necessitating more nodes to investigate the environmental variances adequately and to see these intricate occurrences in greater detail. The study is still conducted to increase the understanding of the energetic behavior of the building (Troi, 2014)
Indoor environmental quality (IEQ)	The levels fixed in each indoor room were identified by audits utilizing thermo-hygrometric (air temperature–air speed–medium radiant temperature–operational temperature–relative humidity–surface temperature), visual, acoustic parametric value, and the resulting estimated comfort parameters (the medium indoor lightning) (Ubertini et al., 2014)	The urban room's thermo-hygrometric well-being measurements revealed values that were comfortable for wintertime (Troi, 2014)
The application of planning passive house buildings (PHPP)		
	The Passive House Planning Package (PHPP) is a user-friendly planning tool for energy efficiency which uses a variety of tested calculations to estimate the heating, cooling, and primary energy requirements of a structure (Ubertini et al., 2014) Bologna climatic information is available in PHPP's most recent version. The calculation's goal, the evaluation of design alternatives, may be accomplished with the available data. When comparing with actual consumption for a specific year, relevant climate data should be used (Pesci et al., 2012)	The implementation of the PHPP analysis produced findings that demonstrated how evenly distributed the energy losses are across the envelope. The non-heated space, exterior walls, and primary components are concentrated in the attic. As a result, there are considerable losses due to infiltration in the ventilation component. Due to the big windows on each external surface, heat transmission losses through the windows are very important. The building's windows allow for natural airflow in the summer, which helps to cool it. The reference component displays a high value. The projected hour of overheating has a low value due to the large value of losses on the entire envelope (Pesci et al., 2012) The results of the calculations related to heating, and consequently for the production of thermal energy in D'Accursio Palace, appear to be incredibly low when compared to the average values obtained in other historic structures located in the Bologna region and taken into consideration by the PHPP calculation (Ubertini et al., 2014)

Table 2 Proposed modifications analysis

Roof restoration	Windows restoration	Insulation for external walls
It is essential to replace the current roof with a ventilated roof that will perform effectively in the summer and insulate the space underneath from the cold in the winter. Two investigated insulating materials are recommended for use: • *The glass wool*: Its characteristics include fire resistance, chemical neutrality, non-absorbency, lightness, flexibility, ease of workability, and age resistance. It establishes the rate of moisture vapor transport, thermal and acoustic insulation, and mechanical compression resistance (Fig. 20) • *The wood fiber*: Design to allow an extremely high thermal wave displacement that will protect the interior spaces from the summer's heat	The existing windows and window frames are to be replaced as part of the intervention, by examining the results produced by the application of various windows solutions whether it was triple glass or double-layer insulating glass. The use of solar control and low emission glass, or glass with high-quality solar factors, has been considered in each case (Fig. 21)	Two intervention hypotheses that each propose replacing the old plaster layer with a thermo-insulating plaster or a conventional lime-based plaster have been considered to manage the current thermal fields associated with deteriorating infra-brick cement layers The thermo-insulating plaster that is being used is an entirely natural and environmentally friendly compost called "Diathonite Evolution," which is composed of "mixed thermo-plaster fiber reinforced with lime, diatomic powders, and hydraulic binding agent," and which ensures an improvement in the conditions of indoor comfort both in the winter and the summer (Dabouh, 2020)

Fig. 17 Raw radargram of "Sala Urbana," acquired with a 900 MHz antenna along a 16.73-m-long East–West survey line at a distance of 4.53 m from the South wall *Source* (Troi, 2014)

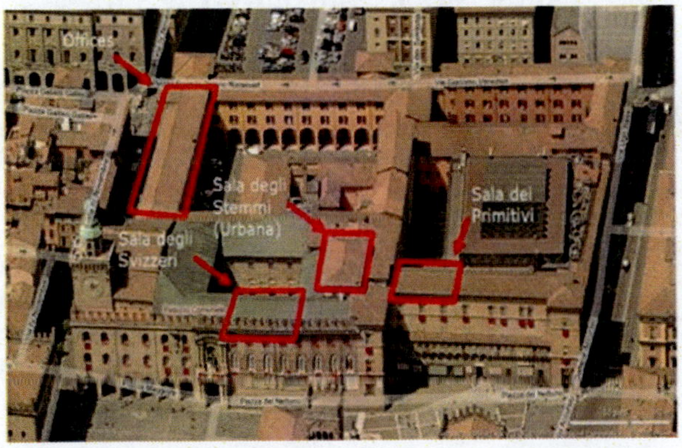

Fig. 18 East façade, aerial view of Palazzo Deaccession *Source* (Troi, 2014)

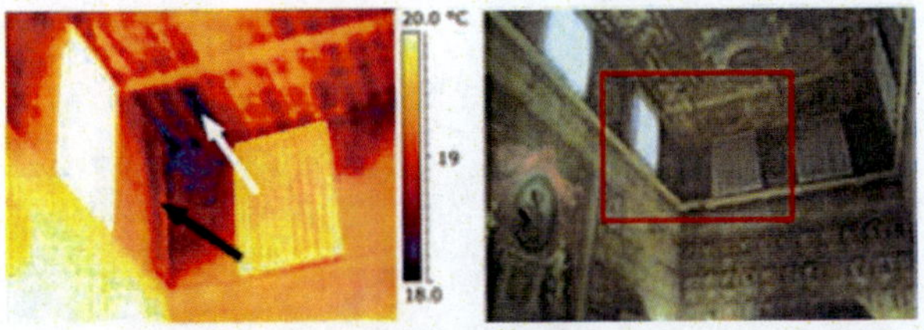

Fig. 19 Example for IRT investigation diagram the white arrow points toward an area that is damp or that would otherwise lack insulation. Although a similar trace is also seen on the opposite side of the same wall, the black arrow points at the South wall and that indicates a cold area that is likely caused by water leaks. *Source* (Ubertini et al., 2014)

- **Proposed modifications**

(D) **Conservation works (implementation)**

The intervention, which is represented in Fig. 22, has occurred according to the phases identified in the proposed solutions determined based on architectural analysis and experimental evaluations. Initially, it was essential to construct a scaffold along the main elevation and a connecting walkway before the maintenance work, which was initiated in April 2013, could begin. Additionally, an internal scaffold structure was built to support the room's entire painted ceiling and prevent the painting layer from falling.

- **Roof substitution**

It was determined that replacing the roof would be sufficient to achieve a ventilated roof shown in Figs. 23 and 24. Applying insulating materials based on wood fiber was important in this condition. Due to the numerous old chimneys that deteriorate the structure represented in Fig. 25, the consolidation of perimeter walls during the intervention by integrating new brick has been conducted as illustrated in Fig. 26. Moreover, curb metal with an anti-seismic function around the walls was installed. Epoxy resins and metal bars were used to strengthen a timber truss as shown in Figs. 27 and 28.

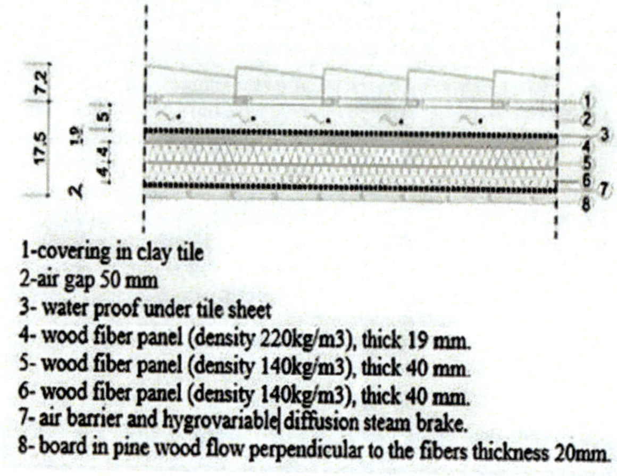

Fig. 20 Glass wool roof layers (*Source* (Troi, 2014))

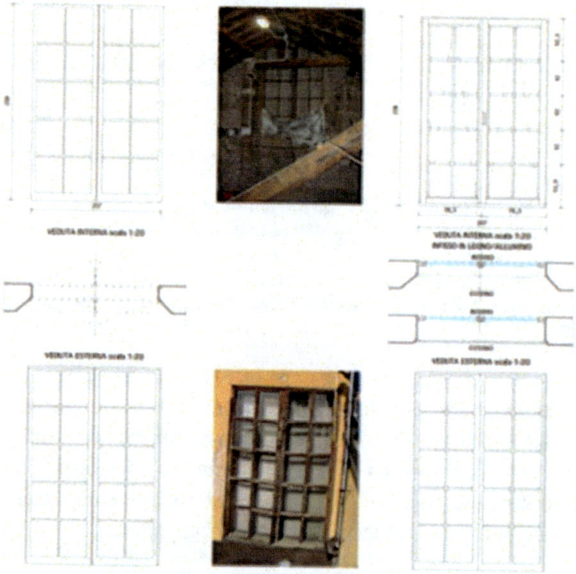

Fig. 21 Windows details *Source* (Ubertini et al., 2014)

- **The substitution of the windows**

East and west-facing windows: In order to increase performance, the same window design was recreated but with double glazing and a cavity (16 mm) filled with argon gas. The changes made to the exterior windows can be observed in Fig. 29, depicting their appearance both before and after the interventions were implemented.

North-facing windows: It is essential to rebuild the frames and install selective glass to reflect sunlight and protect the frescoes from damage while retaining heat during the winter.

South-facing windows: Replace the profiles with a single frame that preserves the same form but uses glass to replace them, particularly regarding the thermal break and the seals (Pesci et al., 2012).

After the interventions that were conducted in the previous stages, the control, automation, and motorization system artificial lighting system experiment was carefully examined to create a clear picture of the project and implement the appropriate corrective actions.

2.2 Technology Innovation and Entrepreneurship Development Center (TIEC), Cairo University, Giza Governorate, Egypt

2.2.1 Building Description

(A) Site context

Technology Innovation and Entrepreneurship Development Center is located in Ben Elsarayat, Dokki, Giza Governorate, as shown in Fig. 30. By providing entrepreneurship and innovation education programs to develop the capacity of young people, TIEC aims to spread the culture

Fig. 22 Interior and exterior scaffoldings for the conservation process *Source* (Troi, 2014)

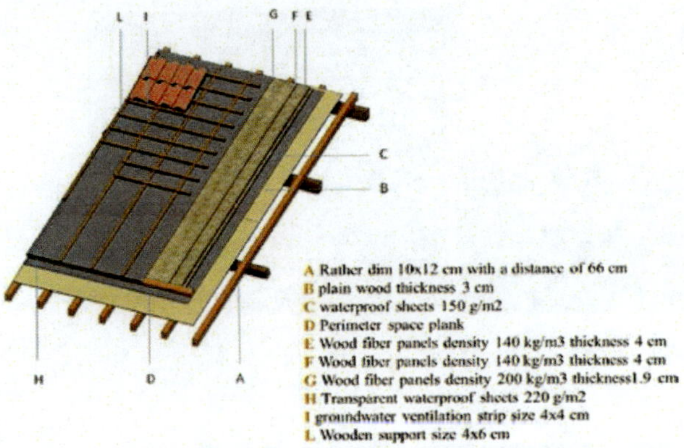

Fig. 23 Illustration of the roof layers (*Source* (Troi, 2014))

Fig. 24 Decay of the roof's wooden structure *Source* (Ubertini et al., 2014)

Fig. 25 Masonry consolidation interventions and historical chimneys *Source* (Ubertini et al., 2014)

Fig. 26 Consolidation of wooden truss components using steel bars and resins *Source* (Ubertini et al., 2014)

Fig. 27 Consolidating perimeter walls with a metal curb *Source* (Ubertini et al., 2014)

of innovation and entrepreneurship among Egyptian university students interested in creativity in the ICT field and aspire to be future entrepreneurs. Therefore, it is located nearby Cairo University in order to serve as a source of knowledge that encourages students to be innovative. As a consequence, ICT entrepreneurs, investors, mobile software developers, and entrepreneurship coaches and mentors are present. The Technology Innovation and Entrepreneurship Center (TIEC) provides a co-working space where they can work from the center, form various partnerships, and attend training sessions and workshops the center offers to innovators and entrepreneurs (Contractors, 2022).

Fig. 28 Establishment of roof insulation and replacing the wooden beams *Source* (Ubertini et al., 2014)

Fig. 29 Exterior windows before and after the interventions *Source* (Pesci et al., 2012)

Fig. 30 Location of the second case study—TIEC (*Source* Google Maps, goo.gl/maps/bR3sKJ5QKFewfKGS8, Edited by Author)

(B) **Architecture and building description**

The building, which was constructed in the nineteenth century, is considered a heritage value with distinctive characteristics and a unique style. The 4120 square meter building has been refurbished into a technology innovation center while preserving its historical value (Egypt's projects maps, 2022). The project's main priorities are adaptive reuse and heritage preservation. The building comprises six buildings and two exterior cafeterias, with the main entrance established recently. Between buildings one and two, the right tower contained stairs. On the other hand, the left tower, located between buildings two and six, consists of stairs, two elevators, and other services. The building layout analysis is illustrated in Figs. 31. Figure 32 depicts the building's appearance before any conservation work was carried out (Contractors, 2022).

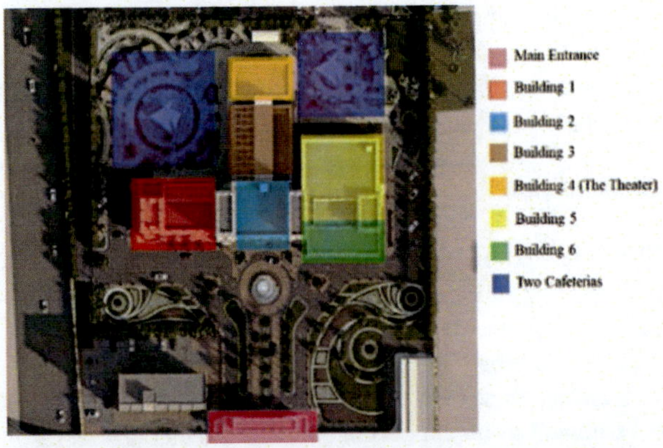

Fig. 31 Building layout (*Source* (Contractors, 2022) Edited by Authors)

Fig. 32 Building before the conservation work (*Source* (Contractors, 2022))

(C) **Historical background**

After the opening of the Suez Canal in 1869, Greeks, Italians, Britons, and others ventured to Egypt in anticipation of new economic opportunities. In 1897, this building was established by a group of businessmen from Belgium who established The Pyramid Brewery (Societ' e Anonyme Brasserie 'de Pyramides) in Cairo, which connected to a parent organization in Brussels. Representatives who live in Egypt and others who continue to reside in Belgium are the directors of the boards of directors for this corporation. The factory was nationalized in 1962 (Foda, 2014). By 1996, the government has decided to privatize the Al-Ahram Beverage Company (ABC). In 2002, the Dutch Heineken Group agreed with Al-Ahram Beverage Company ABC to acquire Al-Ahram Beverage Company at the disposal of the current management, and the management would be transferred to Heineken by January 2006. After that, the company's factories moved to the first industrial area at Obour City (Nederlands et al., 2002). Hence, the building became abandoned. It was reported in 2014 that the land of the beer factory was granted to Cairo University, affiliated with the Ministry of Communications and Information Technology, by Resolution No. 921 of 2014 concerning the heritage building's maintenance, restoration, and rehabilitation. Then, The Arab Contractors Company, Palaces, and Antiquities Maintenance Department carried out the development work of the building. The work on this project has continued to this day (Contractors, 2022). The building stages have been concluded, as shown in Fig. 33.

(D) **The area of intervention that have been selected**

Within the adaptive reuse process, in addition to the restoration work, some renovations and additions have been completed in the majority of the buildings. The conclusion of these updates is as follows:

- **Functional Area 1**: Preservation, reuse, and renovation work: Building one, four, and six.
- **Functional Area 2**: Preservation, reuse, and renovation, in addition to raising the main building`s floors: Building two, three, and five.
- **Functional Area 3**: New construction: The main entrance, the two cafeterias, and the exterior landscape, whether it is hardscape or softscape (Contractors, 2022).

2.2.2 The Building Documentation

The building had been abandoned for a long time. Therefore, there are negative impacts associated with it. Whatever the reasons for project abandonment, it generally affects the building structure. The majority of interior finishes and components require replacement or repair.

Fig. 33 Construction phases of the building (*Source* The Arab Contractors, Edited by: The Authors)

(A) **Diagnosis**

Classifying the type of building's structure and its different implementation details, which include:

- The building envelope's characteristics and maintenance requirements. Contemplate the preservation of its heritage features.
- Utilize both the previously used space and the newly required space.
- Illustrate the traditional architectural and decorative features of the building.

(B) **Proposed solutions**

- Conservation and restoration of the exterior stone facades, interior ceilings, and walls.
- Establishing the mechanical and electrical network, in addition to the installation of new lighting fixtures.

2.2.3 Methodology

(A) **Pre-intervention status**

The inspections and analyses that were done to identify the problems and disadvantages of the building are summarized in Table 3.

(B) **Conservation works (Implementation)**

- **External Facades conservation:**

The load-bearing stone masonry building has been abandoned and disused for a long time. Therefore, the walls' structure has deteriorated. Moreover, it has low strength and insufficient internal cohesion. As a result, it has low durability. In order to conserve the external façade, the following steps must be followed:

Mechanical cleaning of the stone facades

After a condition survey is finished, cleaning is frequently one of the first tasks to be accomplished in order to remove harmful materials from the surface (Doehne, 2014). So, the condition of the underlying stone could be obvious. Therefore, determine what additional conservation measures might be necessary. Mechanical cleaning using sandblasting is conducted in this building which is shown in Fig. 34, as compressed air aims to effectively hit the surfaces with the sand material that is specifically utilized for this process. The stone surface is cleansed of all harmful substances in this technique (Wilson, 1984).

Consolidation

The injection is the strengthening process that is carried out in building facades using equipment such as rubber tubes or needles of various sizes. Firstly, the exterior façade is divided into points horizontally and vertically. The distance between each point is equal to the other and is estimated to be approximately half a meter. Then, fill in the approved specific reinforcement material by injecting it through the cracks and gaps within the surfaces, as shown in Fig. 35. The process is stopped when the injection material leaks, and then, the leak is removed immediately. Additionally, identical replacement original stones have been used to replace the deteriorated stones as a part of the consolidation process and to preserve the original appearance of the building (Nabil, personal communication, July 27, 2022).

Surface coating (Protection)

After the joint fillings shown in Fig. 36a and b, the building has been coated in the same original color, composition, and fabric. Then, the final transparent protective finishing coat was applied, which has no modifications to the stone aspects, as illustrated in Fig. 37a and b (Gherardi, 2022). Moreover, this coat protects the stone from the causes of decay, such as air pollution, weather, wind, rain, and salts (Fig. 38).

Table 3 Experimental assessment analysis

Technique	Implementation strategy	Findings
Experimental assessment		
Ground-penetrating radar tests for masonry stratigraphy (GPR)	Strengthens subsequent conservation decisions with appropriate information on the local geology, groundwater level, identification of structural components, and modifications to the original building layout (Barraca et al., 2014)	The radar survey investigates the underground. Therefore, tunnels were discovered at different levels and depths, extending under the building and outside the project boundaries (Nabil, personal communication, July 27, 2022)
Infrared thermography (IRT)	Investigate the surface material's characteristics (Pesci et al., 2012)	Information regarding the building was found, gathered, and evaluated during an IRT investigation. A map of the material's interior anomalies can be created by mapping the surface temperature and comprehending the heat. IRT was utilized on the building's external façade to gather data that would not be accessible from visual analysis, such as the materials, status, characteristics, and degree of deterioration (Spodek, 2009)
X-ray computed tomography (CT) analysis	Determines appropriate conservation and restoration techniques (Morigi et al., 2010)	It reveals details on the inner structure and composition of materials of the building, which include natural stone, mortar, brick, and concrete (Kock & Orr, 2021)

Fig. 34 Sandblasting technique (*Source* The Arab Contractors, representative of the executing company, 2022)

Fig. 35 Building during the injection process (*Source* (Contractors, 2022))

(a) (b)

Fig. 36 **a, b** Joint fillings process (*Source* (The Arab Contractors, 2022))

(a) (b)

Fig. 37 **a, b** The building during the last coating layer (*Source* The Authors)

Fig. 38 Windows in building 3 before and after renovation (*Source* The Authors)

- **The substitution of the windows and doors**

Most of the building's windows and doors have been replaced because of their deteriorating condition while preserving the same design. Regarding the windows, use selective glass to reflect sunlight and shield the frescoes from it (Nabil, personal communication, July 27, 2022).

- **The interior conservation process**

Structural interventions

The building structure situation and state were analyzed, and conversion factors were considered, such as temperature, moisture, and the duration of the load, as well as the

Fig. 39 Installing new beams (*Source* (The Arab Contractors, 2022))

Fig. 40 Strengthening of the building structure process (*Source* (The Arab Contractors, 2022))

shape and size of the specimens that were used. When deterioration is observed, relevant mechanisms are identified, and based on theoretical or experimental research, a deterioration model that predicts the structure's performance in the future is implemented. In most sections of the buildings, renovations and structural reinforcement have been constructed. Regarding the left tower, which is located between buildings six and two, the intervention strengthened the building structure after analyzing its structural state, as revealed in Fig. 39 and Fig. 40. Moreover, most of the destroyed ceiling beams in the buildings have been replaced. On the other hand, in the majority of buildings, a process of floor reinforcing took place which is shown in Fig. 41 (Figs. 39, 40, 41, 42 and 43).

Fig. 41 Steel reinforcement for the floor (*Source* (The Arab Contractors, 2022))

- **Interior conservation**

After the building consolidation, the renovation took place to suit the new building's needs, as some interventions have indeed been required to convert a factory into an innovation center. Hence, the bracing and accessory conservation process is considered essential in this project. It must begin with mechanical and chemical cleaning, followed by treatments and preparation to be finished like an original piece, such as the heritage furnace located in building one, as demonstrated in Fig. 42 and Fig. 43. Also, some walls were built while preserving the brick arches and the stone walls after restoration, which are shown in Fig. 44.

Fig. 42 Heritage furance (*Source* The Authors)

2.3 Results

Depending on the methodology used in both case studies, previous building analysis, and conservation work, the evaluation process may be carried out using the following procedures, which is concluded as follows in Table 4.

Fig. 43 Brick arch, which is located ahead of the main gate (*Source* The Authors)

Fig. 44 Interior stone walls and brick arches after conservation works (*Source* The Authors)

2.4 Analysis

This section will illustrate the case study assessment using SWOT analysis regarding the adaptive reuse benefits. In this analysis, the STEEP (environmental, economic, social, political, and technological) analysis would be involved, which is defined as an assessment of the different influences on an organization with the aim of using these data to guide strategic decision-making (Szigeti et al., 2011). The working methodology and STEEP analysis were merged, and the results of the SWOT analysis were divided into five categories: political (PO), social (SO), technological (TE), economic (EC), and environmental (EN) aspects, as illustrated in Figs. 45 and 46. This combined assessment serves as a decision-making framework for better visualizing the scenarios' driving forces for urban and territorial development (Anna, 2022).

3 Results and Discussion

According to the previous case studies' comparative analysis, the framework is formulated in four stages to achieve a preliminary map to be followed to prepare the guidelines for heritage building conservation.

3.1 The First Stage—Data Collection

It is essential to assemble all the information about the building and site that is to be conserved before performing any physical interventions in the design approach for a potential conservation project. The original documents are essential (Hesham, 2018).

3.2 The Second Stage—Pre-Intervention

Before beginning any conservation work, the pre-intervention step involves the practice of locating and documenting architectural defects using digital and photographic documentation. The process of observing the building to gather and record data based on analysis of the finished surfaces and the structure, whether the structure is visible or invisible, is the essence of pre-intervention, to determine the building defects.

3.3 The Third Stage—Intervention/Implementation

The method for maintaining the heritage value is chosen when the vulnerability in the historic buildings has been identified. Before the restoration work can begin, some requirements are fulfilled, which will prepare the proposal or method of statement on the technique for the restoration and preservation of heritage buildings. In order to begin any conservation work, it is essential to clean the surface, make the surrounding area clearer, and reinforce any buildings with a temporary framework. Then, the conservation work will take place with suitable intervention.

3.4 The Fourth Stage—Evaluation

After the evaluation of preserving the architectural value, the evaluation of adaptive reuse is developed through the SWOT analysis, which is merged with the STEEP analysis to examine the consequences of the previous experimentation following the implementation of the project methodology.

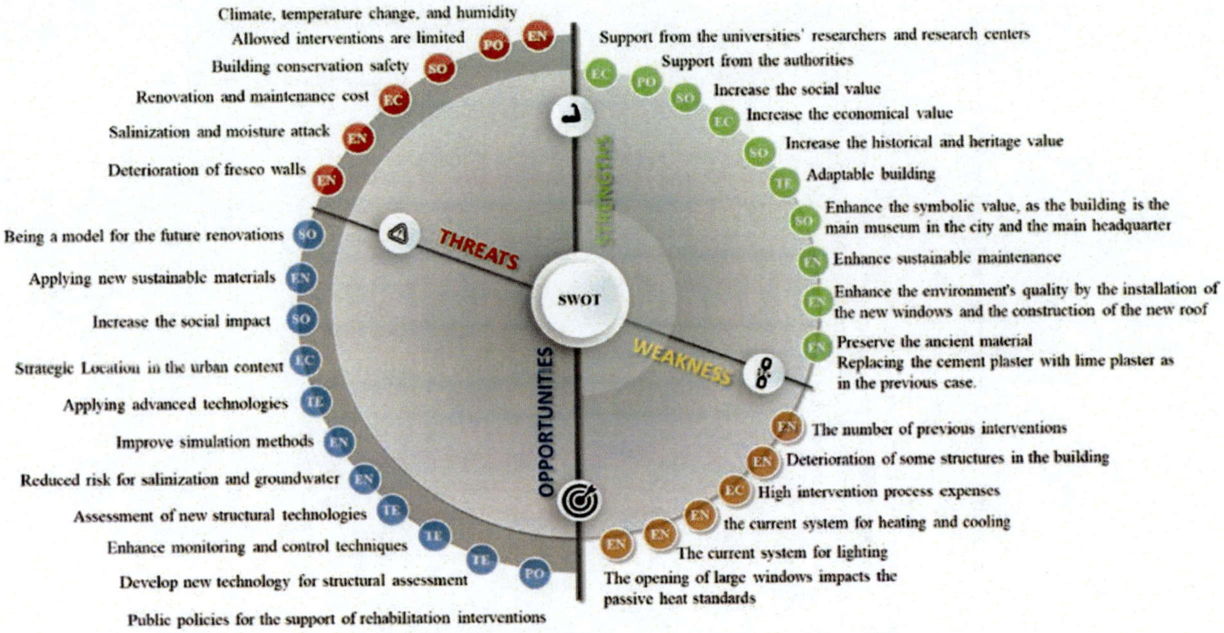

Fig. 45 Palazzo D'Accursio—SWOT analysis combined with STEEP analysis for the adaptive reuse building (*Source* The Authors)

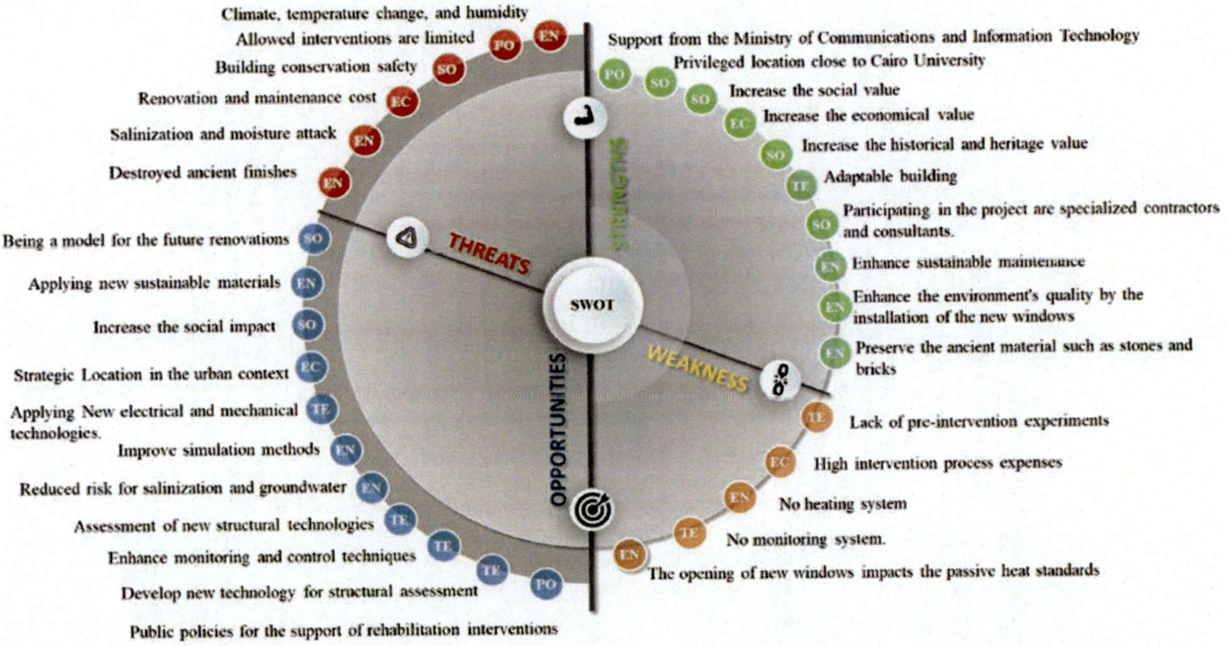

Fig. 46 Technology innovation and entrepreneurship development center—SWOT analysis combined with STEEP analysis for the adaptive reuse building (*Source* The Authors)

Table 4 Comparison between the two case studies results

Case studies	Case study 1 Palazzo D'Accursio	Case study 2 Technology innovation and entrepreneurship development center (TIEC)
Evaluation of the architectural value		
	In order to preserve the appearance of the heritage building, the new plaster, which has a brighter appearance since it is made of lime-based mortar and natural pigments used, essentially as it was done in the past. Hence, this preserves the historical image in a physical way	Preserve the structure's historic architectural original details and decorations. Moreover, the innovative function is compatible with the building's system and new space needs. Also, the physical stability has been enhanced concerning the building's identity
Evaluation of the adaptive reuse targets		
Socially	The interventions respected the importance of the building and its strong identity. After the renovation, the building benefits society and meets regional needs	The building improves living conditions and quality and satisfies regional needs. The closeness of such a building to Cairo university encourages Egyptian university students who are enthusiastic about innovation in the field of information and communication technology to develop a culture of innovation and entrepreneurship. The center for technology innovation and entrepreneurship provides a common workspace for entrepreneurs in various fields
Economically	The economic aspect has concluded with the symbolic and strategic location of the building. A wooden truss is integrated into the site as part of the historical value. The large "healthy" part of the wooden construction was preserved, while the damaged part was removed. During the conservation process, the cost was not taken into consideration	The cost has been rationalized during the conservation process
Environmentally	Building renovations lead to saving more energy, as the result of the previous intervention and the experimental evaluation that was monitored. The findings indicate a 16% decrease in electrical consumption and an increase of 6% in air conditioning consumption (Troi, 2014). Recycling or using fewer materials has been achieved, as the lime plaster has been replaced with cement plaster, which is regarded as a respectful and conservation-minded solution. The fresco restoration, roof insulation, structural wall consolidation, and artificial lighting system all have a high degree of reversibility. On the other hand, the consolidation was conducted with natural lime-based mortars, and acrylic resins were added	The building uses artificial lighting and ventilation, which results in high energy consumption. On the other hand, the building restored the deteriorating construction while restoring its original status of external and interior finishes. It is also easily accessible to users because it is close to Cairo university, vehicles, and pedestrians Goals have been achieved, including recycling and using fewer materials

4 Conclusion

The framework indicators combine numerous quantitative and qualitative indicators. Therefore, it is appropriate at international and national levels. The framework could be concluded as follows in Fig. 47a–d.

The conservation of cultural heritage requires knowledge and understanding of those resources and the history that they reflect. Hence, the main contribution of the research was to conserve the quality of heritage buildings through the use of adaptive reuse features in an accurate and sequential order. Therefore, after the comparative analysis, which observed the differences between the international and national case studies and shed light on the building analysis through the SWOT with the merger of STEEP, consequently, a methodical approach is required for conservation work. From there, the significance of the framework is established through four steps, which are followed by the assessment and then the final report.

It is proposed that the conceptual framework for conserving cultural heritage be used as a guideline for conservation practitioners in Cairo. Also, more work should be undertaken by different sectors to develop the framework of the guidelines for the field's future development.

(a)

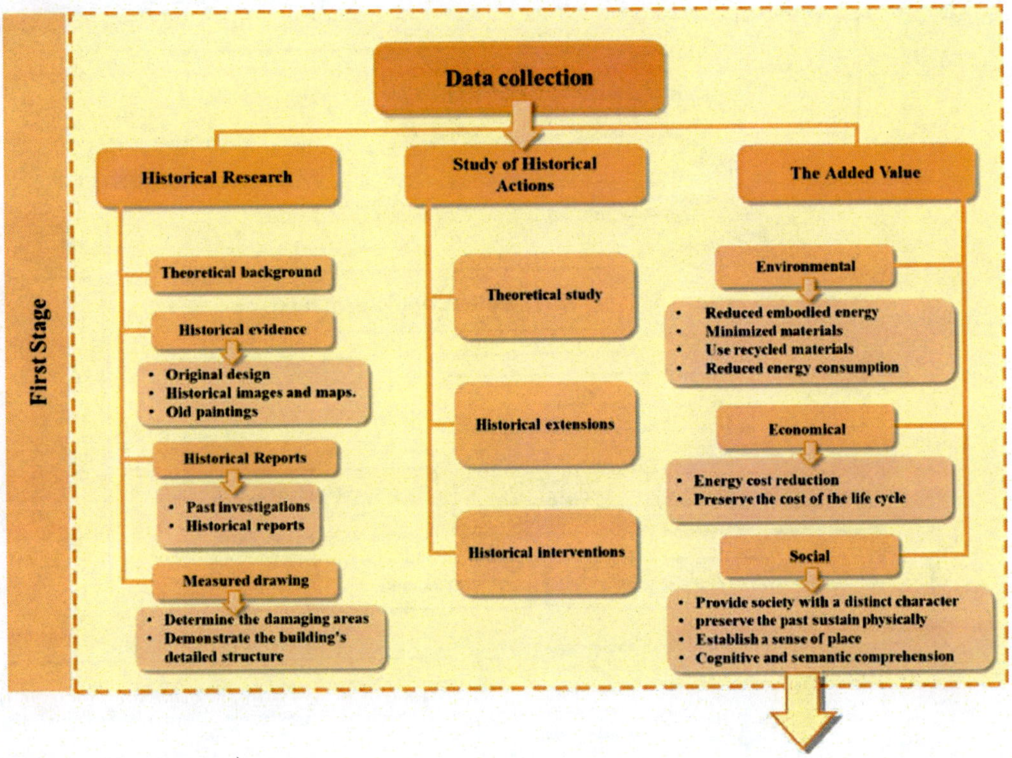

Fig. 47 Guideline framework (*Source* The Authors)

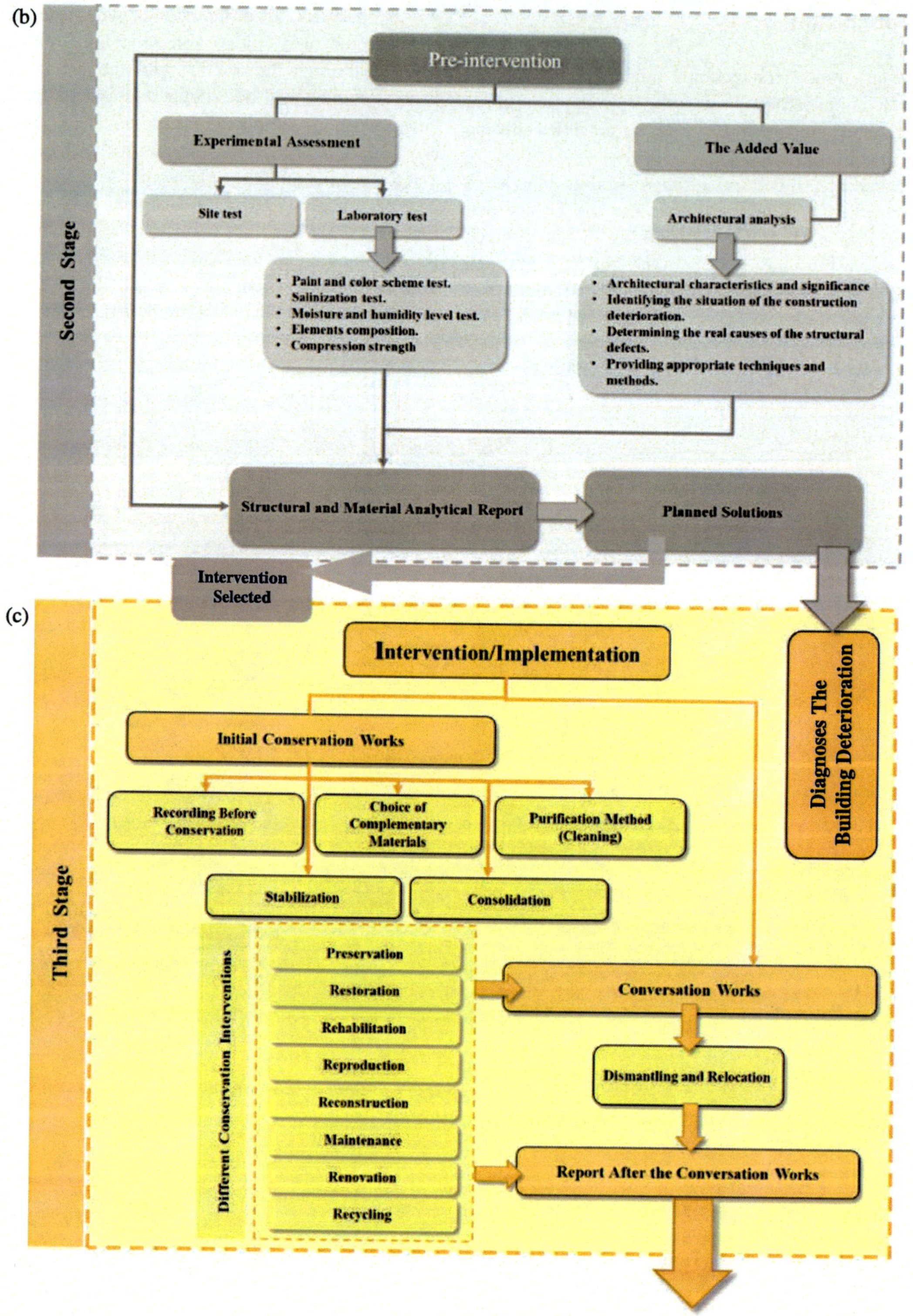

Fig. 47 (continued)

(d)

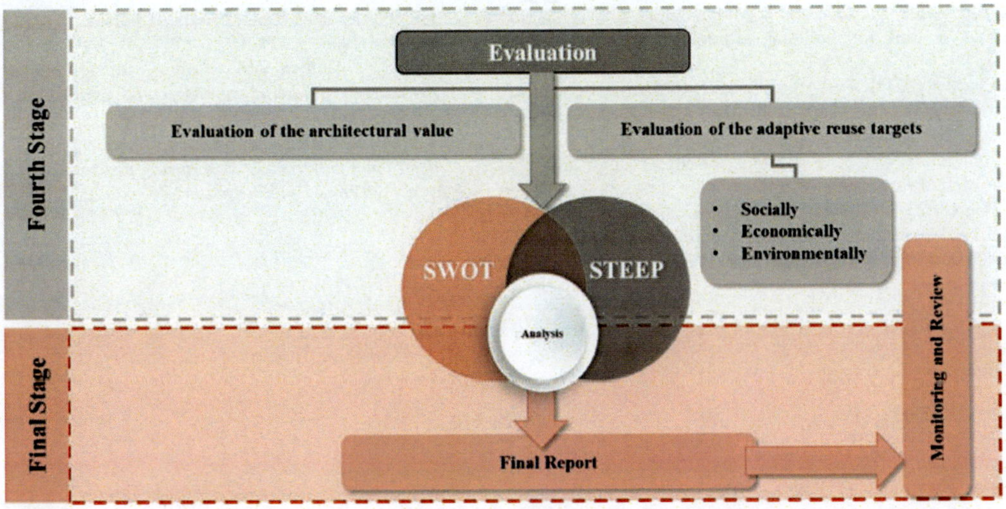

Fig. 47 (continued)

References

Aga Khan Trust for Culture. (AKTC). (2013). Urban conservation and regeneration of heritage areas in Egypt Aga Khan Darb Al-Ahmar project model.

Ali, O., Mansour, Y., Elshater, A., & Fareed, A. (2022). Assessing the identity of place through its measurable components to achieve sustainable development *10*, 137–157. https://doi.org/10.13189/cea.2022.101407

Al-mukhtar, M. M. A., & Hadi, T. H. (2019). A monitoring system using wireless sensor network (Issue June 2014). https://doi.org/10.22401/JNUS.17.2.29

Al-sakkaf, A., Zayed, T., & Bagchi, A. (2020). A review of definition and classification of heritage buildings and framework for their evaluation. Research Gate, August.

Amin, S. K., Maarouf, N. M., & Ali, S. S. (2012). Sustainable development of cultural heritage via anti weathering nanoparticles material *6*(6).

Anna, F. D. (2022). What advantages do adaptive industrial heritage reuse processes provide ? An econometric model for estimating the impact on the surrounding residential housing market.

Barraca, N., Almeida, M., Octopetala, D., Varum, H., & Matias, M. (2014). The use of GPR in the rehabilitation of built heritage Tu Olym 05 the use of GPR in the rehabilitation of built heritage (Issue January 2015). https://doi.org/10.3997/2214-4609.20141998

Bergamini, W. (1976). Guida ai monumenti, alla storia, all'arte della città Bologna, Tamari.

Bologna, O. T. I. S. (2022). Collezioni Comunali d' Arte (Municipal art collection) official tourist information site of Bologna.

Dabouh, I. (2020). A framework for sustainable conservation of architectural heritage using nanotechnologies Khedivian city as a case study.

Doehne, E. (2014). Stone conservation: an overview of current research (Issue February 2011). https://doi.org/10.2307/3179804

Egypt's projects maps. (2022). Creativity development center project in Bein Al-Surayat—Egypt projects map.

Foda, O. D. (2014). *The pyramid and the crown: The Egyptian beer industry from 1897 to 1963* (vol. 46). https://doi.org/10.1017/S0020743813001323

Frogé, E., Richards, P., Sperling, L., Reyes, L., & Kiambi, D. (1999). Courier rethinking the mix UNESCO COURIER (Issue August).

Gherardi, F. (2022). Conserving stone heritage.

Gupta, A., & Prakash, A. A. (2018). Conservation of historic buildings. *International Journal of Engineering Research, 7*(1). https://doi.org/10.5958/2319-6890.2018.00087.9

Hall, F., & Maria, G. (n.d.). Palazzo d' Accursio. 1–2.

Hanso, B. (2016). *Conservation of historic buildings* (vol. 4).

Hesham, E. S. (2018). Boundaries of areas of heritage value in greater Cairo.

Hubert, H. W. (1993). Der Palazzo Comunale von Bologna, Bohlau.

Kock, T. D., & Orr, S. A. (2021). X-ray micro-computed tomography for heritage building materials (Issue m). *Royal Society of Chemistry*. https://doi.org/10.1039/d1ay90154k

Morigi, M. P., Casali, F., Bettuzzi, M., & Brancaccio, R. (2010). Application of X-ray computed tomography to cultural heritage diagnostics (Issue September). https://doi.org/10.1007/s00339-010-5648-6

Nederlands, C., Soci, L., Boulangerie, A., Soci, L., Brewery, A. C., Heyndrickx, A., Brewery, C., & Brasserie, A. (2002). Les entreprises belges en Égypte.

Pesci, A., Bonali, E., Galli, C., & Boschi, E. (2012). Laser scanning and digital imaging for the investigation of an ancient building: Palazzo d' Accursio study case (Bologna, Italy). *In Journal of Cultural Heritage, 13*(2). Elsevier Masson SAS. https://doi.org/10.1016/j.culher.2011.09.004

Solla, M., Lorenzo, H., Rial, F. I., & Novo, A. (2012). Ground-penetrating radar for the structural evaluation of masonry bridges: Results and interpretational tools. *29*, 458–465. https://doi.org/10.1016/j.conbuildmat.2011.10.001

Spodek, J. (2009). Application of infrared thermography to historic building investigation (Issue May 2015). https://doi.org/10.1080/13556207.2009.10785040

Supreme Council for Planning and Urban Development. (2008). Coordinating civilizations and foundation standards for heritage buildings, areas of distinguished value, and urban development for higher planning. Of the approved Law No. 119 of the executive year 2008 and its regulations.

Szigeti, H., Messaadia, M., Cesi, G., & Majumdar, A. (2011). STEEP analysis as a tool for building technology roadmaps (Issue October).

The Burra Charter. (2013). The Australia ICOMOS charter for places of cultural significance. Australia ICOMOS incorporated international council on monuments and sites.

The Arab Contractors. (2022). The project of conservation and rehabilitation of innovation and entrepreneurship development Center at Cairo University.

Troi, M. F. F. (2014). 3encult. Efficient energy for Eu cultural heritage. D 6.2 documentation of each study case CS2 Palazzo d' accursio/ municipal palace, Bologna (Italy) delivered at M42 European commission (Issue 260162).

Ubertini, F., Giuliani, M., Lollini, R., & Henze, G. P. (2014). *An energy modeling approach to analyse historical building performance*.

Wangkeot, K. (2002). *Monumental challenges: The lawfulness of destroying cultural heritage during peacetime*, (p. 225).

Wilson, F. (1984). *Building materials evaluation handbook*.

Preserving Monumental Hospital Heritage While Providing Efficient Health Services. The Case of the Goyeneche Hospital in Arequipa Peru

Edith Suarez-Malaga and Carlos Zeballos-Velarde

Abstract

The preservation and defense of a heritage building, including hospitals considered historical monuments, are essential to maintain and strengthen the identity of a community. On the other hand, it is crucial to provide quality health services for the population, especially in countries of the Global South, where resources are scarce and development opportunities are not frequent. But, what happens when both interests collide? Is it possible to maintain a unique built heritage and, at the same time, meet the basic health needs of the population?. This research presents the case of the Goyeneche Hospital in Arequipa, the best example of Neo-Gothic architecture in the city, built with a local white and pink ashlar stone, and, at the time of its construction at the beginning of the twentieth century, one of the most advanced in Latin America, but which has seriously deteriorated over time. On the other hand, many voices are calling for its demolition to allow the construction of a modern hospital in its place, since it is located in the city's downtown area. This study proposes solution alternatives for both perspectives, beginning with a detailed architectural and typological historical analysis of the Goyeneche Hospital. Then, we offer a perspective to the functional demands required by this facility and the management problems that hinder its development. Finally, an architectural restoration project is proposed that reconciles the preservation of the hospital heritage with the development of efficient and useful health infrastructure for the population by means of the adaptive reuse concept.

E. Suarez-Malaga · C. Zeballos-Velarde (✉)
Department of Architecture, Universidad Nacional de San Agustín, Arequipa, Peru
e-mail: czeballos@unsa.edu.pe

E. Suarez-Malaga
e-mail: esuarezma@unsa.edu.pe

Keywords

Heritage preservation · Hospitals · Health infrastructure · Gothic revival · Adaptive reuse

1 Introduction

1.1 The Importance of Recovering Monumental Hospitals

Historic hospitals are part of the cultural and architectural heritage of a city. At the end of the 19th and the beginning of the twentieth centuries, a process for improving health was developed in various countries, imbued by the principles of Hygienism, which resulted in the construction of several hospital centers. However, over time, these complexes often became obsolete because they could not meet anymore the needs that a health facility required, both in terms of infrastructure capacity for an ever growing population, the arrival of new health technologies, and the use of new sustainable practices in construction.

Particularly in cases of the Global North, many of these historical hospitals have been abandoned, building more modern ones in other locations, which meet current health standards. In some cases, the academic literature proposes the adaptive reuse of these abandoned hospitals, in order to provide them with other uses more linked to culture or similar aspects and reuse the historic buildings instead of demolishing them (Diana et al., 2022; Sicignano et al., 2022). However, in countries of the Global South, where the lack of health infrastructure is noticeable, it is important to reassess the heritage character of these historic buildings without giving up their use as a hospital.

Such is the case of the Goyeneche Hospital in Arequipa, Peru, a building made in an elaborate Neo-gothic style, unique in South America, and which has been deteriorating over the years. However, its role was crucial during

M. F. Shahidan et al. (eds.), *Urban Narratives: Exploring Identity, Heritage, and Sustainable Development in Cities*, Advances in Science, Technology & Innovation, https://doi.org/10.1007/978-3-031-48517-6_17

the COVID-19 crisis, supporting the fight against this pandemic, despite its many limitations (Agenda para el Desarrollo de Arequipa, 2021). The COVID-19 pandemic highlighted the enormous lack of health infrastructure in Peru. At the peak of the pandemic, Peru became the country with the highest number of deaths per million in the world despite the strict quarantine imposed, and an important factor in this tragedy was the lack of appropriate health infrastructure. In the case of Arequipa, which also suffered terrible peaks in deaths from this disease (Universidad Católica San Pablo [UCSP], 2020) it is crucial to substantially improve the hospital offer, without renouncing the historical legacy of this UNESCO World Heritage city.

1.2 Concepts of Adaptive Reuse

Heritage buildings provide communities an important sense of identity and their preservation also include social, cultural, and economic aspects. On the other hand, historic buildings are often located in central areas, where land is very expensive. Then there is the debate whether to preserve a building or demolish it to make way for a new one and, if preserved, whether it should be used for its original function or repurposed. The decision to reuse a building is determined by its location, state of conservation, aesthetics, historical values, and market trends (Bullen & Love, 2011; Konsta, 2019).

The enhancement measures needed to preserve or reuse a historic building are usually expensive. The structural reinforcement of a building that was constructed with inefficient technology or the conditioning of old structures with current technologies usually exceeds the cost of constructing a modern building that does comply with safety and comfort regulations. Therefore, adaptive reuse interventions are usually carried out by public institutions or by solvent private enterprises that can cover the costs of restoration and reinforcement. In these latter cases, however, these companies will prioritize profitable functions that do not necessarily have to do with the original use of the historical building. It is common to see, for example, churches whose structure has been preserved, but they were converted into libraries, restaurants, or gyms, functions completely different than the one they were originally created. In this sense, Mısırlısoy and Günçe (2016) propose that when it is necessary to adapt heritage buildings to different functions, both the new use and the interventions must preserve the originality and architectural character of the building so as not to give erroneous or missing information for future generations.

In the specific case of historic hospitals, various proposals envisioned the adaptive reuse of abandoned health structures, some for purposes different from the original use and others proposing cultural functions or related services, which allow them to be associated with those of health, thus improving the quality of life of the communities (Mısırlısoy & Günçe, 2016).

In this sense, Sicignano et al. (2022) propose the adaptive reuse of monumental hospitals in Italy, such as the cases of "San Giacomo degli Incurabili" and "Carlo Forlanini" in Rome, and the hospitals of "Gesù e Maria", "San Gennaro dei Poveri" and "Santa Maria del Popolo degli Incurabili" in Naples. In these cases, after a historical analysis, they propose a regeneration of these buildings as health facilities at the community level, previously suggesting strategies for structural reinforcement and improvement of environmental efficiency.

Our study has a similar approach since it proposes the demolition of obsolete Modernist style health structures that lack aesthetic or historical value to accommodate a contemporary facility that provides modern health services and complies with the regulations against earthquakes while making sustainable use of resources. At the same time, it proposes the adaptive reuse of the monumental buildings of the old hospital, accommodating compatible health functions, while offering some associated cultural services. All this takes into account specific measures to ensure the preservation of a historical Neo-gothic heritage with many unique features. Additionally, it proposes the recovery of the hospital public space for the local community, integrating it into the surrounding urban spatial system.

2 The Site

2.1 Historical Background

Arequipa is the second city in Peru and the capital of the province of the same name, which has a population of 1,080,635 inhabitants (Instituto Nacional de Estadística e Informática [INEI], 2018). The city, sited at an altitude between 2300 and 2400 m and flanked from north to east by a chain of high mountains, is crossed by the fertile valley of the Chili River. It was founded in 1540 by Spanish conquistadores and its historic center is a UNESCO World Heritage Site (Zeballos-Velarde, 2020).

After a devastating earthquake in 1868, the deterioration of the existing San Juan de Dios hospital, and the abandonment of the order of the Camillians (in charge of the sanitation in the city), health care in Arequipa was in crisis. At that time, hospitals were groups of hospitalization rooms, with basic kitchen and laundry services (Llerena et al., 1996). For this reason, on December 31, 1871, Dr. José Sebastián de Goyeneche y Barreda, who was archbishop of Arequipa and Lima, made a testamentary donation of 150,000 pesos for the construction of a "hospital for the poor" in Arequipa. Subsequently, the Beneficencia (Charity Board) of Arequipa ceded a 55,000 m^2 plot of land at the end of Siglo XX Boulevard for the construction of the hospital (Zeballos-Velarde, 2022).

The architectural design of the hospital was drawn up in Neo-Gothic style by the French engineers Gaillard and Ponsot, and adapted by the local scholar Pedro Paulet (1874–1945). Construction works began on July 1, 1905, carried out by engineers Oscar López and Julio Arce; the masonry work was under the direction of the engineer Manuel Cuba, and the construction of the roofs was entrusted to the Italian architect Giuseppe C. Vialardi (Municipalidad Provincial de Arequipa [MPA], 2019).

After seven years of construction, on February 11, 1912, the Goyeneche Hospital was inaugurated, with a built area of 37,822 m2. With 780 beds and various specialties it was one of the most modern in South America, build in Neo-Gothic style using local materials, in particular a local volcanic stone called ignimbrite. However, destructive earthquakes in 1958 and 1960 caused serious damage to its structure and for this reason, on February 12, 1960, Mayor Jorge Vásquez, together with the Charity Board, decided to demolish the beautiful Neo-Gothic chapel, greatly damaging the image of the emblematic hospital. On June 23, 2001, another earthquake damaged the infrastructure of the hospital, being restored in 2002. The last addition to the complex is the new emergency services built in 2009 (Zeballos-Velarde, 2022).

2.2 Urban Relations

At the beginning of the twentieth century, the Plan for the Rectification and Widening of roads in Arequipa was developed, considering the location of a new hospital, facing the new boulevard Siglo XX. The colorful Neo-Gothic chapel, brought in parts from France, was a remarkable landmark. The triangular profile of the chapel's gable was accentuated by the perspective of the trees that flanked the road.

2.3 Architectural Features

2.3.1 Formal Composition

The building takes several characteristics of the Neo-gothic style such as the horizontal, symmetrical, and repetitive organization. However, it differs from the European and North American Gothic Revival style due to its adaptation to local materials, such as the reinforcement of the roofs with rails and the use of ignimbrite combined with a reinforced concrete structure. The white ignimbrite was initially covered and painted in yellow for better chromatic integration with the chapel, which was covered in marble. The hospital is characterized by having (a) ogival-shaped openings (windows and doors), organized sequentially and repetitively, (b) crowning with a gable, especially in the kitchen and the chapel, and (c) railings that run over all the terminals of the roofs in the form of small columns interspersed with decorations of leaves and flowers (Llerena et al., 1996). In addition, the division into three elements in the facades is notorious: cornice, body, and plinth. Likewise, the corners frame the volumes with smooth stone quoins.

The most notable volume of the hospital was the chapel, which maintained a symmetrical composition. The facade was hierarchized by a triangular gable that was crowned by a statue and pierced by a circular rose window. The entrance portico, ogival in shape, was emphasized by archivolts. On the sides, the volume was flanked by two minarets crowned by pinnacles (Fig. 1). Conversely, the hospital pavilions are one-level constructions, which receive lateral solar light using sequences of pointed windows. This style is also observed in the doors of the consulting rooms and waiting rooms.

Over the years, other buildings were added in International Modern style, which had nothing to do with the style or proportions of the original building. The controversial demolitions after the earthquakes and the abandonment of many of the original elements have brought this beautiful building to a critical condition, where the risk of collapse of its structure compromises the physical integrity of the users, as well as its heritage.

2.3.2 Function

Originally the complex was organized around a large elongated central yard on whose axis both the chapel and the kitchen were aligned. On both sides, the elongated pavilions were organized, linked by the consulting rooms. After the

Fig. 1 Postcard of the hospital's façade. *Source*: Carlos Rodriguez (n.d)

demolition of the chapel, the access square was enlarged. Functional needs to provide auxiliary services led to the creation of new rooms that linked the pavilions at their other end but without an integrating circulation between them. As a result, a series of closed patios were generated between the pavilions. Subsequently, more additions were made and several facilities were built, but in a different style from the original hospital. Thus, to the original hospitalization areas, laboratories, services, emergency, and administration areas were added (Zeballos-Velarde, 2022). Currently, the hospital has a huge infrastructure deficit to supply the demand of users and the health personnel it houses.

2.3.3 Spatial Characteristics

The central yard was and still is the main spatial organizer of the complex. Additionally, other smaller patios are generated between the pavilions and before other hospital services, mostly following the same orientation, which allows these units to have natural lighting and ventilation.

Originally, the spatial organization of the pavilions was simple but clear: the elongated hospitalization rooms were articulated by a perpendicular corridor, to which the consulting rooms and waiting rooms located between the pavilions were also linked. The spatial analysis shows that the places of greatest integration are the contact areas between the corridor and the exit of the pavilions, as well as the spatial hierarchy of the rooms (Fig. 2a). Later additions such as services and other environments have ended up distorting the spatial clarity of these environments (Fig. 2b).

3 Materials and Methods

The methodological flow used for the adaptative reuse proposal results from a combination of methods investigated in the literature and the observation of techniques frequently used by local practitioners. The method includes five steps: documentary research, the definition of criteria for surveying the building, the survey of the building in its current condition, the definition of criteria for the proposal, and the development of the proposal itself (Fig. 3).

Many authors carry out different data collection methods before developing an adaptive reuse project. For example, Djebbour and Biara (2020), used three different instruments for this purpose: semi-structured interviews, a documentary analysis, and an evaluation grid. As for the definition of how to choose criteria to survey a building, the Harris matrix has been used for stratigraphical analysis in archaeological and historical studies (Barros García, 2004; Davies, 1993; Doneus et al., 2022). Similarly, different methods have been discussed for the identification and monitoring of pathologies in buildings (Gonçalves et al., 2017).

Each of the five steps described in the methodology will be explained in detail below.

3.1 Criteria for the Documentary Investigation

In order to obtain the necessary historical information for the study, there are no previous studies to be used as a

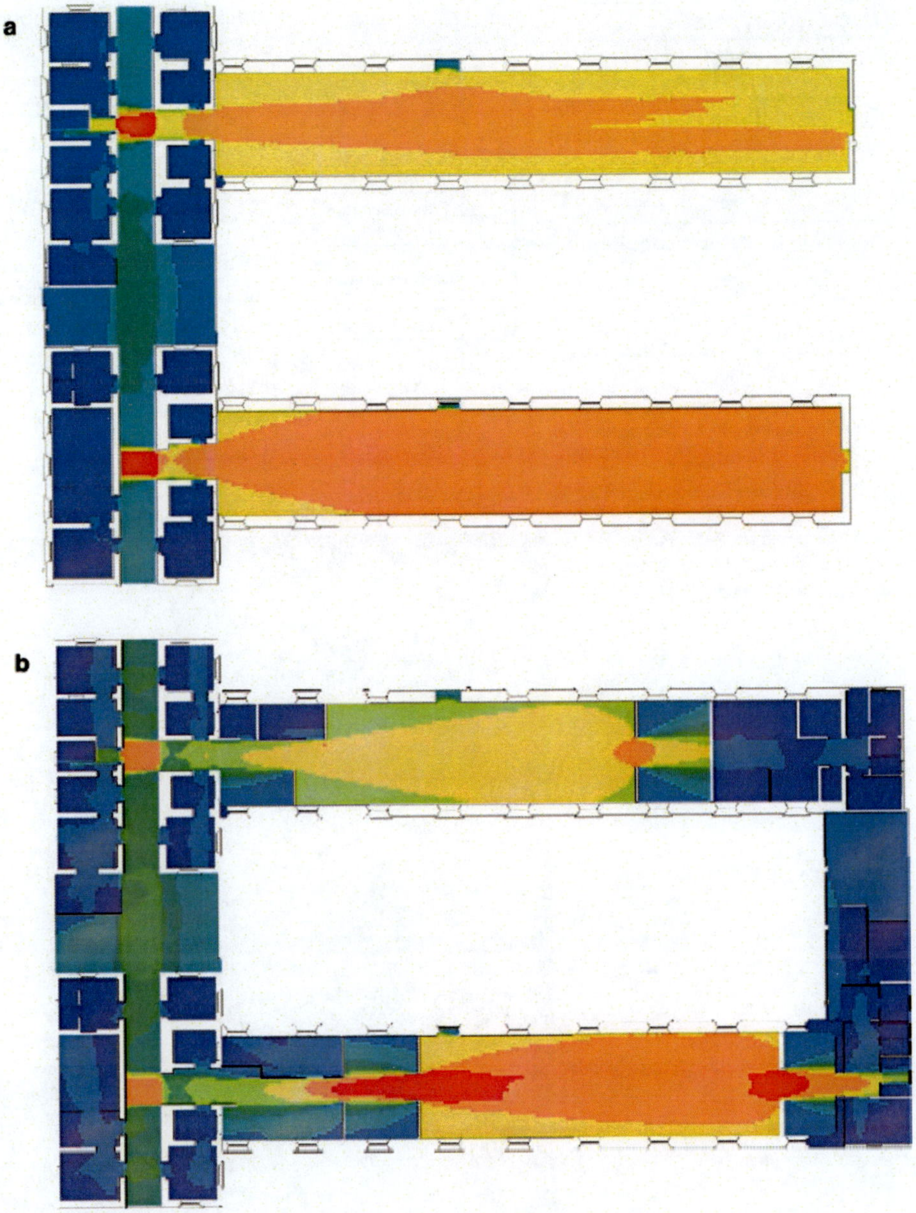

Fig. 2 **a** Spatial integration of one of the pavilions in its original layout. Graph by Carlos Zeballos. **b** Spatial integration of one of the pavilions in its current status. Graph by Carlos Zeballos

reference. However, information published in newspapers and magazines of the time has been compiled: the newspapers Diario El Deber (1910, 1912), Diario La Bolsa (Gibson, 1912); the magazines Revista Variedades (Palma, 1910, 1912), Revista Actualidades (1905), and also the Regulations of the Goyeneche Hospital (Sociedad de Beneficencia de Arequipa, 1920).

Letters between the designer Pedro Paulet and the builder of the Goyeneche Hospital were also reviewed, which helped to understand various details within the

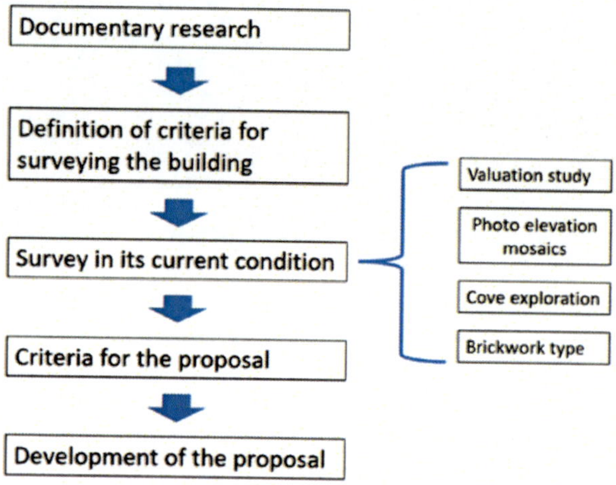

Fig. 3 Methodological flowchart. Graphic: Carlos Zeballos

complex, these were obtained from the National Library of Peru (Fig. 4). Likewise, the first photogrammetric aerial photographs in Arequipa, taken in 1944 by the National Geographic Institute (IGN), were used as a source (Fig. 5). Old photos of the hospital were also obtained, thanks to the collaboration of collectors (Fig. 6).

Finally, it was essential to revise an old plan of the complex, printed on acetate, which, although it is not the original plan, is a survey carried out by the Arequipa Public Charity (Beneficencia Pública de Arequipa), an entity that oversees the social development of the city. The plan shows some constructions that ratify the stratigraphy presented, as well as provided the necessary guidelines and dimensions for the reconstruction proposal of the chapel. (Fig. 7).

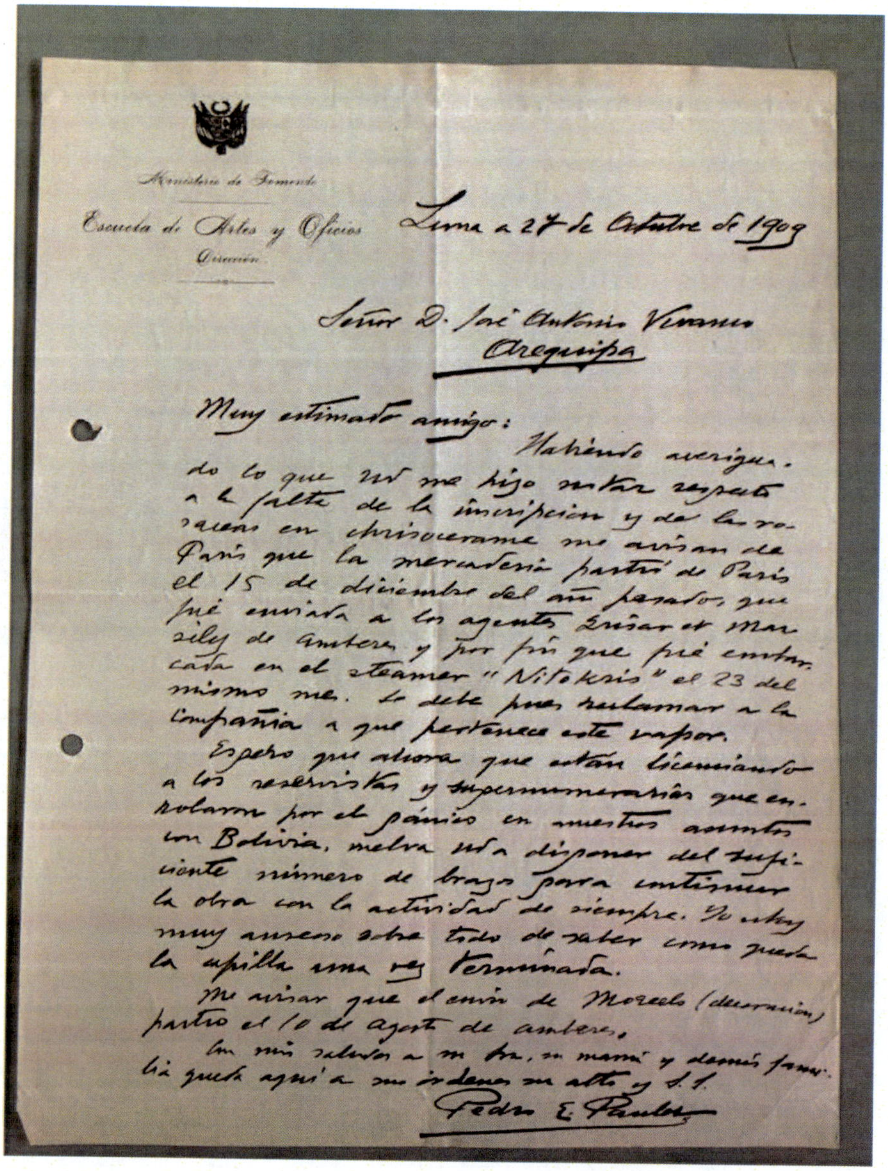

Ministerio de Fomento
Escuela de Artes y Oficios
Dirección

Lima a 27 de Octubre de 1909

Señor D. José Antonio Vivanco
Arequipa

Muy estimado amigo:

Habiendo averiguado lo que Ud me hizo notar respecto a la falta de la inscripción y de las rosáceas en chrisocerame me avisan de París que la mercadería partió de París el 15 de diciembre del año pasado, que fué enviada a los agentes Luisar et Marsiley de Amberes y por fin que fué embarcada en el steamer "Nitokris" el 23 del mismo mes. Se debe pues reclamar a la Compañía a que pertenece este vapor.

Espero que ahora que están licenciando a los reservistas y supernumerarios que enrolaron por el pánico en nuestros asuntos con Bolivia, vuelva Ud a disponer del suficiente número de brazos para continuar la obra con la actividad de siempre. Yo estoy muy ansioso sobre todo de saber como queda la capilla una vez terminada.

Me avisan que el envío de Morecels (decoración) partió el 10 de agosto de Amberes.

Con mis saludos a su Sra, su mamá y demás familia queda aquí a sus órdenes su atto y S.S.

Pedro E. Paulet

Fig. 4 Letters from Pedro Paulet to Antonio Vivanco, builder of the hospital. *Source* Paulet (1909)

Fig. 5 Aerial photo of Arequipa from 1944. *Source* Instituto Geográfico Nacional [IGN] (1944)

Fig. 6 Ancient Goyeneche hospital. *Source* Revilla (n.d)

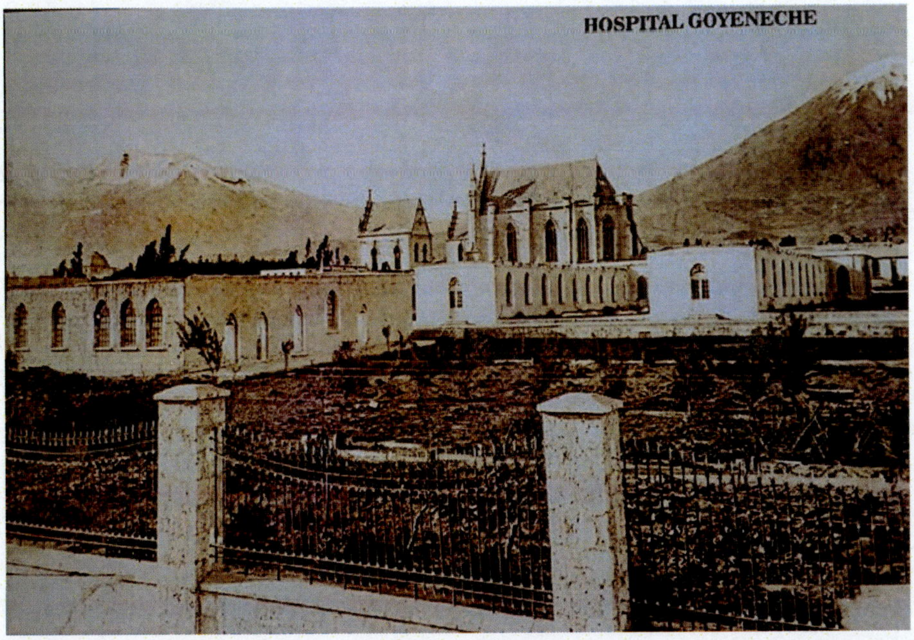

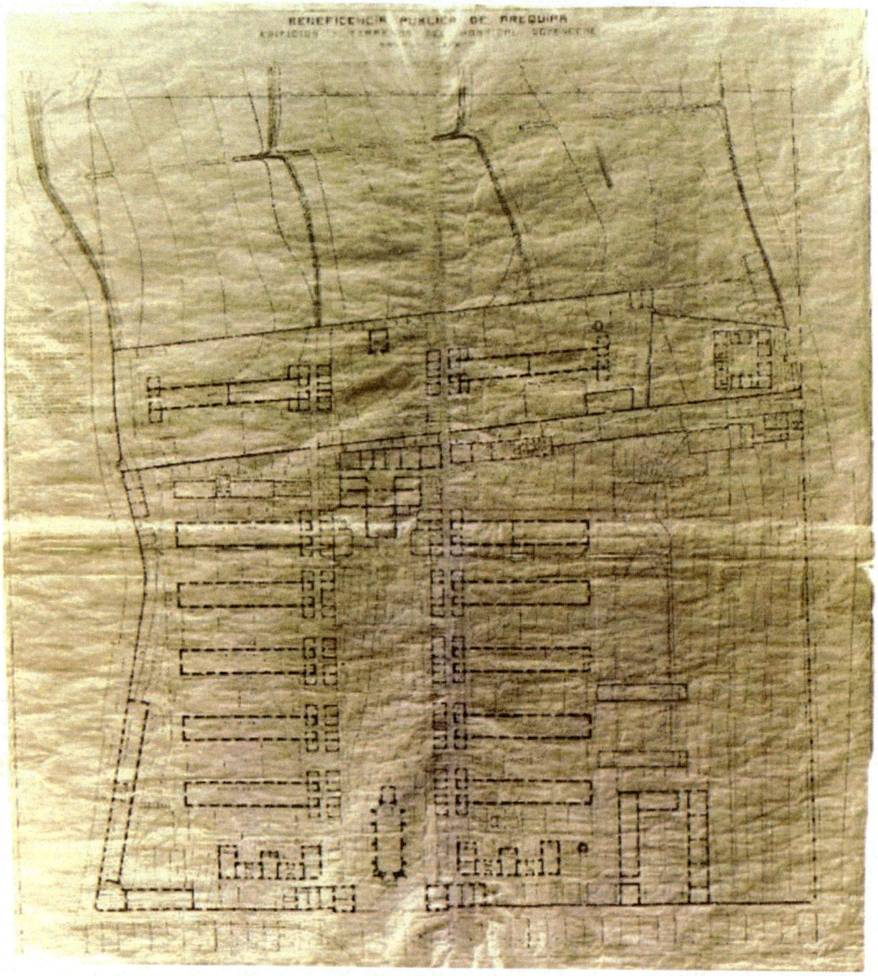

Fig. 7 An old plan of the Goyeneche hospital. *Source* Cordero (n.d)

3.2 Criteria for the Survey of the Building

To carry out an analysis of the current situation of the hospital, the following guidelines were followed:

(a) Identification of the construction phases, reconstructions, and extensions, considering the entire complex as well as each of the buildings. In this way, it has been possible to obtain the construction stageability, from its inauguration to date. For this purpose, stages were identified in a floor plan considering each stratigraphic unit (Fig. 8).

The chart of stratigraphic units (Fig. 8) served not only to arrange each of the buildings in the complex according to their chronological development, but also, according to a Harris matrix, to assign a particular phase, the date of construction, and the architectural style of each building (Fig. 9). This method allowed a better assesment of their conservation and restoration.

(b) Classification of the architectural components and masonry, through photo elevations that show the current state of each of the buildings.
(c) Recognition of the different architectural pathologies, within the most frequent typologies.
(d) Identification of the types of ad hoc intervention and the pathologies and according to the information found in the photographic documents, letters, and plans.

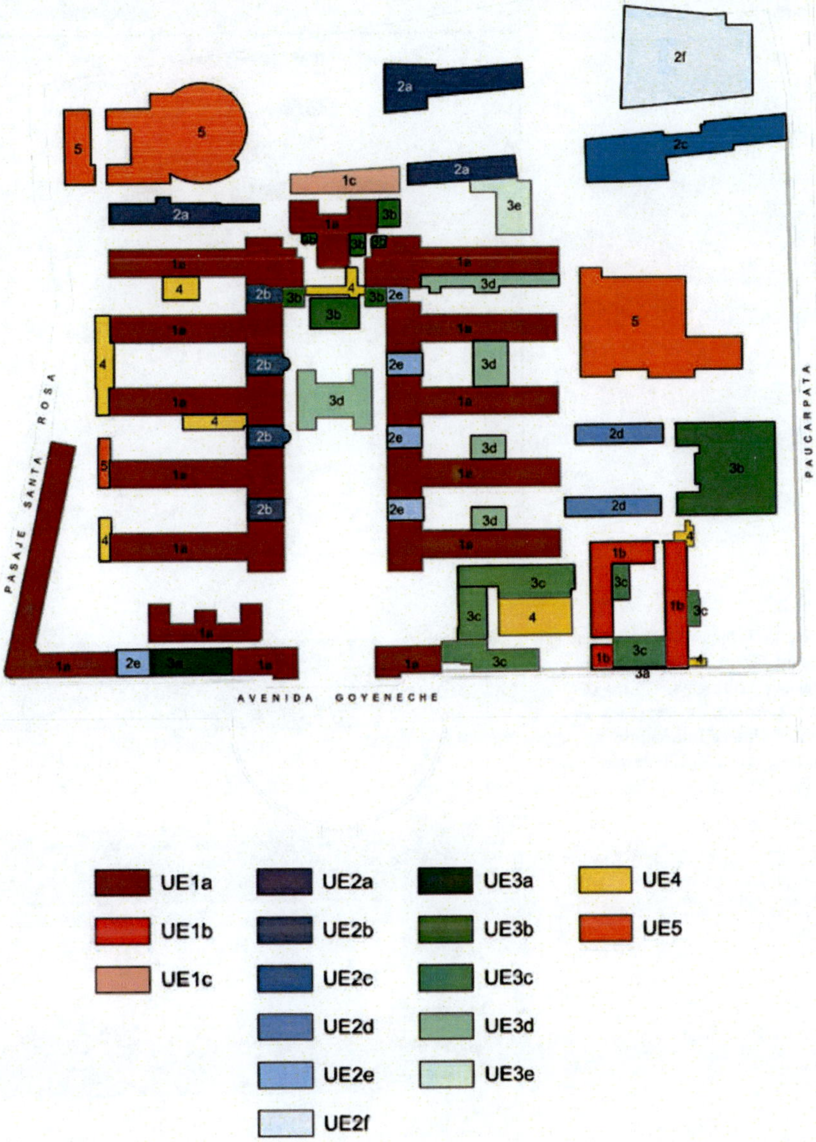

Fig. 8 The overall plan, stratigraphic units. Graphic: E. Suarez

3.3 Survey of the Hospital in Its Current Status

Valuation study

A valuation study was carried out through a record of historical, symbolic, artistic, urban, architectural, technological, environmental, and instrumental values. This study helped to determine the categorization of the property as a first, second, or third order of value; resulting in the permanence or not of each building.

Among the different varieties of constructions in the hospital, two types of buildings are shown as the most emblematic: the longitudinal pavilion and the hydrotherapy module.

Fig. 9 Harris matrix. Graphic: E. Suarez

HARRIS MATRIX			
STRATIGRAPHIC UNITS	PHASE	PERIOD	ARCHITECTURAL INFLUENCE
1a 1b	PHASE 1	1912-1917	NEO-GOTHIC
1c	PHASE 1	1917-1940	ECLECTIC
2a 2b 2c	PHASE 2	1940-1944	ECLECTIC
2d 2e	PHASE 2		TRANSITION
2f	PHASE 2		TRANSITION
3a	PHASE 3	1944-1962	LATE NEO-GOTHIC
3b	PHASE 3		TRANSITION
3c 3d	PHASE 3		MODERN
3e	PHASE 3		MODERN
4	PHASE 4	1962-1978	MODERN
5	PHASE 5	1978-2016	CONTEMPORARY

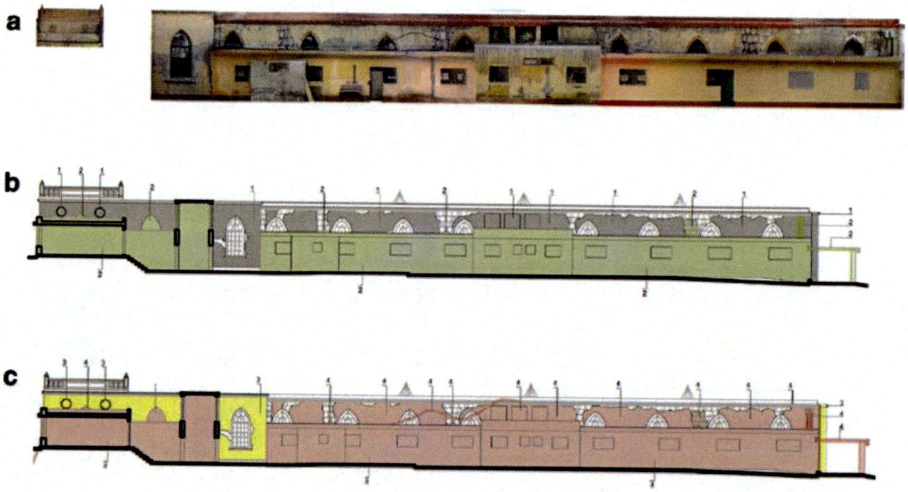

Fig. 10 **a** Typical pavilion photo elevation. Graphics: E. Suarez. **b** Plan of pathologies. 1. Loss of matter. 2. Additions. Graphics: E. Suarez. **c** Plan of Interventions. 3. Reintegration. 4. Liberation. Graphics: E. Suarez

Photo elevation mosaics

As a methodology for collecting information in each pavilion, photo elevation mosaics were prepared, which are photographs taken in small segments to avoid the typical distortion of the images and also to improve reliability in identifying each of the existing pathologies (Fig. 10a).

In this way, various architectural pathologies could be found, which were classified into four groups that were the most recurrent: additions, loss of matter, ruptures, and chromatic alterations; the first two being the most frequent in most pavilions. Among the cases of material loss, they were found mainly in the covering of the ashlar walls, which are decohesive, as well as in the different architectural details:

balustrades, original carpentry, and in the small mosaics of the medallions. As for the additions, the shown pavilion present the highest incidence of built-in brick within the entire complex; another form of addition is shown in the covered original openings, in addition to boxes, electricity, and water pipes, among others (Fig. 10b).

Once the pathologies in the pavilions were identified, they were solved in such a way that they resembled their original status as much as possible. For this, in cases of loss of material, the reintegration of wall coverings, openings, and details is proposed. Regarding the additions of built-in brick, it is proposed to release the surfaces of these intruding materials, leaving the original wall exposed again (Fig. 10c).

One of the most important buildings, both because of its location, as it stands as a finishing point on the central axis that runs through the complex, and due to its shape, is the Hydrotherapy module. It has been complex to be able to carry out its photographic survey, as currently there are several buildings attached to this module, and even more complex to be able to merge the pictures digitally so that serve as a work tool. There is evident damage to the structure, caused by the earthquakes and subsequent poor interventions that have modified both the exterior image and the interior space. This building has gone from being hydrotherapy originally, then the kitchen (that's why it has a brick chimney), and is currently a medicine store (for this an intermediate concrete slab was added) (Fig. 11a).

Hydrotherapy is the only module that has not been structurally reinforced and, due to time and constant earthquakes, it has deteriorated considerably, to the point of almost collapsing. Many pieces that were about to fall are now kept in storage in the same hospital. Over the years, elements have been added and they have deformed the particular morphology of the building; the most notorious are the external lateral additions, such as the lateral chimney, and the internal slab that divides the space in two, modifying the original height of the three openings of the. As for the loss of material, as shown in Fig. 11b, it happened above all in the covering of the walls, which have disappeared in the lower part of the building, in the carpentry of the main façade, in the mosaics of the 4 medallions and the upper gabled balustrade.

Having identified the pathologies in the module in the restoration study, we proceed to free all the additions found, the lateral modules, and the upper chimney; as well as the intermediate slab in a way that allows recovering both the interior space and the main spans. This allows the reintegration of the elements that were lost such as wall coverings, details such as carpentry, medallion mosaics, balustrade, and nomenclature. As indicated, the entire structure of the building is also consolidated through reinforcements and necessary additives (Fig. 11c).

Exploration Coves

Once the pathologies and their corresponding interventions have been identified, we proceeded to determine the originality of the coatings on walls, floors, and carpentry. This is achieved through an exploratory study, which showed each of the interventions carried out throughout the building's 111 years, important changes suffered due to deterioration and damage caused by earthquakes, renovations, expansions, and maintenance which have influenced its external image (Figs. 12a and b).

As a criterion for the location of the exploration coves, three representative buildings have been chosen to carry out stratigraphic surveys or coves. They have been located in places where they can be preserved without altering the general image of the complex after the intervention, taking into account the following criteria:

1. That the walls preserve traces of color or old paint to the naked eye.
2. That they did not have many previous interventions.
3. That the state of integrity of the sector of the cove is complete.
4. That the sample is representative of more than one piece of ashlar. The study was able to determine the different pictorial layers and the plaster or finish applied on the walls.

In this way it was possible to find the color, the plaster, and the original finishing of the pavilions, such as the typical plaster that characterizes the colonial houses in the city, using mortar composed of lime, sand, and calcium hydroxide.

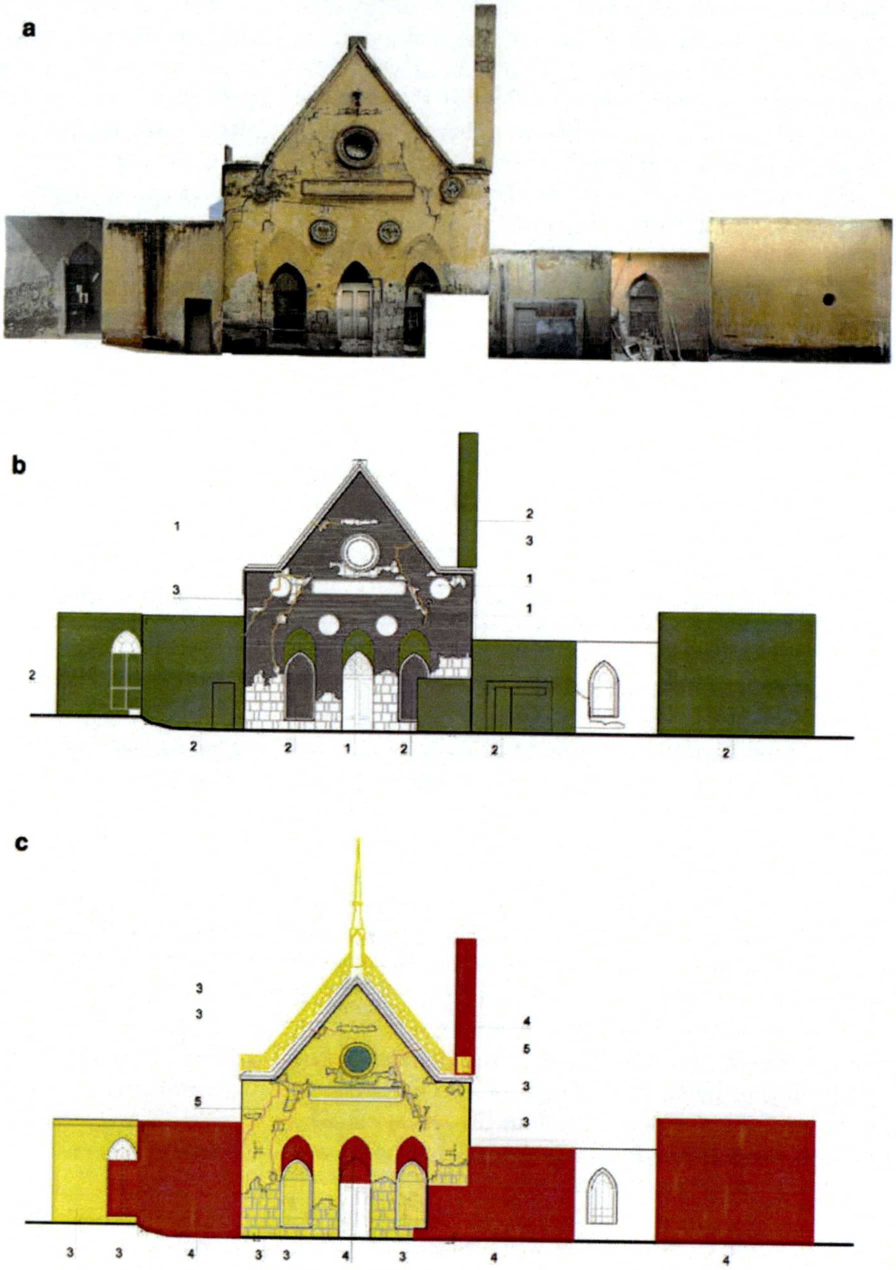

Fig. 11 a Hydrotherapy module photo elevation. Graphics: E. Suarez. b Plan of Pathologies. 1. Loss of matter. 2. Additions. Graphics: E. Suarez. c Plan of Interventions. 3. Reintegration. 4. Liberation. Graphics: E. Suarez

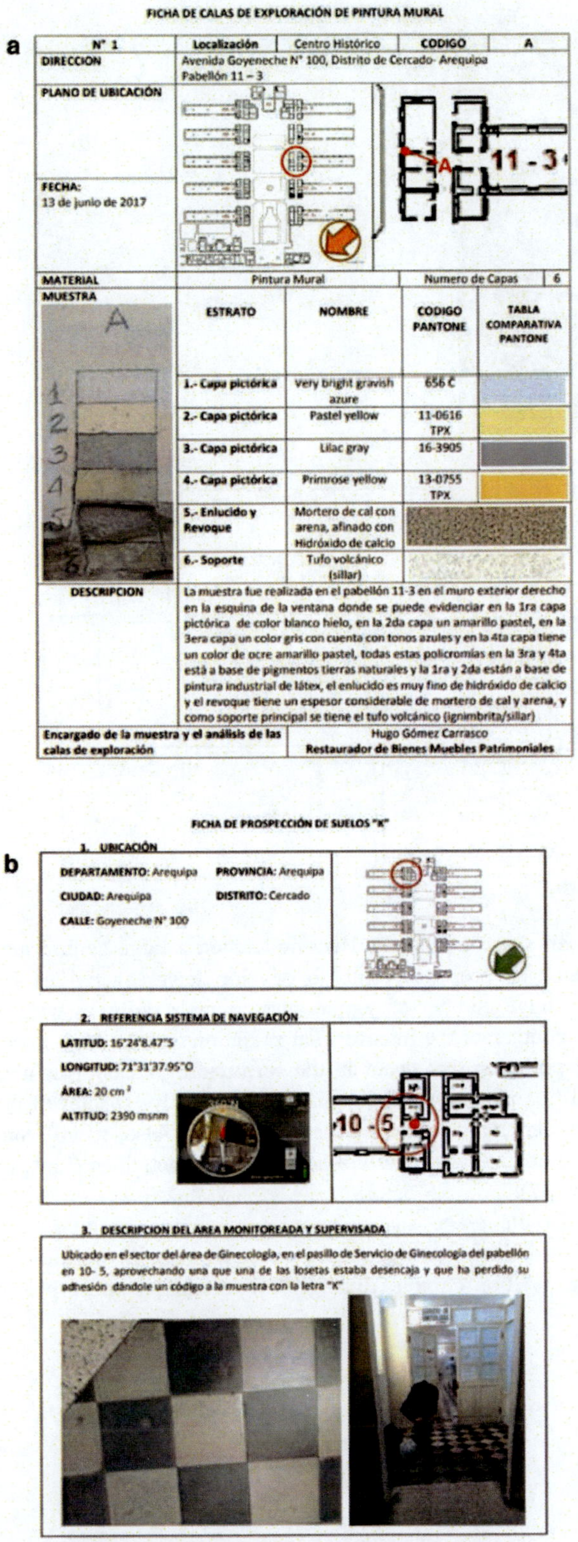

Fig. 12 **a** Wall exploration cove. *Source* Gómez (2017). **b** Floor exploration cove. *Source* Gómez (2017)

Brickwork type

The brickwork type has been considered since, according to this study, the original proposal took into account the level of precision in the finishing of the walls to decide whether a wall would be exposed or covered.

Figures 13a and b show an elaborate brickwork, following a typical pattern in ashlar constructions in the city, with the possible intention of leaving these walls exposed.

However, on other occasions, such as in Figs. 14a and b, the brickwork did not have a meticulous finishing, using rather ashlar pieces of various sizes and shapes, covering the interstices with amorphous pieces. As a result, the wall was subsequently covered with a coating.

3.4 Criteria for the Proposal

The criteria taken into account for the proposal are based on the unique characteristics of the hospital complex:

(a) Urban commitment: the spatial configuration of the hospital complex and its visual and symbolic connection with the perpendicular axis that faces it (Goyeneche Boulevard) are prioritized, removing the structures that conflict with this general order.
(b) Symmetry: this particular characteristic was mainly used in the original design and is preserved within the whole complex as well as on each of its elements (Fig. 15).
(c) Rhythm: a predominant characteristic in all the openings, doors, and windows of the complex.
(d) Stylistic qualities: the unique Neo-Gothic style models brought from France are restorered, as well as their proportions and location give it a singular character. All the openings are ogival, except for the screen at the entrance to each pavilion and the windows of the rear pavilions.
(e) Spatial richness: the scale of the internal scale dimensions is proportional to the scale of the whole complex, freeing any element that divides its integrity and trying in most cases to respect and enhance the spatial value of these internal spaces.
(f) Construction evolution: the different structures of the hospital have been identified and categorized according to their year of construction.

4 Proposal

The most important criterion regarding the restoration of the old hospital has been to recover each of its unique qualities and characteristics. However, in terms of the

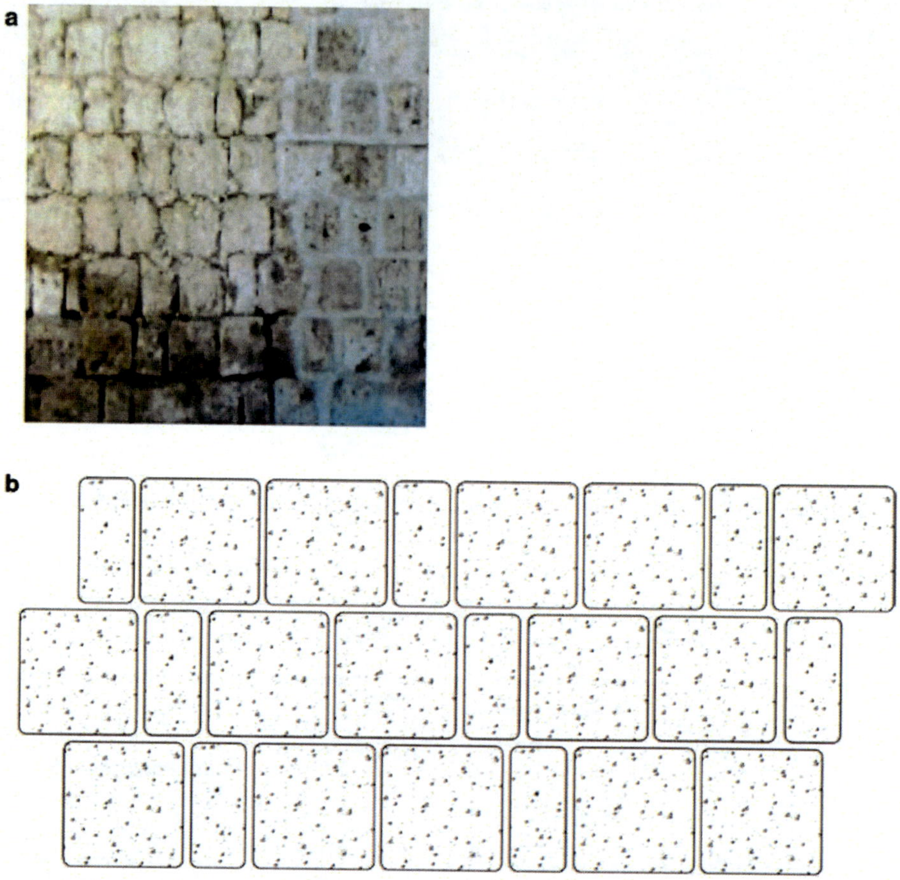

Fig. 13 **a** Ashlar brickwork on the side walls of pavilions. **b** Brickwork scheme. *Source* Author: Edith Suarez

recovery of the ancient elements that no longer exist, the relevance of reconstructing a building in Neo-Gothic style was evaluated, stylizing the proposal with contemporary elements.

4.1 Proposal for Urban Spaces and Landscaping

To the same extent, the relationship between both the main urban space and the spaces between the pavilions was evaluated. Taking into account that this complex has the largest open space within the Historic Center, a large central garden was proposed, both to develop leisure activities for the large number of people who circulate daily, as well as to counteract the intense solar radiation and the high levels of pollution that occur in the surrounding areas. This area will be planted with jacaranda trees, a native species of the city, arranged following the central axis. Also, it will contain other spaces that will house cultural activities (Figs. 16 and 17).

In the green spaces between pavilions, gardens with curved lines were proposed to counteract the rigidity of the pavilion complex, using different shades and types of

a

b
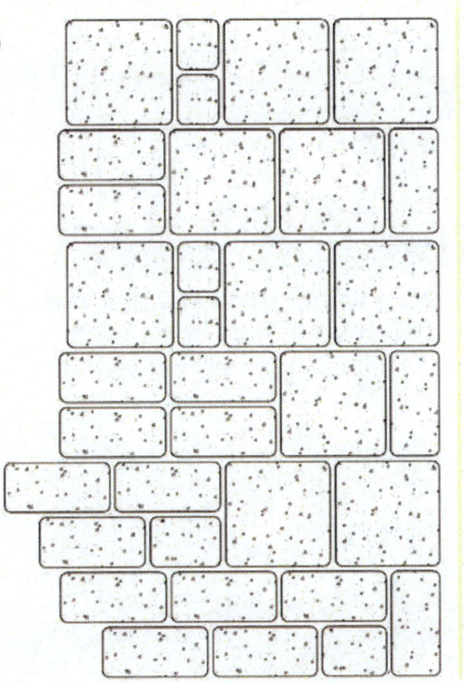

Fig. 14 **a** Ashlar block on the main façades of pavilions in front of the main garden. *Source* Author: Edith Suarez. **b** Brickwork scheme. *Source* Author: Edith Suarez

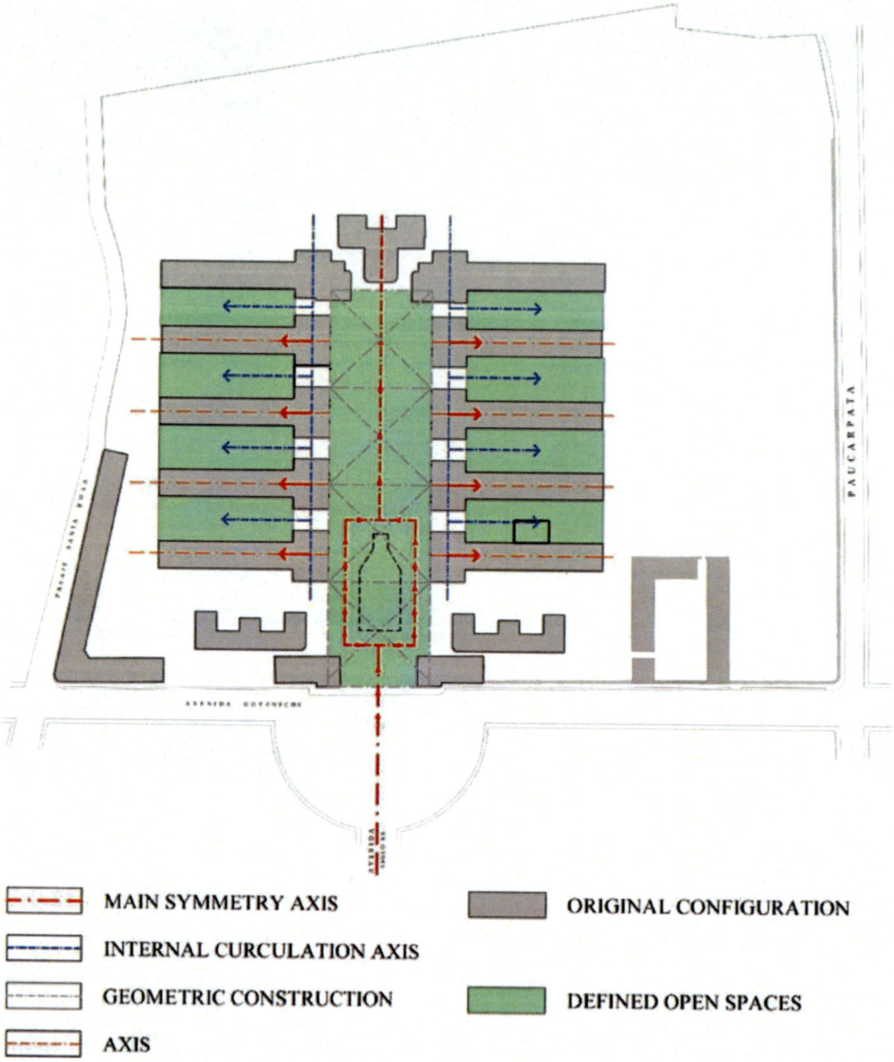

Fig. 15 The layout of the hospital complex showing the spatial and symmetry criteria. *Source* E. Suarez

mostly native flowers; promoting greater biodiversity necessary for the bee population, as bees are crucial in the ecosystem service of pollination and in food production (Fig. 18a–c).

4.2 Proposal for Restoring the Buildings

Once all subsequent additions were removed and the missing parts were restored, it was necessary to decide on the coatings and finishes of each of the surfaces, for which the

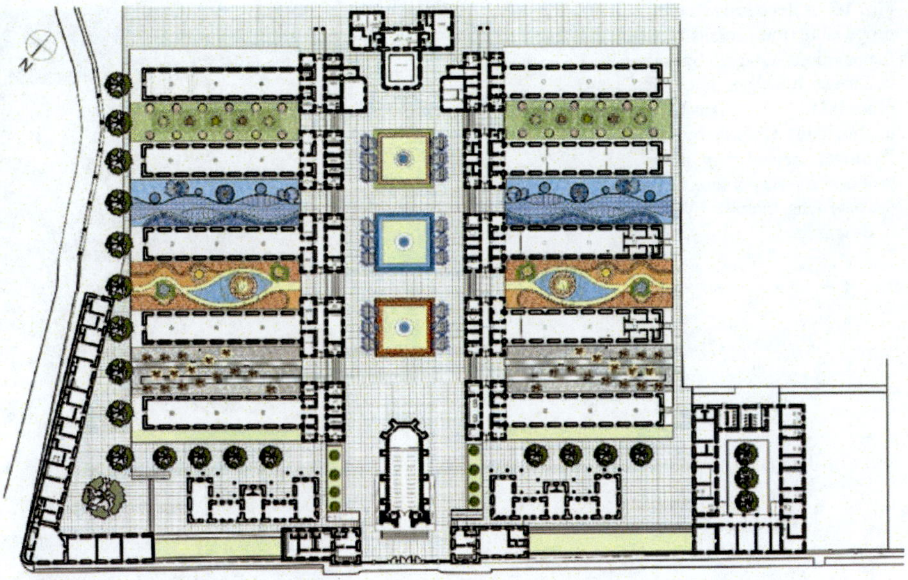

Fig. 16 General landscaping plan. Graphic by E. Suarez

Fig. 17 Central garden proposal. Graphic by E. Suarez

Fig. 18 **a** Rendering of the proposal for the gardens between the pavilions. Graphic by author: E. Suarez. **b** Rendering of the proposal for the gardens between the pavilions. Graphic by author: E. Suarez. **c** Rendering of the proposal for the gardens between the pavilions. Graphic by author: E. Suarez

a

b

c

Fig. 19 The final proposal for typical pavilions. **A** Ochre and lime coating. **B** Casing and stool in exposed ashlar. **C** Ivory-colored metallic carpentry. **D** Medallions. **E** Coverage between pavilions **F** Proposed carpentry **G** Visible ashlar wall **H** Ocher stool. **I**. Ochre color casing. Graphic by E. Suarez

Fig. 20 The final proposal for the hydrotherapy module. **A** Ochre and lime coating. **B** Casing and stool in exposed ashlar. **C** Ivory-colored metallic carpentry. **D** Medallions. **E** Mosaic **F** Archivolta **G** Clock **H** Balustrade. **I** Pinnacle. Graphic by E. Suarez

so-called exploration coves were made in each of the typical buildings (Fig. 19).

According to the stratigraphic exploration coves carried out in the pavilions, all the lateral coatings were made of ashlar and on top of it cement was added, which was not compatible with the ashlar, for which it was proposed to leave the walls exposed without coating, revealing the construction rigging. However, in the heads of the pavilions,

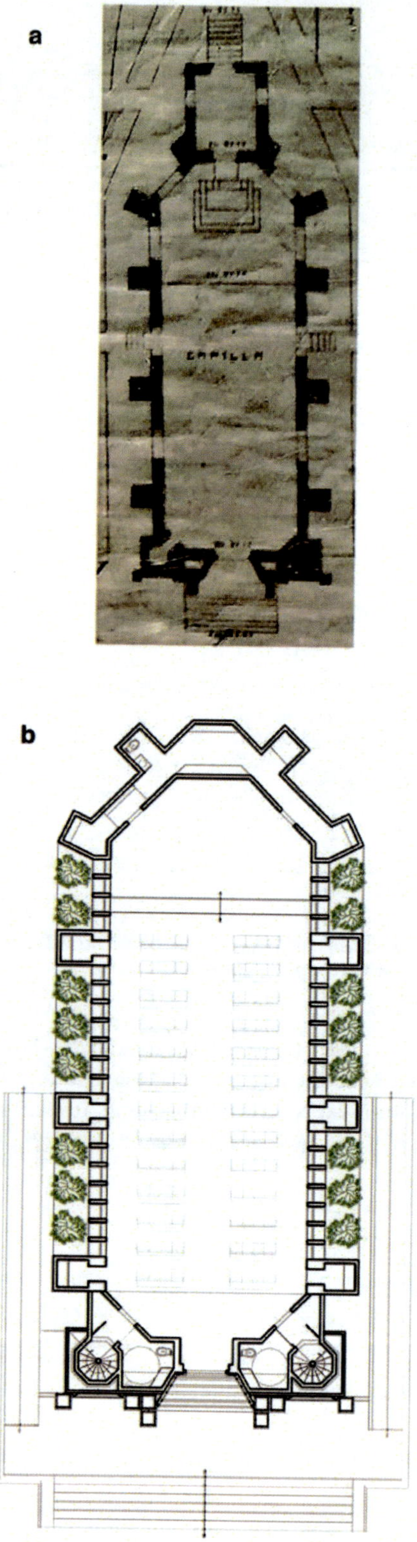

Fig. 21 **a** An ancient plan of the chapel. Graphic: Miguel Cordero. **b.** Proposal for the new chapel. Graphic by author: E. Suarez

the stratigraphy shows a finish of lime and sand, painted with yellow ochre. Therefore, it is proposed to recover them as they were originally, also considering that the rigging on these surfaces was built to be hidden.

The finishes proposed in the hydrotherapy module were also subject to the study of coves, whose stratigraphy showed that, like the case of the pavilions, the walls were covered with lime and sand, and painted with yellow ochre; therefore, the recovery of both finishes is proposed. This finding shows that all the walls facing the large central space were covered and painted in the same way, as a form of homogenization; different from the treatment of the other walls facing less relevant spaces. In the same way, the finishes of missing details are reintegrated, such as carpentry, balustrades, medallions, etc., as significant elements within the location of the module within the monumental axis (Fig. 20).

In the case of the chapel, the main tool for its reconstruction was a survey plan of the entire complex, which shows the building as it was before its demolition (Fig. 21a). In addition to this, considering its reconstruction 110 years later, where the requirements and conditions in our time are different from the initial ones, it is decided to rebuild the structure that forms the building, that is, the main and rear façade with all the stylistic details of the only Neo-gothic example in the city, but this time built in white concrete veneered in ashlar, obtaining an exterior appearance similar to the original (Fig. 21b).

However, on the lateral sides, the presence of buttresses is maintained, but they are not solid as in the original case but become confessional booths and/or meditation spaces, illuminated from above with a pink light. Between each buttress, a sequence of screens with a gradient of blue tones is proposed, from which the whole can be seen from the elevated part where the chapel is located. The colors of the proposed glasses evoke the colors of the Virgen del Consuelo, the patron saint assigned to the chapel (Figs. 22 and 23).

5 Conclusions

This research proposes a methodology that allows the recovery and improvement of a deteriorated historical hospital, maintaining its health functions, but enhancing its heritage and monumental values, which contributes to improving the quality of life of the inhabitants while reinforcing their identity.

The research is based on a historical study of the hospital, as well as an urban analysis of its influences and relationships. Then, criteria are proposed for the survey of the building and, finally, strategies are discussed for the enhancement of the deteriorated historical pavilions,

Fig. 22 External view of the chapel. Graphic by author: E. Suarez

Fig. 23 Interior of the chapel. Graphic: author: E. Suarez

complementing them with new buildings that are integrated into the monumental complex.

References

Actualidades. (15 July, 1905). *El nuevo hospital de Arequipa* (p. 120). Actualidades Revista Ilustrada.

Agenda para el Desarrollo de Arequipa. (2021). *Balance del Sistema de Salud Frente al COVID-19, Arequipa 2021*. Retrieved 09 28, 2022, from https://ucsp.edu.pe/wp-content/uploads/2021/11/foro-tecnico-balance-sistema-salud-covid19-arequipa.pdf

Barros García, J. (2004). The use of the Harris matrix to document the layers removed during the cleaning of painted surfaces. *Studies in Conservation, 49*, 245–258. https://doi.org/10.1179/sic.2004.49.4.245

Bullen, P., & Love, P. (2011). Adaptive reuse of heritage buildings. *Structural Survey, 29*(5), 411–421. https://doi.org/10.1108/02630801111182439

Cordero, M. (n.d.). [Image from his photo collection]. Obtained in Arequipa, in 2018.

Davies, M. (1993). *The application of the Harris matrix to the recording of standing structures* (pp. 167–180). Practices of Archaeological Stratigraphy.

Diana, L., D'Auria, S., Acampa, G., & Marino, G. (2022). Assessment of disused public buildings: Strategies and tools for reuse of healthcare structures. *Sustainability, 14*(2361). https://doi.org/10.3390/su14042361

Djebbour, I., & Biara, R. W. (2020). The challenge of adaptive reuse towards the sustainability of heritage buildings. *International Journal of Conservation Science, 11*(2), 519–530.

Doneus, M., Neubauer, W., Filzwieser, R., & Sevara, C. (2022). *Stratigraphy from topography II. The practical application of the Harris matrix for the GIS-based Spatio-temporal archaeological interpretation of topographical data* (pp. 223–252). Archaeologia Austriaca.

El Deber. (18 June, 1910). *Capilla en construcción*. El Deber, British Library, Library, EAP726/1/1/21/167. https://eap.bl.uk/archive-file/EAP726-1-1-21-167

El Deber. (July 28, 1912). *Capilla construida*. El Deber, British Library, EAP726/1/1/21/207. https://eap.bl.uk/archive-file/EAP726-1-1-21-207

Gibson, C. (February 12, 1912). *Inauguración del hospital Goyeneche*. La Bolsa. Arequipa.

Gómez, H. (2017). *Informe de calas del hospital Goyeneche*. Arequipa.

Gonçalves, J., Mateus, R., Silvestre, J. D., & Vasconcelos, G. (2017). Survey to architects: Challenges to inspection and diagnosis in historical residential buildings. *3rd International conference on preservation, materials and rehabilitation of historical buildings and structures* (pp. 3–10).

Instituto Geográfico Nacional. (1944). [Aero photography]. Obtained in Lima, in 1999.

Konsta, A. (2019). Built heritage use and compatibility evaluation methods: Towards effective decision making. *Built Environment, 191*, 237–242. https://doi.org/10.2495/STR190201

Llerena, R., Muñoz, P., & Rojas, M. (1996). *Hospital Goyeneche: Nueva Estructura. Thesis*. Universidad Nacional de San Agustín.

Mısırlısoy, D., & Günçe, K. (2016). Adaptive reuse strategies for heritage buildings: A holistic approach. *Sustainable Cities and Society, 26*, 91–98.

Municipalidad Provincial de Arequipa. (June, 2019). *Hospital Goyeneche*. Retrieved from Gerencia del Centro Histórico: https://gcentrohistoricoaqp.blogspot.com/2019/06/138_11.html

Instituto Nacional de Estadística e Informática. (2018). *Censos Nacionales: XII de Población, VII de Vivienda y III de Comunidades Indígenas*. Retrieved from Instituto Nacional de Estadística e Informática: https://www.inei.gob.pe/media/MenuRecursivo/publicaciones_digitales/Est/Lib

Palma, C. (June 18, 1910). *Capilla en Construcción* (p. 209). Variedades.

Palma, C. (March 2, 1912). *Inauguración del Hospital Goyeneche* (p. 209). Variedades.

Paulet, P. (1909). [Letter to Antonio Vivanco]. Retrieved form the National Library of Peru, 2019.

Revilla, J. (n.d.). [Image from his photo collection]. Obtained in Arequipa, in 2019.

Rodriguez, C. (n.d.). [Image from his photo collection]. Obtained in Arequipa, in January, 2020.

Sicignano, C., Diana, L., Marmo, R., & Polverino, F. (2022). Abandoned places, complexes and parts of cities. Regeneration and enhancement of monumental hospitals in the historic centre of Naples. Vitrubio. *International Journal of Architecture Technology and Sustainability*, *7*(1), 78–91. https://doi.org/10.4995/vitruvio-ijats.2022.17489

Sociedad de Beneficencia de Arequipa. (1920). *Reglamento del Hospital Goyeneche*. Arequipa.

Universidad Católica San Pablo. (2020). *COVID-19: Exceso de muertes en Arequipa y Perú es peor de lo que parece*. Retrieved from San Pablo informa: https://ucsp.edu.pe/covid-19-exceso-de-muertes-en-arequipa-y-peru-es-peor-de-lo-que-parece/

Zeballos-Velarde, C. (2022). *Arequipa Moderna y contemporánea I. 1868–1940. Una historia urbano arquitectónica*. Universidad Nacional de San Agustín.

Zeballos-Velarde, C. (2020). *Atlas Ambiental de Arequipa*. Universidad Católica de Santa María.

The Impact of Islamic Legislation and Identity on Formulating Environmental and Social Sustainability Indicators for Residential Neighborhoods in Al Diriyah Heritage Area

Noha Kassab, Elsayed Amer, and Faisal bin Sulaiman

Abstract

Heritage neighborhoods in Arab countries form a stereotyped image of architecture whose design was influenced by Islamic legislation, this helped it acquire an identity that distinguished it in form and content. These legislations are a crucial component of social life and the passed-down customs and traditions that continue to be practiced today. Moreover, the forms and designs of heritage neighborhoods evolved to accommodate climatic and environmental factors to mitigate their adverse effects on users. To do this, the study intends to offer insight into how laws and legislations affect neighborhood sustainability from an environmental and social perspective. In order to reach the most important factors (legislation and related architectural and urban elements), the research employs a methodology based on documenting and analyzing the content of what was written from the literature about the subject, and then analyzing a case study from the heritage neighborhoods in Al Riyadh, the capital of Saudi Arabia, by observing, meeting the citizens, and determining whether the principles of sustainability for residential neighborhoods are applied and how.

Keywords

Diriyah · Sharia sources · Muslim cities

N. Kassab (✉) · E. Amer · F. bin Sulaiman
College of Architecture and Planning, Architecture and Built Environment Department, King Saud University, Riyadh, Saudi Arabia
e-mail: 442204359@student.ksu.edu.sa

E. Amer
e-mail: eamer@ksu.edu.sa

F. bin Sulaiman
e-mail: fsouliuman@ksu.edu.sa

1 Introduction

The forms and contents of residential neighborhoods are linked to many factors that contribute to its output. The location, physical characteristics, geographic boundaries, society, culture, the era in which those neighborhoods emerged, the technologies available at that era, and the external influences of the peoples and civilizations that passed through them (Duany & Plater-Moudon et al., 2006; Zyberk, 2008) are just a few of the factors that affect the forms and contents of these residential neighborhoods. The traditional architecture that emerged in Najd may represent and embody the reality of architecture that was not subject to influence but rather originated from human gatherings with an inherent Islamic culture, a harsh desert climate, a predominance of mud and wood as local building materials, and a security reality that necessitates tightly closed designs and fortification, if necessary, in addition to a life that depends on agriculture (Al-Qarni, 2002). As a result, the typical traditional residential neighborhoods in the Najd area appear in the form of residential communities with a compact fabric, where the houses' walls are compact and adjoining back-to-back to open on the other side to the internal courtyards designated for each housing unit. Those neighborhoods were spread next to the agricultural areas and contained special facilities for raising poultry, livestock, and other animals spread in that area (Aina et al., 2013). On the other hand, the neighborhood mosque with its mud structure with substantial walls and restricted exterior access serves as the hub of the community, hosting Friday speeches and prayers as well as a gathering space where neighbors may socialize and share knowledge and daily news. The roads of the neighborhood are narrow and winding, punctuated by arches that widen and narrow in accordance with the symbolic and spatial importance of the dwelling. The human presence has continued in the traditional neighborhoods in the Najd region since its inception to the present time which indicates the sustainability of

M. F. Shahidan et al. (eds.), *Urban Narratives: Exploring Identity, Heritage, and Sustainable Development in Cities*, Advances in Science, Technology & Innovation, https://doi.org/10.1007/978-3-031-48517-6_18

those neighborhoods in terms of environmental and social aspects (Al-Jarallah, 2020). Literary studies have highlighted the importance of the religious component in promoting concepts that support social sustainability, as well as the legislative source, and its retroactive impact on the manner of construction and the method of social interaction, and thus it affects directly or indirectly environmental and social sustainability (Jabbar & Al Ahbabi, 2011; Aina et al, 2013; Al-Sayyad, 2016).

2 The Problem of the Research

Although some studies and research talk about the relationship between the provisions of Islamic legislation and the principles of sustainability in residential neighborhoods, these studies did not eventually reach the formulation of local indicators that are added to the principles of global environmental and social sustainability. At the same time, the local authorities in the Kingdom of Saudi Arabia are striving to achieve the sustainability of heritage sites and their surroundings in line with the Kingdom's Vision of 2030 and are working to develop and rehabilitate the entire Wadi Hanifa region, which is a supervisory area for the historic Diriyah, monitors efforts were hired for this matter in addition to calls for research and studies to be carried out for it. Therefore, the idea of this research stems from the problem related to the scarcity of previous research in the field on one hand, and the need of the local authorities in the city of Riyadh to be concerned and more involved with the development projects in the heritage areas in Diriyah and Wadi Hanifa on the other hand.

The study's main objective is to derive environmental and social sustainability indicators in the supervisory area of Diriyah, drawn from the sources of Islamic legislation.

3 Methodology of the Research

To reach the previous goals, the research follows the descriptive approach (content analysis) for literary studies based on the main books that dealt with the issue of organizing and forming the Islamic city in its early appearance, passing through contemporary writers, and reaching the writings of well-known writers in the Kingdom of Saudi Arabia—the research area—both through University thesis, literary articles, and printed books. Also, the documentation approach is adopted and its content is analyzed through (the derivation of rulings from Islamic law) by inferring from the indexed dictionary of the Book and the Sunnah.

To reach the formulation of local indicators of environmental and social sustainability, the research follows the case study method and works on analyzing the urban, architectural, and organizational content and comparing it with the content used in the Islamic city to be able to highlight the direct and indirect impact of Islamic legislation in the research area. Most aspects of the architectural form that are consistent with the principles of social and environmental sustainability emerge from the analysis of urban and architectural content, which arise through the analysis of local codes and the degree to which the community is attached to the architectural symbols and which must be preserved. To reach the relationship between the sources of legislation and the indicators of sustainability, the research approaches the comparative analysis of the global and local indicators of environmental and social sustainability, after which a final matrix is made to highlight the intersections between those indicators and the sources of legislation.

4 Literary Studies

4.1 Research in the Study Area and Similar Areas

About the mechanisms of achieving social sustainability in the traditional urban structure, specifically the traditional city center of Kadhimiya. Researchers (Jabbar & Al Ahbabi, 2011), started from the hypothesis that says: the traditional urban structure bears the elements of social sustainability as it is derived from the teachings of Islamic law, and the field of research was determined by studying a number of the elements of social sustainability: the sustainability of the kinship system, the sustainability of the Waqf system, and the sustainability of the religious factor.

On the subject of the sustainability of the kinship system, he identified the variables in kinship and its impact on the formation of the urban body—kinship as social support—kinship as a reason for living in the place—kinship as the basis of the strength of the relationship in the neighborhood. As for the axis of religious sustainability, he identified the variables in the following: controlling social behavior—strength social relations—Strengthening social cohesion and increasing interaction between parts of the city—Strengthening social relations—Social participation—Strengthening the sense of belonging—Increasing communication and interaction to and from religious places. Finally, in the aspect of endowment sustainability, the following indicators were studied: Social justice—strengthening social ties and relations and social solidarity.

Then it became clear that the religious factor played the most prominent role in the sustainability of society through the statistical results, which made the researchers recommend re-designing contemporary cities based on Islamic principles to ensure social sustainability.

A researcher (Abu Ali, 2000), when studying sustainable development in traditional architecture in the Kingdom of

Saudi Arabia (Hejaz region), developed a model for evaluating traditional buildings in terms of sustainability and relied on LEED evaluation criteria. The model shows that there are strengths such as the orientation of the building, the use of local labor, lighting, and natural ventilation, and the use of environmentally friendly materials derived from them. However, it demonstrates some flaws and weaknesses in the language of traditional architecture, such as the ineffective use of rainwater collection systems, ventilation techniques like the courtyard and the mantle, and lack of benefit from wastewater recycling processes. The buildings under study achieved similar results, confined between the silver and standard levels. In general, the utilization of renewable energy was ranked first in terms of standards compatible with sustainability, while the rationalization of water use came at the last place. this fact triggered the researcher to look for the architectural elements that contributed into raising this evaluation, such as reliance on natural energy in lighting, the use of straws, wooden Mashrabiyas, and others.

Towards an integrative theoretical approach to sustainable urban design in Saudi Arabia, and the value of geographical design, (Aina et al., 2013) foresees that when adopting urban development, it is necessary to resort to the basic principles and the social and cultural background of the traditional form. There should be a framework focused on the consideration of traditional concepts for the appropriate integration of these concepts into principles of sustainable urban design. The traditional urban and legislative elements that interacted to form the urban form and spatial structure were incorporated into the design process. Such as the institution of Hesba, endowment and land allocation, in addition to traditional energy-saving techniques and symbolic manifestations.

Authors (Al-Saadi & Al-Hamwi, 2000) refute the provisions of construction in Islamic jurisprudence and make them into two chapters, the first related to the construction and repair of the building, and the second related to the provisions of neighborhood and partnership. It collects evidence about the regulation of construction and the provisions of construction, repair, and demolition, as well as the selection of suitable land for cultivation that no one owns and granting permits of construction on it, in addition to the identification of buildings to be built such as mosques and private homes. These buildings should be safe from the wind drifts, and protected from the sun. Even though Islam urged the demand for livelihood and the expansion and decoration of buildings, but at the same time it had forbidden using silver and gold vessels and crockeries and recommended moderation in consumption. As for the issues of social relations, a general basic rule prevailed, which is not to harm (specifically by smell). This rule governed all construction, demolition, restoration, repair, addition, subscription, neighborhood, and endowment works and their repair. This affected the openings, protrusions, surfaces, and doors.

A study (Al-Kinani, 2009) shows that the Islamic city is governed by legislation that fully complies with LEED standards and requirements. The Islamic city was founded on principles related to the importance of choosing the site, linking it to neighboring roads, bringing benefits to it, addressing the environmental impacts of buildings, centralizing services, separating uses in markets, and excluding industries. The polluted areas must be separated from the city, emphasizing the unity of neighborhood, interdependence and organic planning with functional hierarchy and taking into account the movement of circulation and transportation within the city and the classification of streets. Urbanism in the Islamic city has also taken care of bringing in cultivable water, rationalizing its consumption, separating it from waste and distributing it fairly with the necessity of agricultural and green spaces, whether for investment, shade or Decoration as an aesthetic value. Other concepts and principles that are compatible with the requirements of LEED sustainability, are the concept of revival, environmental protection, and the concept of Hima, which corresponds to the preservation of natural areas and the redevelopment of neglected sites. Undoubtedly, innovation and flexibility are two features of the Muslim legislator, which is compatible with the current principles of sustainability requirements.

Referring to a previous study by the researcher on the relationship of the urban form with cultural and social sustainability (Kassab, 2021), the narrow corridors and external patios in the Mamluk city in Tripoli, Lebanon, are considered a reason for consolidating the social relationship between the residents as a gathering area with a compulsory character, as well as the case in Najd representing each of the Hosh, the interior and the narrow external corridors that bring together the people of the neighborhood. Al-Barahat, the neighborhood mosque, and the Najd decorations are landmarks with social symbolism and an important local identity.

As a result of the foregoing case, the same question arises, what is the Islamic city and how did the legislation contribute to its revival?

4.2 The Concept of the Islamic City or Muslim Architecture

The emergence of the Islamic city and the legislative impact in its beginnings

Medina is considered the first Islamic city after the Prophet Muhammad, may God's prayers and peace be upon him, migrated to it, leaving Mecca, where he and his Muslim companions could no longer practice their rituals there, manage their affairs, and carry out their mission. The idea of a city in which Islamic legislation prevails and where the citizens live in harmony and peace within their society began from Medina. The citizens have their religious, commercial, and residential facilities, in which there is no harm, no danger, no animosity or fanaticism, where the center of governance is justice, science and healing. As for the location of the city, it is located between the Harat and has an abundance of water and fertile soil for agriculture and pasture. As for the principle of shura and partnership between members of the community, it was manifested in maintaining the sheds, such as the shed of Bani Sa'idah, the shed of al-Rayyan, and others as places for the tribe's meeting for consultation. The mosque formed the first nucleus of the city, then soon the houses spread in the vacant lands at the behest of the Messenger, may God bless him and grant him peace.

Referring to the book "Futuh Al-Buldan", Al-Balathri (1901) transmits the correspondence that took place between Utbah bin Ghazwan and the Caliph Omar bin Al-Khattab, in this passage, he discusses the ideal location for a new city, emphasizing its accessibility to firewood, water, and pasture. He also suggests using firewood for heating and as a building material. On this foundation, the city of Basra in Iraq was built. Reeds were used for the construction of the main mosque and the homes; however, the usage of reeds caused a fire to start in the structures, which prevented them from making the structures larger. Later he gave the orders to create the largest streets with forty cubits wide, the medium ones with twenty cubits wide, and the alleys to be seven cubits wide, in addition to allocating a large area in the middle of each quarter for horse stables.

Accordingly, and as a result to the orders that came before the Caliph of the Muslims' speech served as a key source of legislation for the Islamic state, blending religious and secular planning. Perhaps the need for dwelling near natural resources to ensure the sustainability of human life, as well as modest construction to avoid wasting space, are what are employed to create benefits and eradicate harm. However, due to the mosque's prominence in terms of law, ideology, and politics, there is little question that its existence is what causes the community to first develop around it.

4.3 Arab Islamic City Planning

Historians and theorists have different perspectives on the identity and planning of the Islamic city, but according to AL-Yaqoubi (Wadeh, 1890), Madinah Al-Munawara was formed as a city after connecting its structure among the plans which were initially scattered, and it got distinguished because of the Noble Prophet's Mosque. As for Al Kufa city, according to Al-Buraqi (1332 AH-2003 AD), there are two plans for Kufa: one is for the Yemeni people from the east and the other is for the Nizar people from the west while the remaining space is left as a courtyard for the mosque and the Emirate's home. And according to Al-Qazwini (682 AH), Cordoba is a prosperous city with the "Al-Jami," one of the biggest mosques in Islam. As for the other cities inhabited by Muslims in various parts of the world, the cities differed from one region to another. For example, military construction in Andalusia, commercial urbanization with khans, and al-Qaysar in Granada, and Mamluk cities in Egypt and the Levant. And if this leads to anything, it would be to functional flexibility that the Islamic city expands from worldly aspects according to what the need requires, while keeping the essence of the constant important things within this flexibility. This statement might be very similar to what was stated by Ibn Khaldun (Ahmed, 1999) in his explanation of the principle of urban planning, where he indicates that urban planning is a phenomenon related to society, its structure, population origins and distribution, and the changing locations and functions performed by those nations in those cities according to time and place. Emerging from that perspective, (Mousa & Habib, 2013) believe that the urban identity is a mixture between constants and values on one hand and temporal and spatial variables on the other. He finds that there are many factors that affected cities in terms of the distribution of blocks in line with the security, economic, social, and political needs, and in terms of the external and internal shape of the buildings, which characterizes them with an architectural identity that reflects the thoughts and taste of the community. In his opinion, the Arab Islamic cities are represented by the ideas derived from the principles of Islamic Sharia (the constants) on one hand, and the urban environment applications on the other, which is the outcome of the interaction between those (constants) and the temporal, spatial, social, economic and political variables. Therefore, the urban form of the Arab and Islamic cities differs from one place to another and constitutes the identity of that place.

Ibrahim (1996), believes that there is an uninterrupted debate about defining the characteristics of the Islamic city, due to the political changes that the cities inhabited by Muslims have witnessed. Ibrahim adds that it is also unfair to describe the Islamic city as the city with the old fabric that no longer suitable for pedestrians in addition to the struggle of the traditional Islamic architectural form of stagnation in the face of the technological attack adding that society must deal with appositive manner even in a post violation, wars and geopolitical changes which played a major role in changing the urban form of the Islamic city and its impact on civilizations in the places of conquests.

As a result, the Islamic city is a set of contents that include politics, economics, and society and is determined by the Islamic faith and the reflection of all of this on the urban and architectural features of the city and is not a fixed form repeated in all regions. Accordingly, all the beforementioned sources agree that the Arab Islamic city is the product of two things: the first is not fixed, in which blocks, architectural shapes, and urban relations between them were formed based on the spatial and temporal situation, and the second is fixed, which is the values that govern society and urbanization all together, and they are the sources of Islamic legislation. Fig. 1.

In general, to build the mosque for prayer in the middle to get to know all its inhabitants: it is the first thing that is planned in the Islamic city, and to estimate its markets to achieve the needs closely, and not to combine opposites population Many authors discuss the prerequisites for the establishment of Islamic cities. According to Al-Balathri (1901), while choosing a location, leveled land should be chosen, the air quality should be examined, and any potential dangers should be avoided. While rerouting, it's important to collect water without oppression and gauge the size of the streets to improve circulation. Since the mosque is the first thing that is planned in an Islamic city, it makes sense to construct a mosque for prayer in the city's center which will help citizens to get to know each other and in order to closely meet the needs of the residences, it is crucial to assess the markets. Rather than mixing populations with opposing viewpoints and to avoid conflicts it is necessary to keep each ruler with his entourage in close communities while the opposite parties on the other side of the city. In addition to surrounding the city with a fence for protection from outer attacks: the fence may be a wall or a trench, as in the time of the Prophet, or a natural boundary such as water, a valley, or others, in addition to the method of closed doors. Finally, it is necessary to bring in skilled and knowledgeable people for self-sufficiency.

Al-Hathloul (1994), in his comprehensive and innovative book, talks about the Arab-Islamic city, its genesis, and its urban characteristics, beginning with Medina, passing through the provinces of Basra, Kufa, and Fustat, and then the Abbasid cities of Baghdad and Samarra, to reach at the end of the analysis to the conclusion that these cities are similar in terms of planning, which is summarized in the presence of the mosque as the nucleus in that city in addition to the Emirate House nearby, while the markets are spread around the mosque. In addition to the cities that were growing in the form of plans or neighborhoods, each plan inhabited by a group of individuals linked by a tribal relationship whose residents were settled in feudalism, where the internal planning of the houses is up to the individuals themselves. According to the narrations, it is worth noting that the streets that were planned in those cities were widening and narrowing according to the need and function of the near buildings, and the winding roads came as a natural result of the passage between the plans that quickly multiplied with time to connect after they were almost scattered. As for land uses, markets have spread in Arab and Islamic cities, and residential areas have been separated from industrial areas according to the "do not harm or be harmed by others" rule, which governs many aspects of the organization in Islamic cities. As for the architectural blocks, many residential styles appeared in the Arab city, including the open courtyard, the covered courtyard, and the Mashrabiyas, all of which guarantee the principle of privacy, which is the basis of social and Islamic life in those cities.

It should be noted that many Arab and Islamic cities were safe cities because of their defense mechanisms presented by either walls, guardrails, or gates. The cities took a semi-radial shape with the distribution of roads while the streets were also centered around the mosque and contained Barahas that were distributed throughout the neighborhoods and called al-Oqad.

Many other models around the Islamic world follow the same pattern, with the knowledge that these cities have grown, and developed and their population with clearer features and a tighter urban fabric. Figures 2, and 3.

Moreover, Al-Hathloul addresses a very important issue, which is Islamic legislation and its impact on the urban formation of Arab Islamic cities. Where the role of the judge and those under him among the trustees and the muhtasib is highlighted in the implementation of God's commands in those human gatherings according to what the sources of Islamic legislation came with, namely the Noble Qur'an, the Sunnah, the consensus of the imams, and the custom

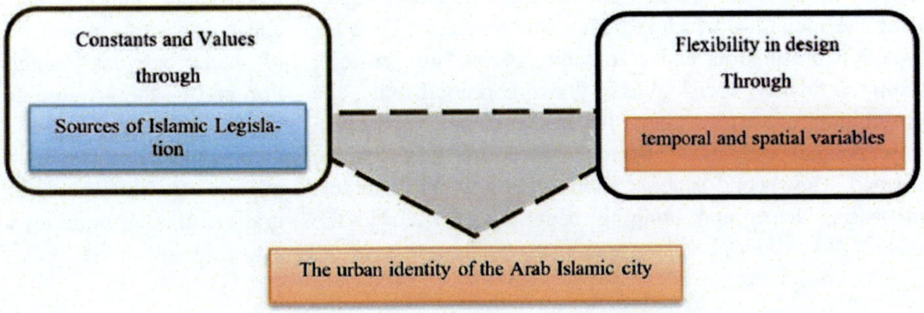

Fig. 1 Summary of definitions of the formation concept of the Arab-Islamic city. *Source* The author

Fig. 2 Damascus—a lane plan with a gated system on streets and windowed alleys. (Al-Hathloul, 1994, p. 94)

that spreads in the Islamic society in a way that does not contradict a legal text.

As for the houses, the urban forms have taken multiple models, some of which are with the inner courtyard, and some are with the clearness, which is the heavenly opening above the main hall like the modern skylight, and some of them are open to the outside in the image of a Mashrabiya. Fig. 4.

It can be concluded from the literature review that Islamic principles ruling and organizing the Islamic city can be summarized in four categories: first is about organizing the social life which includes: The Hesba, endowment (waqf), Consultation, and community partnership, custom, family solidarity, social solidarity, neighborhood unity, the prohibition of extravagance in the resource, Motivation to work and productivity, Principle of do not harm nor be harmed. The second is about site selection which includes: preventing harm and bringing benefits, air quality, the presence of freshwater, bringing water without misery, bringing in the scholars and the skilled, supplier presence, and protecting the natural environment. The third is about urban design which includes: fences, compact tissue, organic layout, and functional space scale, narrow, winding, and walkable streets, street types, street design, metaphors, bouquets, trees, and fabrics, al Barahat, the mosque, public squares in mosques, market, centralization of services, industries outside housing, horse stalls, taking into account the movement of transportation (at that time, animals), yard landscaping, landscaping, death protection, building dams and constructing ponds, entropy, sweep rainwater, division of water between neighborhoods, gutter, reuse of ablution and washing water, common drainage channels, public baths, wells, the forehead. The fourth and the final is about architectural design which includes: privacy and visibility, passive construction, skylights, windows, and openings, natural ventilation, building with mud, stone, and wood, inner

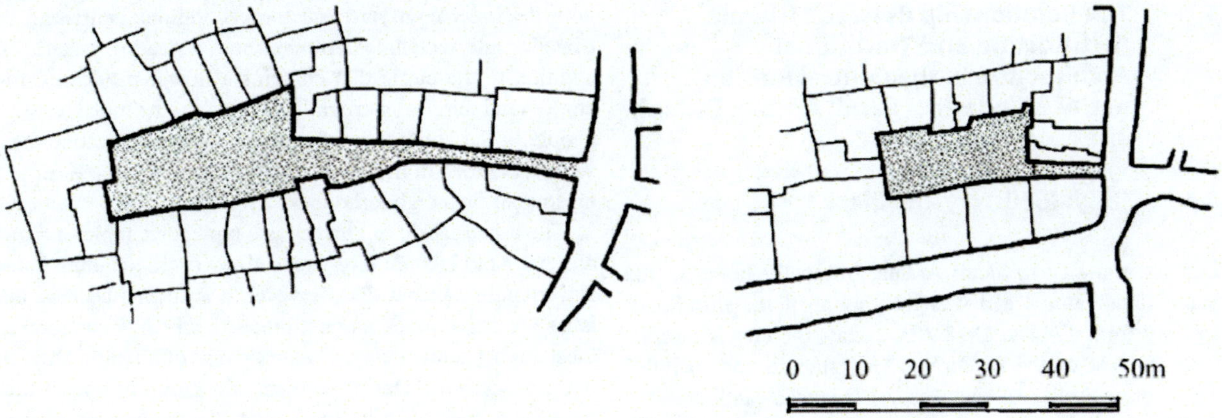

Fig. 3 Medina The courtyard is a semi-private space used for celebrations, events, and children's play. (Al-Hathloul, 1994, p. 100)

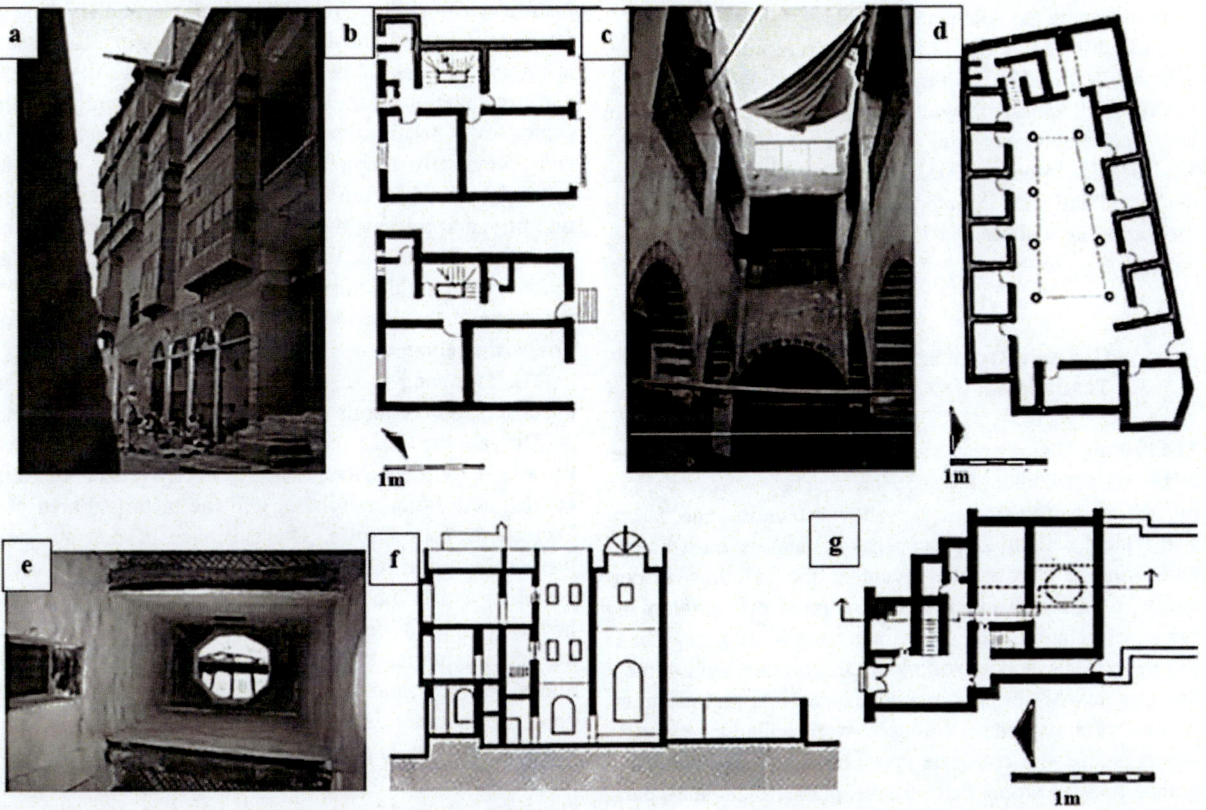

Fig. 4 **a** The outside of a house with a Mashrabiya in Medina, **b** its plan, **c** the inside of a house with a courtyard in Medina, **d** its plan, **e** the inside of a house with a hall, **f** its section, **g** its plan. (Al-Hathloul, 1994, pp. 116–118)

courtyard, hole shapes, introducing natural lighting, providing shadows, minimum and geometrical or plants decorations, function, modesty in construction, don't get stuck in the building, respecting the heritage in construction, reuse of animal dung, reusing dead trees, solid waste collection, innovation in the means and methods of construction.

5 The Relationship Between Islamic Architecture and Traditional Architecture in the Central Region of the Kingdom of Saudi Arabia, Diriyah as a Model

5.1 Diriyah, History and Place

Diriyah is a city in Saudi Arabia, originally home to the Saudi royal family, and was the capital of the first Saudi dynasty from 1744 to 1818 AD. Today, the city is considered the seat of the Al-Diriyah Governorate and includes neighboring villages such as Al-Uyaynah and Jbeilah. At-Turaif neighborhood in Diriyah was declared a World Heritage Site by UNESCO in 2010. Diriyah (24 0 44′.13″N, 4634′.08″E) is located on the banks of Wadi Hanifa, at an altitude of 700 m above sea level, at the level of the central Najd plateau in the Arabian Peninsula.

Diriyah was founded in the fifteenth century in the year 850 AH 1446 AD. The region was the site of many pre-Islamic civilizations as well as Islamic settlements, and in later centuries emerged as the most powerful state along Wadi Hanifa in central Arabia. Diriyah owes its success to the fertile and well-irrigated fields along the valley and its tributaries as well as the long-standing trade relationship with Al-Ahsa to the east and south of the Arabian Peninsula.

5.2 Diriyah Architecture and Urbanism: Traditional Urban Design in Diriyah

The historic Diriyah consisted of several neighboring settlements on the banks of Wadi Hanifa. They were separated by natural features, such as valley tributaries and farms. The typical neighborhood consists of densely built houses with winding streets. The layout of the buildings is perfectly aligned with the prevailing wind direction, allowing for air circulation. The streets are narrow, allowing shade for pedestrians and providing a comfortable environment for residents. Major buildings such as palaces and mosques, as well as most ordinary houses, were built in a century-old technique and consisted of adobe with foundations and pillars made of stone. Palm trunks, palm fronds, and, palm leaves covered with a layer of mud were used to build roofs and ceilings, as well as stairs leading to the roofs of buildings. It contains ten heritage neighborhoods, most notably Al-Turaif. These neighborhoods include mosques in central neighborhoods and are surrounded by defensive walls or agricultural insulators. They also include some urban elements specific to the nature of the neighborhood, such as schools, farms, palaces, and others. The architectural style of the traditional buildings prevailing there is the house with the inner courtyard and the decorations consisting of triangles and rectangles arranged in a repeating pattern, in addition to the circle that contains a flower inside and triangles and circles surrounding it from the outside. The people in the heritage neighborhoods used to practice their social activities in the outer squares and meet in the central mosque of the neighborhood.

The urban fabric of Diriyah is a typical example of a traditional Arab Islamic city, especially with its winding lanes and irregular outer walls. However, it is special because the lanes are crisscrossed and not blocked, which is common in most Arab-Islamic cities. The urban unit of Historic Diriyah is the block: each cluster consists of a group of houses that have been positioned in a way that ensures maximum protection. Each of the clusters clump together to form public spaces as well as semi-private interior spaces. Arcade-type mosques are common and are found in various locations in historical assemblies. While the outlines of the block are tentatively defined, the division of each group into houses is arbitrary. This should be left to the responsive development and urbanization of these vacant sites, following the same principles of Islamic law and regional customary rules that were previously adopted in Diriyah to develop its original urban fabric. Thus, the right of ownership and the protection of visual privacy, during the development of the neighboring houses built in succession, are the main factors that make up the architecture and are therefore responsible for the shape of the external walls of the houses, their orientation, and openings.

The following is an urban analysis of some of the traditional neighborhoods in Wadi Hanifa area, specifically Al-Diriyah, Fig. 5 shows a part of Al Turaif district, One of the World Heritage Sites, which perfectly reflects the urban context and urban morphology in the historical area of Al Diriyah in wadi Hanifa. The existence of the mosque as the starting point of the neighborhood is very clear. At the periphery exists the farms, and the fence, while the royal houses which are considered as residential buildings form an organic fabric, and compacted urbanism with narrow alleys between mud buildings, connected with nodes, and contains an inner courtyard. Moreover, some special features are existing in the district and match the Islamic urban elements such as Al-Waqf (endowment), and the fresh water source represented by the stream of Wadi Hanifa. Figure 5.

As a result, the Diriyah area coincides with Islamic architecture with its fixed and variable parts, so that Islamic law, with its provisions and customs, casts its shadows on urban formations, buildings shapes and openings, and the elements of Islamic urbanism appear in the presence of the mosque as a basic nucleus for the formation and rest between the dwellings and the compact fabric surrounded by the edges and the fences that provide Safety and security.

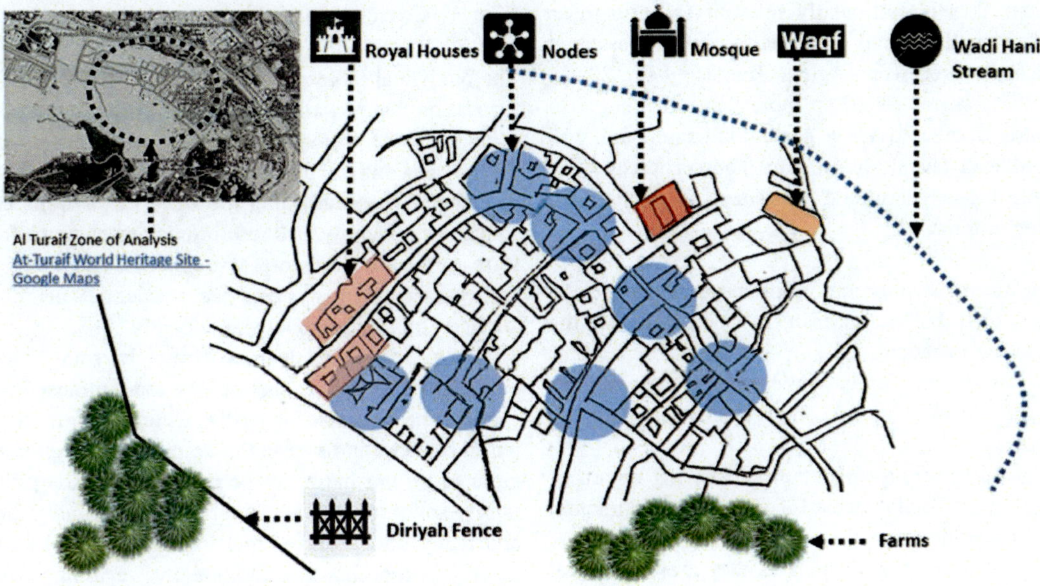

Fig. 5 Al Diriyah heritage site urban context analysis. *Source* the author

Below, in the Tables number 2, 3, 4, and 5 at the appendix we highlight the most important sources of legislation related to the regulation of architecture and urbanization and the extent of the mandatory nature of these sources based on jurisprudence evidence (where Man. means mandatory order named in Quran or Sunnah, and H. I. means highly important to be applied, and Flex. means flexible to do it or not) in addition to their relationship to the traditional urban situation in Wadi Hanifa, which constitutes a reference for urban development in that region.

5.3 Local Environmental and Social Sustainability Criteria in Wadi Hanifa and Their Intersections with Islamic Legislation

Sustainability indicators deal with the local content of the areas under study. These indicators differ from one geographical area to another, as well as from a built urban area to another; There are indicators of sustainability at the level of a single building, as well as indicators at the level of the neighborhood and others at the level of the city as well as the region. The formulation of indicators in general and on the neighborhood level, in particular, is a subject of disagreement—according to previous literary studies—as it is divided into two stages: The first stage is the deduction in the pre-implementation period, which mostly depends on the tools of assessing the neighborhood sustainability, which is known in the acronym (NSA: Neighborhood sustainable Assessment), and the second stage is the evaluation of the existing content and depends on observation and inspection of the applicable field by making deductive reports on sustainability indicators from reality, recording and comparing them, measuring the extent of their impact on sustainability, and adopting them at the end of the reports. Since the study here attempts to derive sustainability indicators for the neighborhoods to be designed in the future in the supervisory area of Diriyah, the Neighborhood Assessment (NSA) tools are the most famous and effective method in such studies. (NSA) is a class of impact assessment tools aimed at promoting urban sustainability by establishing a standard for assessing and certifying the sustainability of urban neighborhoods (Sharifi, 2013). These tools began to emerge in the early 2000s, inspired by environmental impact assessments of urban developments and certification tools for green buildings. While these systems have been developed for local use, attempts to understand and address these indicators of heritage are still limited today (Berardi, 2013). While the two globally prominent Neighborhood Sustainability Assessment tools, LEED-Nand D, BREEM-C, a discourse analysis of tool documents and a conversation between the theoretical advances of the fields of critical heritage and sustainability studies show that while heritage-related aspects are present in both non-governmental agencies, heritage is being reintroduced. as a matter of protecting primarily the physical expressions of culture while ignoring the natural elements and related immaterial practices. Both involve an ontological and ideological problem in the way of dealing with the local context. This is particularly relevant to heritage as approached, visualized, and incorporated into assessment and certification processes (Aguiar Borges et al., 2020). Therefore, the research relied on more than one source to create a matrix in order to derive general indicators that will be compared with Islamic legislation. Those sources were selected from the neighborhood assessment tools based on the following criteria:

1. Choosing the most important, most prominent, and the first assessment tool to appear in the world to ensure the validity of the selection.
2. Choosing assessment tools applied in different regions around the world to ensure comprehensiveness in the research.
3. Choosing assessment tools in the neighboring region of the Kingdom of Saudi Arabia for similar environmental and social conditions.
4. Choosing local assessment tools as they are an integral part of the legislation.
5. The following are the tools that the research relied on:
   a. International Neighborhood Assessment Tool: LEED-ND, LEED Cities and Communities, BREEAM Communities, CASBEE-UD, Building for Life, HQE^2R, Sustainable Renovation of Buildings for Sustainable Neighborhoods, The Neighborhood Sustainability Framework in New Zeeland (NSF), and One Planet Living Tool.
   b. Regional Neighborhood Assessment Tool in the Gulf Countries: Pearl Community Rating System Tool, GSAS district, and infrastructure,
   c. Local tools, charters, and codes: Saudi code for green buildings, Mostadam code, and Wadi Hanifa code.
6. Classifying the codes content into 3 categories or dimensions: Environmental, social and related to heritage.
7. Make Data addition to the same category.
8. Rename the indicators fields and target it as the goal of the related dimension.

Figure 6 helps to summarize the above methodology and elaborates its findings.

The local criteria deduced from environmental and social sustainability assessment tools are now linked with the principles of Islamic legislation in one matrix as shown in Table 1, and as detailed in the Appendix, and thus the links, effects, and interrelationships appear, which is the thrust of this research. The bold underlined principles are the most repetitive titles that maches many targets of social and environmental sustainability dimension. Table 1.

6 Conclusion

The analytical tables showed that the principles of Islamic legislation that prevailed in the Islamic community and contributed to the formation of the Islamic city by choosing its location and shaping its architecture, these legislations contributed to establishing the rules of environmental and social sustainability, as these provisions came to fully conform to the standards of environmental and social sustainability assessment systems for residential neighborhoods applied globally, regionally and locally.

Some principles deduced from Islamic Sharia have emerged as constants that reflect their utmost legitimacy, such as the principle of Hisba, which is currently embodied in the laws in force in the Wadi Hanifa region, the principle of do not harm nor be harmful, and the principle of family solidarity, privacy, public space around the mosques and many others have direct impact on social sustainability in neighborhoods. Additionally, regarding site selection, the importance of water and air quality seems to be the two most important principles in the formation of the Islamic city. In terms of design, the presence of the mosque stands out mainly in the formation of Islamic architecture, in addition to the design that provides a sufficient amount of privacy as a prerequisite that cannot be neglected. On the other hand, the principles of the design of the Islamic city, which are derived from Islamic legislation, showed full conformity with the local standards of environmental and social sustainability, such as saving the resource, using local environmentally friendly materials, not being extravagant or wasting in the use of resources, separating industries from the places of residence and many more, but the indicators that constituted a valuable addition On the criteria of sustainability were the endowment system, the principle of shura, community, and family solidarity, and neighborhood unity so that these principles contribute greatly to creating a diverse, safe, and active society. The mosque again plays a prominent role in the urban and social interdependence of the neighborhood and is considered the strongest link in the formation of the neighborhood. And the connection with the culture, heritage, spatial and spiritual symbolism of the community in those areas. Since the Najd architecture located in the research area in Wadi Hanifa met all its criteria with the formation of the Islamic city, those principles derived from Islamic Sharia, which are indicators of environmental and social sustainability, it is necessary to include these indicators when formulating local sustainability criteria in that area.

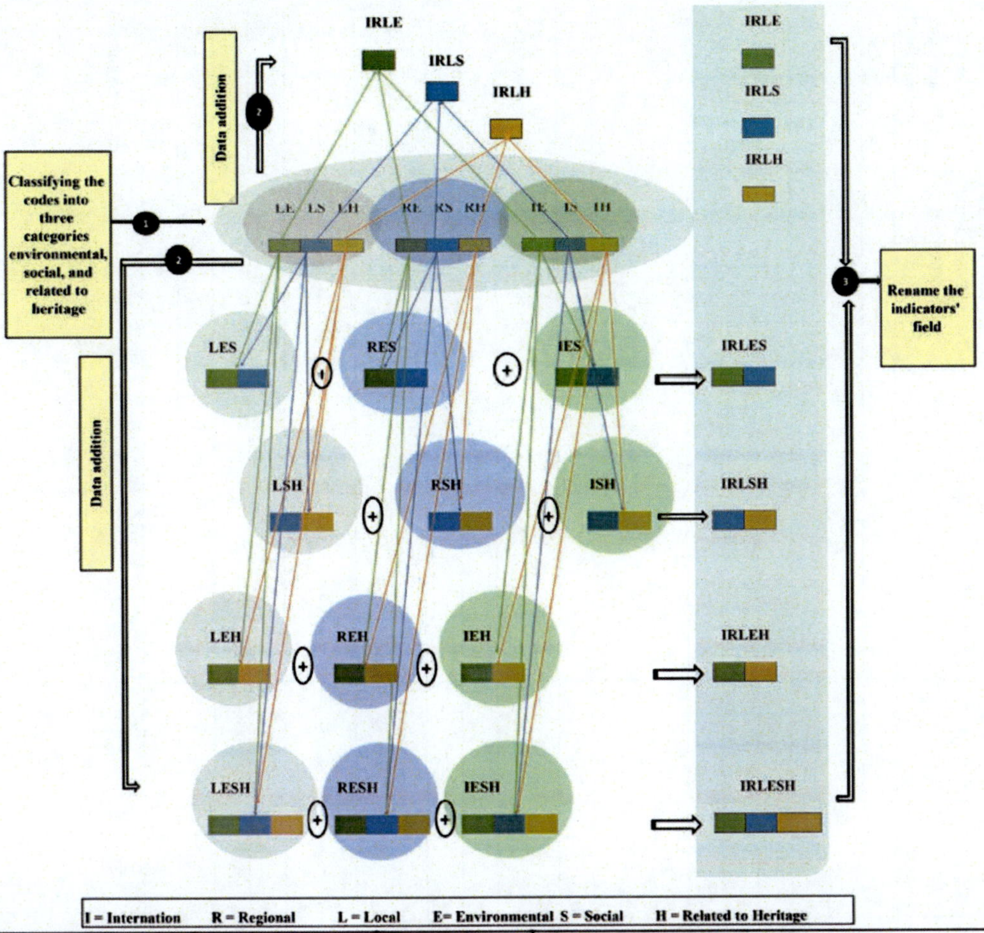

Colors of the dimensions	Symbol	Goals related to the dimension
	IRLE	Conserving resources and clean energy
	IRLS	A diverse, safe and active community
	IRLH	Respecting heritage in designs
	IRLES	Clean Infrastructure
	IRLSH	Respecting culture in designs
	IRLEH	Preserving the site and the natural environment
	IRLESH	Innovating in dealing with the local environment

Fig. 6 Methodology and findings of extracting local sustainability criteria tools in wadi Hanifa. *Source* the author

Table 1 Islamic legislation and local environmental and social sustainability dimensions and goals in wadi-Hanifa matrix

Principles of Islamic city regulations		Local environmental and social sustainability dimensions and goals in Wadi Hanifa						
		IRLE	IRLS	IRLH	IRLES	IRLSH	IRLEH	IRLESH
		Conserving resources and clean energy	A diverse, safe and active community	Respecting heritage in designs	Clean infrastructure	Respecting culture in designs	Preserving the site and the natural environment	Innovating in dealing with the local environment
Administrative and social organization	Hesba		•					
	Endowment (waqf)	•				•		•
	Consultation, and community partnership		•					
	Custom			•		•		
	Family solidarity		•					
	Social solidarity		•					
	Neighborhood unity		•					
	The prohibition of extravagance in the resource	•						
	Motivation to work and productivity		•					
	Principle of no harm, no making cause of harm		•				•	•
Site selection conditions	*Preventing harms and bringing benefits*		•				•	•
	Air quality		•				•	•
	The presence of freshwater		•				•	•
	Bringing water without the misery		•				•	•
	Bring in the scholars and the skilled		•				•	•
	Supplier presence		•				•	•
	Protecting the natural environment		•				•	•
Urbanism	Fences		•					
	Compact tissue		•	•				
	Organic layout and functional space scale			•				

(continued)

Table 1 (continued)

Principles of Islamic city regulations		Local environmental and social sustainability dimensions and goals in Wadi Hanifa						
		IRLE	IRLS	IRLH	IRLES	IRLSH	IRLEH	IRLESH
		Conserving resources and clean energy	A diverse, safe and active community	Respecting heritage in designs	Clean infrastructure	Respecting culture in designs	Preserving the site and the natural environment	Innovating in dealing with the local environment
Urbanism	Narrow, winding, and walkable streets		•	•				
	Street types			•				
	Street design			•				
	Metaphors			•				
	Bouquets, trees, and fabrics			•				•
	Al Barahat			•				
	The mosque		•	•		•		
	Public squares in mosques			•				
	Market			•				
	The plan			•				
	Centralization of services			•				
	industries outside housing		•	•				
	Horse stalls			•	•			
	Taking into account the movement of transportation (at that time, animals)	•		•	•			
	Yard landscaping				•			
	Landscaping				•			
	Death protection						•	•
	Building dams and constructing ponds	•			•			
	Entropy	•			•			
	Sweep rainwater	•			•			
	Division of water between neighborhoods	•			•			
	gutter	•			•			
	Reuse of ablution and washing water	•			•		•	
	Common drainage channels				•			
	Public baths				•			
	Wells				•			
	The forehead				•			

(continued)

Table 1 (continued)

Principles of Islamic city regulations		Local environmental and social sustainability dimensions and goals in Wadi Hanifa						
		IRLE	IRLS	IRLH	IRLES	IRLSH	IRLEH	IRLESH
		Conserving resources and clean energy	A diverse, safe and active community	Respecting heritage in designs	Clean infrastructure	Respecting culture in designs	Preserving the site and the natural environment	Innovating in dealing with the local environment
Architecture	Privacy and visibility				•			
	Passive construction				•			
	Skylights, windows, and openings				•			
	Natural ventilation				•			
	Building with mud, stone, and wood				•			
	Inner courtyard				•			
	Hole shapes				•			
	Introducing natural lighting				•			
	Provide shadows				•			
	Decorations				•			
	Function				•			
	Modesty in construction				•			
	Don't get stuck in the building				•			
	Respecting the heritage in construction				•			
	Reuse of animal dung				•			
	Reusing dead trees				•			
	Solid waste collection				•			
	Innovation in the means and methods of construction				•			

Source The author

Acknowledgements The authors would like to gratefully thank the Deanship of Scientific Research and Researchers Support and Services Unit at King Saud University. The authors would like to extend their gratitude to Professor Dr. Saleh AlHathloul for his support and making easy benefitting from his personal library. In addition, the authors also extend their grat-itude to the anonymous reviewers for their constructive and profound comments.

Appendix

See Tables 2, 3, 4 and 5.

Table 2 Islamic legislation importance and its applications in wadi Hanifa traditional urbanism In terms of Administrative and social organization

Principles of organizing the Islamic city design			The degree of importance of the legal ruling			Traditional urban design matching the principles
Administrative and social organization						
Principle	Description	References	Man	H. I.	Flex	Yes/No, with Explanation
Hesba	The Muhtasib, the judge, and his trustee's assistants had the main role in the politics and organiza-tion of the Islamic city according to the laws of Sharia. The first person to practice the calcula-tion in Islamic history was the Messenger of God, may God's prayers and peace be upon him. He used to walk in the markets and forbid cheating and belittling in weight and measure. His companions and caliphs followed his path. After them	Quran: (And let there be among you a nation inviting to the good and enjoin-ing what is right and forbidding what is wrong). (Al Omran, verse 104)	•			Yes, it is currently embodied in the Royal Commission for Riyadh, which issues codes and indicators, issues laws, and controls their implementation
Endowment (waqf)	Endowment buildings and projects are widespread in Islamic cities, where the Islamic religion urges the importance of the endowment and considers it a good deed for which the individual is rewarded even after his death	Sunnah: the messenger, Mohamad, said to Umar, urging him to endowment and charity: (If you wish, you withhold its origin and give it as a charity, then Umar should give it in charity, because its origin is not sold, bequeathed, or gifted). (Al-Nisaburi, 2006)			•	Yes, the legal rulings were and still are the motive and the motivation for the transactions between individuals in the area of Wadi Hanifa, especially since the religion that prevails there is Islam as customs and traditions govern the societies in that region
Consultation, and community partnership	The sheds were the centers of the meeting of the notables of the society, in which the fateful mat-ters were discussed, and some of them are still present in Medina, where the Islamic community must consult among themselves to agree on the matters that concern them	Quran: (And their matter is consulta-tion among themselves, and from what We have provided them they spend) (Al-Shura verse 38)		•		
Custom	It is what people are accustomed to and familiar with, if it is not corrupt, and not bad at the same time	Quran: (Take pardon and enjoin custom and turn away from the ignorant) (Al-Araf verse 199) Sunnah: Abdullah bin Masoud said: (What the Muslims see as good is good with God)		•		
Family solidarity	It appears on several levels, as it begins with the family and good relations with the parents, then honoring the parents is a duty, and their disobedience is a major sin	Quran: (Either one or both of them reach old age with you, do not say to them "Uf" and do not reprimand them, but say to them a kind word). (Al-Esra Verse 23)	•			
social solidarity	Some many hadiths and verses stipulate cooperation and interde-pendence in the Islamic society between Muslims and each other and all people	Sunnah: (Whoever relieves a Muslim of distress of the distresses of this world, god will relieve him of distress of the distresses of the day of resurrection, and he who makes it easy for a Muslim will be easy for a Muslim, god will make it easy for him) (Al-Nisaburi, 2006)			•	

(continued)

Table 2 (continued)

Principles of organizing the Islamic city design			The degree of importance of the legal ruling			Traditional urban design matching the principles
Administrative and social organization						
Principle	Description	References	Man	H. I.	Flex	Yes/No, with Explanation
Neighborhood unity	The Islamic religion is keen to be kind to neighbors, and this reinforces the principle of coexistence within one neighborhood	Sunnah: Prophet Mohamad said: (Gabriel still recommended me to the neighbor, until I thought he would bequeath him) (Al-Bukhari, 2002; Al-Nisaburi, 2006)		•		
The prohibition of extravagance in the resource	Despite the importance of ablution in Islamic law, the Shariah urges not to waste water and to preserve grace	The prophet, may god bless him, passed by Saad while he was performing ablution, and he said: (What is this extravagance, O Saad? He said: Is ablution extravagant? Prophet said yes, even if you were on a flowing river) (Al-Bukhari, 2002)		•		
Motivation to work and productivity	Al-Shariah urges productivity and local effort that leads to self-sufficiency and a sense of belonging	Prophet Mohamad said: no one has ever eaten better food than what he eats from the work of his hand (Al-Bukhari, 2002)			•	
Principle of no harm, no making cause of harm	A rule that established the movement of architecture and urbanization in the Islamic city and included all aspects of social life	Prophet Mohamad said: Whoever is harmful, god will harm him, and whoever is difficult, god will make it difficult (Al-Tabarani, 1994)/Do not raise your building above its construction, so the wind will block it	•			

Note Man. Stands for mandatory, H. I. stands for Highly important, and Flex. Means there is flexibility to apply or not
The author

Table 3 Islamic legislation importance and its applications in wadi Hanifa traditional urbanism In terms of site selection conditions

Principles of organizing the Islamic city design			The degree of importance of the legal ruling			Traditional urban design matching the principles
Site selection conditions						
Principle	Description	References	Man	H.I.	Flex	Yes/No, with Explanation
Preventing harms and bringing benefits	Bring the city to higher ground so as not to be vulnerable. Providing sources of food and clothing by linking cities and countryside, trade routes, or sea routes	This was mentioned by Ibn al-Rabi in his talk about the Islamic city and the conditions for its construction (Al-Rabee, 885) Quran: (And whoever saves a life, it is as if he had saved all mankind) (AL Maida, verse 32.)	•			Yes, the residential areas overlook the outskirts of Wadi Hanifa and are surrounded by farms and water borders
Air quality	Choosing places with good air	Al-Qazweni mentioned in his talk about Sana'a, indicative of the validity of its air, that meat remains there for a week and does not spoil, and he mentioned about Toledo that from the goodness of its soil and the gentleness of its air, the bugs remain in its landfills for seventy years that do not change. Change (Waziri, 2004)	•			Yes, Wadi Hanifa enjoys clean air compared to the rest of the cities of Riyadh thanks to the presence of green barriers, open spaces, and diversity of terrain

(continued)

Table 3 (continued)

Principles of organizing the Islamic city design			The degree of importance of the legal ruling			Traditional urban design matching the principles
Site selection conditions						
Principle	Description	References	Man	H.I.	Flex	Yes/No, with Explanation
The presence of freshwater	Water is the basis of life/purity is a condition of worship and	Quran: and we made from water every living thing (Al-Anbiya, verse 30)	•			Yes, Wadi Hanifa area contains groundwater extracted through wells
Bringing water without the misery	Availability of a nearby water source	(Waziri, 2004)	•			
Bring in the scholars and the skilled	To bring in the skilled and the people of knowledge for self-sufficiency	This was mentioned by Ibn al-Rabi in his talk about the Islamic city and the conditions for its construction (Ibn AlRabi, 885)			•	Yes, Diriyah in Wadi Hanifa was the seat of the state, from which scholars, preachers, and reformers emerged
Supplier presence	The possibility of the derived Meera: It includes the provision of food from the inside and the neighboring regions (the importance of the commercial era and the relationship of the site with the countryside and the region)	(Waziri, 2004)		•		Yes, Wadi Hanifa range is linked to the neighboring areas by the Northern Ring and Western Ring Lines
Protecting the natural environment	Avoid building in agricultural areas, yards, roads, or nature reserves	The Messenger, Mohamad, legislated the possession of the land that was left to those who would revive it, and Umar ibn al-Khattab stipulated that it be fenced with a fence and he only owns it by reviving it, such as bringing water to it, removing the surplus from it or cultivating it. Whoever encloses a wall on land, it is his (Narrated by Ahmad and Abu Dawood under the authority of Jaber)		•		Yes, Wadi Hanifa code includes provisions related to the preservation of the natural environment

The author

Table 4 Islamic legislation importance and its applications in wadi Hanifa traditional urbanism in terms of urban design

Principles of organizing the Islamic city design			The degree of importance of the legal ruling			Traditional urban design matching the principles
Urban design						
Principle	Description	References	Man	H. I.	Flex	Yes/No, with explanation
Fences	To be surrounded by a fence for protection: the fence may be a wall or a trench, as in the time of the prophet, or a natural border such as water, a valley, or others, in addition to the method of closed doors	Quran: my lord, make this country safe and provide its people with fruits (Surat Al-Baqarah, verse 126)/Building doors to streets and lanes (Othman, 1988)	•			Yes, as previously mentioned in Fig. 5
Compact tissue	The convergence of buildings to represent a single architectural block that resists harsh climatic factors	Al-Qadi (2000)			•	
Organic layout and functional space scale	Dividing the city into parts where all basic services are available	Waziri (2004)			•	
Narrow, winding, and walkable streets	The Islamic city includes a network of winding alleys and impermeable roads	Al-Qadi (2000)			•	
Street types	The street, the lane, and the alley from the widest to the narrowest, from 4 m to one and a half meters	Waziri (2004)			•	
Street design	Orienting cities in hot areas from north to south so that the facades of buildings are not exposed to sunlight	Othman (1988)			•	
Metaphors	They are the ends of narrow streets with a wide area that helps in storing moderately cold air at night and prevents its leakage with the wind at night	Allam (1977)			•	
Bouquets, trees, and fabrics	It was used to shade the lanes	Al-Kinani (2009)			•	
Al Barahat	The squares that arose with time as a result of closing the alleys produced al-Barahat	Al-Hathloul (1994)			•	
The mosque	The mosque is built for prayer in the middle to get to know all its people: it is the first thing that is planned in the Islamic city	Quran: (And set your faces at every mosque and call upon it, being sincere to it in religion). (Al-Araf, verse 29)	•			
Public squares in mosques	It was used to hold social, cultural, and religious events (Ibrahim, 1996)	Quran: "O children of Adam, take your adornment at every mosque." (Al-Araf, verse 31)		•		Yes, as previously mentioned in Fig. 5
Market	Its markets are estimated to meet the needs closely, and those markets are organized to ward off the harm from the organization of the craft according to their nature	Markets are estimated to obtain the needs. Al-Hathloul (1994)	•			Yes, around the mosques. (Affairs, 2002)
The neighborhood	It does not combine the opposites of the population: the neighborhood design appeared with the Messenger, may god bless him and grant him peace	Al-Hathloul (1994)			•	Yes, as per (Affairs, 2002) and (Al-Hathloul, 1994)
Centralization of services	Dividing the city into parts where basic services are available	Al-Hathloul (1994)			•	
Industries outside housing	Keeping industrial places away from residential places to prevent harm from audio, visual, or health pollution (Al-Hathloul) Keeping markets away from residential neighborhoods and distributing them qualitatively so that each craft has its market isolating crafts and industries that produce fumes or smoke	Al-Hathloul (1994)/Same rule: no harm, no causing harms	•			

Principles of organizing the Islamic city design			The degree of importance of the legal ruling			Traditional urban design matching the principles
Urban design						
Principle	Description	References	Man	H. I.	Flex	Yes/No, with explanation
Horse stalls	The Islamic city included stables or climates for animals that used to take a place in the barahat or spaces in every plan Moreover, getting horses and training them is very important in Islamic legislation	Al-Hathloul (1994) Sunnah horses are tied to their good intentions until the day of resurrection) Al-Nisaburi (2006)			•	No
Taking into account the movement of transportation (at that time, animals)	Islamic urbanism took into account the issue of not narrowing the road, obstructing movement in it, or causing harm to pedestrians	Al-Kinani (2009)			•	Yes, because this is an Islamic rule which believe in and obey Wadi Hanifa citizens
Yard landscaping	In Islamic cities, internal and external courtyards spread, and some of them contained water fountains, and others were planted				•	No
Landscaping	Found in Sana'a				•	No
Death protection	Prohibition of possessions so that the void remains permissible for the cultivation of pastures and the grazing of livestock				•	No
Building dams and constructing ponds	To save rainwater			•		No
Entropy	A piece of cloth is placed on the roof with a hole through which rainwater pours into the pool				•	No
Sweep rainwater	It was removed from the roads by order of the Muhtasib				•	No reference
Division of water between neighborhoods	People are equal in watering and allowing those who pass through their land to irrigate for a limited period that does not exceed it	People are partners in three things: pasture, water, and fire Al-Hafez (1986)	•			No reference
Gutter	To drain rainwater	Al-Kinani (2009)				Yes, everywhere in wadi Hanifa traditional buildings. Affairs (2002)
Reuse of ablution and washing water	Establishing channels that reach the farms with the force of gravity, but this water is used before reaching the farms for washing and ablution	Al-Kinani (2009)			•	No reference
Common drainage channels	Each neighborhood has a joint channel that reaches the main channel. The rules of jurisprudence have defined the subscription system to it	Al-Kinani (2009)			•	Yes. (Affairs, 2002)
Public baths	It spread in some Islamic cities, and water was passed to it and drained	Al-Kinani (2009)			•	No
Wells	Wells are erected away from drainage channels	Al-Kinani (2009)			•	Yes, Wadi Hanifa is having many wells. Affairs (2002)
The forehead	To drain waste and be cleaned by specialists periodically	Al-Kinani (2009)			•	No reference

The author

Table 5 Islamic legislation importance and its applications in wadi Hanifa traditional urbanism in terms of architecture

Principles of organizing the Islamic city design						
Architecture						
Principle	Description	References	Man	H. I.	Flex	Yes/ No, with explanation
Privacy and visibility	Islamic urbanism took into account the issue of the occurrence of consideration with special consideration to ensure the sanctity of homes and their residents	Al-Kinani (2009)	•			Yes, as shown in Fig. 5
Passive construction	By passing air in homes, through streets, and in voids, and directing the building inward			•		
Skylights, windows, and openings	By providing internal and external openings such as patios, evacuations, and Mashrabiyas				•	
Natural ventilation	Several types of natural ventilation methods have spread, the most famous of which are the sheds and the inner courtyards		•			Yes. Affairs (2002)
Building with mud, stone, and wood	Environmentally friendly and heat-insulating local materials				•	
Inner courtyard	Plant it or design the fountains inside it				•	
Hole shapes	Small openings that look inward or are protected by Mashrabiyas to provide privacy and thermal insulation				•	
Introducing natural lighting	Opening inwards, using the inner courtyard to allow lighting and ventilation of the building's interior units				•	
Provide shadows	Through the design that opens inward, as well as the presence of protrusions and railings and their varying heights, as well as the roofing of corridors and streets	Waziri (2004)			•	Yes. Affairs (2002)
Decorations	Geometric shapes engraved on marble or plaster where there are no exaggerations and no human drawings	Sunnah: (Angels do not enter a house in which there is an image/if you must do something, then make trees and things that have no soul (Al-Nisaburi, 2006)			•	Yes. Affairs (2002)
	The residential buildings were functional, flexible in design, and used in a way that ensured privacy and met different needs	Flexibility to change the design, even in the Prophet's Mosque. When the Qiblah changed, the Messenger, may God bless him and grant him peace, turned around and changed as he turned the Qiblah wall, and all mosques followed him, as well as the modifications that the mosque witnessed and is still witnessing, and the addition of women's doors in ancient times and to this day		•		Yes. Affairs (2002)
	There was no extravagance or exaggeration in the design, as well as the materials were simple and were prepared from materials found in the local environment	ElDin (1997); Waziri (2004)			•	Yes. Affairs (2002)

(continued)

Table 5 (continued)

Principles of organizing the Islamic city design						
Architecture						
Principle	Description	References	Man	H. I.	Flex	Yes/ No, with explanation
	The buildings were not high, commensurate with the width of the streets and the human size, and provided shades	Al-Kinani (2009)			•	Yes, as shown in Fig. 5
	By building houses with what is common and known and used in every region alike according to the availability of materials				•	
Reuse of animal dung	For splinting walls and thermal insulation				•	No
Function	For the bishop				•	Yes, in construction of all traditional building in Wadi Hanifa Affairs (2002)
Modesty in construction	For heating or fuel	Waziri (2004)			•	No
Don't get stuck in the building	Inventing types of mechanical machines for lifting large weights with little efforts, such as kohl, perm, sieve, shin, and crew, in addition to the rollers that are used to lift stones	Al-Kinani (2009)/wisdom is the lost property of the believer, so wherever he finds it, he has more right to it). (Al-Qazwini, 682 AH)			•	No

References

Affairs, M. (2002). *Urban heritage in the kingdom of Saudi Arabia.* King Fahd National Library.

Aguiar Borges, L., Hammami, F., & Wangel, J. (2020). Reviewing neighborhood sustainability assessment tools through critical heritage studies. *Sustainability, 12*(4), 1605. https://doi.org/10.3390/su12041605

Ahmed, F. Q. (1999). *Planning thought when Ibn Khaldun.* Unpublished doctoral thesis.

Aina, Y. A., Al-Naser, A., & Garb, S. B. (2013). Towards an integrative theory approach to sustainable urban design in Saudi Arabia: The value of geodesign. *Advances in Landscape Architecture.* https://doi.org/10.5772/55888

Al-Balathri, A. B. (1901). *Futuh Al Buldan.* The Arab Press.

Al-Bukhari, A. A. (2002). *Sahih Bukhari.* Dar Ibn Katheer.

Al-Buraqi, H. 1332 AH-2003 AD. History of Kufa. Haidari Library.

Al-Hafez, A.-M. (1986). *Sahih Al Targheeb.* Islamic office of publishing and distribution.

Al-Hathloul, S. B. (1994). *The Arab Islamic city, the impact of legislation in the formation of the urban environment.* King Fahd National Library.

Ali, N. B. (2000). *Sustainable development in traditional architecture in the Kingdom of Saudi Arabia (a case study—The Hijaz region).* Umm Al-Qura University.

Al-Jarallah, Y. Y. (24 May, 2020). *The neighborhoods revival initiative... An analytical view of the development of the city of Riyadh.* Retrieved from Al Riyadh: https://www.alriyadh.com/1822781

Al-Kinani, H. (2009). *Sustainability in Islamic architecture.* Master's thesis. King Saud University.

Allam, N. I. (1977). *Arts of the middle east in Islamic ages* (2nd ed.). Dar Al Maaref.

Al-Nisaburi, A. A.-H.-H. (2006). Sahih Muslim. Al-Azhar—The path of the Turks: Dar Al-Bayan Al-Arabi.

Al-Qadi, S. (2000). The inner courtyard as an architectural element in the compact urban fabric and its role in achieving a comfortable residential environment. The inner courtyard conference in the Arab City (Trends towards the twenty-first century). Al-Baath University.

Al-Qarni, M. B-F. (2002). *Urban heritage in the Kingdom of Saudi Arabia.* Ministry of Municipal Affairs and Village, Riyadh.

Al-Qazwini, Z. B. (682 AH). *The effects of the country in telling the humanity (Athar AlBilad).* Dar Sader.

Al-Rabee, S. A.-D. (2010). (885). *Owner's behavior in managing kingdoms* (1 edn.). Dar Al-Athria for publication and distribution. Retrieved Retrieval date.

Al-Saadi, A., & Al-Hamwi, O. (2000). *Rulings of Imran in Islamic jurisprudence, a comparative study (unpublished doctoral dissertation).* Damascus University.

Al-Sayyad, N. A. (2016). Towards an environmentally friendly strategy to revive the heritage area in Jeddah. *Journal of Engineering Sciences, Assiut University*, 191–205.

Al-Tabarani, A.A.-Q. (1994). *The great Lexicon*. Dar Al-Sumaie.

At-Turaif: Diriyah gate development authority. At-Turaifl Diriyah gate development authority. (n.d.). https://dgda.gov.sa/our-destinations/at-turaif.aspx

Berardi, U. (2013). Sustainability assessment of urban communities through rating systems. *Environment, Development and Sustainability*, 1573–1591.

Duany, A. M., & Plater-Zyberk, E. (2008). *The neighborhood, the district and the corridor*.

ElDin, A. K. (1997). *Environmental treatments in planning Islamic cities and designing their buildings*. Research conference "Interbuild" Cairo. Cairo.

Ibrahim, A.-B. (1996). *Urban characteristics of the Islamic city. Seminar on the principles and criteria for classifying Islamic heritage buildings and cities and how to preserve them* (pp. 1–9). Islamic Capitals and Cities Organization.

Jabbar, S., AL Ahbabi, S. (2011). Mechanisms of achieving social sustainability in the traditional urban structure case study: The center of the traditional city of Kadhimiya. *Engineering Journal*, (3), 133–155.

Kassab, N. (2021). Transformation of the relationship between neighborhood & sociality space and land use in Tripoli Lebanon. *Journal of Urban Culture Research*, 50–64.

Moudon, A. V., Lee, C., Cheadle, A. D., Garvin, C., Johnson, D., Schmid, T. L., Weathers, R. D., & Lin, L. (2006). Operational definitions of walkable neighborhood: Theoretical and empirical insights. *Journal of Physical Activity and Health*, *3*(s1). https://doi.org/10.1123/jpah.3.s1.s99

Mousa, N., & Habib, M. (2013). Factors affecting the planning of the Arab Islamic city. *Journal of the College of Education for Girls for Humanities* (13), 164–196. https://search.mandumah.com/Record/621480

Othman, M. A. (1988). *Islamic city*. World of Knowledge.

Sharifi, A. M. (2013). A critical review of seven selected neighborhood sustainability assessment. *Environment Impact Assessment*, 73–87.

Wadeh, A. (1890). *The countries (Al Buldan)*. Brill.

Waziri, Y. (2004). *Islamic architecture and the environment*. World of knowledge series.

Creation of Sense of Place in Interior Architecture for Adaptive Reuse of Heritage Buildings as Identity Continuation

Faten Issa and Ghada M. R. Al Slik

Abstract

In the lifetime of a heritage building associated with the cultural or political history of the country, new requirements will arise to reuse it for a different function serving new users. In addition, certain quality improvements might be required such as adding new systems or carrying out interior design modification. These new changes will introduce a new sense of place that will have an impact on the users and the identity of the heritage building. The creation of a new sense of place of interior architecture for the adaptive reuse heritage buildings could be categorized into three different levels. Firstly, buildings serve the same function but are improved in terms of interior design. Secondly, buildings serve the same function but require certain changes to their interior architecture, and this might change a bit the sense of place. The third category relates to the buildings that are changed functionally and require major interior architecture modification (mainly adaptive reuse), which will introduce a new sense of place. This paper emphasizes the third category and analyzes the concept of the sense of place created through the act of design that applied to the case study of 'Al Faw Presidential Palace Complex' in Baghdad. Part of this complex was adapted in 2019 to accommodate the facilities of the new American University of Iraq in Baghdad (AUIB). The importance of considering the new sense of place for the adaptive reuse act of such valuable buildings comes from the fact that there are hundreds of massive palaces and presidential facilities built in Iraq in the period from 1979 to 2003. These palaces require tremendous interior architectural changes and adaptation to serve new functions, mainly cultural ones. Creating a new sense of place is a major factor that should be addressed, taking into consideration the value and strength of the old sense of place of these properties. By creating a new cultural function that introduces a new sense of place, the identity continuation of the place will be achieved and reflected in the identity and cultural continuation.

Keywords

Interior architecture · Sense of place · Adaptive reuse · Heritage

1 Introduction

During the last two decades of the twentieth century, hundreds of presidential palaces were designed and constructed in different cities of Iraq. The palaces and their huge facilities are unique in their architecture and interior design. They represent an important era of Iraqi modern history (regionalized international architecture) that cannot be separated from the two major Gulf wars, the 12 years of economic sanctions, as well as the internal political conflicts.

When the previous regime came to an end in 2003, most of these buildings were abandoned due to the damages they experienced during the war, and the massive budget needed for their maintenance and adaptation. In addition, there is an absence of actions and decisions to be made when dealing with these private properties, perhaps due to their connection to a period when these buildings are being constructed while the people of Iraq were facing difficult years of wars and economic sanctions.

F. Issa (✉)
Department of Architecture, College of Engineering, University of Baghdad, Baghdad, Iraq
e-mail: faten@finalfix.ae; faten.taleb2104@coeng.uobaghdad.edu.iq

G. M. R. Al Slik
Theory and Conservation of Architecture, Department of Architecture, College of Engineering, University of Baghdad, Baghdad, Iraq
e-mail: ghada.alslik@coeng.uobaghdad.edu.iq

M. F. Shahidan et al. (eds.), *Urban Narratives: Exploring Identity, Heritage, and Sustainable Development in Cities*, Advances in Science, Technology & Innovation, https://doi.org/10.1007/978-3-031-48517-6_19

Places that have meaning develop people's sense of belonging through their involvement in certain events that occurred in these places. In this regard, it is important to mention the philosophical input of Martin Heidegger, the first philosopher of the twentieth century who linked the place to the people who interact with it through his book *'Being and Time'* published in 1927. Heidegger (1927) states that "when space is intuited formally, the pure possibilities of spatial relations are discovered" (p. 147). He explains more about the connection between the concept of belonging and involvement, stating that "belonging somewhere has an essential relationship to involvement" (Heidegger, 1927, p. 420). By applying this principle to our case (The Presidential Palaces in Iraq), people's interactions and connections that make them attached to the place and develop positive feelings towards it are not seemed being existed in these properties.

Due to the years of negligence and the occupation of the American military troops of some of the buildings, many of their interior elements are damaged. Al Faw Presidential Palace Complex has experienced damage to its interior elements such as the ceilings, finishes, and furniture (Fig. 1a: d).

Since 2003, only a few palatial residences have been given a second lease of life, mainly as military bases or public administrations, and more rarely as museums. In Baghdad, two palaces recently housed the Presidency and the Prime Minister's offices. Other palaces lie empty for many reasons, mainly due to the high cost of renovation.

The Presidential Complexes in Iraq are considered historical buildings, but they also represent a certain heritage era of the country. Considering proper renovation of these palaces can benefit the Iraqi people and the economic development of the country in many aspects. A similar thinking approach was mentioned by Chambers (2006) who claims with respect to the changes in his hometown that "these general uses of heritage are not in and of themselves necessarily bad or injurious things, and they can have quite beneficial results in terms of economic development, public education, and encouraging people to care more responsibly for the experiences of others and for their environment" (p. 37). The act of adaptive reuse of any heritage/historical building should be approached carefully to protect it and at the same time to reflect its value, in order to create spaces that will have a positive impact on their new users. Chambers (2006) states that "creative reuse allows us to save and protect our heritage while exploring its value as a resource" (p. 12). He further explains that this creation allows us to reconsider our cultural needs and aspirations in a new vision and thinking and turns them into advantages.

The preservation of historic buildings by adaptive reuse act through new functions is the right method to ensure their survival and prevent them from negligence. In addition, it will benefit the environment and save embodied energy to make it more sustainable compared to new construction (DEH, 2004).

Canter (1977) asserts that "a place is the result of relationships between actions, conceptions, and physical attributes" (p. 158). The place is not only a physical setting, but rather it is a relationship with the people occupying it and their behaviour within it. Therefore, creating a positive sense of place through the adaptive reuse act should consider the physical characteristics as well as the human needs in terms of function and perception.

The quality of life that a place could generate is explained by Keunhye Lee (Fig. 2). She refers to the body's engagement with interior space that provides a spatial experience and results in an emotional connection (Lee, 2022). Moreover, Lee (2022) adds that "emotions generated through sensory experience within space also recall memories of past space, which embodies interior experiences" (p. 6). This experience towards the interior space represents many factors (Fig. 3) such as culture, memory, and emotions as well as materiality (Lee, 2022). As adaptive reuse in our case study involves the interior space more than the exterior of the building, it is important to take into consideration the people's perception and how they experience the place and interact with it.

Creating interaction, engagement, and community within the used space enhances a sense of belonging and brings a healthy and balance feelings. As a result, it provides a perfect human environmental experience.

According to the above, considering the adaptive reuse concept and applying it to the presidential palaces in Iraq are essential. However, this act is a big challenge for designers/interior architects when they deal with such a sensitive situation. In this regard, Lorraine Farrelly refers in her essay in Brooker and Weinthal (2018b) to the challenge facing interior architects and designers when it comes to the act of transforming what exists and making it relevant for our contemporary use, claiming that it is an opportunity to think creatively about the space reuse, the materials, and the structure (p. 139).

This paper will go through the changes that Al Faw Palace Complex in Baghdad has experienced over four decades, and it determines how the sense of place is affected during the preservation process.

Fig. 1 a and b Damage of the ceiling and chandeliers, Al Faw Main Palace in 2003. Photos by the authors, 2022. c and d Damage of the finishes, Al Faw Main Palace in 2003. Photos by the authors, 2022

2 Paper Methodology

The paper uses a descriptive approach method based on the theories and analysis determined from previous literature and research, considering the main variables (interior architecture, adaptive reuse act, and the sense of place). It tends to clarify how to deal with the function changes concerning the heritage buildings that represent a certain historical era. The method involves observations and surveys, explores certain indicators, and applies them to the case study of Al Faw Presidential Palace in Baghdad, in an attempt to determine their impact on the act of adaptive reuse of interior architecture and the sense of place.

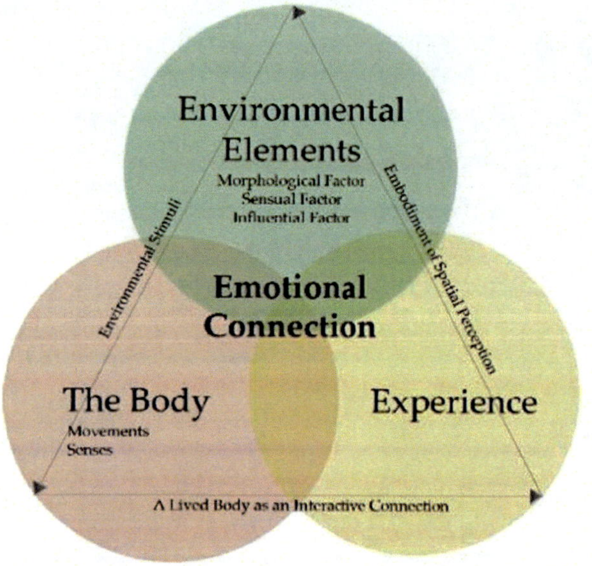

Fig. 2 Embodiment of spatial experience for emotional connection. Reference: Lee (2022)

3 Main Variables

3.1 Interior Architecture

Brooker and Weinthal (2018a) state that "the discipline of interior architecture and design can be considered to be substantial and expansive; primarily because of the many areas it can affect and engage with" (p. 2). They explain that this notion comes from the fact that the interior is a complex of many principles that involve physical structures as well as mental ones, and accordingly, many factors affect the perception of the interior as a whole structure, which is shaped by the space use and its history amongst many other forms (Brooker & Weinthal, 2018a).

Perolini (2014) explains "When we create interiors we express a collection of desires and reflections that live within us, and we express them in a material sense" (p. 1). Interior architecture bridges the act of architecture and interior design to develop spatial spaces. It provides engagement between the environment and the people using the space and interacting with it and goes beyond the physical

Fig. 3 Phenomenological understanding of the body. Reference: Lee (2022)

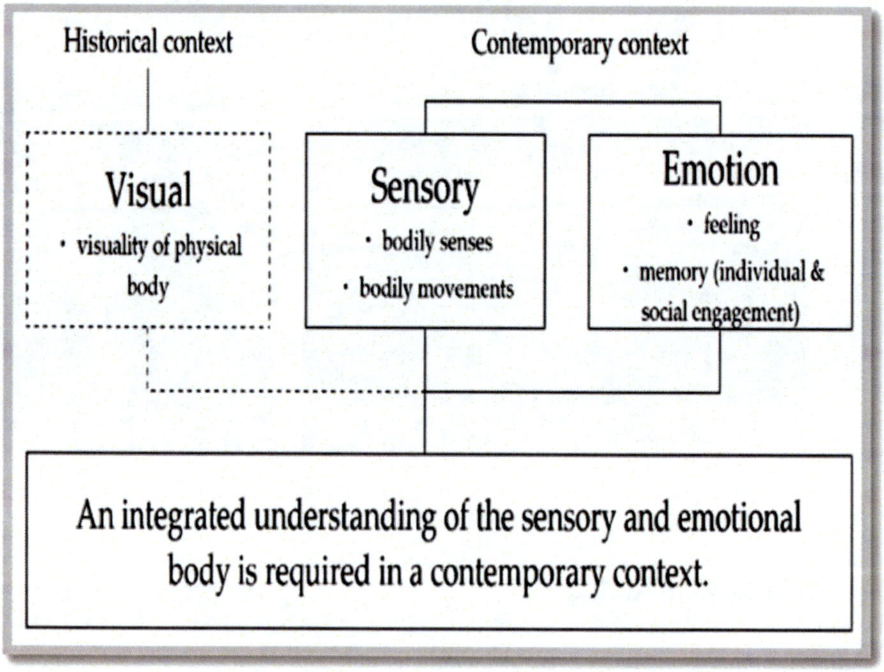

Fig. 4 Three concepts of functionalism. Reference: Nussbaumer (2009, p. 23)

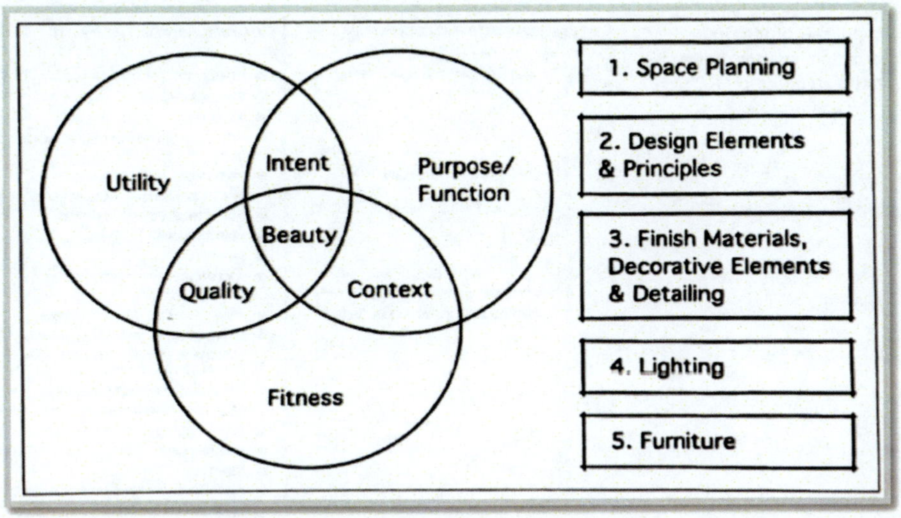

boundaries of architecture. Hay describes interior architecture as an approach to design alterations in existing buildings (Hay, 2007) and not far from this direction. Brooker and Stone (2007) define interior architecture as "a discipline concerned with the remodelling of existing buildings, it bridges the practices of interior design and architecture, often dealing with complex structural, environmental and servicing problems" (p. 126).

Thus, interior architecture combines the responsibilities of interior design with the involvement of architecture when it comes to structural modification and adaptation acts.

Accordingly, interior architecture could be considered as an act that deals with the physical characteristics/components and the perception of society.

3.1.1 Physical Characteristics/Components

In interior design, giving a space its own identity and personality refers to as "interiority", and according to Charles Rice, interiority can be achieved by interrelating the interior elements such as furniture, colours, lighting, and finishing amongst other spatial elements, with the building envelope (Rice, 2006). By saying that, he refers to the physical elements.

One of the most important aspects of architecture and interior design is the function of the designed space and all other aspects associated with it. This concept reformed the theory of what is called *'Functionalism'* which was widely used (as a term) since the beginning of the twentieth century with the rise of the modern architecture movement. In this context, Nussbaumer (2009) describes this term and theory, referring to the functionalism concepts "The three key concepts of functionalism are function, fitness, and utility, beauty is an additional concept that is integrated into all of these concepts" (p. 22). This theory is explained further in Fig. 4 which shows the relationship between these three key concepts and the integration of beauty into them (Nussbaumer, 2009).

The physical characteristics/components of the place, which are essential for any interior space, could be categorized into three different categories based on the act of design. It is determined according to the design act that might involve interior decoration only, expands to an interior design act, or covers the act of full interior architecture (Fig. 5).

In their book *'Place Attachment'* Altman and Low (1992) state that not only physical components matter to place, defining place as "a space that has been given meaning through personal, group, or cultural processes" (p. 5).

3.1.2 Perception of Society

Rasmussen (1959) states "It is not enough to see architecture, you must experience; you must observe how it was designed for a spatial purpose and how it was attuned to the entire concept and rhythm of a specific era" (p. 33). In this context, some signs might influence the environmental meaning and perception (Rapoport, 1977–1990). These signs are summarized in Fig. 6.

Gosling et al. (2018) mention in their essay published in Brooker and Weinthal (2018a) that "environmental psychologists suggest three ways of perception: probabilities-functionalism, affordances, and collective properties" (p. 283). They further explain that the influential approach of the first way (probabilities/functionalism) is based on the work accomplished by Egon Brunswik who connected the perception of people and their understanding of a particular setting to what he called 'The Lens Model' (Brunswik, 1956) as shown in Fig. 7. The second way (affordances) is defined by Gosling et al. (2018) "Affordances are functions

Fig. 5 Physical characteristics/components of interior decoration, interior design, and interior architecture. Reference: by the authors, 2022

INTERIOR DISCIPLINE	COMPONENTS - ASPECTS
INTERIOR DECORATION	• FINISHES (SOFT/ HARD) • FURNITURE • TEXTILE • ARTWORK (PAINTINGS, ACCESSORIES,ETC.)
INTERIOR DESIGN	• SYSTEMS IMPROVEMENT (AC, LIGHTING, SOUND SYSTEM, SANITARYWARES AND FITTINGS) • FINISHES (SOFT/ HARD) • FURNITURE • TEXTILE • ARTWORK (PAINTINGS, ACCESSORIES,ETC.)
INTERIOR ARCHITECTURE	• FUNCTIONAL ADAPTATION • STRUCTURAL MODIFICATION (ADDING OR REMOVING STRUCTURAL ELEMENTS, CHANGING SPACES' AREAS, SCALE, DIMENSIONS, ETC.) • SYSTEMS IMPROVEMENT (AC, LIGHTING, SOUND SYSTEM, SANITARYWARES AND FITTINGS). • FINISHES (SOFT/ HARD) • FURNITURE • TEXTILE • ARTWORK (PAINTINGS, ACCESSORIES,ETC.)

Fig. 6 Signs that may influence the meaning and perception of the environment, Summarized from Rapoport (1990: 106–107 and 1977: 229–230). Reference: Birer and Adem (2021, p. 16)

Theme	Sub-Theme	Examples
Physical Elements	Sight	Form Features of spaces Light and shadow Greenery Age Age Order Perceived density Protection and maintenance Topography Place, Location
	Sound	Quality and changes of sound
	Scent	Quality and changes of scent
Social Elements	People	Culture, age, gender
	Activities & Uses	Density, transportation; functions
Temporal Differences of Various Kinds	Objects	Signs, advertisements, cuisines
	Long Term Changes	Change, continuity and stability
	Short Term Changes	Type of uses (day and night, days of week), intensity of use over time, rhythms of activities

of an environment that are instantly detectable by a perceiver as useful for a particular purpose" (p. 283). The third way to be considered is (collective properties), which are attributes and characteristics of a setting that made the perceivers compare the details of the environment and stimulate their interest in the setting in general.

Hence, the interaction between the person and his physical environment in addition to his interaction with other individuals will be reflected in his/her behaviour (positive or negative), which will vary according to the design of the interior setting provided by the architect/designer.

3.2 Adaptive Reuse Act

There is a variety of different terms that are used to indicate the practice of changing the function and architecture of existing buildings. Plevoets and Van Cleempoel (2019) identify four terms concerning this act: Renovation, Adaptation, Alteration, and Remodelling (Fig. 8). They explain that when referring to the discipline and concept of working with existing buildings, 'adaptive reuse' is the most suitable term, which has come from the words 'adaptation' and 'reuse'. The adaptive reuse act refers

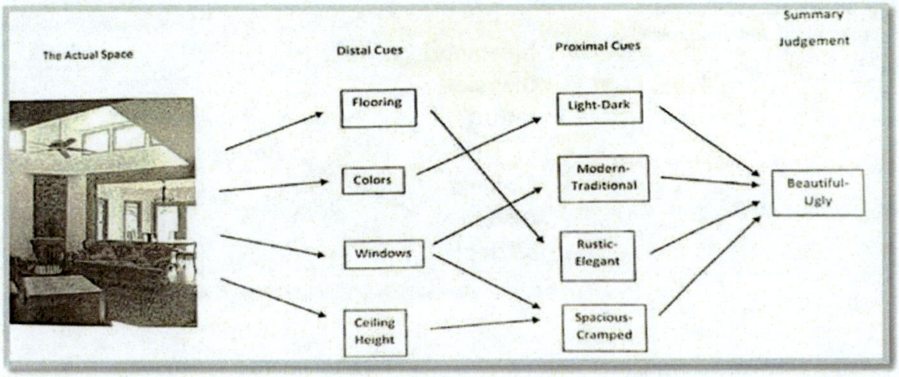

Fig. 7 Lens model for an interior. Reference: Robert Gifford in Brooker and Weinthal (2018a, p. 283)

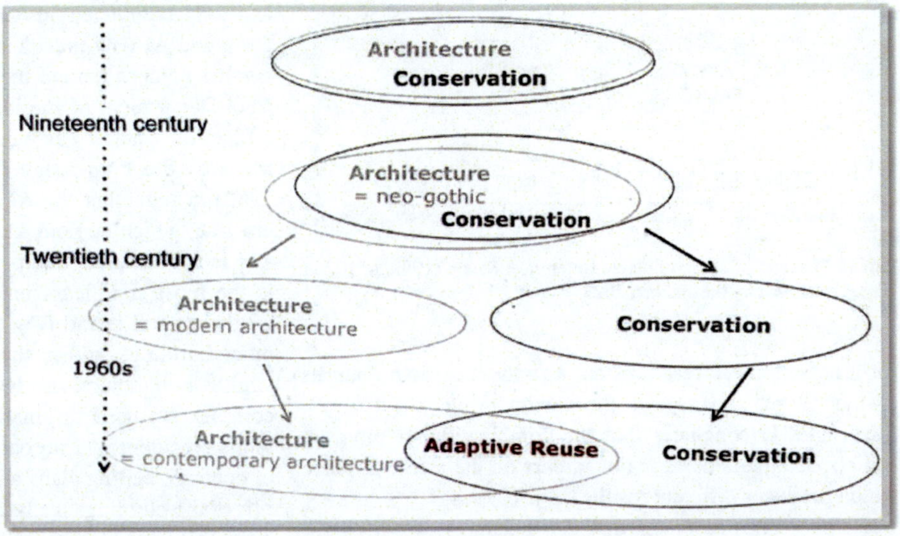

Fig. 8 Adaptive reuse as a discipline in its own right. Reference: Plevoets and Van Cleempoel (2019, p. 16)

to the act of changes that involve physical and functional components.

Tuan (2018) has quoted "History has depth, and time bestows value" (p. 191), stating that by preserving the value of the place, we preserve our identity. Due to the importance of the historical place value, Plevoets and Van Cleempoel (2019) mention that the quality of any project that went through an adaptive reuse act is based on two main facts: the new design intervention that relates intelligently to the future, versus the conservative or restorative interventions. The adaptive reuse act of a heritage/historical building should be approached carefully and in a creative way that protects it, reflects its value, and at the same time gives it a new meaning. Fiorani et al. (2017) claim that "in the field of conservation theories, the relationship between potential current use and the historical building itself was initially measured by looking at the historical distance between the present and the date of creation of the existing architecture" (p. 119). They explain that the relationship between the potential new function/use of the historical building is initially measured by the distance between the present and the time of the creation of such a building (Fiorani et al., 2017). Therefore, the design act associated with adaptive reuse creates a harmony of form and function and generates a new spirit in contrast with the negative history of such a building. Moreover, it is important to generate a new positive spirit and successful function that serve the new needs through the act of interior architecture, taking into consideration the preservation of the original character of the historical building and the character of the place.

3.3 Sense of Place

Steele (1981) states that "the sense of place can refer not just to a person's sensation, but also to whether or not it makes sense as a place for human use" (p. 203). He offers more explanation of the factors and characteristics associated with the sense of place (Fig. 9) claiming that "a sense of place is the pattern of reactions that a setting stimulates

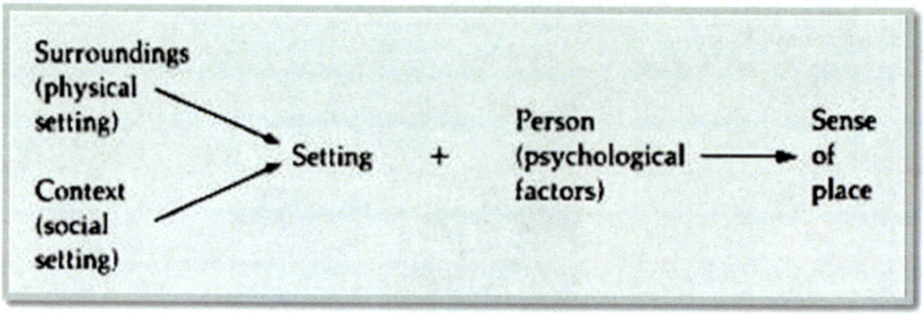

Fig. 9 Aspects affecting the sense of place. Reference: Steele (1981, p. 12)

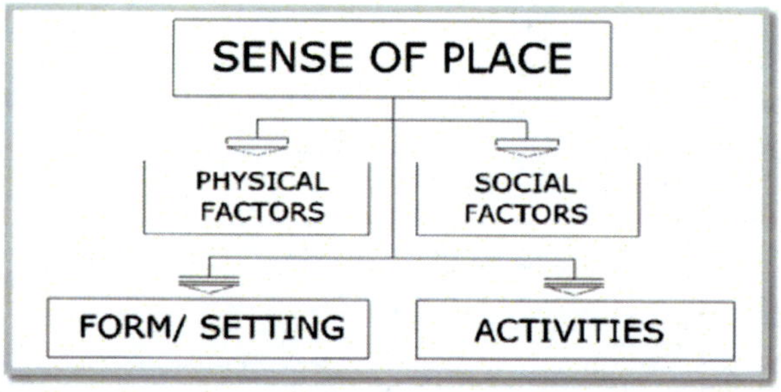

Fig. 10 Factors of sense of place. Reference: Steele (1981, p. 102 & 184). Developed by the authors, 2023

for a person, these reactions are a product of both features of the setting and aspects the person brings to" (p. 12). Steele (1981) concludes that the combination of physical and social factors has a major impact on the sense of place which produces different results (Fig. 10).

Relph (2008) mentions that the meanings of places are not considered as a property of the physical settings, activities, and objects although they are linked to these aspects, and they are rather considered as a property of human experience and intentions. He states that "meanings can change and be transferred from one set of objects to another, and they possess their own qualities of complexity, obscurity, clarity, or whatever" (p. 47).

According to the mentioned variables and indicators, the next chapter will go through the case study of Al Faw Palace Complex in Baghdad, reviewing and analyzing the changes applied to its buildings based on the new activities.

4 Al Faw Palace Complex in Baghdad—Iraq

4.1 The Presidential Palace (1991–2003)

Al Faw Palace Complex was built near Baghdad International Airport, and it is surrounded by a huge artificial lake diverting the water from the Tigris River. The complex covers an area of almost half a million square feet (Kelly, 2021). The construction which was started in 1989 and was completed in 1991 consists of many buildings, the grand Al Faw presidential palace, numerous villas, and smaller palaces around the lake. According to the previous Iraqi Department of Engineering Affairs of the Presidential Office, the palace was built to commemorate the sacrifices made by the Iraqi Army during the Iran–Iraq War (1980–1988) in regaining the Al Faw peninsula located in southern Iraq, which marked an end to the eight years of fighting and is named after this peninsula. The main palace which is the most significant building of the complex is located on the central island (Fig. 11a and b), in addition to many other buildings within the complex site that have almost similar architectural and interior styles. The buildings of the complex are used originally to host the guests of the previous president of Iraq occasionally. The structure consists of concrete ceiling slabs and brickwork walls, and the exterior of the buildings is cladded with travertine stone brought from the northern Iraqi city of Mosel. The interior design of the complex shows a massive use of aesthetical elements which could be seen through the ornamental design and details on the false ceilings and walls, most of which are handmade by Iraqi skilled craftsmen.

The Palace Complex, like many other similar buildings, was built and supervised by the previous Iraqi Department of Engineering Affairs of the Presidential Office which represents the authorities during the previous regime in Iraq. It reflects the legacy of a luxurious life. As a result, it has introduced a sense of place engaging power, lavishly, relaxation, and a show of strength and control from the side of its user/users.

4.2 The Victory Base (2003–2012)

During the invasion of Iraq in 2003, the American troops targeted many of the palaces built during the previous regime in Iraq. The American bombers hit the bridge of Al Faw complex which connects its main palace with the surrounding buildings (Williams, 2007), causing damage to

Fig. 11 **a** Al Faw palace complex. Reference: AUIB website, 2022. **b** Al Faw Palace Complex location and master plan. Reference: Aboutblank company website, 2019

a

b

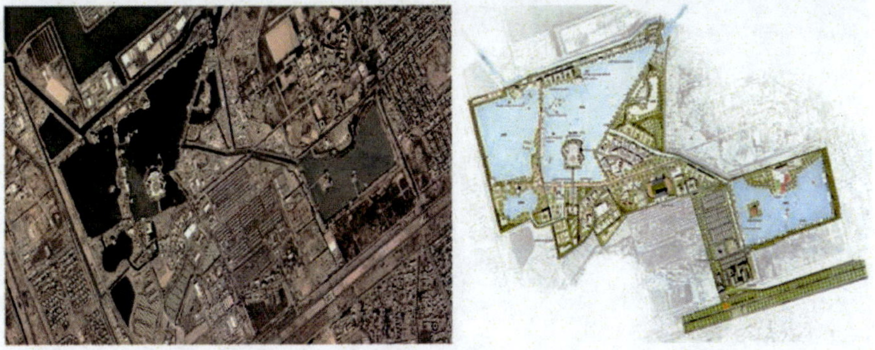

the palace and the other buildings. The complex then has become the US Army's Command Headquarters located at the heart of Camp Victory (Fig. 12). "Unlike other architectural crown jewels belonging to the regime, Al Faw Palace witnessed changes to its function, and renamed 'Camp Victory' to house 42,000 military personnel and 20,000 civilian staff" (Tarzi, 2018). In 2011, Jim Loney, a journalist from Reuters mentioned in one of his articles that the top US war leaders used one of the villas as their residence within the base, describing it as a mansion consisting of 20 rooms and covering an area of 25,000 square foot. He added that Al Faw Palace, a 450,000 square feet edifice of 62 rooms including 29 bathrooms, is used by the US forces as their war operations' centre (Loney, 2011).

Richard Mosse, an Irish Photographer based in the USA who visited the palace in 2009, explained that "the effect is like an ironic form of camouflage, making the perilously foreign seem all the more familiar and habitable a kind of military twist on postmodern interior design" (Mosse, 2009). He described the interiors of some areas within the palace that have been transformed to accommodate the soldiers, mentioning that the soldiers who occupied the palace seemed pretty interested in its original function (Fig. 13b), adding that the soldiers looked more bonded and happy with their job compared to the other troops he visited and met in massive military bases in Iraq, which were built at the desert for a military purpose (Mosse, 2009). This could be considered as a rare transformation of the sense of place which has not been changed much, although it was expected to be different (due to the new function and occupants of these buildings).

The physical components of the interiors are changed to suit the needs of the new occupants. During a site visit to the palace in July 2022, the authors documented and photographed the buildings where changes of many areas to create a place suitable for the new function are witnessed. New AC system and ducts are installed without paying attention to the existing finishes. New washrooms and bathroom fittings are added, and new partitions are installed to accommodate offices. The main halls are converted to conference rooms, multi-purpose halls, and auditoriums where a new wooden structure is erected, and a new false ceiling is installed. These changes were not done properly, causing further damage to the interior elements and finishes. However, this is a temporary and rare situation that should not be considered a normal case of an adaptive reuse act (Fig. 13a: f).

Before the end of the year 2011, the US military vacated Al Faw Palace Complex and turned the property over to Iraq. The site was used then by the Iraqi Military troops and

Fig. 12 US soldiers tour et al.-Faw palace. Reuters/the wall street journal. Photo by Shalash (2011)

the function was not changed. The same properties were used by the Iraqi troops for some years before they were vacated and the ownership of the complex was transferred to the new 'Presidential Palaces Authority' in Iraq, which was established at that time to commence the renovation work of the complex. An agreement was signed between this authority and the American University to turn the site into the new American University of Iraq in Baghdad (AUIB).

4.3 The American University of Iraq in Baghdad AUIB (2019 Till Date)

Al Faw Palace Complex was chosen to accommodate the American University of Baghdad. The work commissioned in April 2019 has covered a 2,500,000 sqm of a designed master plan area and a 93,000 sqm of a construction area (American University of Baghdad, Iraq, 2019). The villas are renovated and adapted to suit the new function after carrying out the necessary survey, assessment, and design. Moreover, a new gate was designed and constructed; modification and maintenance work of the internal roads was carried out, in addition to the necessary maintenance work and improvement of the lake and the landscape (Fig. 14a and b).

The buildings/villas surrounding the lake are adapted to accommodate the colleges and other facilities of the university.

As an example, one of the villas (Building No. 65) was adapted to house the College of Arts and Design, despite the damage this building (and so for other buildings) has experienced from inside (Fig. 15a: c), and outside (Fig. 16a: c).

The act of design was carried out to adapt this building to house a new function serving new users. Modification to its interior architecture has included some changes to its structural elements (see the layouts in Fig. 17a and b).

The university's President Michael Mulnix mentioned in an interview by Kullab (2021) from the Associated Press after the university officially opened its doors: "I feel more like a mayor of a big city than a university president" (Kullab, 2021), adding that he feels the presence of the deposed dictator everywhere through his initials that are still etched on the walls, columns, and ceilings, expressing

Fig. 13 a New ducting work, victory base. Photos by the authors, 2022. b Outdoor gym, victory base Photo by Mosse (2009). c Dining room used as a meeting room. Photo by Pupec, 2020. e & f New conference hall—auditorium erected using wooden structure at the victory base. Photos by the authors, 2022. d New partitions for offices. Photos by the authors, 2022

Fig. 14 a New gate of AUIB. Photo by the authors, 2022. b Landscape work at AUIB. Photo by Habbobi (2019)

Fig. 15 a, b & c Interior damage, building 65, Al Faw complex. Reference: Survey done by Kurk Insaat Company (2019). Photos by Habbobi (2019)

Fig. 16 a, b & c Exterior damage, building 65, Al Faw complex. Reference: survey done by Kurk Insaat Company (2019). Photos by Habbobi (2019)

his feeling and how taking that legacy and transforming it into what has been done are a kind of interesting. According to Kullab (2021), the work was funded by an Iraqi investor, and an amount of USD 200 Million was allocated to complete the renovation and refurbishment of the campus.

The exterior of the building is maintained (Fig. 18a and b), while the interior architecture of the building's main lobbies and rooms is being modified as classrooms, offices, and lecture auditoriums to suit the new function (Fig. 19a: f). New systems of AC, Audio, and lighting are being installed, in addition to new washroom fittings, equipment, and different types of furniture (Fig. 20a and b).

According to the site visits conducted by the authors in July 2022, it could be concluded that the adaptive reuse act has changed the buildings' interior to maintain the new function and requirements of the new users. A new sense of place is created by changing the physical components/characteristics of the interior architecture. This was achieved by adding or removing walls and partitions, changing spaces and areas, using new finishes, furniture, and equipment, in addition to the improvement of the systems and the installation of new technology that serve the new function and requirements.

The public and students are interacted with the spaces and the new facilities in different ways. According to the writers and journalists mentioned previously, in addition to some opinions from the public, it was expected that the users will negatively perceive the space by connecting the site to its past. However, the perception of the new main users (the students) has reflected joy, excitement, and

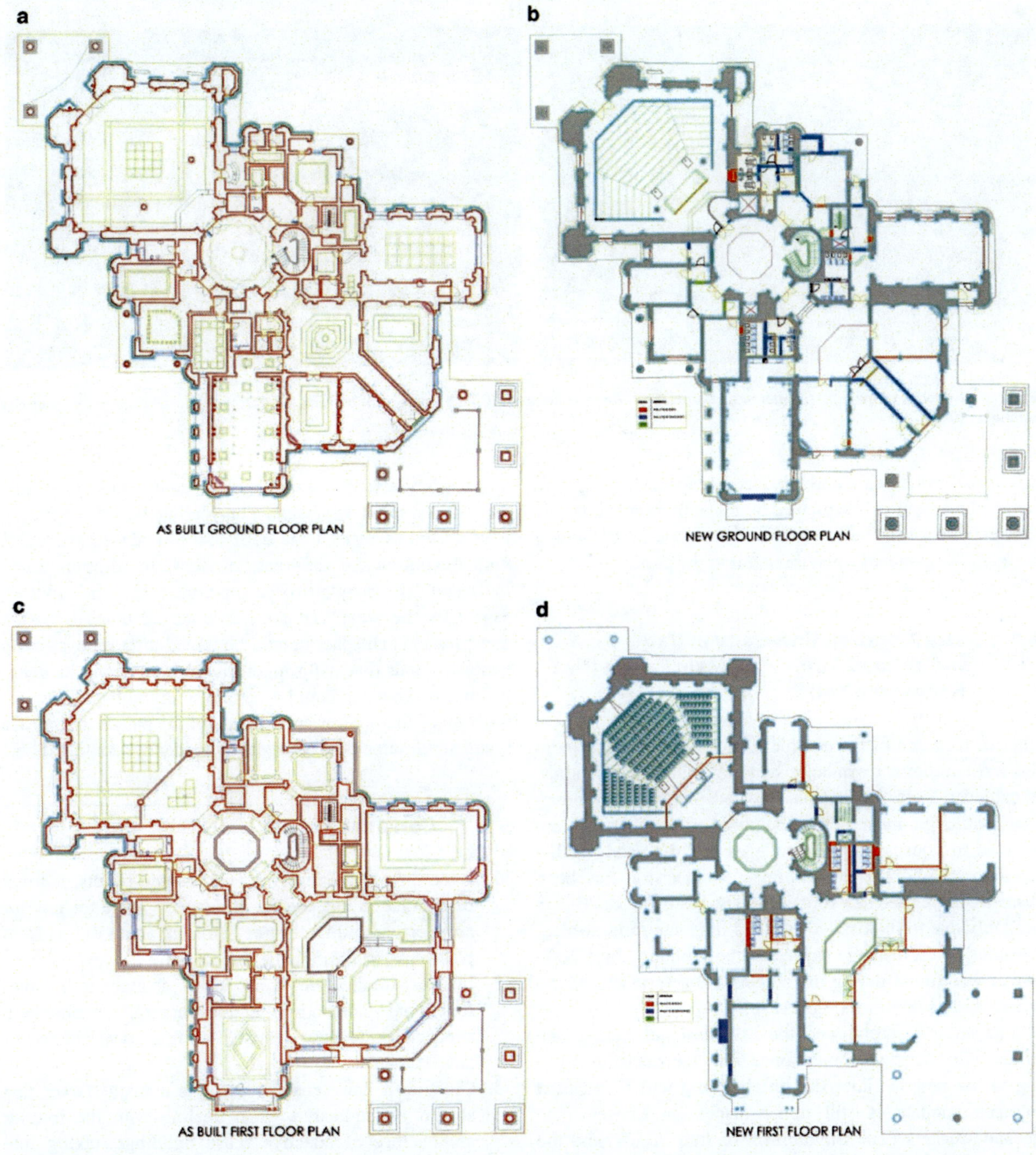

Fig. 17 **a** Ground floor plan of building 65, Al Faw complex. Left: original layout. Right: modified layout. Reference: Habbobi (2019). Updated and drawn by the authors, 2022. **b** First floor plan of building 65, Al Faw complex. Left: original layout. Right: modified layout. Reference: Habbobi (2019). Updated and drawn by the authors, 2022

optimistic vision. The reason behind this is the fact that these students represent a generation who did not experience the past of the previous regime in Iraq as they were born and raised after the end of that reign, and hence, their perception was mostly positive. This concept comes in line with the view of Whitlock and Powers (2008) who mention that "youths want to belong to their community and to their schools, to be endowed with meaningful rights and

Fig. 18 **a** Exterior of building 65 after work completion. Ref. the authors, 2022. **b** Exterior of building 65 during the renovation work. Photo by Habbobi (2019)

privileges, to be heard and adequately represented" (p.15). Some older visitors interviewed by the authors mention that they experience the same positive sense of place, but with a feeling of excitement when linking it to the past.

4.4 The American University of Iraq in Baghdad AUIB—The Main Palace (Not Renovated Yet)

The main palace of the complex, which is the biggest and the most important symbolic building within the campus, is excluded from the first phase of adaptation and renovation. Although most of the buildings surrounding it are adapted to house the AUIB colleges and the newly added laboratories and service buildings, no decision has been made regarding the new function of the main palace. This is probably due to its massive size and the huge fund needed, considering the damage caused to the building since 2003 which was found during the survey and assessment of the palace and its facilities (Fig. 21a: c).

The authors interviewed the lead designer during one of the site visits who mentioned that the decision relating to the new facilities that the building will be adapted to accommodate is still under discussion between the decision-makers (The Government of Iraq, AUIB, and the investors funding the project). He added that the assessment reports and the as-built layouts and drawings are prepared and submitted to the authorities, awaiting the final decision (Fig. 23).

Based on the study conducted by the authors, it is recommended to seek the opinions of experienced architects and interior designers to turn this massive palace into cultural and multi-functional spaces that could serve the university as well as the public. Adapting the palace to accommodate a cultural centre, conference halls, and art workshops is advisable, considering that the building is consisted of many large halls that provide the designers with flexibility and different solutions. In addition, some halls could be adapted to accommodate certain activities, especially the areas that are not damaged much (Fig. 22a and b), and such spaces could be rented after renovation to generate funds that will support the university and its financial needs. This aim could be achieved through proper modification of the interior architecture of the palace to improve it and at the same time to preserve its identity continuation.

5 Conclusion

1. According to the research analysis and results, it could be concluded that the act of adaptive reuse for interior architecture and interior design is all about the sense of place created through the design act.
2. Interior architecture is an act of change; it involves functional adaptation and structural modification needed due to new demands of functional change or new adaptation.
3. Creating a new sense of place is a major factor that should be addressed when dealing with the interior architecture of adaptive reuse buildings, taking into consideration the value and strength of the old sense of place related to the long-term relationship between the place and the people (sensory ones and social ones).
4. Physical characteristics/components of interior design and interior architecture, in addition to the cognitive/perceptual attributes that are generated from them, are considered the main two groups of aspects concerning the sense of place. Both of these groups affect the quality and character of the created place.

Fig. 19 **a** & **b**: One hall is adapted to accommodate a classroom, building 65, Al Faw complex. Photos by Habbobi (2019). **c**, **d**, **e** & **f**: Main hall is adapted to an auditorium. Photos by Habbobi (2019)

a b

Fig. 20 **a** New washrooms, building 65, Al Faw complex. Photo by Habbobi (2019). **b** One lobby adapted for the students gathering area Photo by: Habbobi (2019)

a b c

Fig. 21 **a, b & c** Damage of Al Faw main palace. Photos by the authors, 2022

5. It is necessary to involve interior architects when dealing with the adaptive reuse act, especially for valuable buildings, considering the preservation of the original character of such buildings.
6. Transforming an interior space into a place that people can experience and interact with will improve the quality of the space and increase its sense of place. This act of design does not engage the recent emotions only but also recognizes the previous related memories along with their historical and emotional experiences.
7. Creating a positive sense of place through successful interior architecture will benefit the public by introducing new cultural functions as well as assuring the identity continuation, which will subsequently preserve the identity continuation of the country.
8. The Presidential Palaces are considered historical buildings representing a certain heritage era of the country. These buildings should be preserved and reused to serve the public with new functions that introduce a new sense of place without ignoring the old ones.

a b

Fig. 22 a & b Existing condition of Al Faw main palace. Photos by the authors, 2022

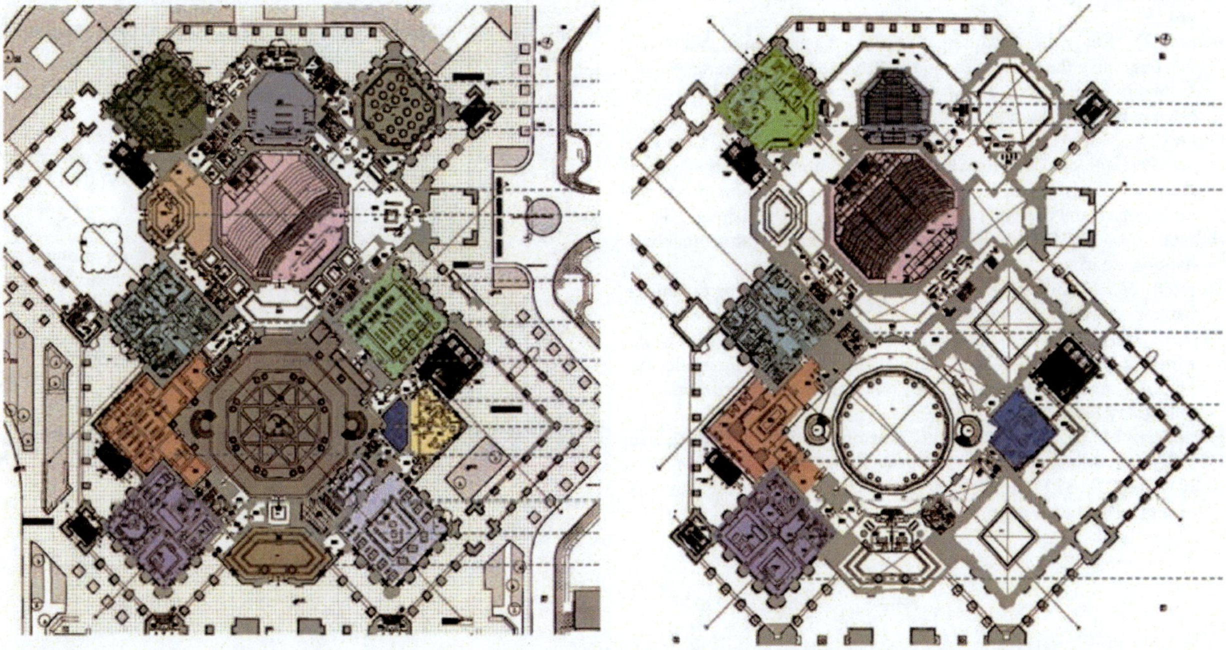

Fig. 23 Ground and first floor layouts of Al Faw palace proposed to AUIB by Manhal Habbobi Consultants, 2019. Ref. Habbobi (2021)

9. Considering the adaptive reuse concept and applying it to these palaces in Iraq are essential. This paper took the case study of one of these palaces and provided an analytical study concerning the changes that Al Faw Palace Complex has experienced over four decades, and how the sense of place has been affected during the preservation process for different functions.
10. Adapting part of Al Faw Palace Complex to accommodate the American University of Baghdad was a successful act. The same concept could be followed to preserve the other palaces which are abandoned since 2003.

References

Altman, I., & Low, S. M. (1992). Place attachment. Plenum Press, London & New York.

American University of Baghdad, Iraq (2019). Aboutblank Website. Retrieved May 15, 2022, from https://aboutblank.cc/Projects/AUIB-American-University-of-Baghdad-Iraq

Birer, E., & Adem, P. (2021). *Role of Public Space Design on the Perception of Historical Environment: A Pilot Study in Amasya.* https://doi.org/10.1016/j.foar.2021.09.003

Brooker, G., Weinthal, L. (2018a). (Third ed). The Handbook of Interior Architecture and Design. Bloomsbury Publishing Plc.

Brooker, G., & Weinthal, L. (2018b). New Occupancy. In The handbook of interior architecture and design (Third ed.). The essay written by Farrelly, L. (p 139). Bloomsbury Publishing, UK.

Brooker, G., & Stone, S. (2007). *Basics Interior Architecture- 01- Form and Structure*. AVA Publishing.

Brunswik, E. (1956). *Perception and the Representative Design of Psychological Experiments*. University of California Press.

Canter, D. (1977). *The Psychology of Place*. Architectural Press.

Chambers, E. (2006). Heritage Matters: Heritage, Culture, History and Chesapeake Bay. The Maryland Sea Grant College.

DEH, Department of Environment and Heritage (2004). Adaptive Reuse: Preserving our past, building our future. A booklet published for the initiative of (Year of The Built Environment). https://www.dcceew.gov.au/sites/default/files/documents/adaptive-reuse.pdf.

Fiorani, D., Kealy, L., & Musso, S. (2017). Conservation Adaption-Keeping alive the spirit of the place adaptive reuse of heritage with symbolic value. European Association for Architectural Education. EAAE, Hasselt, Belgium.

Gosling, S., Gifford, R., Mccunn, L. (2018). The Selection, Creation, and Perception of Interior Spaces - An Environmental Psychology Approach. In Brooker, G., & Weinthal, L. The handbook of interior architecture and design (Third ed.). Bloomsbury Publishing, UK.

Habbobi, M., (2019). Manhal Habbobi Consultants. https://manhal-habbobi.co.uk

Habbobi, M., (2021). Manhal Habbobi Consultants. https://manhal-habbobi.co.uk

Hay, F. (2007). Interior Architecture. In J. Gigli, F. Hay, E. Hollis, A. Milligan, A. Milton and D. Plunkett (eds.). Thinking inside the box. A reader in interiors for the 21st Century London: Middlesex University Press. p. 33–42.

Heidegger, M. (1927). Being and Time, in German, translated to English by Macquarrie, J. & Robinson, E. (1962). SCM Press.

Kelly, E. (2021). 33 Photos of Saddam Hussein's Palace that show the excess of his fallen reign: The Boydell Press.

Kullab, S. (2021). American University hopes to fill higher-end gap in Iraq. Associated Press. https://apnews.com/article/urban-sprawl-saddam-hussein-baghdad-higher-education-iraq-897067824c11c022513d837211193c9a

Kurk Insaat Company (2019). https://kurk.com.tr/en/projelerimiz/3320/

Lee, K. (2022). The Interior Experience of Architecture: An Emotional Connection between Space and the Body.

Loney, J. (2011, Dec. 2nd). U.S. hands main war base, Saddam palaces back to Iraq. Reuters: https://www.reuters.com/article/iraq-withdrawal-base-idINDEE7B10E520111202

Mosse, R. (2009) in Manaugh, G. Saddam's Palaces: An Interview with Richard Mosse.

Nussbaumer, L. (2009). Evidence- Based Design for Interior Designers. Bloomsbury Publishing Inc.

Perolini, P. (2014). Interior Environments: The Space of Interiority. Zoontechnica and Griffith University.

Plevoets, B., & Van Cleempoel, K. (2019). *Adaptive Reuse of the Built Heritage: Concepts and Cases of an Emerging Discipline*. Routledge.

Rapoport, A. (1977). *Human aspects of urban form: Towards a man-environment approach to urban form and design*. University of Wisconsin-Milwaukee.

Rapoport, A. (1990). *History and Precedent in Environmental Design*. Springer.

Rasmussen, S. (1959). Experiencing Architecture. Chapman & Hall, London. Association. Oklahoma City, OK: EDRA Press.

Relph, E. (2008) Third edition - Fist published in 1976. Place and placelessness. Pion, Lt. London

Rice, C. (2006). The Emergence of the Interior: Architecture, Modernity, Domesticity. Routledge, London.

Shalash, S. (2011). US to Hand Palaces Back to Iraq. Reuters, the Wall Street Journal. http://online.wsj.com/article/SB10001424052970203833104577072633810452606.html?reflink=desktopwebshare_permalink

Steele, F. (1981). *The sense of place*. CBI Pub. Co.

Tarzi, N. (2018). Building an American University on grounds of Saddam's palace is heavy on symbolism. https://thearabweekly.com/building-american-university-grounds-saddams-palace-heavy-symbolism

Tuan, Y. F. (2018), 9th edition (first edition published in 1977). Space and Place- The Perspective of Experience. University of Minnesota. Library of Congress Catalog No. 77–072910, USA.

Whitlock, J. & Powers, J. (2008). Places to Be and to Belong: Youth Perceptions of Life in Community. Prevention Researcher, 15(2), 12–15.

Williams, B. (2007). Saddam's Al Faw palace not at all what it seems. NBC NEWS. https://www.nbcnews.com/id/wbna17524022

Designing with Uncertainty—Built Heritage in Circular Economy

Iida K. Kalakoski and Riina M. Sirén

Abstract

Circular economy is rising to general interest. In architectural context, it is mostly considered as a technical problem, solved by engineers, lawyers, and economists. However, architectural heritage and culture are tightly connected, and architecture defines the local milieu, functions, routes, and maintenance. Accordingly, solutions made along circular economy are strongly influencing the human encounter, and vice versa. Circular economy will change the way we consider construction, preservation, demolition, storage, logistics, and repair. Furthermore, the spectrum of aspects of circular economy expresses and creates variety of *uncertainties* discussed in this chapter. The study is based on considerations collected from Finnish architectural heritage professionals and practicing architects. Firstly, the authors have hosted two round table discussions among Finnish professionals on architectural heritage and circular economy. Secondly, the authors have collected responses from architects participating into a "Reimage Buildings" camp focusing on circular economy taking place in Helsinki in September 2022. This article is based on round table discussions and questionnaire replies collected from the participants of the camp. The paper examines material reuse as an expression of circular economy and identifies different perspectives on its relations and potential dissonances with built heritage. The discourse is interpreted through a literature review. According to the dataset, circular economy in architectural heritage is not only a question of technical and economic acts but also of cultural aspects, and it is a reaction to societal challenges or an expression of uncertainty. The paper demonstrates that the premises of the circular economy of buildings require multidisciplinary understanding of the phenomenon.

Keywords

Circular economy · Architectural heritage · Built heritage · Reused construction materials

1 Introduction

Circular economy is rising to general interest. According to a summary study, circular economy is a system that takes into account material flows and their impact to economy and ecology and aims to minimize the use of fresh materials (Kirchherr et al., 2017). Circular economy has versatile fields of implementation in the built heritage, including platform economy and consumption-based pricing models, but the key part of the phenomenon is the material reuse (Huttunen, 2021).

Material reuse has diverse classification methods in conservation theory. Typically, material reuse is categorized by the level of modification. From lightest to heaviest, they are often classified as follows: (a) continued use (repair in place); (b) adaptive reuse (repair to some another use in place); (c) component reuse (as repair to same use in other place or as transformation to something else in other place); (d) material recycling (recycle as blank material); (e) earthworks/energy use; (f) landfill (Huuhka, 2016, 13). In terms of both conservation theory and circular economy, the lightest operations are generally regarded as the most recommended.

In architectural context, reuse is mostly considered as a technical problem, treated by engineers, lawyers, and economists. However, built heritage is an environment both

I. K. Kalakoski (✉) · R. M. Sirén
Faculty of Built Environment, Tampere University, Tampere, Finland
e-mail: iida.kalakoski@tuni.fi

R. M. Sirén
e-mail: riina.siren@tuni.fi

M. F. Shahidan et al. (eds.), *Urban Narratives: Exploring Identity, Heritage, and Sustainable Development in Cities*, Advances in Science, Technology & Innovation, https://doi.org/10.1007/978-3-031-48517-6_20

tangible and intangible, influencing the human encounter, and vice versa. As a part of the built heritage, material reuse defines not only the local functions and routes but also the maintenance, milieu, and aesthetics. Accordingly, when cultivating the material reuse of built heritage, we see that it will change the way we consider construction, preservation, demolition, storage, logistics, repair, and use. Therefore, it is essential to study the whole spectrum of aspects of reuse in built heritage.

In this article, we focus on the cultural perspective of the built heritage of circular economy. The concept of culture may refer to the way of life, especially the general customs and beliefs, of a particular group of people at a particular time; or it can refer to the attitudes, behaviour, opinions, etc., of a particular group of people within the society (Cambridge Dictionary, 2023). In both cases, culture is often originated or attributed to a specific region or location. The uncertainties discussed in this article are connected to both of these interpretations of culture. The broader meaning of the concept is under review, when considering the massive cultural shift as a respond to the ecological crisis. The more specific meaning is needed to work with the tools, such as architecture and circular economy, to cope with the crisis.

This paper discusses the cultural view of material reuse in a Finnish context. As a forum for circular economy, Finland is inclined for the new development since the country is determined by rapid and total paradigm shifts aroused by technological and cultural innovations. (Haapala & Peltola, 2018, 175–177.) The agility is driven by a relatively small and uniform population and an organized society with strong mutual trust. Mainly for these reasons, complex problems are solved at the national level, and accordingly, the decisions are quickly implemented throughout the society. Correspondingly, this applies to architecture and industrial design. The best-known great shift, the reconstruction era after the Second World War, generated many insightful and internationally known phenomena such as Finnish design, architect Alvar Aalto, Puutalo Oy/Timber Houses Ltd, and the general standardization of building parts and building regulations (Vesikansa et al., 2021). All the inventions still affect the Finnish building culture.

Following these examples, Finland is a country where building industry is heavily controlled, the building stock originates mostly from the twentieth century, and most building components are industrially produced. Accordingly, everyday environments for most people are modernist, and the culture has an overemphasized affinity for novelty. In Finland, component reuse is not yet a common practice despite the intuitive reuse practiced by home repairers (Huuhka et al., 2021, 9). Big building volumes are handled either through adaptive reuse, transformation, and additions, or through demolition, ending to energy use or landfill. In terms of the material reuse, the impending cultural change is going to have a considerable imprint in Finnish built heritage.

One publication has been establishing the concept of material reuse in the Finnish building construction discourse. The publisher, Rakennustieto, self-claimed as the "Leading Provider of Construction Information in Finland", also manages the database on Finnish building standards, initiated by architect Alvar Aalto in 1942 (Jääskeläinen, 2017, 308; Rakennustieto [n.d.]). Noticing the publisher's authoritative status, the publication called Circular Economy in Built Environment (in Finnish Kiertotalous rakennetussa ympäristössä) paints an accurate picture of the current vision of the phenomenon in Finland in 2021. By introducing the concept of circularity in terms of building industry, the book aims at motivating practitioners in the industry to promote reuse. The book relates the circular economy with other close concepts such as material's lifecycle, presents circularity projects, and shares information about the facilitation of the demolition phase. In addition to these technical views, there are two chapters about the cultural aspects of the circularity, another one written by one author of this article (Kalakoski, 2021). The publication has served as a starting point for this article and has most likely influenced the interviewed experts' understanding of the concept of circular economy.

1.1 Methodology and Material

This study is primarily not a theoretical consideration but is based on discussions with Finnish built heritage professionals and practicing architects and concerns that were arisen during the conversations. Firstly, the authors have hosted two round table discussions among Finnish professionals on built heritage and circular economy. These events have been recorded and distributed to professional and wider audience through the YouTube channel of Finnish Association of Architects (see Rakennusperintö-SAFA, 2021, 2022). Instead of an academic analysis, the conversations were led to enhance the general and professional awareness on the relations between circular economy and built heritage. However, the discussions prove to be an interesting material also from the academic viewpoint to perceive the way in which architects and heritage professionals approach the theme of circular economy in this time. The keywords "circular economy" and "built heritage" were used in the headings in both of the two sessions, and the participants were led to consider the theme from their interfacial perspective.

To achieve a more versatile picture of the professionals' opinions on circular economy, we accomplished our material through collecting responses from designers attending a "Reimage Buildings" camp focused on circular economy (Kalakoski, 2022). The camp took place in Helsinki in September 2022 and was organized by "The City of Helsinki cluster programme for circular economy in construction", the Ministry of Environment, and Ethica, a consultancy specialized in circular economy. The practice-oriented camp was arranged for building designers, interested in implementing circular economy in their work. A short questionnaire was given to the participants before and after the course, denoted with the keywords "circular economy" and "architecture", but without the concept of "built heritage". The data were collected to study the conceptions of circular economy among the broader scope of planning specialists. Firstly, is heritage acknowledged at all when the concept is not specifically noted in the layout of the questions? And secondly, if heritage is detected, so in what way it is considered.

The first panel consisted of four professionals including an entrepreneur–engineer, an entrepreneur–architect and two authorities from different fields of the museum sector (see Table 1). Since the perspectives varied from independent actors to public officers and from engineer practice to heritage supervision, the event was characterized by the enthusiasm for the collective problem-solving. The second panel consisted of architects working in different professional roles from education to practice and politics (see Table 1). Followingly, the event was characterized by architects' self-reflection and focused on the architectural education and the role of an architect in the practice of circular economy. In this article, the specialists are referred by their abbreviations.

The Reimagine Building camp was attended by a broader group of actors, including interior designers, engineers, planners, and city officers. The two-day camp consisted of a one-day excursion and another day with short lectures and workshops. The group of circular-economy-oriented building specialists is relatively small in Finland, and it is plausible that some of the participants had already got acquainted with the above-mentioned round table discussions.

We sent a questionnaire to the participants before and after the camp. Before the camp, we asked three questions: (1) how the participant had become interested in circular economy, (2) how they had already practiced circularity in their work, and (3) what kind of challenges and potential they see in circularity. After the camp, we asked how the circularity will or could change (4) the practices of architectural design and (5) the principles of contemporary architecture or its "architectural language". Despite the narrow response rate—we received only seven replies to the pre-camp survey and four replies to the after-camp survey—the replies correspond well to the discussions that took place at the camp.

Based on the replies, the participants became interested in circularity mainly for ecological reasons. They were aware of the environmental impact of the building industry and wanted to find ways to make better use of existing buildings and building parts. The current practice to demolish and rebuild was described as extravagant and irritating. When asked about previous experience on circularity in their work, four of seven replies mentioned lectures or studies as an arena of practiced circularity. Two of the participants mentioned that they had worked with reuse of furniture or other pieces of interior design. One participant mentioned the difficulty in applying circularity into the conventional building design project. The deduction of these replies is that circularity is still rather under a theoretical consideration than a common practice. It also seems to be adaptable to furniture more likely than of building components.

Table 1 Table presents the panelists of the two round table discussions held in December 2021 and February 2022

Panelist	Abb	Education	Professional role	Organization
Panel 1				
Robin landsdorff	RL	Architect	Museum authority	Finnish heritage agency
Petri Salmi	PeSal	Engineer	Entrepreneur	Spolia design
Juha Päätalo	JP	Architect	Entrepreneur	Architecture office Päätalo
Anne Uosukainen	AU	Building conservator, construction foreman	Conservation instructor	Regional Museum of Pirkanmaa (Tampere region)
Panel 2				
Anni Vartola	AV	Architect, Dr of Sc	Academic	Aalto university
Panu Savolainen	PaSa	Architect, Dr of Ph	Academic	Aalto university
Pekka Saatsi	PeSaa	Architect	Entrepreneur	Saatsi architects
Henna Partanen	HP	Architect	Politician	The greens in Finland

When considering the results of the study, it must be that the authors of this paper are architects working in the field of built heritage. The authors contributed to the collection of the material by planning the round table discussion, selecting the participators, preparing the questions, and leading the discussion. One of the authors also attended the preparatory meeting of the Reimagine Buildings camp, gave a lecture, and participated into workshops on the camp. As a follow-up, the authors also wrote an article about circularity and built heritage to Finnish Architectural Review (Kalakoski & Sirén, 2021). The round table discussions as well as the replies to the after-camp questionnaires are most likely influenced by the participation of the authors.

Once the material was collected and the discussions were recorded, we transcribed the material and analyzed it in order to identify themes or categories within the discussions and replies to the questionnaires, concerning the cultural aspects of circularity in built heritage. We found out that the main motivation for discussing about circularity was the professionals' concern about the future of the planet. Secondly, the participants required clearer definitions and vocabulary to the field. Thirdly, the participants expressed their concern about the ethics and aesthetics of reuse. Fourthly, the discussions were concerning different professions and their way of adaptation to new practices. There was also discussion about building regulations and other techno-economic aspects of reuse that were limited out from this analysis, since they have been more discussed by experts on other fields (Huttunen, 2021). The above-mentioned categories are further discussed in the following sections.

2 Analyzing the Uncertainties

Based on the previously mentioned materials, we examined the reuse and circular economy in terms of built heritage through identifying different perspectives on their relations and potential dissonances. During the discourses, we discovered that most of important observations were made by expressing uncertainty. The ultimate uncertainty mentioned was the uncertainty about the future of the planet. The ecological crisis with the impoverishment of nature, pollution, and deterioration of living conditions initiates all other uncertainties. It is the number one motivation for considering circular economy in the frame of built environment.

The material reuse was regarded as a key factor to return to the long tradition of sustainable lifestyle. The participants mentioned that the "throw-away culture" has only been a short and exceptional period in the history of humankind (PaSa, JP). The development is the result of the imbalance produced by easily accessible energy and mass production, which has led to the relative high cost of handicrafts. Only from 5 to 6 decades ago, the reuse of building parts was an every-day practice (PeSal). The so-called intuitive reuse is still a common practise among home repairers. When building for oneself, one does not count the number of working hours, if the material is easily available and feasible (AU). Today, it is necessary to expand the field of activity and unlearn the throw-away culture. In the development of circularity, we are rather turning back to good old practices instead of creating something new.

2.1 Filtering Unnecessary Uncertainties—Definitions and Measures

When deconstructing the uncertainties, the central problem is the actual effectiveness of the operation. Decision-making is under pressure since there are lots of new actions to trial in terms of the material reuse. According to the experts, it is difficult to assess the environmental sustainability of the new practices. Are there some values or aspects that we risk losing if we practice the material reuse? The experts shared a common concern on truly affecting the world, instead of indulging in greenwashing (HP). According to them, we are facing a complete reorganization of the construction industry (AV).

Firstly, the experts debated on the unification of terminology. It turned out that there is illegibility in the terminology of material reuse. The speakers discussed the way in which material reuse should be defined in terms of practices like repair, reuse, and relocation, and further on, how the concepts of recycling, upcycling, and downcycling should be defined in terms of built environment. The panelists agreed that recycling should mean maintaining the value of the material in reuse. Upcycling should mean raising the value of the material by reuse, and downcycling, correspondingly, losing the value of the material by reuse. (RL, JP.) According to the experts, most of the present acts called as recycling are actually acts of downcycling.

The building material can be reused either with or without utilizing the aesthetics or stories of the object (AV, HP). When trying to preserve the manifold cultural values in the process of circulation, one option is to perform the building transformation on site. This helps to maintain the qualities related to the surroundings, in particular. It is probable that the heavy transformations with buildings modified beyond recognition are going to grow in number in the future. Nevertheless, the cultural aspects may still be bypassed and should be considered diligently. (AV.) The topic of the risks of heavy transformation has also been discussed in the scholarly literature (Yarrow, 2019, 11).

From the material energy perspective, the value decreases the more you reprocess (Braungart & McDonough, 2008). In this idea, higher level of processing is comprehended as downcycling, i.e., putting material into small pieces and eventually losing them, both material and energy. For example, crashing 30 years old concrete elements to landfill is downcycling as the elements could have a lifespan on at least 100 years (PeSal). In turn, if you can build from reused concrete elements instead of new ones, you can save up to 95% of the carbon footprint (JP).

From the perspective of building preservation, the original building tends to have the biggest value. Relocated building parts have lost a section of their cultural values, because they are no longer part of the original entity. This applies at least if the original building still exists. But if the reused components are the only things remaining from the original building, they may gain cultural value parallel to relics. At least this concerns decorative building parts such as doors and windows—or even handmade bricks or hand carved logs withholding the character of uniqueness and handicraft. (AU, RL.) Instead, with more anonymous or industrial building parts, the value of a component may even rise when reprocessed. A raw board, for example, becomes more valuable when planed. The value rises even more if the transformation is affiliated by artistic capital or handicraft skill. This is due to the value of work and time used, as quite the opposite to the previously mentioned idea of material energy.

Followingly, the simplest component may turn into an art piece if it is treated with artist's, architect's, or artisan's intention (JP). When implemented in a new construction, such elements are often highlighted with a visual contradiction, and therefore, they may appear as "spolia". The term "spolia" primarily refers to the early Christian and medieval architectural reuse of stone building blocks dating from the times of Greco-Roman antiquity. Quite recently, it has also been revisited as a hyponym for the reuse of building components and materials that accentuate experiential values, such as historicity and authenticity (Kalakoski & Huuhka, 2018). In that sense, spolia can be referred to upcycling, since such parts are valued more than others with only tectonic value: the spolias become pieces of decoration or even works of art.

In aiming for generalization of the material reuse, it is also essential to reconsider the concepts of demolition. In the twentieth-century Western age of abundance, we have forgotten not only the practices of reuse but also the way to dismantle buildings accurately. According to the experts, we should understand demolition, not as a practice that results in destruction, but as a practice that allows the collection of usable materials and building components to new uses (LR, AU). Correspondingly, the concept of lifespan should be regarded in a new way. Since we consider buildings as long-lasting environments, we must accept that distinct building parts and construction materials are in different phases of their lifespan (RL). This leads to the fact that all building components should be easily repairable or replaceable. It is likely that in some other use, the worn component can be fully appropriate and valued.

Despite the focus on cultural aspects of circularity, the panelists brought up certain technical issues about the reuse. Most importantly, they were concerned about the logistics. Even if there was an ambition to obtain reuse and there were suitable materials, we also need ways to store and transport the components when and where needed. To facilitate the planning and design phases, we should understand the upcoming material flows. (AU.) We should also have ways to analyze and classify the materials based on their size and condition, etc. Especially the unpredictability of materials challenges the contemporary building industry. With traditional building materials, the risks are quite manageable, since the range of materials is relatively narrow, and the condition of the material can be seen based on the surface. (RL.) However, a prominent part of building components in Finland is industrially produced and has a run of only a few decades. This problem demands for an encompassing cultural shift in the field.

In Finland, used materials are often polluted by negative connotations. As the professionals acknowledged, Finnish culture is defined by a general admiration of novelty (AV). For example, kitchen cabinets are replaced regularly, just because they are not fashionable (PaSa). The ethos is strong also among the architects and it is reinforced by architectural competitions and other representations of good architecture, verified by reviews. (AV.) On the other hand, we can see that the younger generations of architects and architecture students are not so fascinated by the novelty and wow-architecture (PaSa).

Besides the aesthetics, the novelty in architecture is often related to functionality and healthiness. The pursuit for novelty refers to eagerness to afford structures that are as safe as possible. People think that reuse materials are more prone to detrition and microbes than new ones. Fear of old materials is related to the potentially harmful compounds. This implies the need for further study on the young building components in order to recognize the products and find out what they are made of (PeSal). It is the predictability that increases the usability of reused building components. As stated before, we need ways to analyze and evaluate the reusable materials and components to convince people about the safety of the reused material (PeSal). We also need a cultural change that accepts and appreciates circularity. Such a change starts from good examples that are safe, healthy, and aesthetically pleasing. Followingly, the circularity also challenges the production of all new buildings. Buildings should be designed to be easily dismantled later (RL) (see Fig. 1).

Fig. 1 Photo of TKY 3 building, the current JMT 3, which was built in the neighbourhood of Teekkarikylä in Espoo from the bricks of the Soviet embassy building destroyed in the Second World War bombings of Helsinki. Photo by UA Saarinen 1952, courtesy of Finnish Heritage Agency. Reproduced by Creative Commons permission CC BY 4.0

2.2 Sympathizing the Uncertainty—Cultural Ground

Building with multi-aged components is a technical challenge, but it can also provide cultural and aesthetic depth to the buildings, as noticed in recent studies (Huuhka & Vestergaard, 2019; Kalakoski & Huuhka, 2018; Kalakoski & Thorgrimsdottir, 2022). Also, the interviewed professionals expressed that reuse building has a history and a unique character already when just finished (JP). Historical building parts can tell stories and evoke emotions. Since stories are important to people, especially home repairers tend to preserve and emphasize the historical layers (AU). There are at least two perspectives to analyze such eagerness to preserve the layers. Either the layers are claimed to (1) tell about history, of a building, or (2) layers are regarded as valuable for their aesthetic qualities, for example, their rough or romantic looks (AU). Third possible motivation is the circularity as such, the willingness to practice the aesthetics of reuse. Such aesthetic approach to reuse may also be considered problematic if it completely neglects the provenance of the component. Especially, museum authorities were concerned about the stories and the authentic history carried along with the materials and components (RL, AU).

Furthermore, there is a dark side in the heritage reuse: the story market. Building components may be sold only because of their high market value (AU, RL). The value of recycled and detached buildings components may sporadically rise and fall, but the value of the original entity most often decreases when it loses its original building parts. In conservation theory, this is known as the value of original entity (Venice Charter, 1964 §9, 12). While we promote material reuse, there is still a need for traditional building preservation to protect some historical building entities (RL). The question applies not only to historical but to buildings of all ages, since the demolitions are now addressed at the younger building stock (PeSal). And knowing that buildings need regular repair, there must be mandatory regulations to secure the temporary storage of old building parts after the point of origin (AU). This applies to construction parts of all ages (AU, RK) (see Fig. 2).

Fig. 2 Building components in transition. Morkku, Tampere. Photo by Anne Uosukainen, 2020, reproduced by courtesy of Vapriikki archives

It is urgent to find ways to convey stories by recycled material (RL). Basically, there are two key elements: firstly, the reused materials should be recognizable from the new material, and secondly, the provenance of the materials should be indicated. Concerning the building component database, the story should also be included. (RL.) Another kind of step towards the protection of stories is to relocate the reuse components as near as possible. This is also ecological, in terms of energy consumption (PeSal). Accordingly, all the existing construction materials in site should be used at first (JP).

This brings us to the diverse contexts of material reuse. The common heritage is typically built amongst a specific cultural environment. For example, traditional two-storey Ostrobothnian houses in Finland are under pressure, since they stand in a region with a declining population, due to which there is a trend to move a part of the houses to more densely populated areas. Relocation is a way to hold the house in use, crucial to timber heritage. However, it is possible that the new cultural environment is fundamentally different with no background of high log buildings. This is one perspective to be assessed in the transformation design (JP).

In Finland though, most of the building stock is not crafted. The great mass of ordinary buildings is post 1950s with repetitive and anonymous building components. The question is: how could the meanings be transmitted in the reuse of industrial building parts? One answer could be found with diverse building types having different functions. The rare heritage represents the artwork, the culture, and the authorship, while the big industrial mass is used for economic, ecological, and technical purposes. An exception proving the rule: a reused concrete element with a graffiti preserved could be a special attraction (PeSal). Even industrial building parts can obtain considerable value with patina or other traces of past life. The weathering of building materials can be conceived as an important quality transmitting historicity in the built environment (Matero, 2006, 76), but its acceptance amongst the spectators often depends on the character of the weathering as well as on the ageing material.

Thereby, industrial elements can be used in different ways as well. Artistic value can arise from the insightful use or connection of reuse building parts. Even if the parts are similar across the country, combination methods may gradually start to form local traditions. The multilevel nature of the material reuse shows its special quality. Correspondingly, this relates to the meanings of exceptionality. Rarity is a time-and-place-dependant phenomenon, and so is the youth with the lack of nostalgia. Handmade building components are now rare in Finland, not the concrete elements of the 1970s. However, as the elements become rarer, they will take on new meanings.

Hence, the material reuse has an impact on the architectural and artistic aesthetics (JP, PeSaa). The reuse parts influence malleability, usability, and stability of the construction. Now, two trends can be observed in terms of aesthetics. When an old building is repaired with reuse components, the changes are often tried to hide (AU). When reuse components are used as part of contemporary architecture, the distinction of the parts becomes more important. The interest in the building increases even more if you know the origin and previous function of the building part (RL). Over time, as recycling becomes more common, and one reuse part has several previous buildings, the stories get chained if only there is a method to preserve the knowledge. Ioannis Poulios (2010, 182) has argued how the communities are crucially significant in creating the continual process and maintaining the messages on the site.

The interviewed professionals are hesitant whether the material reuse would or should not be manifested in the practiced architectural language. Such was the opinion amongst the participants of the Reimaging Building camp. One participant noted that architecture might become more playful, rough, and enriched with the usage of reused material. Another estimated that reuse could evolve our preferences on what is considered beautiful and good architecture. Despite many challenges, participants also mentioned the possibility of creating more sustainable, versatile, and historically layered built environments. The challenges were seen as a possibility to come up with completely new innovations and architectural expression. One respondent evaluated that combining elements from different proveniences can bring along new aesthetics which represent the new values of the society.

Correspondingly, the professionals around the round table thought that there is a need to increase tolerance towards the peculiarity and unfinishedness to make circular economy mainstream in built environment (PeSaa). It is likely that the expectations for the aesthetics of reuse may vary in function of time. As far as the reuse practice is unfamiliar for people, there might be a need to manifest the reuse aesthetics to increase the awareness of the opportunities of material reuse. (PeSal, RL, AV, HP.) Later, the reuse should not dominate the aesthetics of sustainable architecture (RL). The risk with specific "reuse architecture" is that it might end up as a short-lived trend (AV). The material reuse should not be considered as an ism but as the new normal.

As the panelists pointed out, the architects should always aim at good architecture, and material reuse should not be used as a pretext for poor architecture (AV, PeSaa). This concern arises from earlier bad experiences with eco-architecture referred sometimes as "hairy buildings" (AV). However, there is much more than just restrictions with material reuse, since old building components also inspire for new aesthetics. For example, a new house built from old bricks might have an impressive character, just because of the looks of the bricks. Reuse architecture is a welcome opportunity to implement decorative, or some other way unexpected, elements to contemporary architecture (JP).

2.3 Profession in Transition

One key part in introducing the material reuse is, according to the professionals, the politics. The politics affects the construction industry which has great economical, material and energy impacts with the education, design, logistics, production, and legislation involved. Via the ample affiliates, the politics has societal impacts on people's everyday environments and lives. Due to the high societal impact, the reuse should be studied and encouraged by the politics. Politicians determine the target level and the direction of action. The new product will be used if it is the cheapest option, and for politicians, it is possible to influence the purchase price. (HP.) In addition to regulations, methods of influence include taxation and subsidies, e.g., for repairs (AV).

Material reuse is a work-intensive and knowledge-intensive matter. One uncertainty is involved in the society's interest and ability to generate a sufficient level of knowledge and understanding on new specifications, regulations, instructions, and techniques concerning dismantling, mending, installing, and maintaining the building stock (RL, PeSal, JP, AV, HP). In the circular economy, there is a constant need for a continuous scale of experts specialized in different kinds of reuse methods (AU, JP, RL, PeSal). Correspondingly, there is a constant need for education which develops with experience. Education should be developed to the direction of lifelong learning (AV, PaSa).

One uncertainty lies in the professional appreciations. Since working with virginal new buildings is getting rare, there is no need any more for a generalist hero architect ideal (PaSa, AV). In other words, there is a need for a redefinition of the status of architect (HP, AV). As mentioned above, the first imperative of ecological action is to avoid new construction, if you can use the old one (JP). As an architect, it is better to plan the way that the existing components, buildings, or sites would be transformed as little as possible, and consequently, to make a bigger impact with doing less. Is it sufficient for the professional identity to embrace the principle of minimum intervention? There is a need for an architect to adjust to the role of a facilitator or a group leader.

Is it possible for an architect to refuse to design? (HP). The experts concluded that this is not a real option for an architect, because of the parallel designers. Instead of deciding what to plan, it is necessary to decide how to plan. To study architecture is to study problem-solving (PeSaa) and critical thinking (PaSa). Thus, education provides the appropriate skills for this new time of uncertainty. We must move into an architectural culture where it is valuable and respected to think about carbon footprint and material resources (AV).

Since the camp was targeted for practitioners, they mainly contributed, not unreasonably, to the professional and practice-oriented dimensions of the circularity. When asked about the challenges, they mentioned the general attitude and tendency to keep doing the things "as they always have been done". It was noted that we need not only new distributing platforms but also specific professionals to collect, organize, and market the reuse components and materials. However, we also need to educate practitioners to find professional satisfaction in working with the limitations of reused material and their aesthetic qualities. In terms of the impact of the circularity to the practices of architectural design, the respondents estimated that the whole design process needs to change because of the central influence of the available materials on the architecture. In addition to these, the repair and project-specific modification of the components need to be considered in the project management (see Fig. 3).

3 Conclusions

This article examined professional discourses on circular economy in built environment. We interpreted the widespread interest in circular economy as a reaction to uncertainty caused by the ecological crisis. At the intersection of building heritage and circular economy, there are social and cultural aspects of built environment with its values and attitudes, aesthetics, society and education, economy, production techniques, and measures. The main outcome of our considerations is that the circular economy must be viewed from a cultural perspective and not be overshadowed by economic and technical factors. The bundle of sub-uncertainties defined in the study must be handled as an entity.

According to the conservation theory, the most recommended way of treating a building is to repair it in place. However, it should be assured that even when it is necessary to transform or dismantle a building, its building components are relocated on site or as close as possible. Firstly, the shorter the transport distance is, the less energy it takes. Secondly, the shorter the transport distance, the stronger the dialog between the building and the surrounding building tradition.

Material reuse carries multiple meanings. In the new contexts, the reused parts can form surprising new entities with or without the former cultural heritage. With the stories introduced, the layers of history can produce new significances to the environment. It is also a new kind of task

Fig. 3 Reuse building executed in an everyday manner at Vaksali street in Tartu, Estonia. Photo © Author: Riina Sirén

to design with reused materials. With predefined building materials, designing architects must re-evaluate their professional identities. The art of architecture must be interpreted from the art of creation to the art of collaboration and—as a result—to the art of apposition.

The special characteristic of reuse is that it decreases the need for mass production and mass logistics in society and requires lots of manual work on site. Accordingly, as the benefits of mass production diminish and goods move less, craft becomes more profitable, and people specialize on certain craft sectors. Consequently, materials are treated in specific ways in separate areas, which may lead to the diversification of architectural aesthetics. Hence, material reuse can be seen as a technology for creating local identities. Buildings are considered as steady bearers of identities, and recycled construction parts equate to recycled identities. Accordingly, built heritage and material reuse relate to fundamental needs to continuing something anterior: carry on memory or tradition, hold on the life on earth. Both locality and identity are culturally bound phenomena that alleviate the feel of uncertainty.

As one of the participants of the Reimagine Buildings camp noted, there will be a time when we will no longer consider beautiful anything that is destructive to the environment. By saying this, she combined the ethics, aesthetics, and different dimensions of culture. To cope with the current environmental challenges, we need a cultural shift that allows us to reconsider architecture and the ways we build today. The reuse material policy produces factors that reduce uncertainty and responds to the ultimate uncertainty itself, to the ecological crisis.

References

Braungart, M., & McDonough, W. (2008). *From cradle to cradle*. Penguin Random House UK.

Cambridge Dictionary. (2023). *Meaning of culture in English*. Cambridge University Press & Assessment. Cited 22.2.2023. Available at https://dictionary.cambridge.org/dictionary/english/culture

Finnish Heritage Agency. (1952). Figure 1. Photo by UA Saarinen. JOKAUAS2_1335:9. Reproduced by creative commons permission CC BY 4.0. Available at https://www.finna.fi/Record/museovirasto.E578F22EF39C92987FB0A55856EF6463?sid=3063310762

Haapala, P., & Peltola, J. (2018). Elinkeinorakenne 1750–2000. In P. Haapala (Ed.), *Suomen rakennehistoria. Näkökulmia muutokseen ja jatkuvuuteen (1400–2000)* (pp. 170–209). Vastapaino.

Huttunen, E. (Ed.). (2021). *Kiertotalous rakennetussa ympäristössä*. Rakennustieto.

Huuhka, S. (2016). *Building post-growth. Quantifying and characterizing resources in the building stock*. Thesis for the degree of Doctor of science in architecture. Publication 1414. Tampere University of Technology.

Huuhka, S., & Vestergaard, I. (2019). Building conservation and the circular economy: A theoretical consideration. *Journal of Cultural Heritage Management and Sustainable Development*. Accepted November 2, 2019.

Huuhka, S., Vainio, T., Moisio, M., Lampinen, E., Knuutinen, M., Bashmakov, S., Köliö, A., Lahdensivu, J., Ala-Kotila, P., & Lahdenperä, P. (2021). *Purkaa vai korjata? Hiilijalanjälkivaikutukset, elinkaarikustannukset ja ohjauskeinot*. Ympäristöministeriön julkaisuja, 9. (*To demolish or to repair? Carbon footprint impacts, life cycle costs and steering instruments*. Publications of the Ministry of the Environment, 9).

Jääskeläinen, L. (2017). Hyvä rakentaminen edellyttää sääntelyä. In H. Hautajärvi (Ed.), *Rakennetun Suomen tarina* (pp. 306–309). Rakennustieto.

Kalakoski, I., & Huuhka, S. (2018). Spolia revisited and extended: The potential for contemporary architecture. *Journal of Material Culture, 23*(2), 187–213.

Kalakoski, I. (2021). Rakennusperintö ja kiertotalous. In E. Huttunen (Ed.), *Kiertotalous rakennetussa ympäristössä*. Rakennustieto.

Kalakoski, I. (2022). *Reimagine building inquiry. Replies from the participants*.

Kalakoski, I., & Thorgrimsdottir, S. (2022). Learning from the secondary: Rethinking architectural conservation through barn architecture. *Journal of Material Culture*, OnlineFirst.

Kalakoski, I., & Sirén, R. (2021). Messages from the past. *Arkkitehti, Finnish Architectural Review, 5*(2021), 22–28.

Kirchherr, J., Reike, D., & Hekkert, M. (2017). Conceptualizing the circular economy: An analysis of 114 definitions. In *Resources, conservation and recycling* (vol. 127, pp. 221–232).

Matero, F. (2006). Loss, compensation and authenticity in architectural conservation. *Journal of Architectural Conservation, 12*(1), 71–90. https://doi.org/10.1080/13556207.2006.10784961

Poulios, I. (2010). Moving beyond a values-based approach to heritage conservation. *Conservation and Management of Archaeological Sites, 12*(2), 170–185. https://doi.org/10.1179/175355210X12792909186539

Rakennusperintö-SAFA. (December 17, 2021). *Rakennusperintö ja kiertotalous*. Youtube. https://www.youtube.com/watch?v=X5i0PqW8ij8&t=2105s

Rakennusperintö-SAFA. (February 3, 2022). *Rakennusperintö ja kiertotalous 2—Arkkitehtuurin ja muotoilun päivät*. Youtube. https://www.youtube.com/watch?v=2oI4IWSwM7Y&t=16s

Rakennustieto. (n.d.). Yritys. Site of the publisher *Rakennustieto*. Retrieved August 9th 2023 https://www.rakennustieto.fi/yritys

Sirén, R. (2022). Figure 3. Photo by Riina Sirén. Reproduced by courtesy of Riina Sirén.

Vapriikki archives. (2020). Figure 2. Photo by Anne Uosukainen. Reproduced by courtesy of Vapriikki archives.

Venice Charter. (1964). ICOMOS.

Vesikansa, K., Tidwell, P., & Berger, L. (2021). *New standards: Timber houses Ltd. 1940–1955*. Garret Publications.

Yarrow, T. (2019). How conservation matters: Ethnographic explorations of historic building renovation. *Journal of Material Culture, 24*(1), 3–21.

Development of Cities' Vocabularies

Process of an Architectural Concept Generation: The Revival of a Historical Building: A Case Study of Jifna Castle

Marwa S. Alshanti and Dalia O. Hafiz

Abstract

An architectural design concept represents the core value of the design, and it is the string that keeps all design decisions related and original. It needs to be achieved by understanding the holistic nature of the project before adopting one main idea that will be the essence of the design. Approaching historical buildings needs special effort since several aspects are involved and every design decision needs to be carefully studied and sensible. This paper examines different influences affecting inspirations and approaches to develop the architectural design concept of a historical building, as well as how they can lead to the concept formation. The research methodology is based on a practical application to upcycle a case study of a historical castle that dates back to the crusader era in Palestine. In order to generate an architectural approach for the proposed historical structure, several design aspects were studied, e.g., historical, geographical, spiritual, and incidental. These aspects assure a more comprehensive approach to the design and generated a framework representing a design process aid tool. The progressive framework helped in generating and developing the architectural design decisions while approaching the historical building to maintain and celebrate its values while accommodating recent or existing needs. The proposed framework can be used by designers, archeologists, and buildings preservers to generate inclusive approaches to design concept generation.

M. S. Alshanti (✉)
InSite-Planners/Landscape Architects, Dubai, UAE
e-mail: marwa.alshanti@insiteinternational.com

D. O. Hafiz
Ajman University, Ajman, UAE
e-mail: dr.dhafiz@gmail.com d.hafiz@ajman.ac.ae

Keywords

Concept generation · Upcycling · Sustainable historic preservation · Village of Jifna · Contemporary reuse · Socio-cultural significance

1 Introduction

During the conceptual idea stage of the design process, the design notion is externalized using purely abstract representations. Sketches assist students in the conceptual design phase of the design process by allowing them to visualize the spatial arrangements of shapes and offer solutions to the design problem. During the design process, it enables students to share concepts and potential solutions while cooperating to capture fresh ideas and assess design concepts. Additionally, it aids students in putting their concepts and the significance of spoken statements into a pictorial framework that aids in solving design challenges (Yalcin, 2021).

To develop a concept, various sources of inspiration can be examined throughout the process, resulting in a unique and cohesive design that fits the building's function and context while also telling a story through architecture. Through the case study, several aspects have been studied that influenced the decisions in creating the concept.

1.1 Approach to Historical Buildings Upcycling

Historical structures are important because they represent similar historical periods and offer insight into them. Rather than being demolished, they must be maintained since they are proof of the way of life and culture of the people who lived in or around them. Future generations can learn more about them by examining their location, size, and potential.

Adaptive reuse with a new function is inevitable if the building stock's useful life maintains its function. In order

M. F. Shahidan et al. (eds.), *Urban Narratives: Exploring Identity, Heritage, and Sustainable Development in Cities*,
Advances in Science, Technology & Innovation, https://doi.org/10.1007/978-3-031-48517-6_21

to preserve the historic structure's cultural relevance, the anticipated new use must be appropriate. In order to prevent future generations from receiving inaccurate or incomplete knowledge, the functions, new uses, and interventions must retain the building's originality and architectural identity.

2 Literature Review

2.1 Creation of Concepts

Previous efforts aimed at understanding the architecture concept generation and its phases. Research divided concept generation into two primary phases, namely problem- and inner-sense-driven phases, in concept derivation. The classification was based on two criteria: the basis for concept generation and the capacity that allows concept generation to proceed (Taura & Nagai, 2013). In addition, Taura et al., (2012) examined the patterns in the process that produced creative design ideas when studying the development of creative concepts in design. These patterns represent both explicit and implicit ideas that, when linked at the abstract level, establish a productive thought process that yields original design ideas.

Concept emergence was defined by Cross (1997) as a perceptual act that results in a creative leap or as a link between a design challenge and its solution. This definition helped to clarify the relationship between the two phases. Furthermore, Pressman (2011) proposed a design cycle with three main "I" phases: inspiration, ideation, and implementation, linking concept generation with design processing. Ideation is primarily concerned with concept derivation as a bridge between inspiration and implementation among these stages.

2.2 Adaptive Reuse Process

Architects, designers, buildings conservators, and researchers drew attention to significant gaps in theories around the adaptive reuse of old structures. Three modern approaches to adaptive reuse were identified by their study: the technical approach, the typological approach, and the strategic approach. They observe that when describing the adaptive reuse process, many authors give little consideration to the genius loci of the target building. On the other hand, research categorized as broad studies with a typological or technical approach deals with the genius loci less in-depth than studies with a strategic approach (Plevoets & Cleempoel, 2011).

According to Misirlisoy and Günçe (2016), castles are employed as museums at the level of functional use because of their physical attributes. Because castles are typically constructed for protection, their facades are solid with few openings. Most plan layouts feature an introverted arrangement with a courtyard in the center. Even though the museum's role is physically acceptable for castles, preservation-related factors should be taken into account as well.

Barakat and Goriel (2019) examined the importance of the adaptive reuse process using Qubahan School as a historical case study. This process requires precise and effective labor. The study found that a variety of impact elements, each of which has a different impact on the decision to choose a new use, are involved. The strongest and most significant of these are the site's functional, environmental, cultural, and architectural characteristics. In this approach, the study suggested various appropriate uses that were both in keeping with the original plan and compatible with its origin (passive: museum/gallery; active: café/restaurant).

In addition to other significant factors, Eyüce and Eyüce (2010) noted that one of the main influences on adaptive reuse is spatial configuration. Adaptive reuse is concerned with spatial properties and their relationships in the configurational system, as well as the degree to which a new use is compatible with the need for modifying the original spatial configuration.

3 Problem Statement

While historical structures need to be preserved for their socio-cultural relics of the past, these structure may become unoccupied and lose their purpose during a duration of their lives. Altering a building's purpose in accordance with the needs of its neighborhood and the current context can be a useful way to bring these buildings to life. One of Jifna's (a city in Jerusalem) most significant historical structures is Jifna Castle. It has a significant geographical placement in the city center in addition to its historical and cultural significance.

4 Research Aim and Objectives

The goal of this study is to develop a concept generation framework that emphasizes the sustainable conservation of existing buildings, especially historic structures, in order to enhance the possibility of built-in adaptive reuse. Specific factors were chosen to accomplish this goal, and their application to Jifna Castle was used as a case study to guide the approach to the reuse concept for this building.

Also, this research aims to identify some underlying factors and specific criteria for the creation of historic building adaptive reuse concepts. Additionally, as part of the repair process, to investigate the viability of reusing this historic castle, moreover, explore and examine the variables that contribute to the idea of adaptive reuse of historical monuments, and in the end, provide innovative solutions and be prepared to integrate into and serve the local communities.

5 Methodology

The factors that influence the concept development of reusing Jifna Castle are the main subject of this paper. This study used an analytical survey to examine Jifna Castle's historical, cultural, site, environmental, and architectural features in order to achieve its goal. The survey is based on four major influences as shown in Fig. 1: historical, geographical, spiritual, and incidental. This study provides insight into the significance of the adaptive reuse process as a component of environmental development, social benefits for society, and economic benefits for individuals. In essence, it is a living example of the continuous transmission of traditions and culture to any related location.

6 Case Study Description

The case study performed in this study is the author's thesis project, which is a preservation and adaptive reuse project of an abandoned and underutilized ancient castle (El Burj) located in the Palestinian village of Jifna.

Jifna, as shown in Fig. 2b, is a Palestinian village in the governorate of Ramallah, located 8 km north of Ramallah and 23 km north of Jerusalem (650 m above sea level) as shown in Fig. 2a. The village is one of the oldest in Palestine, with a history dating back to the Bronze Age, and it has gained importance during different eras. The castle belongs to the Crusaders and has historical importance, dating back to the different ages of the Crusaders, Mamluks, Ottomans, etc. It was preserved in 1999; however, the preservation process failed for several reasons.

This project aims to preserve (El Burj) and reuse it, employ it to serve the people of the village, and develop tourism in the village. The proposed function is an exhibit that will protect the history of the building. Respect its worth and use a hotel extension building to support the main idea of celebrating the old history.

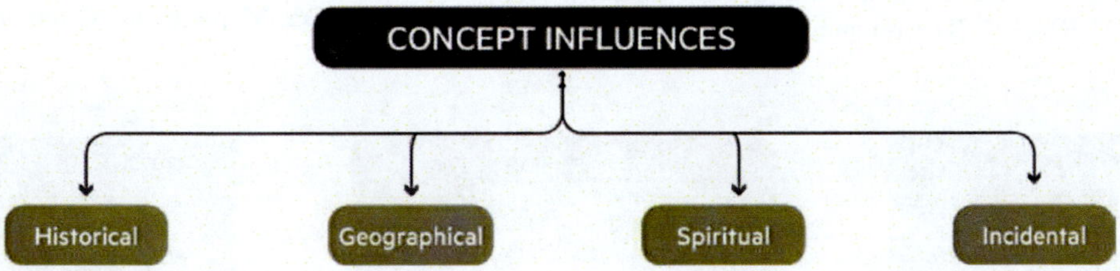

Fig. 1 Concept generation framework (Author, 2022)

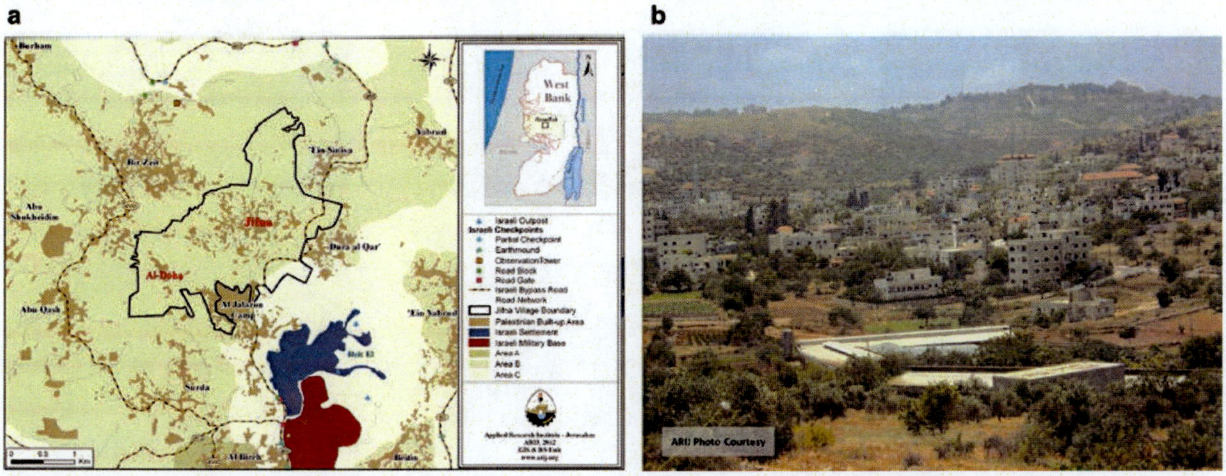

Fig. 2 **a** Jifna Village map (location and borders); **b** aerial view of Jifna (The Applied Research Institute, 2012)

7 Influences

7.1 Historical Influence

A project begins with understanding the story of the building through its history, context, culture, and neighborhood, where it can be used in countless ways to create spaces that are meaningful reflections of their time and place but free from the confines of a dictated historical style.

Jifna history is summarized in Fig. 3. Jifna village was mentioned in the Old Testament of the Holy Bible as "Gophna," which dates back to the time of the Arab Canaanites, but then the village was renamed as Jifna, which means "the vines" in Arabic, because of the many vineyards in the territory of Jifna. Jifna had great importance in different eras; it was the county seat and administrative seat of the regions in the period of Herod, as well as a military base. The Romans chose it to build castles, towers, and Khans for its location on the Roman road, which was built to facilitate the transportation of armies and the establishment of security.

Jifna was mentioned in the Byzantine mosaic map of Madaba, and the importance of this period is shown by the remains and churches that are still present in the village: the church of the Latin sect on the northern side of the village and the remains of the SS Giorg, a Byzantine church (the Church of Al-Khader). The remains of a mosaic-tiled church were discovered on the southern side, and traces of a grape press for making wine were discovered on the eastern side. Other churches were built in the Ottoman era that still exist in the village: the Latin church on the northern side of the village, opposite to the ruins of the old Byzantine church, and the Orthodox church in the southern region, outside the central old village. Jifna has been destroyed seven times, and that was indicated by the Jifna spring that was originally three steps up to the spring of water, but now there are 27 steps down to reach it.

El Burj castle, shown in Fig. 5a–c, is the selected building to be conserved that is located in the historic center of Jifna. The historic center is located on the northern side of Wadi Ashkol (vine valley), which divides Jifna into two parts. It includes almost all of the town's historic buildings, which form the historic traditional fabric of the town: El Burj Castle, which dates back to the fifth century, Virgin Mary Spring, ancient Ahwash (old buildings) shown in Fig. 4; and the old Byzantine Church. Also, there are the remains of two ancient churches, tombs, and an ancient oil press (Riwaq, 2012).

The castle was chosen to be the location of the project for the following reasons:

- Its historical significance and age, which date back to the succession of different eras shown on it in Fig. 6, range from the Crusader, Mamluk, and Ottoman periods.

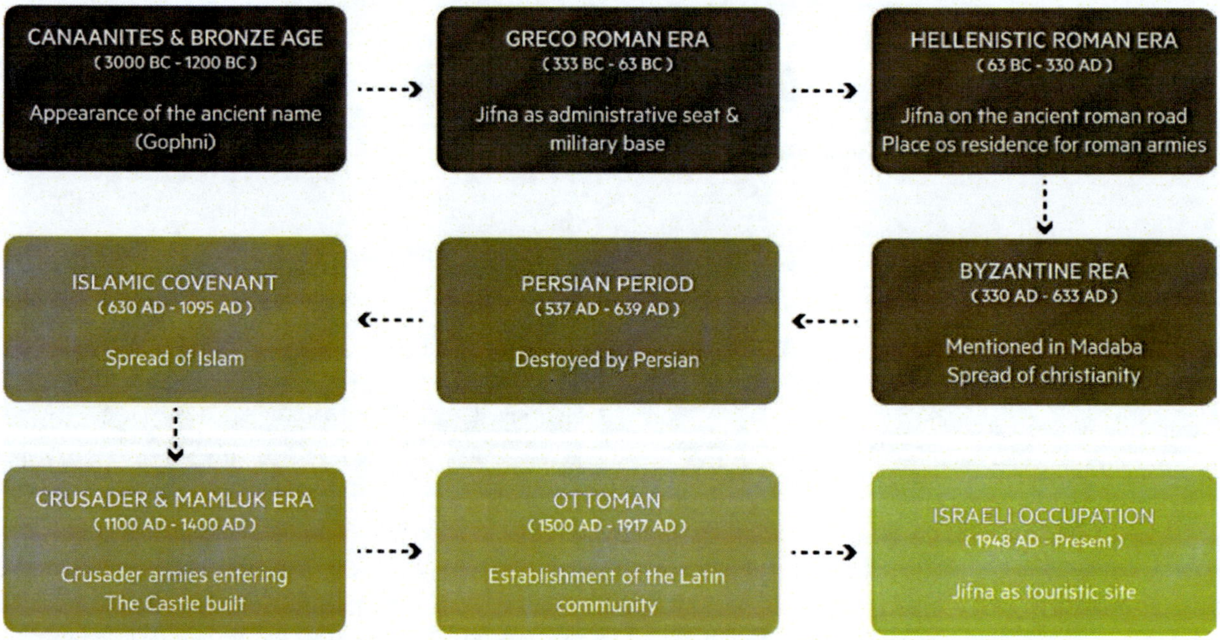

Fig. 3 Jifna's history is depicted graphically (Author, 2022)

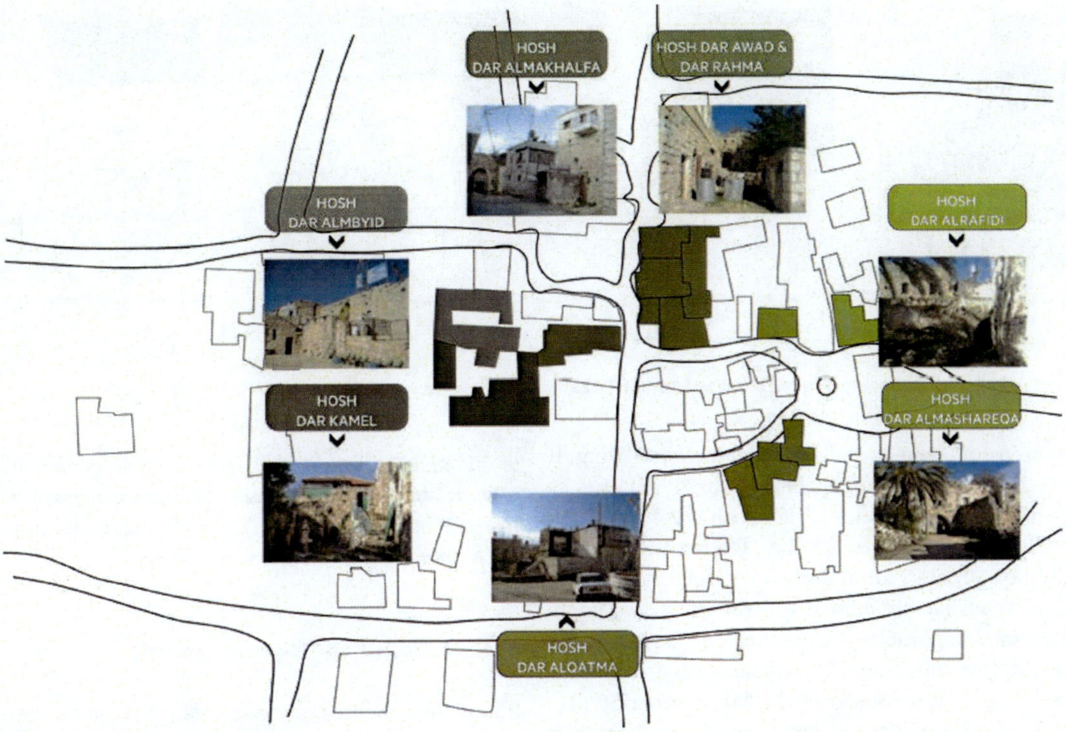

Fig. 4 Hoshes of the old town (Author, 2022)

Fig. 5 Views of the castle, **a** second floor, **b** main entrance, **c** aerial view (Hamdan, 2000)

- The nature of the spaces that make up this building consists of two hoshes,[1] which helped to merge them together and exploit them as a single building, consisting of three floors, which is thus different from the rest.
- It is a convenient location near the Al-Khdir Church and the Byzantine Church, and the state of the surroundings allows for horizontal expansion.

Historical and archeological analyses carried out between 1970 and 1982 indicated that the castle served as a farm and as a hostel for pilgrims traveling along the road running from Acre to Jerusalem via Nablus. In the original site plan, the rooms of the building were constructed off the 2.05-m-thick exterior walls, creating a fortified structure that has a large internal courtyard and an arcade in the east wing, where the main entrance that was built in the Crusader era is located (Zahran, 2004).

Fig. 6 Highlighting the different eras of the building (Author, 2022)

On the basis of a similar hall found in the north wing, it is inferred that the south wing included a large hall used as a stable and pilgrims' dormitory. The north wing contained a large hall, which served as a stable. During the clearing and restoration process, several channels for draining the stable were found. The structure was divided into two floors during the Mamluk pilasters. Stones, capitals, and columns from the old Byzantine church were used to modify the Crusader site for its new use during the Mamluk period.

At the beginning of the Ottoman period, the large hall was transformed. The second floor had collapsed, and an olive oil press was installed on top of the debris. A second press was added later during the Ottoman period, during which the large hall in the south wing was also transformed into an installation for olive oil production. These olive presses indicate that the site was industrial in nature, serving not only the village of Jifna but also the entire territory (Halayqa, 2014). The site continued in use, mostly for domestic purposes, until 50 years ago, when it was abandoned. The alterations made to transform the structure to serve domestic functions that are still evident.

The castle restoration and preservation project began in April 1998 and lasted until 2001, with the goal of revitalizing an important archeological site in Palestinian history. The conservation work on the site concentrated on part of the Crusader manor house complex, which included remains of previous Byzantine structures as well as Mamluk and early and late Ottoman construction. Two olive oil presses were preserved and maintained within the restored buildings and the late Ottoman tower to the left of the main entrance (Hamdan, 2000).

7.2 Geographical Influence

Some are inspired by the unique cultural heritage of the region to tell the story of a place through its built environment.

The village of Jifna is located on the slope of a hill that overlooks a green valley illustrated in Fig. 7b. It is located to the west of the main road from Jerusalem to Nablus and Nazareth, which was the ancient Roman Road as well as the route of pilgrims during the Crusader period from Acre–Nazareth–Nablus–Jifna–Jerusalem. Jifna is a Palestinian village in the governorate of Ramallah, located 8 km north of Ramallah and 23 km north of Jerusalem (650 m above sea level) shown in Fig. 7a.

The village is accessible by the main road, which runs from the west of the village of Jifna and makes its way east to its border with Ein Senya. To get to the historic center of Jifna, take the Abu Jassar roundabout, which is located near the castle (El Burj).

Residents of Jifna village can reach the tower via one of the routes shown in Fig. 8a, depending on their location.

Fig. 7 **a** Site plan contour lines; **b** site sections (top) north–south site section; (bottom) west–east site section (Author, 2022)

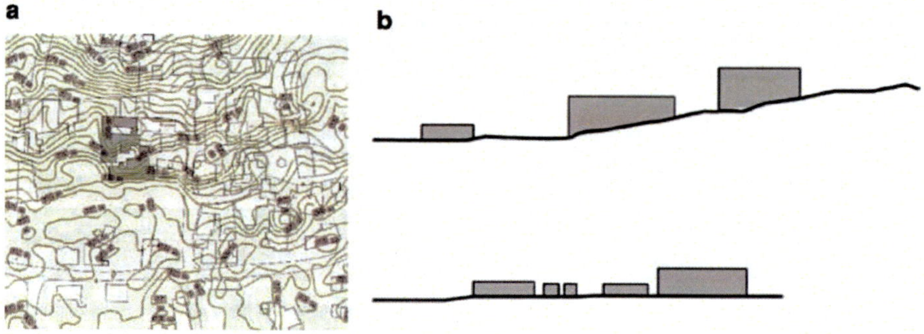

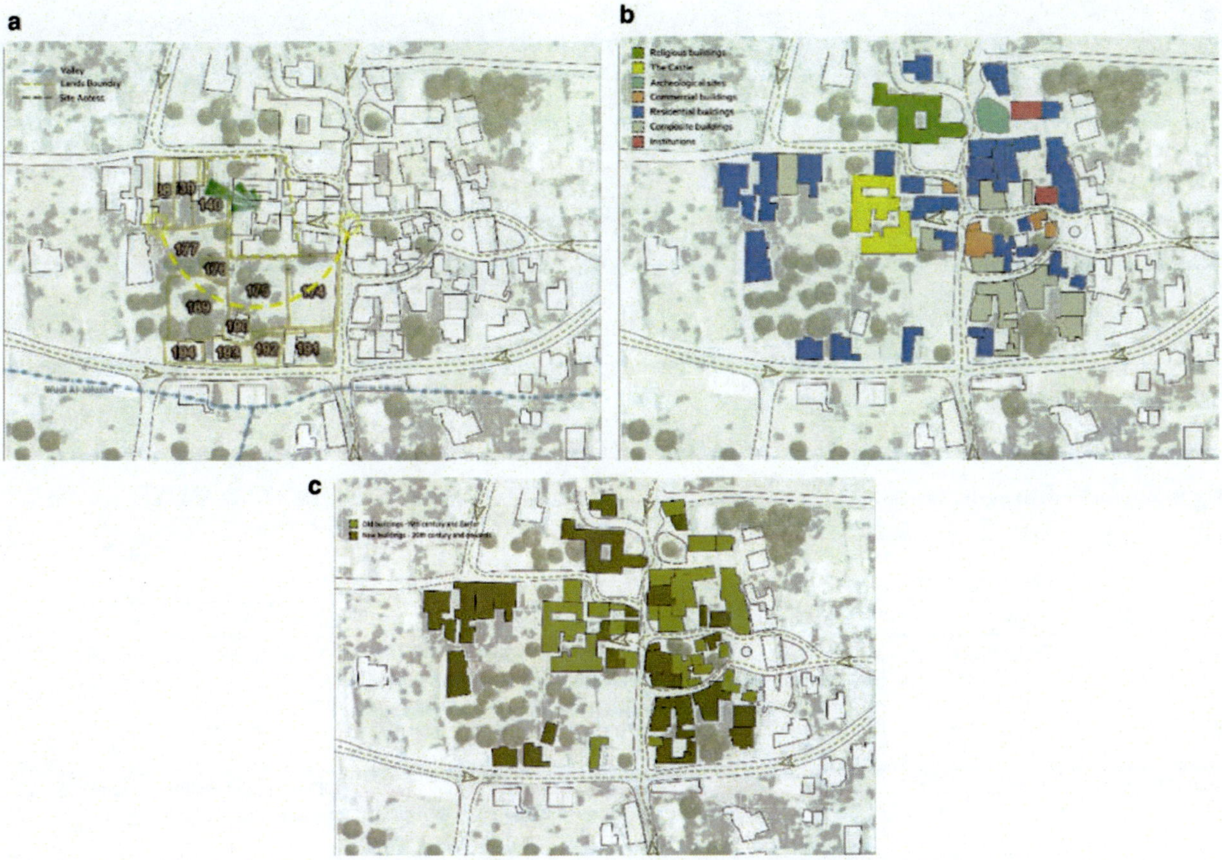

Fig. 8 Urban studies. **a** Sun and wind analysis, **b** surrounding landuses, and **c** surrounding buildings age (Author, 2022)

A result of the context analysis is shown in Fig. 8b and c, and it should be noted here that the castle in the medium where it was found helped to distinguish it, due to the presence of a three-story tower, which is rarely seen in the old hoshes. The castle is located on the edge of the old town, which helps to create a strong visual connection with the building from the main road on the southern side due to the availability of open spaces on the west and south sides, while on the north and east sides it sticks with the old town hoshes.

7.3 Spiritual Influence

Art can be used to enhance the architectural concept and give it a spiritual aspect that will provide a deep meaning that will touch the human soul.

For the El Burj reuse project, as the project is in Palestine and the site is near a settlement, the theme of resistance was taken for the idea of protection and resisting. The art pieces, Temporary Escape in Fig. 9a, Olive Groves #1 in Fig. 9b, and Rakan Artworks in Fig. 9c, which represent resistance were analyzed in Table 1. Table 1 to adapt symbols or modes of resistance and understand how to present a story through art.

7.4 Incidental Influence

Another "art of resistance" painting in the Sharjah Art Museum's permanent collection is the Sabra and Shatila painting shown in Fig. 10a and b which was analyzed in Table 2, which depicts the 1982 war in Lebanon that killed thousands of Lebanese and Palestinian civilians and culminated in the Sabra and Shatila massacre, in which Israel colluded in the killing of between 800 and 3000 civilians (Sinwar, 1984).

8 Framework Effect and Concept Application

Based on suggested framework, the concept generated evolved a multifaceted idea. The framework has informed numerous decisions about the architectural design approach

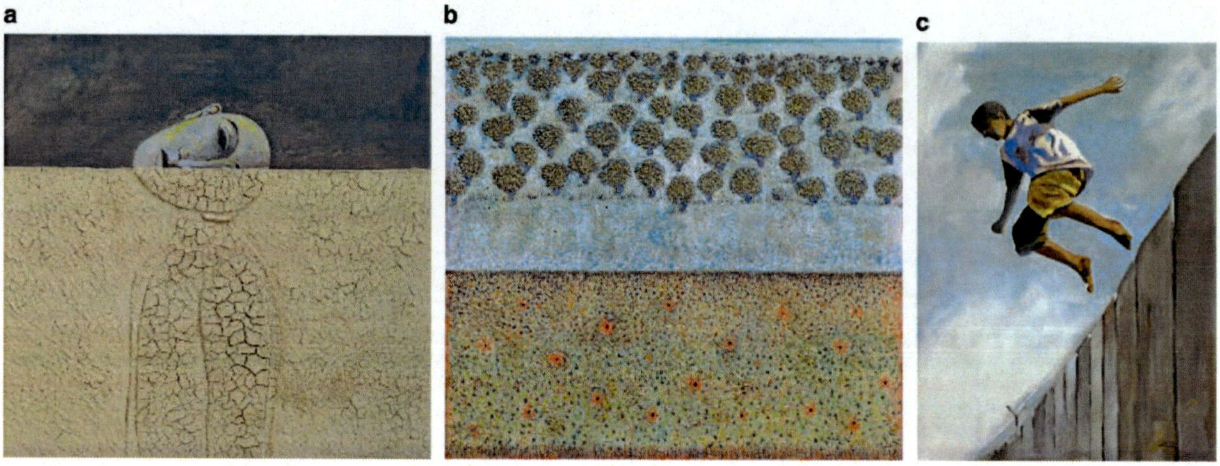

Fig. 9 a Temporary Escape, b Olive Groves #1, and c Rakan Artworks (Zawyeh Gallery, 2020)

Table 1 Artwork inspiration (Zawyeh Gallery, 2020)

Artist	Painting title	Description	Adaptation
Sliman Mansour	Temporary Escape	This artwork expresses the predicament in the current Palestinian reality and the changes that permeate the nature of the Palestinian people by using the tension resulting from the prevailing rough mud cracks that contrast with the areas painted with acrylic on wood	The use of local natural materials
Nabil Anani	Olive Groves #1	By capturing an empty, semi-arid landscape where plants grow sparingly and olive trees spread out over the horizon, this artwork expresses rooting in the ground despite adversity	Olive trees symbolize how to root in the land despite adversity
Khaled Hourani	Rakan	This artwork focuses on the young generation who take upon themselves the challenge of the occupation, as it shows a child jumping high in the air over the apartheid wall without giving it any importance	Proving ownership of the building to the Palestinian people and preserving it for the future

Fig. 10 a, b Bashir Sinwar's Sabra and Shatila, 1984 (Sinwar, 1984)

and elements. It had a great effect on both the design process and outcomes. The generated design concept is more holistic, while understanding and considering all possible geographical, historical, spiritual and incidental aspects affecting the design. These aspects had great impact on the formulation of the memory of the building, the story of the structure, and the culture of the residents as shown in Fig. 11.

The current status of the tower is similar to its pre-restoration status, as many of its sections are currently closed and its outer squares have become dumps, the only difference being the shape of the walls that have been restored and the part that is still in operation, the Palestinian village club, which occupies two rooms of the castle. The situation the village has reached may be due to a number of reasons, including the political situation and the deteriorating

Table 2 Sabra and Shatila painting analysis (Author, 2022)

Analysis	
Content	The artwork is focused on a woman carrying her child in panic and bereavement; bodies of the martyrs are piled up after the massacre, and in the background, the place looks devastated
Colors	Black, white, primary cool colors (blue, green, and brown), and primary warm colors (red). Shade at the top, tint at the bottom. The use of color in its different degrees
Function	Expressing the Palestinian cause, specifically the killings that were carried out based on identity in the Sabra and Shatila massacres
Feeling	It left a tragic black memory and a pain that the passing of days could not erase in our souls
Characteristics	Large mural, oil paintings, and dramatic realism express people's concerns and suffering. The style and the drawing are realistic, with high quality and accuracy. Person and location are central axes in the artwork. Its realism is lyrical, not photographic. Experiences the personality or event to be highlighted, so it shows it inside and not only in external form. Dramatic dimension with an emotional sense
Meanings	The steadfastness of the mother to preserve her son allows him to complete what his ancestors did to resist The use of the colors white, which indicates purity, and blue, which indicates hope Overall express the hope for peace and a better environment for future generations
Adaptation	The art work's composition and how to tell a story that will leave an impression on people Adapting the Kufiya patterns

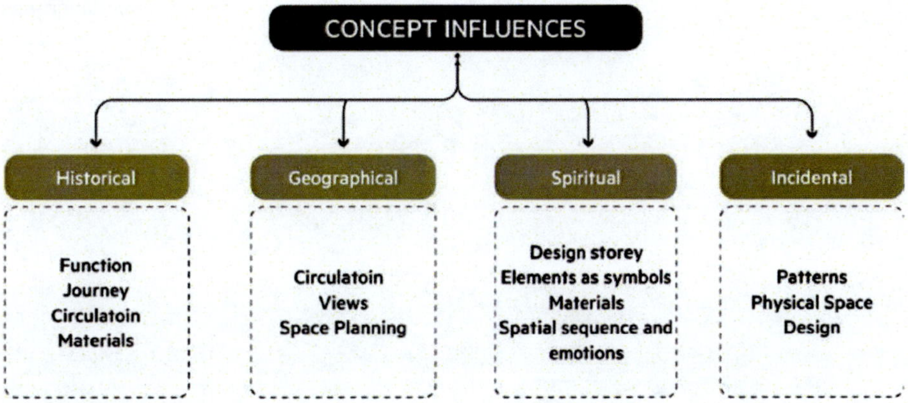

Fig. 11 Architectural design features that were affected by the chosen approach (Author, 2022)

economic situation. Add to that the fact that there are some disagreements between the owners, where each hosh is for an owner.

After all of the factors were considered, the concept began to emerge. The idea is to develop a journey through spaces that creates a narrative story from the history of Jifna and the castle, started with creating a collage of the city in Fig. 12. The main aim of the project was to recreate it with the same historical significance and to serve as an example of Palestine's history and a symbol of resistance.

Fig. 12 Creating a collage helps understand the village's cultural heritage through the built environment (Author, 2022)

Fig. 13 Castle's valuable elements (Author, 2022)

The design creates different experiences through different spaces that will bring back memories of how people lived through the years.

The proposed function for the building is an exhibition that will show the building's history while focusing on the main valuable elements in the castle shown in Fig. 13. Concurrently, an envisioned hotel extension is crafted to harmonize with the main concept, boost the village's economy, and provide tourists with a convenient stay for exploring the historical landmarks.

The site plan in Fig. 14 shows the conceptual design of the project, where all the inspired ideas from the previously discussed aspects are shown and explained in detail.

A. Memorial Path Through the Site to Reach the Castle

A path shown in Fig. 15 that represents eras that Jifna went through using different types of arches was proposed to give the experience and celebrate Jifna's history while

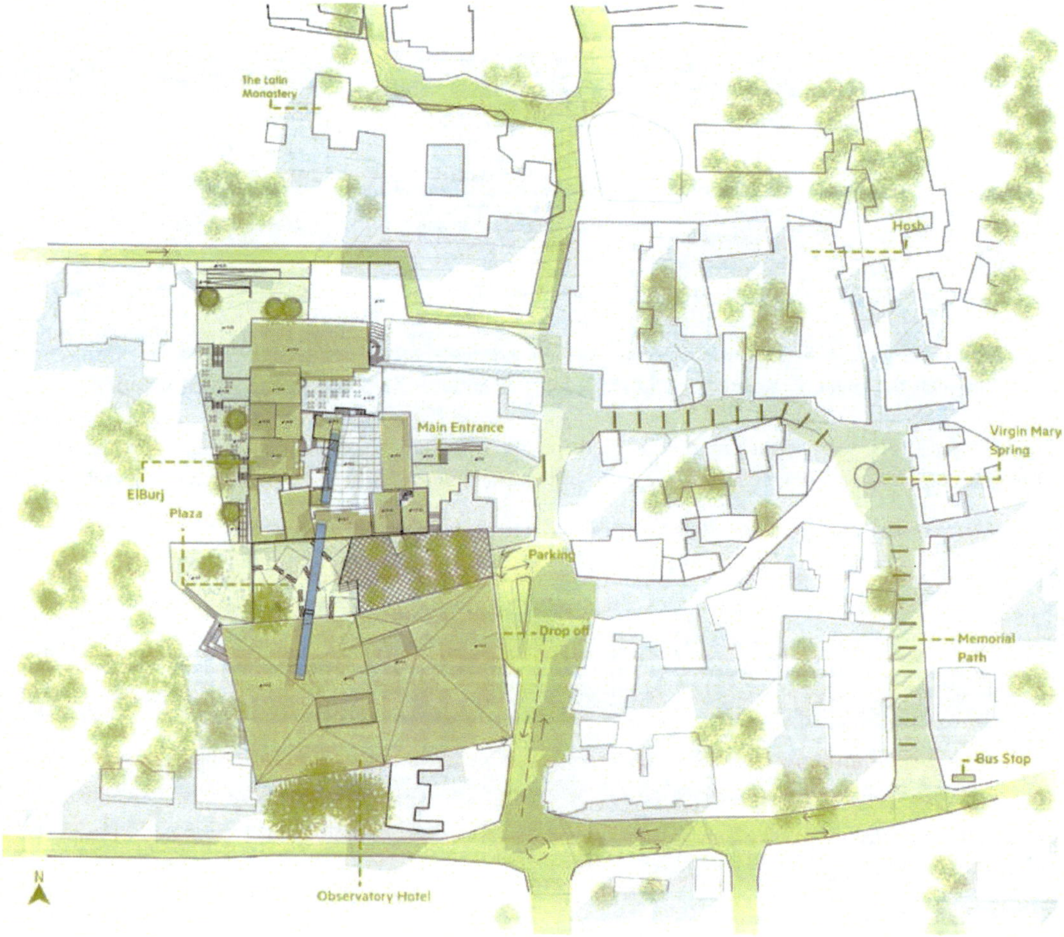

Fig. 14 Proposed site (Author, 2022)

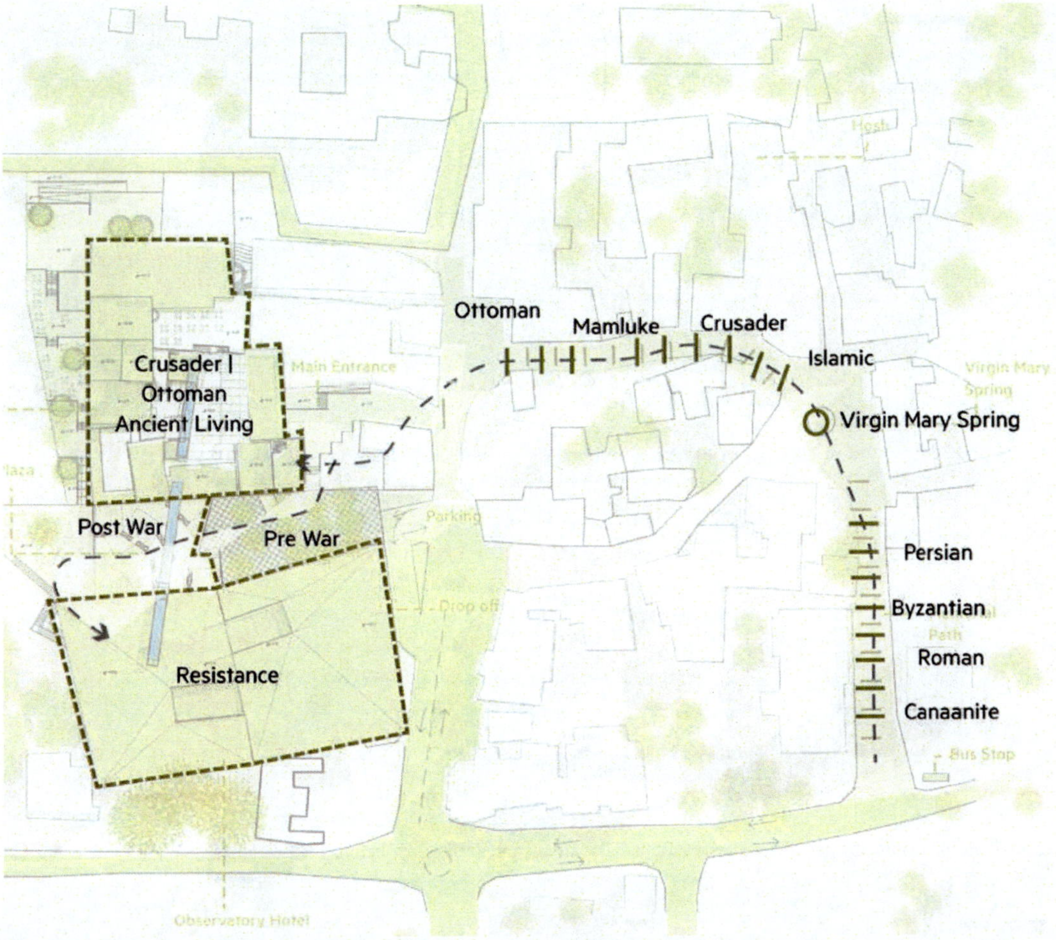

Fig. 15 Journey through history using arches that represent each era (Author, 2022)

going to the castle, which is an ancient monument in the historic center that also shows different layers of history.

B. **Castle, Plaza, and Hotel Extension Journey**

In the old local Palestinian architecture, the use of "hosh" was always present. Keeping this in mind, I designed the castle's extension as a single building separated by a plaza that will serve as both a hosh and a link between the two. The castle represents the ancient living, while the plaza will act as the living pre- and post-occupation, and the hotel will be a representation of the current status and the people's resistance.

C. **Hotel Extension Journey**

The main idea is to connect visitors with all the landmarks that are in Jifna that represent different ancient eras, as sketched in Fig. 16a. For this reason, the building was designed as a journey on-ramp to allow viewing the landmarks from different levels because of the topography of the village, as well as using lines that are oriented to these landmarks to frame the views in each direction to a landmark as illustrated in Fig. 16c. However, the main focus is to create a connection between the extension and the ancient building (El Burj) as shown in Fig. 16b, so that from all levels and sides of the hotel, (El Burj) will be visible to visitors. Also, both buildings are connected through a bridge that passes through the Ottoman wall.

D. **Plaza**

Figure 17 depicts a plaza design that uses the Kufiya pattern regularly for the prewar living zone and irregularly for the postwar zone to demonstrate how life can continue using rubble as a counter monument to design the landscape of the zone.

The plaza will be the design's focal point, connecting both the castle and the hotel, as well as the terraced garden on the west side of the castle, which will be designed to take advantage of the topography.

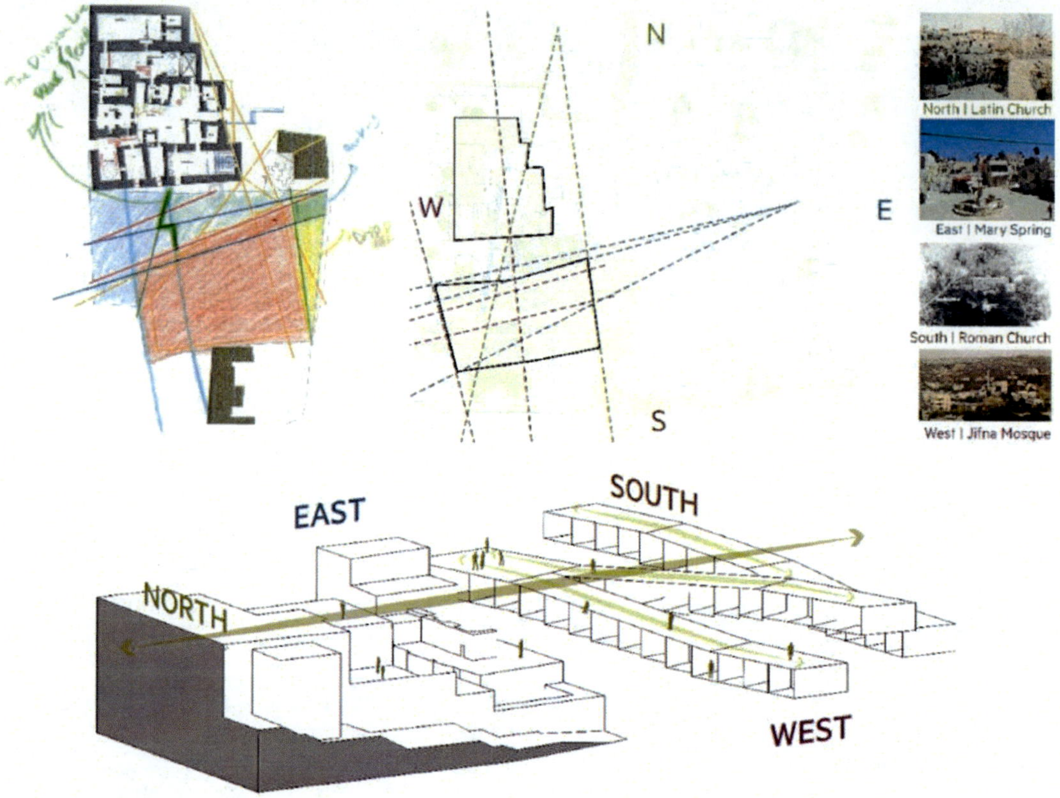

Fig. 16 Design lines based on the landmarks generates the layout of the design. **a** Sketch, **b** 2D implementation, and **c** 3D visualization (Author, 2022)

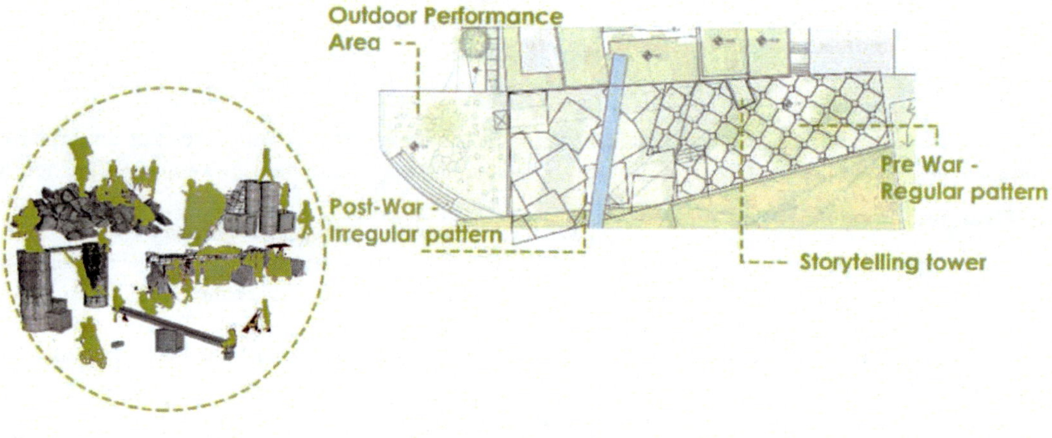

Fig. 17 Proposed plaza conceptual design (Author, 2022)

9 Conclusion and Future Research

In conclusion, the research offers an approach to upcycling an existing historical building through various sources of concept inspiration. The proposed framework can be used to approach the process of reusing valuable buildings, where generated the concept must be a collection of interconnected ideas derived from historical, geographical, spiritual, and incidental settings to be accepted by historians, its residents, and the public.

The proposed framework can guide architects, designers, urban planners, and decision-makers to redesign these buildings without deteriorating their historical and cultural values. The final concept of the case study takes a multidirectional approach where it can bring attention to the village, enhance its touristic value and its memory, and activate its surroundings. El Burj Castle gives Jifna its distinct character and, at the same time, provides a tangible connection to the past. It will show itself as a unique piece that will be given a second life to create a dynamic place and define the character, spirit, and sense of belonging of a space. Future phases of the research will investigate the effect of new Artificial intelligence technologies on the imagination of historical buildings during the early phased of the design process where the concept is being both explored and tested while being generated using full immersive experience made possible with virtual reality and augmented reality technologies and devices. Such application will allow for a fully tested experience during and even before construction.

References

Barakat, A. I., & Goriel, W. A. (2019). Factors affecting the adaptive reuse process for historical buildings in Kurdistan Region. *Sulaimani Journal for Engineering Sciences, 6*(3).

Cross, N. (1997). Descriptive models of creative design: Application to an example. *Design Studies, 18*(4), 427–440. Retrieved from https://doi.org/10.1016/S0142-694X(97)00010-0

Eyüce, Ö., & Eyüce, A. (2010). Design education for adaptive reuse. *International Journal of Architectural Research, 4*(2–3), 419–428. Retrieved from https://www.academia.edu/2317467/Design_Education_for_Adaptive_Reuse

Halayqa, I. (2014). An Olive Press from Jifna. In *A pioneer of Arabia* (Vol. 10, pp. 82–99). La Sapienza.

Hamdan, O. (2000). *Conservation of architectural heritage in Palestine—The Crusader manor house restoration and conservation project in Jifna and the establishment of the Museum of memory in Ein Sinia* (Vol. 3). Gloss Verlage.

Misirlisoy, D., & Günçe, K. (2016). Assessment of the adaptive reuse of castles as museums: Case of Cyprus. *International Journal of Sustainable Development and Planning, 11*(2), 147–159. https://doi.org/10.2495/SDP-V11-N2-147-159

Plevoets, B., & Cleempoel, K. V. (2011). Adaptive reuse as a strategy towards conservation of cultural heritage: A literature review. *WIT Transactions on the Built Environment, 118*, 155–164.

Pressman, A. (2011). *Designing architecture: The elements of process* (1st ed.). Routledge. Retrieved from https://doi.org/10.4324/9780203122174

Riwaq. (2012). *Re-walk heritage: Ramallah highlands trail.* Riwaq. Retrieved from https://www.riwaq.org

Sinwar, B. (1984). *Sabra & Shatila.* Sharjah Art Museum. Retrieved from https://sultanalqassemi.com/articles/sharjah-art-museum-bellwether-of-gulf-art-museums/

Taura, T., Yamamoto, E., Fasiha, M. Y., Goka, M., Mukai, F., Nagai, Y., & Nakashima, H. (2012). Constructive simulation of creative concept generation process in design: A research method for difficult-to-observe design-thinking processes. *Journal of Engineering Design, 23*(4), 297–321. https://doi.org/10.1080/09544828.2011.637191

Taura, T., & Nagai, Y. (2013). Concept generation for design creativity: A systematized theory and methodology. https://doi.org/10.1007/978-1-4471-4081-8

The Applied Research Institute. (2012). *Jifna Village Profile.* Retrieved from http://vprofile.arij.org

Yalcin, Z. O. (2021). Intuition in the design studio: A perspective on student's creativity and design process. *Design Studio, 3*, 37–48.

Zahran, O. F. (2004). جفنا برج في التاريخ. Ramallah, Palestine.

Zawyeh Gallery. (2020). *Palestinian art: Resilience and inspiration*. Retrieved from Zawyeh Gallery. https://zawyeh.net/palestinian-art-resilience-and-inspiration/

"Community of Interest" as a Method of Transformation Toward Integrated Green Community in Egypt

Ghada Assal, Yasser Mansour, and Shaimaa Kamel

Abstract

Globally, economic growth tends to be concentrated in cities. Due to the UN, by the year 2050, around two-thirds of people will live in urban and peri-urban regions. Increasing national wealth and providing the community with hope for a better standard of living are pushed toward massive migration from rural to urban areas. Migration impacts urban environments, infrastructure, and quality of life. Additionally, climate change will also have a negative impact on the environment, which might threaten urban areas. All these circumstances lead to the search for solutions. The focus of the search lies in exploring approaches to reduce pressure on cities while concurrently conserving and enhancing current natural resources. This article is claimed to study successful green communities with common interest examples in rural Egypt. Those communities' interests could help to improve the quality of life for their residents and environmental improvement and may be the reason for reverse migration in the future. The research objectives are to highlight the main features of successful communities and state the influence factors that led communities of interest to achieve positive results. The research uses diagnostic and descriptive methods for studying the communities to set the features by comprehensive analysis to determine the common factors and detect the criteria. Finally, the outcome indicates that the extent of apparent features can be divided into three main elements (economy, environment, and human). Additionally, the suggested factors could be repeated in different spaces, and accordingly, the spaces could be transformed to be developed based on these results.

Keywords

Community of interest · Green community · Transformation · Recycled space

G. Assal (✉) · Y. Mansour · S. Kamel
Architecture Engineering, Ain Shams University, Cairo, Egypt
e-mail: Ghada.Mohammed@hti.edu.eg; ghada.assal@gmail.com

Y. Mansour
e-mail: yasser_mansour@eng.asu.edu.eg

S. Kamel
e-mail: shaimaa_kamel@eng.asu.edu.eg

G. Assal
Higher Technology Institute, 10th of Ramadan City, Egypt

1 Introduction

Oscar Newman released his work "Community of Interest" at the beginning of the 1980s. In his book, he argued about American neighbourhoods and communities' difficulties in the 1960s and 1970s. He observed that the changes in their communities were split up into countless tiny, racially and economically homogeneous suburban communities. Oscar noticed the evident shift in the American community that followed the Industrial Revolution, which resulted in racial and economic segregation that grew over time. What happened was the opposite of the American dream of having communities composed of extended families and friendship networks, age groups, and lifestyles within a few blocks of each other and containing a variety of shared and commercial institutions and stable societies which are harmonious and integrated. The extended families have a unified economic activity. Where they lived in stability and homogeneity, and each age group has an integrative role that gives affiliation to the family and place, providing each other with mutual support. It preserves the surrounding environment and helps in a healthy community shared among them. Oscar observed that it is hard to see extended families in the cities. The idea became just a leftover in the countryside, with the migration of young people to cities for a

M. F. Shahidan et al. (eds.), *Urban Narratives: Exploring Identity, Heritage, and Sustainable Development in Cities*,
Advances in Science, Technology & Innovation, https://doi.org/10.1007/978-3-031-48517-6_22

better lifestyle. Extended families are threatened with being declined (Newman, 1980).

Oscar studied by mixing different types of people in the same neighbourhood, where some groups were either ignorant or unemployed. The experience of mingling with other economic classes and ethnic and racial groupings in the same neighbourhood resulted in poverty among those unfortunate groups. The rate of crime increased as a result. The most prosperous members of this society opt to leave, while the remaining residents struggle with the issue of unstable neighbourhoods and increased crime. They shifted away from the area they believed to be the source of their problems. Therefore, segregation and the tendency to isolate specific groups in the same place because of their differences negatively affected people and space. Some neighbourhoods, to solve those problems, preferred to be pure wealthy, white, black, poor, or middle-class. Each has moved tens of kilometres away from their former residences. People of the unlucky groups that were ignorant or unemployed increased their sense of racism against them. And as a reaction, their attitude pushed others to avoid the spaces they exist in, lifting the space to be neglected instead of participating in various activities inside those spaces, become for transit only—no security control or police, which encourages crime to increase. Additionally, the difficulty of living increases the desire to move from this neighbourhood due to isolation, crime, or the lack of harmony and safety, producing an unsuccessful urban space unattractive to new residents. Finally, all these lead to a lack of belonging from residents toward the place, directly affecting its development. Oscar suggested some strategies that help to form integrated communities based on the same interest (Newman, 1980).

Currently, in Egypt, the reality is very similar to what was discussed previously, even though American society is more diverse, unlike Egyptian society, which has no apparent racial differences. On the other hand, the gap between economic and social communities in Egypt rapidly increased, leading to the same situation that happened in America 40 years ago. Different socio-economic levels in a serious attempt to separate, reaching the point of isolation between each social class from the other. The Hilarious expansion of The Gated Communities (Compounds) is evidence of the existing desires between residents for segregation and isolation. Moreover, in the television advertisements, it is announced very clearly that: "Come and get your unit; all the owners are similar," It is easy to notice in the last few years, especially in the summer, on the northern coast, which, is a natural and environmentally extended coast, and there are no fundamental natural differences in it, and people from the different socio-economic class divided the beach into two halves represent each half of a social group, each trying to stay away and isolate from the other. Several posts on social media platforms describe this situation by joking. They gave a name to each part, "Kind-Coast and Evil-Coast," which shows the inner desire of each socio-economic group to separate and isolate. Referring to what we reached, each group has not attempted to share the same space, even if it was a limited period during summer vacation. In order to prevent the desire to isolate and achieve a successful community due to Oscar's theory that the community must be based on the same interest and to be integrated, this research focuses on examining successful green communities with common interest examples in rural Egypt. Those communities' interests could help to improve the quality of life for their residents and environmental improvement and may be the reason for reverse migration in the future.

2 Research Background

2.1 Community of Interest

Oscar Newman suggests solving the conflict of integration inside the same neighbourhood between different economic classes and ethnic and racial groups, that leads to isolated spaces and people. In his opinion, if there is a common interest between those groups, this would push them toward sharing space and activities. He called this a Community of Interest, grouping people with similar environmental needs: "It is this form of segregation that is the key to the integration & interaction of neighbours of different racial, ethnic, and economic backgrounds" (Newman, 1980). Melvin Webber, in 1963, was the first to present the term "Community of Interest" in his article "Order in Diversity: Community Without Propinquity". He described interest communities as sharing bowling clubs in a particular town, a concert society, or even a group of skilled workers making up a union local in an assembly plant. Webber reflects on, describes, and refers to the important value of interest communities. However, Newman showed an image of bringing together separate communities to develop urban environments, stabilise threatened neighbourhoods, and rehabilitate older cities. Also, the second societal goal is the achievement of racial and economic integration within communities, to plan integrated communities so that the social and physical environments created enable their members to transcend the economic and racial differences between them (Newman, 1980).

Newman tried to put a framework that fulfils the needs through many scales: housing, urban, architecture, and socio-economic. Newman's concept for creating physical communities is based on defensible space and shared lifestyle needs.

- **Beginning with the housing scale, he suggests that** (Newman, 1980):

1. Neighbourhood should preferred to be arranged by clustering similar age and lifestyle groups in small communities of fifty to five hundred dwellings.
2. Denser, smaller in size and scale, and more closely put together with others than suburbs facilitates interaction between different ethnic.
3. He discusses three lifestyles: families with children, elderly, and working adults (single or couple key: no children), and he suggested a design for each type, but it is out of this research scope.

- **On an urban scale, he introduced the defensible space theory illustrated in his book Defensible Space, 1972. The concept of this theory is a way of housing design that gives the residents a sense of ownership, responsibility, and control outside their living spaces.**

Five factors make a defensible space based on Newman's theory (Marshall, 2016; Newman, 1972):

1. Territoriality: the private units are respected.
2. Natural surveillance: physical characteristics allow the residents to motor outside their homes.
3. Image: the physical design capacity gives a feature of security.
4. Milieu: other security tools like proximity to a police substation, commercial area, and bus stations.
5. Safe Next-door Areas—for better security, residents obtain a higher ability to monitor adjoining areas by designing the attached area surrounding the neighbourhood.

- **On an architecture scale, he suggested three functions are required of the new form language of modem architecture** (Newman, 1980):

1. Open plans are preferred to allow the planning and layout of areas to be free, with unity to give a mental image to the community.
2. Use of contemporary materials that were capable of spanning large distances, opening walls to allow light and air, etc., guaranteed a better quality of life.
3. Integrate the resident's opinions in design to express their needs and desires for materials and colours to identify their backgrounds linking the present building to those of the past.

- **On the socio-economic scale** (Newman, 1980):

1. Grouping similar age and lifestyle groups in housing environments is designed to fit their lifestyle needs carefully.
2. A mix of income groups will allow the values and lifestyles of the upper-income group to dominate.
3. The percentage of low income, interests, and the capacities of all the parties are respected and are codified into strict housing and management policies (Quota).
4. Break out of the cycle of poverty by being well educated and providing social opportunity through mutual benefits between all communities.
5. Economic groups invest in public facilities that serve a large and varied populace possible rather than encourage investment in private, small-scale, narrow facilities.

The Output of Community of Interest

- Common interests with mutual benefits.
- Integration and collaboration between different groups.
- Improve the quality of life.
- Respect private zone, ownership, and equality.
- Security and defensible space.

2.2 Green Community

Much research has been attempting to provide a sustainable society in various fields, such as society, economy, and the environment. As sustainability is emphasised, public and socially responsible architecture has gained much interest. In turn, this caused the role of public architecture to be greater than ever due to its existence being tied to the public because sustainability is linked to shared values. It is most important to meet geographical and environmental conditions. Some ecological researchers have adopted the term "urban ecosystem" to identify the qualities of urban areas. That is why vernacular architecture, a style, in which contemporary architecture seeks to find a new design source from traditional and indigenous architecture, is becoming increasingly important (Kim & Kwon, 2018).

The academic community, the broader context of practitioners, and policymakers recognised the need to involve citizens and other stakeholders in the reactivation and urban renewal processes (Arena, 2006; Clark & Wise, 2018; De Carlo, 2013). Several academics and professionals have been committed to developing strategies and frameworks to enable innovative strategies of participatory planning

(Cioffi & Mela, 2011). These theories and practices aim to produce successful recycling actions with increased stakeholder involvement throughout the process to understand peoples' needs concerning where they live, work, or study (Lo Piccolo & Pinzello, 2009; Lusiani & Zan, 2013), which connects what the professionals envision for that place and the community's desires. According to (Loures, 2008) and later (Scaffidi, 2016)), this has led to an increased sense of belonging to the place, especially in the case of brownfields, and ultimately prevents its untimely dereliction. The vast literature and experiences about "commoning" illustrate a change of mindset and the development of a collaborative economy (Scaffidi, 2019).

Social enterprises' efforts positively impact places and local communities thanks to the connection to the local site and its stakeholders (Scaffidi, 2019). Therefore, social innovation (SI) invites reflection on evaluating social benefits (Caroli, 2015; Maiolini, 2015; Moulaert et al., 2005; Moulaert et al., 2005; Phills Jr. et al., 2008), while (Polman et al., 2017) used social innovation (citizen-engaged). As: the "reconfiguring of social practices, in response to societal challenges, which seeks to enhance outcomes on societal well-being and necessarily includes the engagement of civil society actors" (Roka et al., 2020).

Additionally, as mentioned above, two important terms must be well defined, which are *Sustainable communities* and *Community development*:

- Sustainable Communities, as defined by the EPA for CBEP, represent a community effort encompassing areas associated with environmental issue(s). The community may be organised around a neighbourhood, a town, a city, or a region (such as a watershed, valley, or coastal area). Either natural geographic or political boundaries may define it. The critical factor is that the people involved have a common interest in protecting an identifiable, shared environment and quality of life (USEPA, 1999).
- While community development is a process aimed at fostering economic and social progress for the entire community, involving active participation and relying as much as possible on the community's initiative (UN, 1955). "Community development is about building active and sustainable communities based on social justice and mutual respect. It is about changing power structures to remove the barriers that prevent people from participating in the issues that affect their lives" (Hatton-Yeo & Watkins, 2004; Roka et al., 2020). The difference between both terms is that the first describes the existence of the behaviour, while in the second, the behaviour needs to be encouraged among the community.

Many associations and non-profit organizations deliver environmental programs and services that promote sustainable practices and address real-world barriers, aiming to encourage behavior change and action. They operate based on Community-Based Social Marketing principles and practices. Even in communities without green language in their covenants, these initiatives strive to improve members' quality of life by offering choices, lifestyles, services, and efficiencies valued by people (Bray et al., 2014; Green Communities Canada, 2008). Some governments try to encourage people's PEBs to reduce environmental loading. Japan's government launched a campaign from 2010 to 2012 targeting a 25% reduction of greenhouse gases called the "Challenge 25 Campaign" following the statement by then Prime Minister Hatoyama Yukio in 2009 that Japan aimed to reduce its greenhouse gases by 25% from 1990 levels by 2020. It puts a list of target behaviours. Also, the UK Department for Environment, Food, and Rural Affairs (Defra) has developed the framework of PEBs and categorised them based on 12 goals and behavioural areas, such as consumption, food and drink, personal travel, homes and household products, and travel and tourism. Additionally, in the USA, the US Environmental Protection Agency (US-EPA) has also contributed to this initiative. They have emphasized environmental loading reduction and the avoidance of citizens' exposure to toxic compounds as recommended targets (Kurisu, 2015).

The output for *the success of* Green Communities

Summary of the most common green behaviour internationally is recommended for green communities' success (Table 1).

3 Material and Method

The research structure is shown in Fig. 1.

- **Step 1**: The study determines the factors that create succeeded community from the reliable literature (Interest + Green) (Inductive Method).
- **Step 2**: Depending on the case study data and observation, the study analyses the user pattern in real-time and their behaviour according to the factors from the first step (Tunis Village) (Descriptive-analytical Method).
- **Step 3**: Design a comprehensive model/framework to describe the influential main factors that affect successful communities based on output from the second step (Deductive Method).

Table 1 Common green community strategies selected in this research (*source* The authors)

Categories	Behaviour recommended strategies
Energy	Reduce the consumption of energy; interactive techniques
	Change to renewable sources
Materials and waste	Reused (eco-friendly)
	Recycled
	Reducing
Water	Conserving
	Water treatment
	Desalination
Transportation	Transportation provision
	Walking friendly: provide ways for pedestrians
	Biking: provide ways for bicycles
	Using seas/canals/rivers transportation
	Electronic cars—monorails: alternative fuel vehicle
Infrastructure	Providing clean water
	Providing electricity/energy
	Providing drainage
	Providing roads
Quality of life	Affordable
	Provide job opportunities
Human development	Training to raise efficiency
	Education
Policies	Policy development ruler
	Strictly enforced the low

3.1 The Study Area

This study investigates Tunis Village—many factors in this article controlled site selection. The site has to be recycled/transformed from a depleted recourses area to a green community with a common interest, as shown in Fig. 2. Therefore, Tunis Village is an excellent example of community transformation, as the space was dying due to pollution and had depleted recourses to be recycled and developed by projects that gave opportunities for new life. Moreover, it changed the image of the place by providing a new identity with environmental context and sustainable green meaning. The intention is to choose a "ground" that attempts to be part of the city and create better places. Since skin and material have a deep relationship with the environment, they should have durability and outer skin suitable for the regional environment ((Kim & Kwon, 2018)) to form a new community with a mutual benefit effect between the surrounding environment and the people living there. The research was applied by connecting qualitative and quantitative data with qualitative observations in Tunis Village.

Why Tunis Village? The village is located at the south-west end of Lake Qaroon on the way to the Wadi El-Rayan protectorate area. On a -17-m sea level, it overlooks the lower farmland (-35-m sea level). The village has a spectacular view of the desert on the other side of Qaroon Lake and farm fields between them (Radwan, 2016). Despite all those potentials, Fayoum is one of the most central demographically unattractive governorates due to the quality of education, low income, poor infrastructure, and low resident satisfaction, which led residents to emigrate from the governorate (Bakr, 2016).

On the other hand, water pollution is one of the threats that hit the village. In 2016, Abdou et al., (2016) conducted a study to detect toxic heavy metals in Wadi Al Raiyan lakes above the permissible Egyptian limit that had been reported as a hazardous environment. Wadi Al Raiyan lakes are three artificial lakes connected to the agricultural drainage system of El-Fayoum province near Cairo. The water quality of lakes is a dynamic system that depends on weather conditions and polluting parameters. Qaroun Lake, like Wadi Al Raiyan Lake, is contaminated with vast quantities of chemical fertilisers and pesticides. Chemical pollutants can cause severe damage to the lake's ecosystem (Khalifa & El-Khateeb, 2011). Despite all these

Fig. 1 Research structure (*source* The Authors)

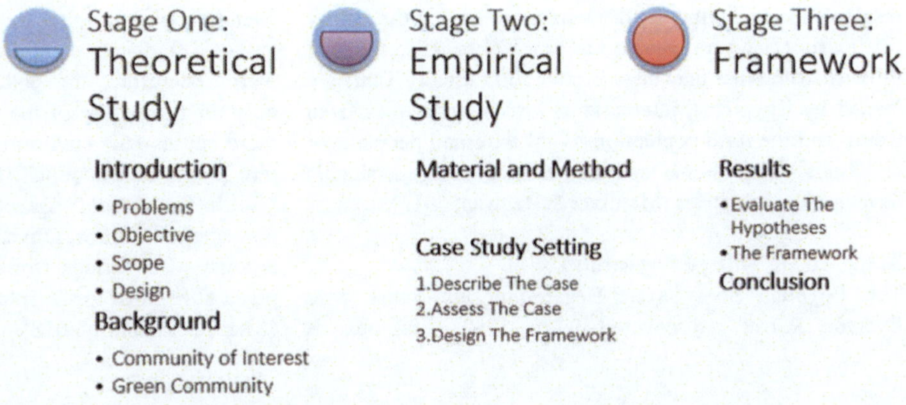

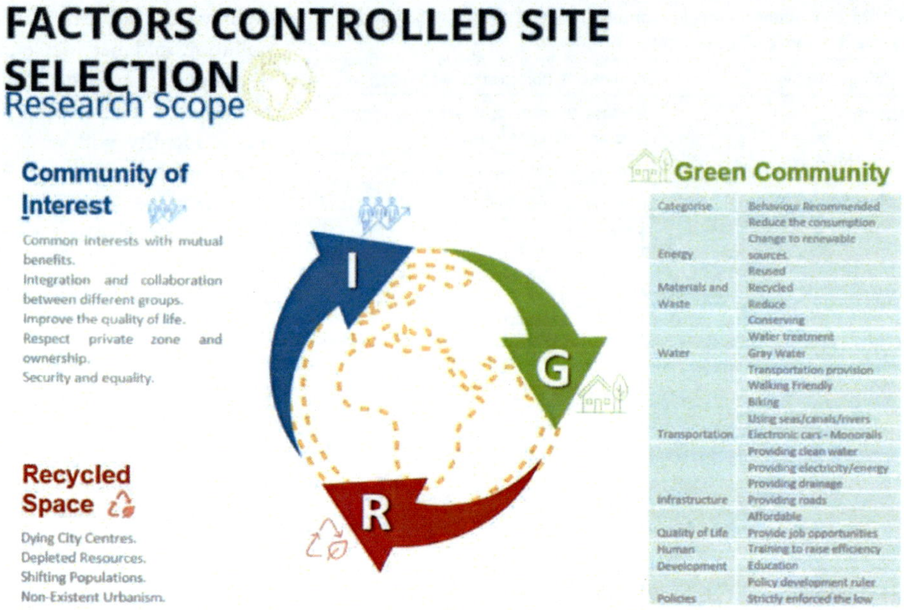

Fig. 2 Factors controlled site selection (*source* The authors)

unfavourable circumstances in Fayoum, Tunis Village enjoys a different situation that deserves to investigate because it is one of the critical points of ecotourism on the map around the world and is widely known. Even many foreigners preferred to live in it and live a modern life.

3.2 Data Analysis

On the way to Wadi Rayan, the small village of Tunis ('izbat Tunis) is in the Fayoum Oasis. The village, which overlooks a scenic view of the edge of the desert on the opposite side of the lake, is located on a hill facing a large saltwater lake. On the southern shore of Lake Qaroun, the village is a farming and fishing community, as shown in Fig. 3 (Mohamed & Refaat, 2019).

3.2.1 Community in Tunis Village

The community comprises a combination of Bedouins and Peasants, representing rural Egypt and life in the desert. The residents are incredibly kind. Despite not having the luxury that exists elsewhere, people are peaceful because they are satisfied with what they have. About 200 farms in Tunis are owned by Egyptians, international artists, academics from Cairo, and the rural population. Eight thousand people now call Tunis Village home, and all are involved in sustainable development initiatives (Mahfouz & Radwan, 2016).

3.2.2 Tunis Village Beginning

The Egyptian poet Sayed Hegab and his Swiss wife Evelyne Porret rediscovered it in 1962 (Mahfouz & Radwan, 2016). After falling in love with the area, Evelyne decided with her husband to build her own country house; in the 1980s, after her divorce from Sayed Hegab, she married the Swiss artisan Michel Pastore, and together they started the pottery workshop there. She built a house inspired by Hassan Fathi-style designs with a domed ceiling design constructed with eco-friendly building materials from mud bricks. She noticed that children in the village were labouring and herding their animals and creating toys out of the mud in the irrigation ditches. The children enjoyed playing at home with her kids and were fascinated with her pottery-making. Evelyne decided to use her workshop as a school to develop the natural skills of the village's children (Khalil & Ibrahim, 2018).

The potter's house was an example for others in Tunis to construct their homes using the same design because it was a masterpiece of eco-architecture. The pottery school rapidly attracted neighbourhood children to learn how to create pottery, inspiring them to begin a potter career and providing lies with money. Because the residents are friendly and hospitable many foreigners, well-known authors, thinkers, painters, and architects bought homes there. Therefore, the potter craft market increased. Not only the hospitality of the residents attracted others to live there but also the environment itself. This environment has rare plants and abundant trees and shrubs, including the "Bottle Brush tree," Spikey Silk Cotton tree, rose bushes, pomegranate shrubs, Olive, Orange, and Lemon trees many arbours with various fruits and flowers. Both people and place gave birth to the intellectual community to establish (Khalil & Ibrahim, 2018).

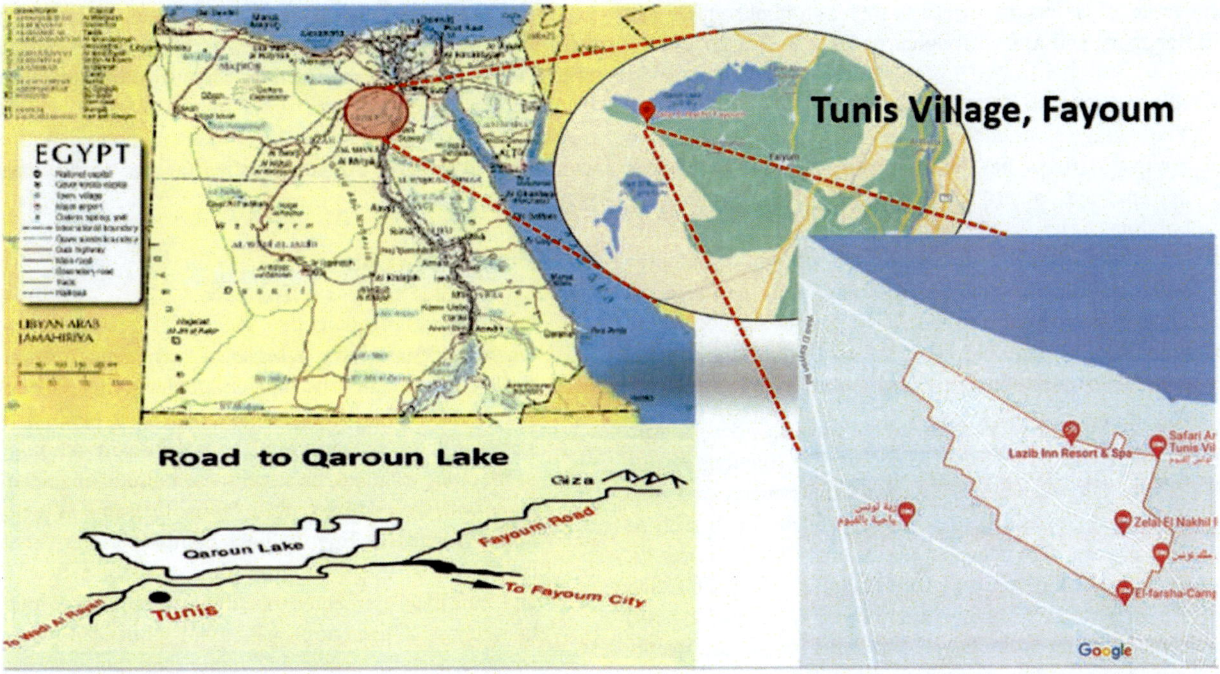

Fig. 3 Tunis Village (*Source* Google Earth, Google Maps. Edited by author)

3.2.3 Community of Interest in Tunis Village

- **Common Interests with Mutual Benefits**
  The village is an obvious example of how individual effort can affect a community. Oscar built in his theory his ideas on government policies, associations, or extended families to work together for more integrations and collaborations. But here, a woman with a dream made changes in the community with her efforts. Nevertheless, pottery was deep inside Egyptian culture. Ancient Egyptians sanctified the Nile, and a thick, rich layer of fertile mud was left behind during the annual flood that receded. They believed that ram-headed Khnum was the god of the source of the Nile, who ensured the regular arrival of the flood and the mud it carried. He was also known as a potter god; he shaped the mud of the Nile on his potter's wheel to create humankind. Such beliefs made ancient Egyptians which have shaped, decorated, and fired the Nile's mud in various forms for various functions ((Hewison, 2021)). Evelyne Porret came to regenerate the craft and taught children how to create pottery.

- **Integration and Collaboration Between Different Groups and Improve the Quality of Life**
  Evelyne Porret initiated the interest by teaching the local children pottery in her pottery school. At the same time, their crafts were sold in shops inside the village, which gave mutual benefits for the space (the village) and the families of those children; between gaining money and opening new fields inside the village, the crafts opened markets. Gradually, the ordinary village turned into a modern pottery centre with an open-air museum of both eco and traditional rural buildings. As some refer to her, today's "students of the Swiss woman" is a new generation of independent potters representing the interaction between humans and their environment. Each maintains a studio and exhibits pottery in national and international exhibitions (Khalil & Ibrahim, 2018). Today Tunis Village has more than 25 pottery workshops and produces more than 1500 art pieces per month. Each piece's price ranges between 20 and 700 Egyptian pounds (Fahmy, 2021). This community attracted investors and entrepreneurs who began to provide the village with different facilities, like a guesthouse and a horse-riding centre.

Moreover, the place also attracted a Non-profit Organisation in 2006. The International Summer Academy inspired the organisation founded by Egyptian artist Mohamed Abla called Fayoum Art Centre in Salzburg, where Abla taught for several years. He hoped to connect artists with international artists. The centre consists of: (several large studios, an art library, living areas, and a communal dining room). Additionally, Abla constructed the first Caricature Museum in the Middle East. This museum holds a vast collection of caricatures by artists such as Saroukhan, George El Bahgoury, and Ahmed Toughan (Ahmed & Hassan, 2020).

- **Respect Private Zone and Ownership**
  The village's architecture can easily divide into two groups. Group has built using Hassan Fathi-style designs, and others have built concrete structure-rise buildings. The first group of locals preferred the Fathi style inspired by a domed ceiling design constructed with eco-friendly building materials. They relate to the pottery industry or tourism. Their houses are located in the village's centre. While on the outskirts of the village, there are concrete structures with unfinished facades.

Hassan Fathi-style designs group used to "design with nature"—vernacular architecture and passive solutions for the hot climate like arcades, cooling towers, inner courts, and wind catchers. People build houses from local materials, such as limestone rocks, clay, and palm tree beams. The roof is domes, vaults, and wooden ceilings (Abd Elrahman, 2020). This kind of design gave the village a unique style. This style attracted numerous people from Egypt and other countries, mainly Europe. They used to enjoy both the vernacular architecture and native traditions of the village (Bassiouny et al., 2021).

On the other hand, the group that preferred concrete can raise the building for the maximum number of floors they need so that all family members can live together and each member can have a private unit. Also, concrete is more durable than how they used to build their old houses. At the same time, they used the non-finished facades to decrease expenses.

- **Security and Equality**
  - Security: the space is self-secured.
    Oscar recommended the defensible space theory, where residents have a sense of ownership, responsibility, and control outside their living spaces (Newman, 1972). The concept has been applied in Tunis Village, but differently; the urban morphology of the village shares the same socio-cultural characteristics with the compact urban fabric, the same as many other villages in Egypt; these compact urban fabric (as shown in Fig. 4), narrow-shaded street networks, and buildings' protected openings were environmental and social responses, which are influenced by the social aspects of private life rather than their geometric shapes (Shehata, 2022).

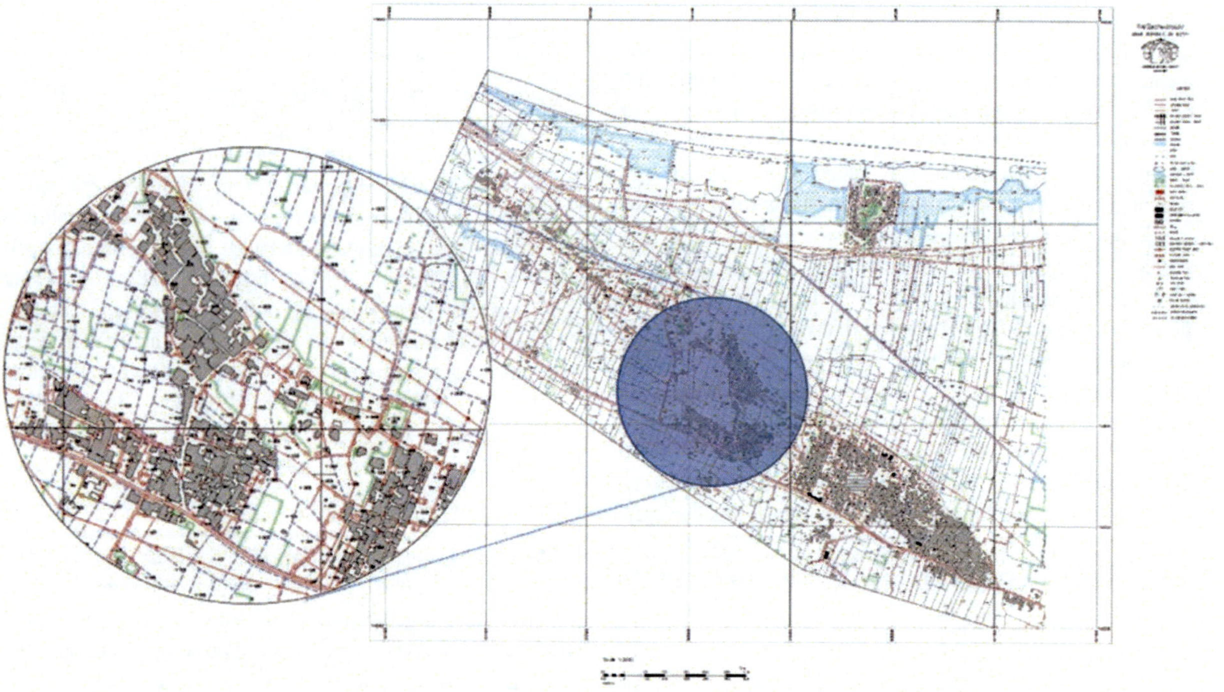

Fig. 4 Tunis Village map (*Source* Egyptian General Survey Authority (2021) Edited by author)

- Equality

In 2019, a study revealed that 92.5% of crafters are men and only 7.5% are women. Long ago, the conservative lifestyle prevented girls and women from learning this profession. Nowadays, there are some changes as many rural families begin to accept learning about their daughters this profession. (Mohamed & Refaat, 2019).

3.2.4 Green Community of Tunis Village

- **Energy**
  - Trials to produce energy through renewable sources include biomass, pre-packaged photovoltaic systems, and solar hot water heat (Abd Elrahman, 2020).

- **Materials and Waste**
  - Materials
    Reusing material (Mahfouz & Radwan, 2016):
    Mud brick walls reform the mud into bricks, and about 200 houses are built using the simple mud brick style, symbolising the rural houses surrounding the environment.
    Reusing palm fronds to form chairs and shades, and roofs.
  - Recycling waste
    Some Garbage Recycling Projects are trying to take place in Tunis (Mahfouz & Radwan, 2016).
    Besides, they tried to shift toward organic agriculture (Abd Elrahman, 2020).

- **Water**
  They used methods for filtering and storing water, such as sand soil strainers and biofilters ((Abd Elrahman, 2020)).

- **Transportation**

  - **Transportation provision**
    Tunis Village is easily accessible from Cairo; the trip takes approximately one hour by car or public transportation.

Using a private car from Cairo-Fayoum (from the pyramids), 20 km before Fayoum City.

There are no direct buses to Tunis. The traveller needs to change between buses and taxis to reach the centre of Tunis (Mahfouz & Radwan, 2016).

- **Walking Friendly**
  The compact urban fabric of the village produced a narrow-shaded street network that allowed pedestrians to move comfortably. Also, all the village's prominent landmarks and essential spots are almost arranged linearly. Most of the main sites are concentrated on the same central axis.
- **Biking: Provide Ways for Bicycles**
  More visitors meant more vehicles in the small village where the vehicles came with car-related problems, like the need for additional parking spaces on what was formerly a farm. Moreover, noise and pollution also increased. As known, short car trips burn more fuel and produce more pollution than longer ones, encouraging small projects to rent bikes for visitors to transport inside the village.
- **Riding horseback**
  There is a horse-riding centre, and many other stables in different spots surround the village. One of the top activities in Tunis Village is riding horseback to explore the village.
- **Using seas/canals/rivers transportation**
  Tunis village is located on Qaroun Lake, so as in any waterfront city, it is common to use the lake for transportation.

- **Quality of Life**
  - Affordable (Mohamed & Refaat, 2019)
    87.5% of crafters stated that their incomes are stable because of their work in pottery.

While 12.5% said that their incomes are only partially stable because the sales volume is affected by the climate of Fayoum, as in the summer, their sales declined.

Pottery crafters stated that their income has increased by 46.5% after working in pottery.

Provide job opportunities:

100% of potters are satisfied with their profession, and 87.5% consider that their occupation has helped the community's unemployment rate decline ((Mohamed & Refaat, 2019)).

The village has many potentials besides pottery markets, tourism fields (hotels and restaurants), and stables for riding horses. Also, there are farms and fishing fields.

- **Human Development**
  - Training to raise efficiency
    Results showed that 52.5% of pottery crafters received their education at Evelyne Porret's "School of Pottery," which she founded in Tunis in 1980. The remaining 47.5% received their education in the private workshops that the first generation of crafters trained by Evelyne established. It takes approximately five years, on average, to learn the profession ((Mohamed & Refaat, 2019)).
  - Education
    The farmers: 32.5% of interviewed had obtained a high school education, 45% were young teenagers

still studying, and only 2.5% had obtained a bachelor's degree.

Of the crafters: 20% had no formal education and were illiterate.

Pottery crafters: 100% indicated that the profession helped them provide better education and health care to their families; they also stated that they feel safe and socially secure working in this profession (Mohamed & Refaat, 2019).

The Egyptian Association implemented an educational project at a nursery school (Radwan, 2016).

4 Results and Discussion

Tunis Village, as a result of Evelyne's efforts, has developed as a destination for ecotourism even though Evelyne herself was not aware of these significant catalysts and transformational leaders for socio-economic change in the local community, environmental, socio-cultural, and economic dimensions (Khalil & Ibrahim, 2018) not only but also the majority of guests and the tourists becoming friendly with the environment. Tunis Village is one of a few places in Egypt that offers Ecolodges, a new trend in the hospitality industry that aims to preserve the environment and society and achieve sustainable travel. Usually, small-scale hotels give opportunities for interaction with the local environment. They are low-impact, nature-based, involve and benefit local communities (Ahmed & Hassan, 2020). All these developments helped to form an integrated green community that started from the main interest of the village, which is the pottery industry.

As previously discussed in the literature view investigation, a study in 2019 interviewed the (local crafters—locale non-crafter and visitors) at Tunis Village. This study confirms the theory of which community of interest can transform a developed community, as shown before Tunis Village; even though it has many potentials to be an attractive point, this happened only after the pottery school started, according to crafters. About 100% of pottery crafters indicated that the profession helped them provide better education and health care to their families; they also stated that they feel safe and socially secure working in this profession. They are satisfied with the work environment of the domain, while 87.5% of them agreed that the job contributed to the reduction of unemployment in the village. Moreover, 90% believed that handicrafts, particularly with pottery playing a prominent role, constitute the primary reason for tourists to visit Tunis Village., as shown in Fig. 5 (Mohamed & Refaat, 2019).

90% of non-crafters from the local community highlighted the transformative value of pottery workshops for Tunis Village. They emphasized through discussion that these workshops shifted the village from its traditional identity as an Egyptian agricultural community to a bustling tourist destination, all credited to the establishment of pottery schools and markets (Fig. 6). However, after the establishment of the pottery school and workshops, tourists began to feel inspired to visit the village. Hotels and other tourism facilities and activities started to exist to meet their needs. Moreover, of the local non-crafters, 93% believe that Pottery workshops improve the village overall by creating jobs in these workshops, hotels, and other tourism-related businesses, which helped the entire village develop. It brought infrastructure that was beneficial to the whole village's people. Additionally, they noted that every family in the village has at least one member who works in pottery handicrafts. All these employment opportunities have inspired their relatives, who previously left the village to seek employment in Cairo, to return to the village and launch their businesses. While 100% of the visitors (tourists) agreed that they had read about Tunis before visiting it, the story of the Pottery school by the Swiss motivated them to see it (Mohamed & Refaat, 2019).

As a result of this study and by examining the theory on Tunis Village (case study), the output is to draw a framework shown in Fig. 7, representing the main factors of transformation toward a developed green community.

Fig. 5 Main reasons tourists visit and stay in Tunis Village (*source* Mohamed and Refaat (2019))

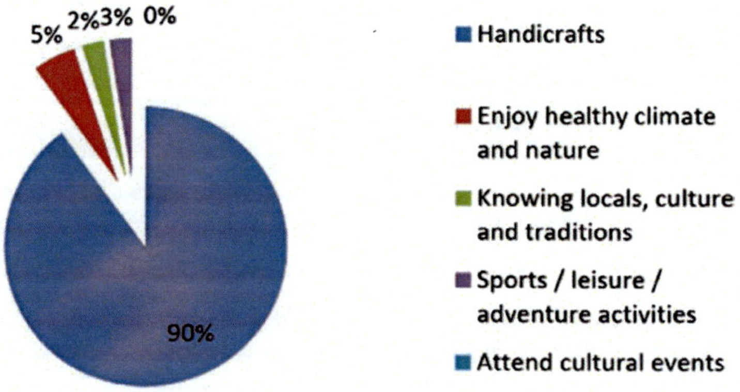

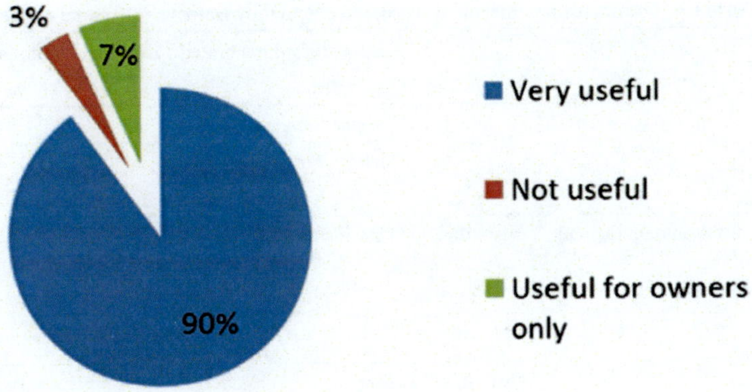

Fig. 6 Impression of locals (non-crafters) about the pottery workshops (*source* Mohamed and Refaat (2019))

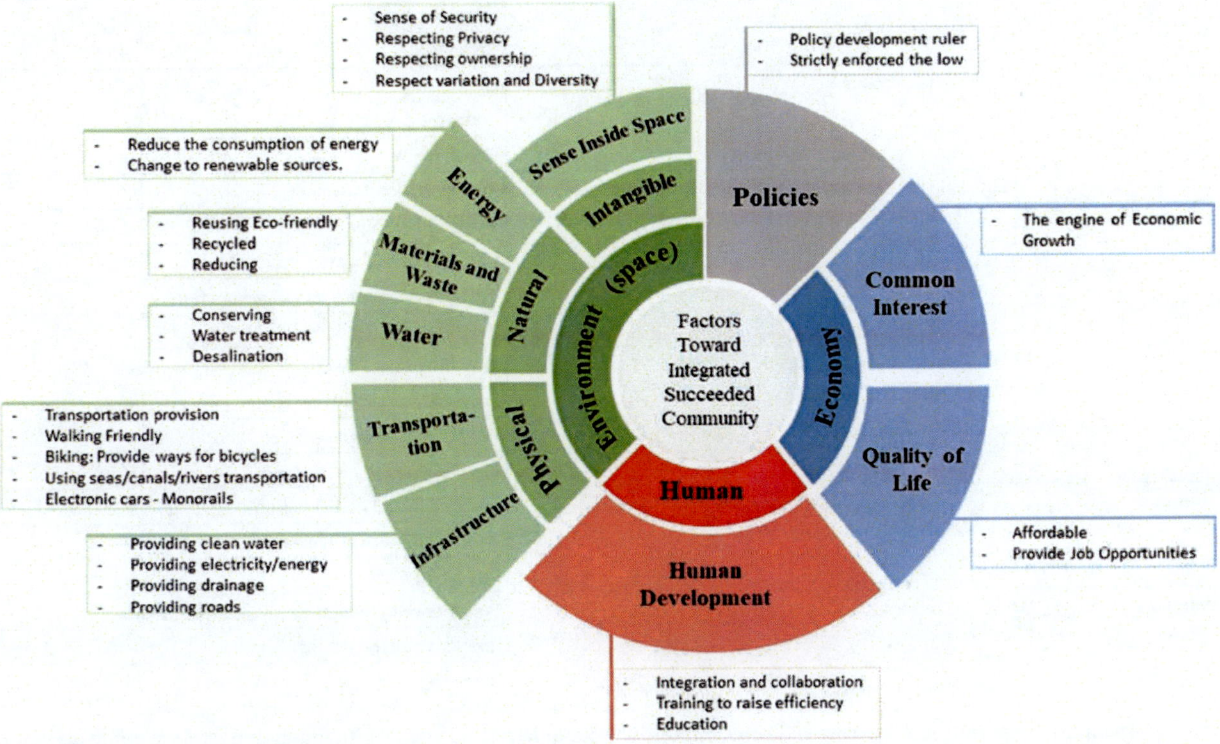

Fig. 7 Factors toward integrated succeeded community framework (*source* author)

Nowadays, no development is without sustainability, and the three main aspects of sustainable development are economy, environment, and human. Each element plays a role in directing this transformation in the way of equilibrium, as shown in Table 2.

In conclusion, the findings highlight the visible characteristics and suggest factors that could be replicated in various settings. As a result, these spaces have the potential to undergo development based on these identified results.

5 Conclusion

Oscar Newman mentioned the factors that led to the unsuccessful neighbourhood in his theory. He thought that the primary factor was a lack of harmony and no integration between residents. He stated the reasons for the disintegration between different age groups and educational and socio-economic classes. This lead to increase discrimination and isolation. Accordingly, the rate of crimes, poverty,

Table 2 Factors toward integrated succeeded community framework (*source* author)

Elements		Factors toward integrated succeeded community		
		Categories	Recommended achievement	Achievement
Economy		Common interest	The engine of economic growth	✓
		Quality of life	Affordability	✓
			Provide job opportunities	✓
Environment (space)	Intangible	Sense inside space	Sense of security	✓
			Respecting privacy	✓
			Respecting ownership	✓
			Respect variation and diversity	✓
	Natural	Energy	Reduce the consumption of energy	–
			Change to renewable sources	✓
		Materials and Waste	Reusing eco-friendly	✓
			Recycled	✓
			Reducing	–
		Water	Conserving	✓
			Water treatment	✓
			Desalination	–
	Physical	Transportation	Transportation provision	✓
			Walking friendly	✓
			Biking: provide ways for bicycles	✓
			Using seas/canals/rivers transportation	✓
			Electronic cars–monorails	✓
		Infrastructure	Providing clean water	✓
			Providing electricity/energy	✓
			Providing drainage	✓
			Providing roads	✓
Human		Human Development	Integration and collaboration	✓
			Training to raise efficiency	✓
			Education	–
Policies			Policy development ruler	–
			Strictly enforced the low	–

and the unemployment rate has increased. The community not being integrated has a negative impact on humans and spaces, as some parts are left neglectable and empty. All these factors affect development. His idea is that if all members of the same community have to join the same interests with mutual benefits, this will make a difference. He put a framework based on many levels for this community of interest. However, his theory did not include sustainable development as a factor of growth and transformation.

Oscar Newman built his theory that governments, quotas, or associations could make changes toward a thriving community. However, one person, Evelyne Porret, made a difference in the research case study in Tunis Village. Even she did not plan for all this transformation as she said: "I never imagined this. There was nothing to imagine, things came about … you don't decide things, they come about on their own". Evelyne Porret (Khalil & Ibrahim, 2018), the project expands even after her death. She invested in children (humans) by teaching and training them to raise their efficiency. Her project exceeded the limit of its walls and attracted many investments to start in the village.

Additionally, her choice to use local eco-friendly architecture for her house and the pottery school gave the space identity as she was an ideal model for her artistic taste spread. She also helped to provide the village with an urban identity or 'macro level' (Cheshmehzangi, 2020) in which the place is urban branding or urban industry that emphasises a place's image. The urban area is frequently branded nationally or regionally, making it known to a larger public group (Ali et al., 2022).

By examining the theory of Oscar on Tunis Village and introducing the factor of sustainability for better outcomes, a framework is drawn for the result, which helps to achieve the transformation of three aspects (economy–environment–and human) toward a thriving green community. The outcome of the community of interest indicates certain features toward an integrated green community, while the suggested factors could be repeated in different spaces. Accordingly, spaces could be transformed to be developed based on these results.

References

Abd Elrahman, A. S. (2020). Cittaslow and applicability in the Egyptian Context: The case of El-Fayoum. In *Architecture and urbanism: A smart outlook* (pp. 371–386). Springer International Publishing. https://doi.org/10.1007/978-3-030-52584-2_26

Abdou, K., Aab, S., Mm, M., Oh, E., & Ia, K. (2016). Assessment of some environmental pollutants in Wadi El-Rayan Lake–Fayoum governorate. *Journal of Scientific and Engineering Research*, *3*(3), 218–137. https://www.researchgate.net/publication/304717875_Assessment_of_Some_Environmental_Pollutants_In_Wadi_El-Rayan_Lake-Fayoum_Governorate

Ahmed, M. O., & Hassan, M. S. (2020). Ecolodges promotion in Egypt: assessing the status quo. *International Journal of Hospitality & Tourism Systems*, *13*(2020).

Ali, O., Mansour, Y., Elshater, A., & Fareed, A. (2022). Assessing the identity of place through its measurable components to achieve sustainable development. *Civil Engineering and Architecture*, *10*(5A), 137–157. https://doi.org/10.13189/cea.2022.101407

Arena, F. G. (2006). Cittadini attivi. Un altro modo di pensare all'Italia (Laterza (ed.)). Laterza.

Bakr, M. (2016). *Population migration out of Fayoum governorate study in population geography* [Fayoum University]. https://www.researchgate.net/publication/309953779_Population_Migration_out_of_Fayoum_Governorate_Study_in_Population_Geography?channel=doi&linkId=5827064308aecfd7b8c07962&showFulltext=true

Bassiouny, Y., Mosaad, G., & Hany, N. (2021). Ecolodging as a solution for sustainable ecotourism development in Al-Fayoum Egypt: Indoor air quality simulation. الدورية العلمية لكلية الفنون الجميلة, *9*(2), 51–31. https://doi.org/10.21608/sjfa.2021.204220

Bray, A., Bunting, J., Fellows, L., Meyer, M. J., Richter, H., & Warren, D. A. (2014). *Best practices: Green communities*. https://foundation.caionline.org/

Caroli, M. (2015). L'innovazione sociale: Caratteristiche chiave, determinanti e princi_pali manifestazioni empiriche. In *Studies and experiences on social innovation* (p. 41:79). Franco Angeli. http://ojs.francoangeli.it/_omp/index.php/oa/catalog/book/129

Cheshmehzangi, A. (2020). *Identity of cities and city of identities*. Springer. https://link.springer.com/book/10.1007/978-981-15-3963-3

Cioffi, D., & Mela, A. (2011). Participatory urban planning. Models and experiences. Carocci.

Clark, J., & Wise, N. (2018). *Urban renewal, community and participation* (J. Clark & N. Wise, Eds.). Springer International Publishing. https://doi.org/10.1007/978-3-319-72311-2

De Carlo, G. (2013). *L'architettura della partecipazione* (S. Marini, Ed.) (Vol. 9). Quodlibet.

Egyptian-General-Survey-Authority. (2021). *Fayoum, dwg Map (Tunis Village)*. https://www.esa.gov.eg

Fahmy, Y. M. (2021). Productive village project in Fayoum governorate as a sustainable development model: A case study of Al Nazla and Tunis Villages. مجلة مستقبل العلوم الإجتماعية, *6*(1), 192–147. https://doi.org/10.21608/fjssj.2021.270514

Green Communities Canada. (2008). *How to grow a green community*.

Hatton-Yeo, A., & Watkins, C. (2004). *Intergenerational Community Development*. file:///C:/PHD/002 REFRENCES/001 BOOKS PDF/InterCommunityDevelopment.pdf

Hewison, R. N. (2021). Fayoum pottery: Ceramic arts and crafts in an Egyptian oasis. 181.

Khalifa, M. A., & El-Khateeb, S. M. (2011). Fayoum oasis between problems and potentials: Towards enhancing ecotourism in Egypt. *Urbenviron Cairo 2011*.

Khalil, S., & Ibrahim, O. (2018). Socio-economic development in Tunis Village: The success factors of a heritage tourism destination. *Minia Journal of Tourism and Hospitality Research*, *6*(2/1).

Kim, S., & Kwon, H. (2018). Urban sustainability through public architecture. *Sustainability*, *10*(4). https://doi.org/10.3390/su10041249

Kurisu, K. (2015). Pro-environmental behaviors. *Springer Japan*. https://doi.org/10.1007/978-4-431-55834-7

Lo Piccolo, F., & Pinzello, I. (2009). Citizens and citizenship. Perspectives, role and opportunities of Agenda 21. Local in the urban context. Palumbo.

Loures, L. C. (2008). Industrial heritage: the past in the future of the city. *WSEAS Transactions on Environment and Development*, *4*(8).

Lusiani, M., & Zan, L. (2013). Planning and heritage. *Journal of Cultural Heritage Management and Sustainable Development*, *3*(2), 108–115. https://doi.org/10.1108/JCHMSD-06-2013-0026

Mahfouz, W. R., & Radwan, I. (2016). The Role of Tunis Village in Fayoum governorate as a cultural tourism model. *Journal of Association of Arab Universities for Tourism and Hospitality*, *13*(1), 209–220. https://doi.org/10.21608/JAAUTH.2016.49696

Maiolini, R. (2015). Lo stato dell'arte della letteratura sull'innovazione sociale. In M. G. Caroli (Ed.), *Modelli ed Esperienze di Innovazione Sociale in Italia* (pp. 23–37). Franco Angeli.

Marshall, M. (2016). *OCR Psychology Student Guide 3: Component 3 Applied psychology - Molly Marshall - Google Books*. Hodder Education. https://books.google.com.eg/books?id=0TjEDAAAQBAJ&pg=PT48&redir_esc=y#v=onepage&q&f=false

Mohamed, M., & Refaat, H. (2019). Rural tourism and sustainable development: The case of Tunis Village's handicrafts, Egypt. *Agrosym 2019*. https://www.researchgate.net/publication/341256173_RURAL_TOURISM_AND_SUSTAINABLE_DEVELOPMENT_THE_CASE_OF_TUNIS_VILLAGE%27S_HANDICRAFTS_EGYPT

Moulaert, F., Martinelli, F., Swyngedouw, E., & Gonzalez, S. (2005). Towards alternative model(s) of local innovation. *Urban Studies*, *42*(11), 1969–1990. https://doi.org/10.1080/00420980500279893

Newman, O. (1972). *Defensible space; crime prevention through urban design*. Macmillan. https://www.ojp.gov/ncjrs/virtual-library/abstracts/defensible-space-crime-prevention-through-urban-design

Newman, O. (1980). Community of interest. *Anchor Press/doubleday*, *18*(1), 52–57. https://doi.org/10.1007/BF02694845

Phills Jr., J. A., Deiglmeier, K., & Miller, D. T. (2008). Rediscovering social innovation. *Stanford Social Innovation Review*, *6*(4), 34–43. https://doi.org/10.48558/GBJY-GJ47

Polman, N., Slee, B., Kluvánková, T., Dijkshoorn, M., Nijnik, M., Gezik, V., & Soma, K. (2017). *Classification of social innovations for marginalized rural areas*.

Radwan, W. (2016). The Role of Tunis Village in Fayoum Governorate as a cultural tourism model. *Journal of Association of Arab Universities for Tourism and Hospitality*, *13*(1), 209–220. https://doi.org/10.21608/JAAUTH.2016.49696

Roka, K., Filho, W. L., Ravazzoli, E., Valero, D. E., Ang, J. C., Diogo Guedes Vidal, N. B., & R. L. M. (2020). *Sustainable cities and communities* (W. Leal Filho, A. Marisa Azul, L. Brandli, P. Gökçin Özuyar, & T. Wall, Eds.). Springer International Publishing. https://doi.org/10.1007/978-3-319-95717-3

Scaffidi, F. (2016). Citizen participation and valuation of heritage. Methodologies of possible application to abandoned industrial spaces. In *Abaco* (Vol. 4, pp. 118–126). https://www.revistasculturales.com/revistas/72/abaco-revista-de-cultura-y-ciencias-sociales/num/86/

Scaffidi, F. (2019). Soft power in recycling spaces: Exploring spatial impacts of regeneration and youth entrepreneurship in Southern Italy. *Local Economy: The Journal of the Local Economy Policy Unit*, *34*(7). https://doi.org/10.1177/0269094219891647

Shehata, A. M. (2022). Current trends in urban heritage conservation: Medieval historic Arab City Centers. *Sustainability, 14*(2), 607. https://doi.org/10.3390/SU14020607

UN. (1955). Social progress through community development.

USEPA. (1999). Compendium of methods for the determination of inorganic compounds in ambient air EPA/625/R96/010a. *United States Environmental Protection Agency*, 21. http://www.epa.gov/ttnamti1/files/ambient/inorganic/iocompen.pdf

Eco-friendly Communities in Recycled Spaces: Environmental Behaviour in Rural Egypt

Ghada Assal, Yasser Mansour, and Shaimaa Kamel

Abstract

Recently, several studies have been conducted to develop a sustainable society, one of which is urban, linked to public and socially responsible principles; thus, sustainability is tied to public values. Consequently, eco-friendly communities might serve as a model of sustainable urbanism and improve the environment's and the general public's quality of life. This research aims to investigate one of the eco-friendly communities in Rural Egypt, SEKEM Farm, a sustainable green urbanism project whose founders established it with a holistic, sustainable development vision with the potential of representing, upon spreading and can positively impact reversing rural–urban migration in the future. The research objective is to set the features of eco-friendly communities of recycled spaces and to determine the factors that led those communities to achieve positive results. In addition, it states the criteria that could help others to repeat the same concept successfully. Methods of both diagnostic and descriptive will be used for setting the features of eco-friendly communities. The empirical study in the research includes site visits for observation and mapping the human behaviour using paper, time-lapse, and video photography; this step aims to analyse the user pattern in real-time. Finally, the outcome indicates a comprehensive model/framework to describe the influential primary factors that affect pro-environmental behaviour. Additionally, based on these results, the suggested criteria could be repeated in different recycled spaces to be transformed into eco-friendly communities.

Keywords

Eco-friendly communities · Sustainable urbanism · Recycled space · Quality of life

1 Introduction

Cities are attractive for migration due to job opportunities and industrialisation located in cities. More than half of the world's population resides in cities of all sizes. Around two-thirds of all people will live in urban and peri-urban regions by the year 2050, the UN predicts. This situation demonstrates the need to enhance urban livability to achieve sustainable growth. As a result, it is vital to alter the way that cities are created and run. Most cities worldwide are expanding significantly, creating megacities—cities with more than 10 million populations. These megacities are overgrowing, especially in emerging nations (El Ghorab & Shalaby, 2016).

On the other hand, industrialisation negatively impacts the environment and human existence, including biodiversity loss, climate change, and deteriorating living and health conditions. Those circumstances made looking for integrated solutions to make development more sustainable even more critical. Some current economic growth strategies are unsustainable because they increasingly rely on exploiting natural resources. Research suggests that a more responsible or sustainable approach to using our finite resources is required. Research indicates a more reliable or sustainable approach to using our limited natural resources is needed (Roka et al., 2020). "Sustainable Development" was first used and defined in 1987 by the

G. Assal (✉) · Y. Mansour · S. Kamel
Architecture Engineering, Ain Shams University, Cairo, Egypt
e-mail: Ghada.Mohammed@hti.edu.eg; G15051382@eng.asu.edu.eg; ghada.assal@gmail.com

Y. Mansour
e-mail: yasser_mansour@eng.asu.edu.eg

S. Kamel
e-mail: shaimaa_kamel@eng.asu.edu.eg

G. Assal
Higher Technology Institute, 10th of Ramadan City, Egypt

M. F. Shahidan et al. (eds.), *Urban Narratives: Exploring Identity, Heritage, and Sustainable Development in Cities*, Advances in Science, Technology & Innovation, https://doi.org/10.1007/978-3-031-48517-6_23

World Commission on the Environment and Development. "Sustainable development," in their definition, is altering people's behaviours to make life easier so that people meet their needs without endangering the capacity of future generations to do the same. The life of humans or non-humans is built on natural resources. Water, air, soil, plants, and animals are some of these resources because irresponsible consumption would risk everyone's security in the present and future (El Ghorab & Shalaby, 2016).

Due to the cost of living, many people accept bad conditions like slums, which are causing the situation to worsen. This translates to declining quality of life and a lack of green space and ecosystem services (Roka et al., 2020). Other strategies must be considered to minimise the pressure on cities and resources. By raising the standard of living in those places and creating sustainable lifestyles, encouraging people to move back towards rural areas may be necessary. Economic growth and equality, environmental protection, and social development are the three acts that must be combined to be achieved (El Ghorab & Shalaby, 2016).

By investigating one of the successful green sustainable communities of SEKEM farm in Egypt, which was founded on depleted land that was transformed/recycled to become a well-known project dedicated to completing sustainable development and social innovation, the goal of this research is to define the characteristics of eco-friendly communities of recycled spaces. SEKEM collaborates with a broad range of organisations in its institutional ecosystem to carry out a variety of projects, from environmental and educational efforts to textile and agricultural companies (SEKEM Association, 2021). This extensive range of initiatives based on the UN's Sustainable Development Goals aids in identifying the elements that caused this community to experience success. Furthermore, by outlining the factors that made SEKEM a leader in sustainable development at the human, organisational, and community levels, where it started on depleted resources land to be transformed into a sustainable green community, this information might be applied to society and act as motivation for others.

2 Research Background

2.1 Eco-friendly Behaviour or Pro-environmental Behaviour

Resource depletion, biodiversity loss, and climate change are some of humanity's major global environmental issues. People react to their surroundings according to their experiences and the information the environment provides them. They both notice it at the same time. The broader conceptualisation of environmental behaviour and society is the study of the mutual interaction between humans and the environment at all scales to enhance the quality of life. EBS emphasises the interaction of sociocultural and environmental elements and physical and biological environmental systems. It functions at the individual, group, societal, and cultural levels. They comprise different environments brought on by these different levels of human activity. These activities can be divided into three categories: functional activity, environmental and physical factors, and cultural and social factors. Emphasises that there is a mutually beneficial relationship between how the physical environment affects human behaviour and how that behaviour affects the environment (Fard, 2014). It is essential to take appropriate measures, including understanding the key drivers and processes behind the behaviour driving them, to determine how these issues will evolve over time and eventually change the system effectively to reduce adverse implications (Klöckner, 2013).

The EBS field also covers studies on environmental intervention, information dissemination, and the application of research to urban policy, planning, urban design, and architecture. Moore also thinks that environment, behaviour, and society have developed to bridge the gap between people, social groups, cultures, and the surroundings in which they live and to use these insights to improve built environment planning and design (Moore, 2006). Therefore, environmental behaviour definition is "all types of behaviour that change the availability of materials or energy from the environment or alter the structure and dynamics of ecosystems or the biosphere" (Steg & Vlek, 2009, p. 309) for either behaviour with an environmentally damaging impact or behaviours that benefit the environment (Gatersleben et al., 2019).

While pro-environmental behaviour has been defined as "behaviour that consciously seeks to minimise the negative impact of one's actions on the natural and built world" (Kollmuss & Agyeman, 2002, p. 240; Gatersleben et al., 2019). PEBs are sometimes called "ecological behaviours," "environment-friendly behaviours," or other variants. One of the PEB definitions related to environmental conservation can be differentiated into two categories: reduction of negative impacts and increase of positive impacts (Kurisu, 2015). The degree of reduction of impacts on the environment by individual behaviour depends on several aspects (Dietz et al., 2009): the impact the behaviour has the number of people who perform the behaviour and the percentage of those people who are willing or able to change the behaviour, referred to as "plasticity".

- **How to Measure Behaviour?**

Initially, it has to be determined that there are two types of measurements: behavioural and environmental impact (Stern et al., 1997).

- **Environmental behaviour**: concerned with analysing goal-directed environmental behaviour. Additionally, it measures the existence of a group of behaviours without considering how those behaviours affect and impact the environment. (Gatersleben et al., 2019).
- **Environmental impact** measurement is more relevant for environmental policy to support the policies' ultimate practical goal, reducing people's ecological footprints and overall environmental impacts. (Gatersleben et al., 2019).

Environmental impact measurement is this article's scope. So to determine the principles and factors (inputs) that affect the impact volume, many studies analysed and categorised them as positive have to be encouraged, and negative footprint has to be reduced. United Nations General Assembly agreed and approved in September 2015 the document "2030 Agenda for Sustainable Development", which contains measures to balance economic progress and protection of the environment. The agenda document consists of 17 Sustainable Development Goals (SDGs) (Roka et al., 2020; United Nations Development Programme (UNDP), 2015). These goals will be used as the key to measuring the environmental impacts on sustainable green communities in rural Egypt using those keys which include climate change, economic inequality, innovation, sustainable consumption, peace, and justice, among other priorities (UNDP, 2015). The 17 SDGs are:

- Goal 1: No Poverty.
- Goal 2: Zero Hunger.
- Goal 3: Good Health and Well-Being.
- Goal 4: Quality Education.
- Goal 5: Gender Equality.
- Goal 6: Clean Water and Sanitation.
- Goal 7: Affordable and Clean Energy.
- Goal 8: Decent Work and Economic Growth.
- Goal 9: Industry, Innovation, and Infrastructure.
- Goal 10: Reduced Inequalities.
- Goal 11: Sustainable Cities and Communities.
- Goal 12: Responsible Consumption and Production.
- Goal 13: Climate Action.
- Goal 14: Life below Water.
- Goal 15: Life on Land.
- Goal 16: Peace, Justice, and Strong Institutions.
- Goal 17: Partnerships for the Goals.

2.2 Egypt and the Sustainable Development Strategy

Africa is currently experiencing a rapid rate of urbanisation that is only second to Asia. It is estimated that Africa's urban population may triple between 2010 and 2050, with the urban population increasing from 395 million in 2010 to 1.339 billion in 2050 (United Nations, 2014; Bello-Schünemann & Aucoin, 2016). A continent with over 50 countries, well endowed with natural resources such as mineral ore, wildlife, fisheries, fertile soils for agriculture, and a youthful population, suffered from socioeconomic and political problems, especially poverty, deindustrialisation, and climate change (Roka et al., 2020). By 2050, the urban population will comprise 55% of the continent's total population, and 21% of the world's projected urban population will correspond (Güneralp et al., 2017). Muggah (2016) described emerging African cities as fragile and weak to socioeconomic and environmental shocks (Matamanda & Nel, 2020).

Egypt is a part of Africa; in the holistic development process to attain economic and social success and environmental sustainability, the sustainable development strategy is an important step. To provide a competitive, balanced, and diversified economy with a balanced ecosystem for attaining sustainable growth and improving the quality of life for Egyptians. Additionally, over recent years, renewable energy development has become a priority for Egypt. Due to its geographical location and climate, Egypt has an average level of solar radiation between 2000 and 3200 kWh/m^2 per year. As a result, Egypt has significant potential for developing solar energy applications (Assal et al., 2018). Egypt was one of the nations that embraced "Vision 2030" (The Ministry of Environment, 2017). The State of the Environment Report 2017 is the latest report published by The Ministry of Environment. The study is centred on addressing globally and regionally agreed-upon targets, notably the 2030 Sustainable Development Goals and the 2063 Africa Development Strategy. These frameworks aim to foster Africa's prosperity, ensure its health and position it as an influential global power. The research aligns with these international objectives and complements Egypt's national agenda encapsulated in 'Vision 2030' and the Egyptian Sustainable Development Strategy. (The Ministry of Environment, 2017). As part of Africa's Sustainable Development Strategy 2063 and following the UN Sustainable Development Goals 2030, these goals are to make everything possible for Egypt to participate actively in the global environment, marked by dynamic and consecutive sustainable developments. The state's goals for sustainable development policies place a high priority on the following:

- To implement the principles of collaboration with all governmental and non-governmental organisations, the corporate sector, and scientific research to achieve the state's goals for sustainable development.
- To contribute to the transformation to a green economy.

- To start the consumption and manufacturing that is sustainable.
- To use strategies and programmes to combat climate change.
- To integrate the environmental factors into strategic plans.

3 Material and Method

3.1 Case Selection

The study is interested in examining eco-friendly communities of recycled spaces. Based on the definitions of eco-friendly community and recycled space, the study chooses SEKEM sustainable development since 1977 as a case for studying. This study uses diagnostic and descriptive methods, eliciting the criteria that could help others repeat the same concept successfully.

Among the many successful communities, Why SEKEM? The advantage of SEKEM society over similar societies is that it is a successful case replicated in other communities with the same concept.

3.2 Data Sources

The study relies mainly on two primary data sources for the case study, the documents as the archives and reports of SEKEM and the site inventory.

- **The documents**: SEKEM's official Website publishes reports and periodicals on projects and events in this community. These will be cited during this study.
- **Site inventory**: The authors visited the place during several events, in addition to the ethnography of the community, for five days, including working days and days off, to document individuals' behaviour in different circumstances and situations. Digital cameras were used to take images documenting the scenes, outdoor activities, and life in the community. The scene choices were based on the facial details of users, their attitudes, and the atmosphere in the space. Recording notes are also used to rely on for assessment.

3.3 Methods Setup and Analysis

Since the purpose of this study is present a comprehensive model/framework to describe the influential primary factors that affect pro-environmental behaviour, the study achieved that through the following steps:

- **Step 1**: The study determines the factors that create Succeeded Community from the reliable literature (inductive method).
- **Step 2**: This step depends on the case study data and observation. The study analyses the user pattern in real time and their behaviour according to the factors from the first step (descriptive-analytical method).
- **Step 3**: Design a comprehensive model/framework to describe the main influential factors affecting pro-environmental behaviour based on output from the second step (deductive method).

4 Case Study Setting

4.1 Recycled Spaces

It is simple to assert that the term "recycling space" is uncommon after extensive investigation into its existence. It is frequently used with many definitions; some researchers used it for areas where recycled or reused wastes were rebuilt. According to Allen (2016), modern architects create contemporary hubs for waste disposal and reuse using recycled and salvaged materials and innovative façades meant to inspire. A 2020 article titled "Recycled Cities: How Circular Design Shapes Urban Life" further described the recycling sector by stating that circular design is motivated by systems that self-manage and regulate, taking both time and whole life cycles into consideration. Roberto D'Arienzo's Ph.D. thesis outcomes are published (defended in 2015) (Bocquet, 2019). He referred to the garbage situation that occurred in Naples and recycling as a solution on an urban scale when the city 'developed a "zero-hectare" strategy aimed at "recycling" the urban space within its bounds.' Introducing a zero-hectare policy in line with the zero-waste approach, densification, and "urban recycling" was advocated in a 2004 urban planning document by a municipal committee selected in Naples (Waine, 2018).

Others used recycling space to describe the green movement transformation for public spaces (Lalbakhsh, 2012); the author discusses "The Impact of Recycling Urban Space in Sustainable Development in Developing Countries" in his article. He asserted that bringing back open, green space into urban life is the basic concept of space recycling. Due to the multiple uses of these spaces, each individual is responsible for maintaining their own space in social interactions. Others take the idea of recycling a step further, applying it to something incorporeal, like the development and renewal of the space's previous image, as Timothy A. Simpson's (1999) study needed an anthropological examination of the history and culture of Ybor City through the work and experiences of two local artists. The musical technique of "sampling" can be used as a metaphor to explain how artists appropriate and reuse old photographs and artefacts to create new works of art. The analysis demonstrates how the artists challenge and retell the prevalent story of Ybor City's gentrification (Simpson, 1999). Utilise recycling as a renewal or rebirth of the area or city (Gligorijevic, 1997). Urban recycling, according to her definition, is the process of transforming urban areas, structures, or entire historical towns to preserve the memory of the cities and raise the level of their urban culture (Gligorijevic, 1997). Hilary Powell examined Jacques Tati's 1967 film "Playtime" and its constructed city set on the outskirts of Paris as an example of the negotiation of urban space in his paper Recycling Junk-space: Finding Space for "Playtime" in the city in 2005. Its prescribed "instructions for use in this paper recycling appeared as expanding of the space." (Powell, 2005).

In contrast, recycling spaces are called abandoned places like brownfields that must be reactivated with creative activities that alter the resource and its urban–rural environment (Scaffidi, 2019). The local community, legislators, and the organisation managing the recycling site all play active roles in the interaction between social innovation and recycling procedures (Scaffidi, 2019). This study borrows the definition of "recycling the space" from the corresponding book, "Recycling Spaces." The book examines how the public sphere might be used to provide answers. How to revitalise ageing urban cores, build meaningful neighbourhoods in quickly expanding cities, offer economic and social value to resource-poor settings, and adapt to the ongoing changes in urban lifespan. The solution provided in the book is to design environments that encourage emotional investment in a location's long-term survival. Sustainable places are those that people identify with it. This book's enlarged definition of sustainability included four categories for those recycling areas (Schwartz-Partners et al., 2012):

- **Dying City Centres**: These spaces can no longer rely on their primary industries for survival; they lost their vitality as a shrinking population and a depressed economy led to vacant. These spaces need to regenerate by finding new ways to cultivate a stable population, diverse economies, and vibrant urban life.
- **Depleted Resources**: Resource extraction sites are paradoxically the world's most and least valued. Once the last ounce of gold, coal, oil, or gas is extracted, the perceived value of the land disappears with the last resource, leaving scarred, toxic land, and devastated economies. Populations are forced to either move away or live in the ruins of the past. These spaces need reinvention: new life, purpose, and identity.
- **Shifting Populations**: When populations leave one place, they arrive in another. As some cities grow and diversify with international immigration and rural migration from within their countries, others are left with diminished populations and unbalanced demographics. These cities must adapt. Rapidly expanding urban areas must adjust to accommodate higher densities and multiple cultures, while shrinking cities must diversify their offers to reverse out-migration trends.
- **Non-Existent urbanism**: Recent "from-scratch" cities and rapid cities and rapid sub-urban sprawl raise questions about how to activate vital cities where there is no current population density, no urban context, and no discreditable centre. In these cases, an urban activity must be generated. Destinations must be created. Identity is forged, and the sense of place intensified to introduce urban life where it previously did not exist.

In this study, a desert area in Bilibis, Egypt, will serve as the primary space that will be researched SEKEM, recycled, or transformed from dead space due to pollution and exhausting resources. Ibrahim Abouleish founded SEKEM Initiative in 1977 as an agricultural company, making it the first company in Egypt to apply biodynamic farming techniques to revitalise the soil. Over time, the space gets regenerated. SEKEM evolved into a project that combines a diverse mix of agro-industrial businesses and non-profit organisations that share the same values as Dr Ibrahim—added a new identity with an environmental context and a sustainable green meaning, changing how people see the location. The goal is to select a "ground" that tries to integrate into the city and provide better places. To create a new community that positively impacts the local environment and the inhabitants, materials and skin should be durable and have an outer skin appropriate for the local environment (Kim & Kwon, 2018).

The "Alternative Nobel Prize" was awarded to SEKEM in 2003 for its "enterprise model for the twenty-first century", which linked corporate success with the social and cultural growth of society via an economy of love. SEKEM is widely hailed as an "Egyptian organic pioneer." So SEKEM is the case study because it is well known for its organic food and integrated green communities. Their sustainable development strategies adhere to the SDGs and Egypt's national sustainable development strategy (Egypt's Vision 2030), which has a holistic outlook by integrating cultural, ecological, economic, and social life (SEKEM Association, 2021).

4.2 Data Analysis

4.2.1 The Space

This research's main scope is to investigate the impact of eco-friendly behaviour inside green sustainable spaces by observing behaviour through some activities. OECD, in 2013, listed some activities that can reduce cities' environmental impacts: land-use planning–Transportation–Buildings–Energy–Wastewater.

Land-use planning (Fig. 1):

- **The farm** (Fig. 2) is divided by the main road into two parts, the left side of the road is an extension that contains only farming areas and cattle ranches, and the place also includes a small plant for animal food production.
- **The gate** of SEKEM: The place's control point is well secured. The security did not allow anyone to enter the SEKEM gate only if he/she had permission by informing the gate of his/her name and date of the visit in advance. The medical centre can be found immediately after the farm's entrance, and the main farm itself is divided into many parts.
- **The medical centre** (Fig. 3): Provides care to the local community. General practice and family medicine, paediatrics, internal medicine with gastroenterology, cardiology, dentistry, ear, nose, and throat, minor and general surgery, neurosurgery, urology, orthopaedics, ophthalmology, and obstetrics and gynaecology are just a few of the departments of medicine that it houses. (SEKEM Report, 2016) However, many other departments the community may need are still not available. It serves employees with medical insurance (eco-health) and others with paid service.
- The houses of the owners of the farm also exist in the area behind the medical centre. (Private Zone Isolated by buffers zone of trees), the workers and employees there respect the family privacy very well and prevent any obtrusive to being near this zone.
- **Schools** (Fig. 4): After the medical centre, offices existed that work five days a week, the offices are not big enough with all facilities needed. And then the school area is one of the essential projects inside SEKEM that

Fig. 1 SEKEM land use (*source* (E,G.S.A) (Egyptian General Survey Authority, 2021): Edited by: Author)

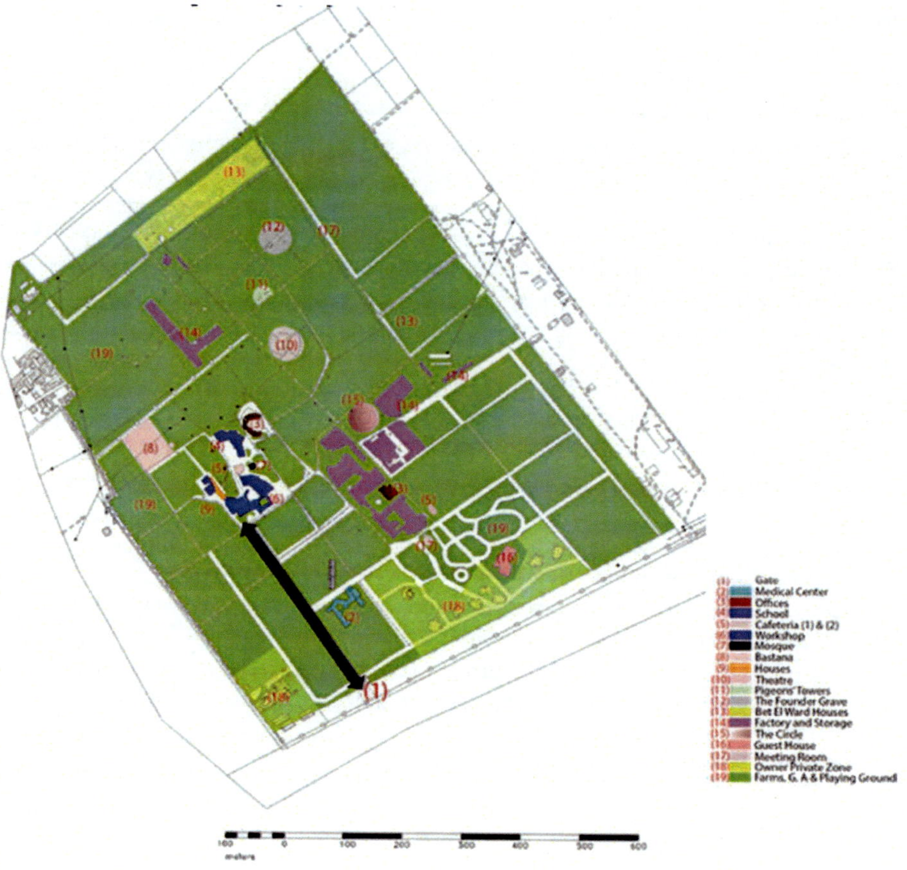

Fig. 2 SEKEM farm extension (*source* Author)

(a) (b)

Fig. 3 Medical centre (*source* By author)

are open to all communities around the farm and teach the students different languages and art—the output of the schools is alums with technical and artistic skills and sound knowledge, which consists of mainstream school: (primary, preparatory, and secondary), also nursery and kindergarten for the employees' kids while work. Moreover, technical education has workshops (Fig. 5a) in different departments for their students and others who joined vocational training centre. It contains several workshops in various categories, including a machine shop (metalwork), a carpentry shop, a knitting workshop, a ready-made clothing workshop, a plumbing workshop, an electronics workshop (including a computer maintenance workshop), and a welding workshop (with a particular focus on solar energy systems). The regular curriculum combines eurhythmy, crafts, theatre, dance, or music lessons. Working and learning are related at all ages and developmental stages. The workshops and theatre of the SEKEM School are among its spacious and well-appointed facilities. The workshops

Fig. 4 SEKEM School (*source* By Author)

already have the necessary equipment to meet basic demands but still need more sophisticated and technologically advanced equipment.

- **Special education**: Fagr: (Fig. 5b) for special needs (disabled), the students in this class trained them to make handicrafts and artworks, and they also sold all their products and showed samples of their products.
- **Community school**, Class El Ghroub: for child labour, to solve this problem, they provided the educational part coupled with an opportunity for the students between 12 and 14 to gain income, especially in the agricultural sector through light work during the chamomile harvest (those children are called "Chamomile Children"). Between the school and workshops building, there is a restaurant, a big kitchen attached to the dining room; it provides all students with free meals; the dining area is also open to everyone and serves three healthy and organic clean and tasty food with affordable price, at the right of the school is the mosque (Fig. 5c), the mosque has a unique design with easy access to all the people on the farm. At the same time, the playgrounds on the left, which are large, contain many sports, but the balls and usable facilities are not always available. Also, partly attached to the school above and beside the Nursery and KG: there are houses (Fig. 6) for the employee at affordable prices; in general, it is a good unit, but there are not enough spaces inside the apartment. And no air conditioning on hot days.

(a) (a) (c)

Fig. 5 **a** An eco-energy workshop, **b** Artwork of the special needs student's work, **c** The Mosque (*source* By Author)

- **The factories**: After this area, the factories are present, containing several factories such as herbs factory, an ISIS packaging factory, a textiles factory called Nature-Tex, and a dug factory called Atos. The factory area is designed with a large circular area in the middle where everyone working on the farm meets every morning as a SEKEM tradition called "The Circle" (Fig. 7). They are eco-friendly factories that produce sustainable products but can be modified technically to increase production and lower power consumption.
- Separated by green farming areas, on the left of the factories area, there is an area at the end of the farm that includes a big roman theatre (Fig. 8a) that hosts any celebration. The tomb/grave of Dr Ibrahim Abouleish (Fig. 8c–e), the founder of SEKEM farm, a unique building with a cross form, gives a spiritual feeling to the visitors; people often visit it to pray for him. Beside the grave is a group of houses called (Beit El Ward). The managers' families also accommodate some Abouleish family members (the farm's founders).

Fig. 6 Employees' houses (*source* by Author)

Fig. 7 Circle surrounded by the factories (*source* Author)

Fig. 8 **a** Open Roman Theatre, **b** The Pigeons' Tower, **c** The Grave Entrance, **d** The Grave Headstone, **e** The visitors' waiting area seats arranged in circle form (*source* Author)

- There are also **hotels** for visitors called the **guesthouse** (Fig. 9a) for foreign delegations who come for collaborations, the building is overall in good condition but needs more room service, and AC and bugs from the farms can be found sometimes. Moreover, a **building's meeting room** (Fig. 9b) consists of a big hall, library, and music room. The meeting hall has great rooms with a cosy feeling for meetings with bathrooms and facilities needed, while the library and music room contain plenty of musical instruments and books and are located inside the green court of the building.
- The gathering nodes: There are three gathering and meeting points. The most important one is the circle: The space is used for a unique tradition that people inside SEKEM must follow: all stand in a circle, crossing hands to signify equality and start the working day

Fig. 9 **a** Guest House, **b** Meeting Room (*source* Author)

with positive energy. This stand is repeated every day and once on the weekdays (Thursday) at the end of the day (Fig. 7).

- The same tradition is repeated inside the school open court, all students and teachers in the morning, same as (Fig. 10) The Circle. While the entertainment area is called **Bastana** (Fig. 11a, b), and it is considered the busiest area on the farm; in the morning, children can be found playing and working on their exercises, and at night people prefer to stay there. It has a fireplace, fixed stone seats, and a small cafeteria and toilets attached to this space. Everyone living on the farm on "Wednesday" meets, has food, and plays music around the fire.

- **Transport:**
  - Workers and farmers usually live in the villages around the farm in Bilbies.
  - Busses are the primary way of travelling from and to the farm.
  - Some workers and farmers use motorcycles and bikes to travel to the farm; parking for motorcycles and bicycles (Fig. 12a, b) is found at the farm's gates.
  - People are moving inside the farm mainly by walking around. Some travel using bicycles, and also, travelling using donkeys and horses is present but rare.
  - Bicycle parking in front of the mosque is considered a meeting node (Fig. 12a, b).
  - Passages between facilities are well-designed because there are few shortcuts outside the passages.
  - The passages/walkways between buildings and facilities are mainly tile/bricks (Fig. 13a) between facilities or (Madak) as shown in (Fig. 13b) and (Fig. 14a–c) between buildings and facilities.
  - Cars are not allowed inside many farm areas; there is parking after the main gate for visitors.
  - People living inside the farm, like the owners and workers, have cars to travel outside the farm. Several of their vehicles are electric and hybrid cars.

Fig. 10 SEKEM School, Main Classes Building and Courtyard where all students stand in a circle holding hands in the morning (*source* By Author)

(a) (b)

Fig. 11 Bastana: **a** at day, **b** at night (*source* By Author)

(a) (b)

Fig. 12 Workers use motorcycles (**a**) and bikes (**b**) and park them in unique parking spots in SEKEM (*source* Author)

- Usually, passages and roads have long trees from both sides that drop shadows on the road (Fig. 15a, b).

- **Buildings**:

The architecture in SEKEM is unified in style; all building is minimal with local material and vernacular soul. The coloured also unified the main dominating one is white for the facades, while only the windows and doors could be coloured, with blue, red, and yellow (Fig. 16a). Most buildings' heights are ground and first floor; some are only one (Fig. 16b). Architectural style: Environmental architecture employs passive strategies, utilizing fresh air for indoor ventilation. Windows are often designed for self-shading, achieved through wall thickness or sun breakers. Additionally, emphasis is placed on maximizing natural daylight through the strategic design of facades with generously sized windows for illumination. The windows are distributed in many directions to allow bigger daylight to enter the inner space as much as possible (Fig. 16c). Also, they used vernacular local materials as shown in Fig. 16d.

- **Energy**:
  - Energy consumption:

    Farm buildings are designed to reduce energy consumption by decreasing the number of heaters or AC and using efficient daylight instead of electricity.

The management of SEKEM usually increases the awareness of energy usage for the workers and people inside the farm and organisation.

Factories are not consuming much energy as they use solar energy for many facilities, and the factories have low carbon emissions as they use electrical machines or manual processing.

(a) (b)

Fig. 13 Shows the different types used on walkways (Captured by: Author)

(a) (b) (c)

Fig. 14 Streets made of compact soil (Madak) for either walking or car ways (Captured by: Author)

(a) (b)

Fig. 15 Shaded streets inside SEKEM (Captured by: Author)

- Renewable energy (Fig. 17a, b):

  A significant linear Fresnel solar reflector system is found beside the medical system.

Photovoltaic PV solar panels can be found in various places inside the farm.

  A flat plate-collector water heaters system is in the drugs factory (Atos) to heat the water using solar power.

- **Waste**:
  - Reducing:

    There are plenty of trash cans everywhere on the farm that separate the different types of waste.

The management of SEKEM usually increases the awareness of people inside the farm to separate and recycle their waste.

By observation, all spaces are clean, and people are awarded to put waste in trash cans by default.

  They used paper bags instead of plastic in factories.

  - Recycling:

    The farm facilities have a deal with waste recycling companies that take the separated wastes from the farm and recycle them.

    The use of organic wastes to form organic compost fertiliser is derived from organic compost sources, cattle manures, poultry droppings, and domestic sewage (Singh, 2012). This kind of fertiliser is used to get rid of wastes and, at the same time, recycle them to form a useful product by using worms and organic wastes to produce this natural fertiliser. This idea is spread outside SEKEM

Fig. 16 **a** Architecture style, minimal, coloured windows, and white facade. **b** Shaded buildings by sun breakers. **c** Depend on daylight by increasing the number of windows with big openings. **d** Local materials in the ceiling (*source* by author)

Fig. 17 **a** An extensive linear Fresnel solar reflector system, **b** PV at roofs (*Source* By Author)

borders, so it is easy to recognise small projects to create this kind of fertiliser on the way to SEKEM farm.

- Reusing (Fig. 18a–e):

  The landscape over the farm is built from reusing materials like wheel tires on playing grounds.

  At Bastana, the space is furnished by wooden seats, tables, and even the ping-bong table are reusable materials.

- **Water**:
  - The farm depends mainly on wells water.
  - A wastewater treatment plant exists on the farm to recycle and reuse wastewater for irrigation.
  - SEKEM owns a mineral water factory outside the SEKEM farm; the drinking water is present in various places as mineral water dispensers.
  - The irrigation is clean as there are no chemicals, and they use compost fertilisers.
- **People**:
  - There is a diversity of people in the place; various foreigners are working alongside Egyptians, well-educated or not, kids and elders, and low socioeconomic class and owners, not only healthy people but also disabled, all work in good harmony, integrated, and with space. They have mutual benefits and collaboration.
  - People who live and work on SEKEM farm are quiet and friendly people who are very friendly with strangers and each other.
  - Women are working in every possible job on the farm, and people are okay with that.
  - People are very loyal to the place and respect the site's owners.
  - The founder of the place is very respected by the people there. After his death, his grave is considered a sacred place. People feel nostalgic for him and pride because they worked and dealt with him during his lifetime.

Fig. 18 **a** Trash cans everywhere, **b** using paper bags instead of plastic, **c** using wheels tires in the landscape, **d** ping-bong table, **e** wooden seats and tables (*source* By author)

- People know each other and have many activities together, making them like a big family.
- The population's capacity inside the farm is low. The place is always enough for people.
- The place has a family atmosphere which gives the feeling of security.

4.2.2 The Impact on the Environment

Numerous associations and non-profit environmental organisations offer environmental programmes and services about sustainable practises and methodologies that address the obstacles and motivators to behaviour change and action. Based on the tenets and methods of community-based social marketing even in neighbourhoods without any mention of the environment in their covenants, some governments (in Japan, the USA, the UK, and others) make an effort to encourage individuals to use PEBs that lessen environmental loading. It created a framework for PEBs, created a list of target behaviours, and categorised them according to aims and behavioural domains like consumption, food and drink, personal travel, houses and household goods, and travel and tourism (Kurisu, 2015). The most common and efficient target goals are collected in Fig. 19 to be used to measure the impact of PEBs on the environment of SEKEN using a data sheet and annual report of SEKEM published every year on their website (www.sekem.com) by comparing the statistics of SEKEM Annual Report to the years 2016 and 2021 guided with the factors listed below.

5 Results and Discussion

In his book Life between Buildings, Jan Gehl (2012) made a compelling case for the importance of considering the quantity of these activities concerning the quality of the urban environment and space. He separated the three types of outdoor activities in public areas into social, voluntary, and necessary. Each type of activity has different demands on the physical environment. He suggests that only required tasks should be performed in unappealing outside spaces. When outdoor places are of high quality, vital activities generally take place at the same time (Gehl, 2012) (Fig. 20).

- **Necessary activities**: include those that must be done, such as going to work or school and awaiting a bus.
- **Optional activities**: that is, the attendees are only there if they want to do so and if the circumstances—time, location, etc.—allow for it, such as sitting, enjoying life while standing, or walking to obtain fresh air.
- **Social activities**: They require other people to be present in public areas. Children at play, greetings and small talk, group activities of all types, and so on are examples of social activities.

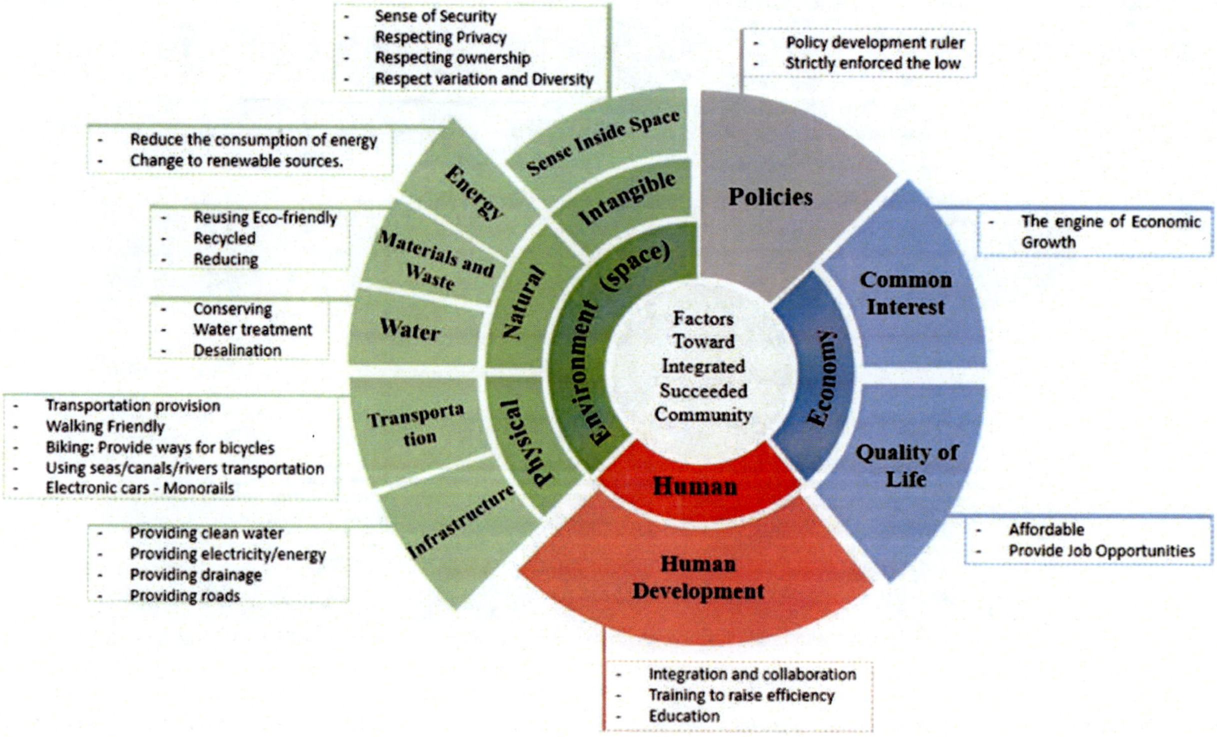

Fig. 19 Towards succeeded community framework (*source* Author)

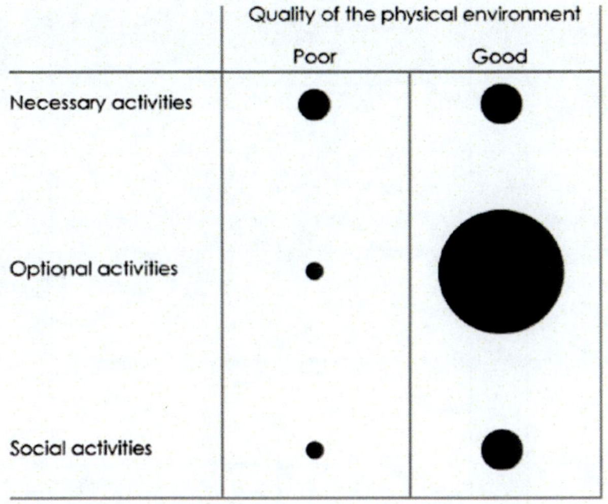

Fig. 20 Relationship between the quality of outer space and the incidence of external activities (*source* Gehl, 2012)

5.1 The Relation Between Activities and the Quality of Space

Applying Gehl's suggestions to SEKEM Farm's various spaces during the site visit yielded the following observationresults for space and activity (refer to Table 2): Approximately 59% of the activities within SEKEM are categorizedas Necessary, around 27% are classified as Social Activities, and the remaining portion falls under Optional. Thisoutcome can be attributed to SEKEM being an investment project for sustainable development. As shown in (Fig. 21a, b). Generally, the payment investment concept is linked with economic benefits and money returns. While sustainability encourages the absence of unused or underused spaces, this indicates that the project fits the users' primary needs and that the users are adapted to the space's capabilities. By comparing the activities with the quality of the space, we found the rate ranged between good and satisfying, this also may give the answer during the ethnography in the space, and by observation, most people there have feelings of loyalty to SEKEM, and they are proud to belong to the place. The loyalty feeling may be because the quality of life inside the work environment is much better than in other surrounding locations.

5.2 The Most Influence Factors

The results of analysing the outputs of the SEKEM rates and statistics (Table 1) show that how they deal with waste and use eco-friendly materials are the most successful part of their goals. Measuring their rates shows achievement by 62% in five years. While total net revenue is in the second level, it will rise in 2021 by 60% more than in 2016. Although the year 2016 witnessed local currency devaluation. Additionally, in 2019, the Corona pandemic swept the world, and most countries were locked down. A partial lockdown occurred in Egypt in the first half of 2020. Despite all these factors negatively affecting the economy, SEKEM profits have increased.

Additionally, workers' and employees' salaries increased, with the lowest compensation raised by 53% in 5 years, and medical insurance for workers expanded to cover 100% instead of 35% in the past years, and this confirms the sustainable development policy inside SEKEM. Water conservation and using fossil water is at the next level at 47%, then human development by 20%, while energy is the lowest with only a 7% increase between the two years. Table 3 shows the rate of growth, which shows the factors that are the most influenced towards a successful green community, while Fig. 22 shows priorities of influenced factors for developing a green community based on a SEKEM case study. Finally, the outcome shows the extent of apparent features the suggested factors could be repeated in different spaces, by using PEB model shown in Fig. 23, and accordingly, the spaces could be transformed to be developed based on these results.

Fig. 21 **a** Space activities, **b** Space quality (*source* by author)

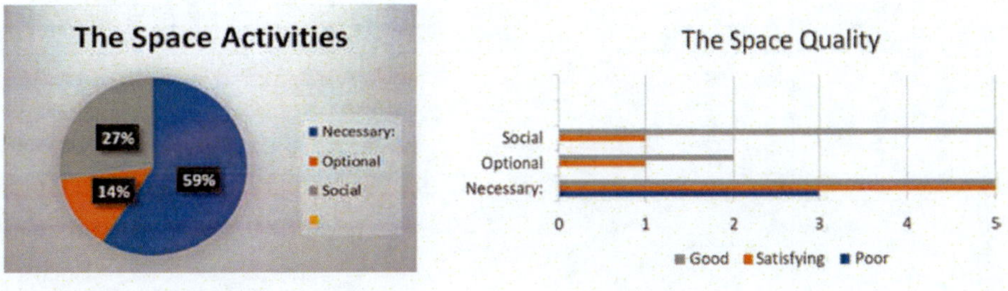

Table 1 Achievement rate of SEKEM impact statistics (*source* SEKEM annual report, 2016 and 2021 edited by author)

Elements		Factors Towards Integrated Succeeded Community		Extensive Performance Evaluation Framework		Year		Evaluation %	-VE impact		Positive improvement				Total % of Achievement
		Categories	Recommended achievement	Indicator	Unit	2016	2021		(0): (-1)	From 1:25%=1	From 26:50%=2	From 51:75%=3	From 76:100%=4	More than 100%=5	.=sum. of positive impacts points/ (5*(no. OF elements) %
Economy		Common Interest	The engine of economic growth	Total net revenue	m/year	376	619.4	64.73				3			60
		1.2 Quality of Life	Affordable	Minimum salary paid, excluding non-monetary benefits	LE/ month	1300	2000	53.85				3			5
				Average amount of non-monetary benefits per employee per month	LE/ month	130	963	640.77						5	
				The ratio of The highest to the lowest annual full-time salary	Ratio	1:31	1:38	22.50		1					
			Provide job opportunities	Total number of employees	count	1,700	1,904	12.00		1					
Ecological	Intangible	Sense Inside Space	Sense of security												
			Respecting privacy												
			Respecting ownership												
	Natural	Energy	Reduce the consumption of energy	Total electricity consumption (grid, diesel & renewables)	MWh	4,439	6,938	56.30	−1						7
			Electricity												
			fuel	The total amount of gasoline consumption	KL	731	601	−17.78		1					
			Change to renewable sources	Share of electricity consumption from renewable sources	%	2.7	24	21.30		1					
			Installation of solar panels												
			Installation of solar water heater												
			Introduction of wind power												

(continued)

Table 1 (continued)

Elements		Factors Towards Integrated Succeeded Community		Extensive Performance Evaluation Framework		Year		Evaluation %	-VE impact		Positive improvement				Total % of Achievement
		Categories	Recommended achievement	Indicator	Unit	2016	2021		(0): (-1)	From 1:25% = 1	From 26:50% = 2	From 51:75% = 3	From 76:100% = 4	More than 100% = 5	. = sum. of positive impacts points/ (5*(no. OF elements) %
		Materials and Waste	Reusing eco-friendly	Eco-friendly packaging material	Ton	1045	2177	108.33						5	62
			Use reusable bags/bottles instead of disposable ones												
			Recycled: Increase recycling (and segregation)	Share of recycled non-organic waste	%	42	73	73.81				3			
				Share of recycled organic waste	%	66	62	-6.06		1					
				Share of packaging material input	%		62								
			Reducing using plastic	Total plastic waste	Ton	25	18	−28.00			2				
				Total paper waste	Ton	44	37	−15.91		1					
				Total glass waste	Ton	71	18	−74.65				3			
				Total organic waste	Ton	493	167	−66.13				3			
		Water	Conserving	The total amount of water usage for agricultural	KM3	2,405,438	4,510	−99.81					4		47
			Avoidance of water running												
			Repair all leaks												
			Water treatment	Share of waste-water recycled and reused	%	100	100	0.00		1					
			Grey water												
			Rainwater												
			Desalination	Fossil water	%	54	77	42.59			2				
	Physical`	Building	Does the design of the house provide thermal comfort, or it needs AC, fans or heaters?												

(continued)

Table 1 (continued)

Elements	Factors Towards Integrated Succeeded Community		Extensive Performance Evaluation Framework		Year		Evaluation %	-VE impact		Positive improvement				Total % of Achievement
	Categories	Recommended achievement	Indicator	Unit	2016	2021		(0): (-1)	From 1:25% = 1	From 26:50% = 2	From 51:75% = 3	From 76:100% = 4	More than 100% = 5	. = sum. of positive impacts points/ (5*(no. OF elements) %
		Do houses spaces, areas and dimensions suitable for living?												
		Does rent fees reasonable or appropriate for the service provided?												
	Transportation	Transportation provision	Available or not	%	100	100	0	0						0
		Use of public transportation												
		Walking Friendly	Available or not	%	100	100	0	0						
		Biking: Provide ways for bicycles	Available or not	%	100	100	0	0						
		Or riding horses												
		Using seas/canals/rivers transportation	Available or not	%	0	0	0	0						
		Electronic cars - Monorails	Available or not	%	0	0	0	0						
	Infrastructure	Providing clean water	Available or not	%	100	100	0	0						
		Providing electricity/energy	Available or not	%	100	100	0	0						
		Providing drainage	Available or not	%	100	100	0	0						
		Providing roads	Available or not	%	100	100	0	0						

Table 1 (continued)

Elements	Factors Towards Integrated Succeeded Community		Extensive Performance Evaluation Framework		Year		Evaluation %	-VE impact		Positive improvement				Total % of Achievement
	Categories	Recommended achievement	Indicator	Unit	2016	2021		(0): (-1)	From 1:25% = 1	From 26:50% = 2	From 51:75% = 3	From 76:100% = 4	More than 100% = 5	. = sum. of positive impacts points/ (5*(no. OF elements) %
Human	Human Development	Integration and collaboration	Worker aged above 60 yrs. old	%	0	3	3.00		1					20
		Training to raise efficiency	Average yearly hours of training per employee	hrs/Y	99	121	22.22		1					
		Education	Total number of students	no	303	544	79.54				3			
		Respect variation and diversity	Female employee	%	22	24	9.09		1					
		Loyalty	Long-term employees (working 7 + years at SEKEM holding)	%		49								
Policies		Policy development ruler												
		Strictly enforced the low												

Table 2 Represents the relation between activity done in different spaces and the quality of the same space (*source* By Author)

Land use as in Fig. 1	Space elements	Function	Activities			Quality of space		
			Necessary	Optional	Social	Poor	Satisfying	Good
1	Gate	Secured—controlling point	①			❶		
2	Medical centre	Serving community	①				❶	
3	Office buildings	General administration	①			❶		
4	Schools	Serving community	①					❶
5	Cafeteria	To provide all students with free meals	①					❶
6	Vocational training centre	workshops	①				❶	
	Nursery and kindergarten	For the employees' kids	①					❶
7	Mosque	For praying		①				❶
	Art workshops	For special-needs-students	①				❶	
8	Bastana	Night gathering			①			❶
	Playing area	Playing grounds			①		❶	
10	Houses group (1)	For employees	①				❶	
11	Open theatre	For social events			①			❶
12	Pigeons' towers	For the breeding of doves			①			❶
13	The founder grave	Stated as a landmark and monumental building			①			❶
14	Bet El ward houses	For the owners' and managers' families Bet El Ward	①					❶
15	4 factories	• Lotus: Organic herbs and spices • ATOS Pharma: natural medicines and healthcare products • ISIS Organic: herbal teas • NatureTex: Organic cotton textiles	①				❶	
16	Circle space	Gathering point	①					❶
17	Storage hangers	To store products and goods	①			❶		
18	Guest house	Hotel for the foreigners		①			❶	
19	Meeting room	For discussing and press		①				❶
	Library and music room	For culture development			①			❶
20	Owner private zone	Private villas with buffer trees for isolation						

Table 3 Most influence factors for succeeded green community in rural

Factors towards integrated succeeded community		The total rate of achievements (%)
Economy	Economic growth	60
	Affordable	50
Ecological	Energy	7
	Materials and waste	62
	Water	47
Human	Development	20

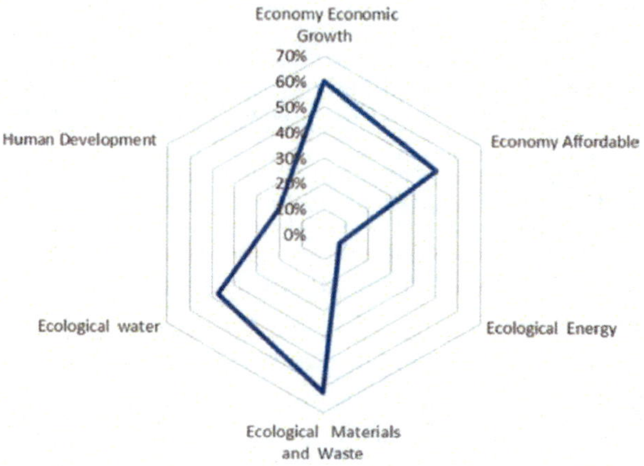

Fig. 22 Priorities of actions for developing a green community based on the SEKEM case study (*source* Author)

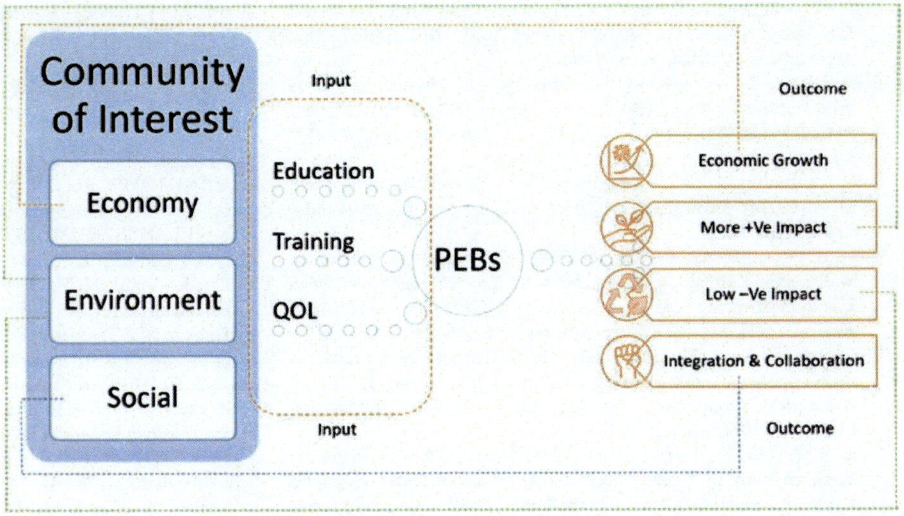

Fig. 23 PEB model in SEKEM (*source* by Author)

6 Conclusion

At the beginning of this research, there were several concerns: If the study of PEBs could exist in rural Egypt, due to the prevailing impression that this type of behaviour is contrary to the nature of daily life among the Egyptians, specifying that these concepts come from developed countries and that the human being there is more luxury and know well about the threat of climate changes. And if PEBs in rural, it will be individual, and few people will follow. Even by knowing the existence of a greener society like SEKEM promoted for their products, as they are 100% organic, there are some worries that it may be just for advertisement for the place and no green community exist. These concerns decreased to a minimum level after visiting the farm plot. It seems a clean place, with no garbage or plastic bottles on the walkways and no unpleasant odours. They realised sustainability concepts and other terms by dealing with the workers/employees inside. After investigating the reasons, SEKEM was founded by Dr Ibrahim Abouleish on a sustainable basis, based on the principle of "community of interest" (Newman, 1980). Planned communities exist to improve members' quality of life because they offer choices, lifestyles, services, and, most importantly, efficiencies that people value (Bray et al., 2014; Green Communities Canada, 2008; Useful Community Development, 2021).

The SEKEM farm was founded as an agricultural investment project to "restore and preserve the vitality of the soil and food as well as the biodiversity of nature" via sustainable, organic agriculture and to support social and cultural development in Egypt. From the beginning, the founder had an unwavering vision and goals to create a green community. He built a system and state policies to solve any problems (Abouleish et al., 2005). To preserve his project and his ideas, people who will help produce this must be aware and know enough about sustainable developments and their goals. In SEKEM, this happened on two levels, daily awareness level, by daily meeting in the CIRCLE, while education and training do the same function but in long terms and target different group ages by introducing sustainable development from the beginning of the educational journey.

Thus, creating a successful green community with a positive impact on the environment in rural Egypt is feasible under specific conditions encompassing human development (both individual and community), environmental considerations, and economic factors. Ultimately, the observed outcomes reveal the degree of apparent features, suggesting that these factors could be replicated in various spaces. Consequently, spaces can undergo transformation and development based on these findings.

References

Abouleish, I., Cardwell, A., & Kirchgessner, M. (2005). *SEKEM : A sustainable community in the Egyptian desert.* Floris Books. https://www.wellspringbookshop.co.uk/books/abouleish-ibrahim/sekem-a-sustainable-community-in-the-egyptian-desert/

Allen, E. (2016). 12 rad recycling centers that go green in style. https://www.architecturaldigest.com/gallery/recycling-center-design

Assal, A. M., Atia, M. R. A., & Shaaban, S. (2018). Performance of solar dish Stirling engine systems under Egyptian operating conditions. *Engineering Research Journal, 160*(0), 79–104. https://doi.org/10.21608/erj.2018.139543

Egyptian General Survey Authority (EGSA). (2021). Bilbeis dwg map (SEKEM). https://www.esa.gov.eg/

Bello-Schünemann, J., & Aucoin, C. (2016). African Urban Futures—ISS Africa. Institute for Security Studies (ISS). https://africaportal.org/publication/african-urban-futures

Bocquet, D. (2019). Métabolismes urbains. De l'hygiénisme à la ville durable : Naples 1884–2004, Roberto D'Arienzo, Genève : MétisPresses, 2017. *Flux, N 115*(1), 74. https://doi.org/10.3917/flux1.115.0074

Bray, A., Bunting, J., Fellows, L., Meyer, M. J., Richter, H., & Warren, D. A. (2014). Best practices: Green communities. https://foundation.caionline.org/

Dietz, T., Rosa, E. A., & York, R. (2009). Environmentally efficient well-being: Rethinking sustainability as the relationship between human well-being and environmental impacts. *Human Ecology Review, 16*(1), 114–123. https://www.jstor.org/stable/24707742

El Ghorab, H. K., & Shalaby, H. A. (2016). Eco and Green cities as new approaches for planning and developing cities in Egypt. *Alexandria Engineering Journal, 55*(1), 495–503. https://doi.org/10.1016/J.AEJ.2015.12.018

Fard, H. R. (2014). Evaluating spatial behavior in the urban public space of Kadıköy Square. https://www.semanticscholar.org/paper/Evaluating-Spatial-Behavior-in-the-Urban-Public-of-Fard/038c64edaa7ba02fd1e26a21399a2a0fce0ee43f

Gatersleben, B., Joye, Y., Berg, A. E. van den., Midden, C., & Ham, J. (2019). *Environmental psychology* (L. Steg & J. I. M. De Groot, Eds.) (2nd ed.). Wiley.

Gehl, J. (2012). *Life between buildings using public space* (J. Koch, Ed. & Trans.). Island Press.

Gligorijevic, Z. (1997). Urban recycling: A way to save the character of the cities. In *Structural studies, repairs and maintenance of historical building* (Vol. 29). WIT Press.

Green Communities Canada. (2008). *How to grow a green community.*

Güneralp, B., Lwasa, S., Masundire, H., Parnell, S., & Seto, K. C. (2017). Urbanization in Africa: Challenges and opportunities for conservation. *Environmental Research Letters, 13*(1), 015002. https://doi.org/10.1088/1748-9326/aa94fe

Kim, S., & Kwon, H. (2018). Urban sustainability through public architecture. *Sustainability, 10*(4). https://doi.org/10.3390/su10041249

Klöckner, C. A. (2013). A comprehensive model of the psychology of environmental behaviour—A meta-analysis. *Global Environmental Change, 23*(5), 1028–1038. https://doi.org/10.1016/J.GLOENVCHA.2013.05.014

Kollmuss, A., & Agyeman, J. (2002). Mind the Gap: Why do people act environmentally and what are the barriers to pro-environmental behavior? *Environmental Education Research, 8*(3), 239–260. https://doi.org/10.1080/13504620220145401

Kurisu, K. (2015). Pro-environmental behaviors. *Springer Japan.* https://doi.org/10.1007/978-4-431-55834-7

Lalbakhsh, E. (2012). The impact of recycling urban space in sustainable development in developing countries. *APCBEE Procedia, 1*, 331–334. https://doi.org/10.1016/J.APCBEE.2012.03.054

Matamanda, A. R., & Nel, V. (2020). Sustainable urbanization in Africa: The critical enablers and disablers (pp. 738–751). https://doi.org/10.1007/978-3-319-95717-3_119

Moore, G. T. (2006). Environment, behaviour and society: A brief look at the field and some current EBS research at the University of Sydney. *Environment, Behaviour & Society.*

Muggah, R. (2016). Development specialists must get to grips with fragile cities | Devex. https://www.devex.com/news/development-specialists-must-get-to-grips-with-fragile-cities-87538

Newman, O. (1980). *Community of interest.* Anchor Press/Doubleday.

Powell, H. (2005). Recycling Junkspace: Finding space for 'Playtime' in the city. *The Journal of Architecture, 10*(2). https://doi.org/10.1080/13602360500114874

Roka, K., Filho, W. L. ., Ravazzoli, E., Valero, D. E., Ang, J. C., Diogo Guedes Vidal, N. B., & R. L. M. (2020). *Sustainable cities and communities* (W. Leal Filho, A. Marisa Azul, L. Brandli, P. Gökçin Özuyar, & T. Wall, Eds.). Springer International Publishing. https://doi.org/10.1007/978-3-319-95717-3

Scaffidi, F. (2019). Soft power in recycling spaces: Exploring spatial impacts of regeneration and youth entrepreneurship in Southern Italy. *Local Economy: The Journal of the Local Economy Policy Unit, 34*(7). https://doi.org/10.1177/0269094219891647

Schwartz-Partners, M., Waugh., E., & Waldhei, C. (2012). Recycling Spaces: Curating Urban Evolution. Oro editions.

SEKEM Association. (2021). SEKEM 2021. https://www.sekem.com/wp-content/uploads/2022/05/SEKEM-Report-2021-1.pdf

SEKEM Report. (2016). SEKEM | Annual Report 2021. https://www.sekem.com/wp-content/uploads/2022/05/SEKEM-Report-2021-1.pdf

Simpson, T. A. (1999). Recycling urban spaces. *Western Journal of Communication, 63*(3). https://doi.org/10.1080/10570319909374644

Singh, R. P. (2012). *Organic fertilizers : types, production and environmental impact.* Nova Science Inc. https://www.researchgate.net/publication/274896698_Organic_Fertilizers_Types_Production_and_Environmental_Impact

Steg, L., & Vlek, C. (2009). Encouraging pro-environmental behaviour: An integrative review and research agenda. *Journal of Environmental Psychology, 29*(3), 309–317. https://doi.org/10.1016/j.jenvp.2008.10.004

Stern, P. C., Dietz, T., Ruttan, V. W., et al. (Eds.). (1997). *Environmentally significant consumption.* National Academies Press. https://doi.org/10.17226/5430

The Ministry of Environment. (2017). State of the environment 2017 Arab Republic of Egypt summary for policymakers. https://www.eeaa.gov.eg/portals/0/eeaaReports/SoE-2017/EgyptSOE2017-SPMEnglish.pdf

United Nations. (2014). 2014 revision of the World Urbanization Prospects | Latest Major Publications—United Nations Department of Economic and Social Affairs. DESA. https://www.un.org/en/development/desa/publications/2014-revision-world-urbanization-prospects.html

United Nations Development Programme (UNDP). (2015). The sustainable development goals (SDGs). www.undp.org

Useful Community Development. (2021). Green communities address livability, climate change. https://www.useful-community-development.org/green-communities.html

Waine, O. (2018). Recyclable city or disposable city? A century of urban waste in Naples. Metropolitics.

Gentrification Study and Preservation Efforts Semarang Old Town

Mila Karmilah, Ardiana Yuli Puspitasari, and Syarifah Atia

Abstract

The term of gentrification is often associated with rehabilitation, revitalization, rejuvenation, and improvement concerning area's environmental quality. It is believed that environmental quality improvement generates investments such as commercial activities to Semarang Old Town. In addition, such improvement alters the land use and social class, resulting from gentrification in Semarang Old Town. This study aims to observe the conducted efforts in order to preserve Semarang Old Town and the development of cultural tourism, and to determine whether these efforts affect the gentrification in the old town area. The method of this study is a qualitative approach, conducted with interview and observation techniques to observe changes and to determine the change's impact by distinguishing physical, economic, and social aspect in the area. The results of the study analysis reveal that gentrification occurs from the transformation of environmental characteristics which is deemed necessary as a means of adaptation for society. In particular, the impact of gentrification is determined by both good and bad connotations, affecting physical, social, and economic aspects.

Keywords

Gentrification · Impact · Preserve · Semarang Old Town

M. Karmilah (✉) · A. Y. Puspitasari · S. Atia
Urban and Regional Planning Department,
Faculty of Engineering, Sultan Agung Islamic University,
Semarang, Indonesia
e-mail: mila.k@unissula.ac.id

A. Y. Puspitasari
e-mail: ardiana@unissula.ac.id

1 Introduction

The phenomenon of gentrification has been the discussion in America and Europe, leveraging into the urban topic in Indonesia whose presence is inevitable, along with the increasing urbanization. Gentrification refers to the process of land-use change which transforms middle-to-lower community settlements into middle-to-upper community settlements (Medha, 2017). In addition, the term of gentrification refers to an activity conducted to improve the area's image by improving the area's quality without causing physical differences in the area (Prasetya, 2006). Referring to various studies on gentrification, the phenomenon of gentrification could be related to efforts, conducted to rejuvenate, revitalize, rehabilitate, and improve the region's environment quality. Semarang Old Town area (*KKLS*) is an acknowledged area that has a history, preserved since 2015. As such, various types of activities are performed in the form of improvements in physical, social, and economic aspects. Prior to conducting the conservation efforts, Semarang Old Town area experienced various problems which are diverse and complex, including building, environmental, governance, and social problems (Puspitasari & Ramli, 2018).

On the other hand, Semarang Old Town area is also memorable due to illegal activities such as homeless people living in empty buildings. Worsening the situation, public transportation stops, street vendor activities, and cockfighting activities decrease the function of the area. Hence, improving the environmental quality of Semarang Old Town area is performed by structuring and improving the area by handling public transport stops, street vendors, homelessness, and cockfighting activities in order to improve the image of the region.

In particular, physical improvements are performed by repairing and utilizing various existed buildings into buildings with new functions such as for commercial activities. Previously, several buildings in the Semarang Old Town

M. F. Shahidan et al. (eds.), *Urban Narratives: Exploring Identity, Heritage, and Sustainable Development in Cities*,
Advances in Science, Technology & Innovation, https://doi.org/10.1007/978-3-031-48517-6_24

area were vacant and damaged due to unknown ownership of the land (Harani et al., 2016). The landowner repaired and leased the building to investors later utilized for commercial activities. Meanwhile, some buildings were left with the unknown owners.

According to Paula and Wakhidah (2020), it is reported that there is a relationship between gentrification and revitalization. This study reports that the revitalization which was carried out did not affect on the lives of people living in the old city area, even this revitalization resulted in the displacement and eviction for people who are unable to adapt.

The objectives of this study are (1) to identify preservation efforts in the Semarang Old Town area; (2) to identify the social and economic implications of gentrification.

2 Review Literature

2.1 Gentrification and Impact on Heritage Site

Opinions regarding gentrification phenomenon are typically divided from the viewpoints of policy makers/government, researchers, affected communities, public (commentators), and developers. In determining the impact of gentrification, various stakeholders must be involved to conclude the cost and benefit from a comprehensive point of view (Rahardjo, 2017). The impacts of gentrification are basically grouped into both negative and positive connotations (Arkaraprasertkul, 2018). As such, the consequences affect economic and social aspects (Hancock & Algozzine, 2006).

Gentrification initially introduced by Ruth Glass in 1964, due to changes or improvements in housing leading to the transformation of society's social class. However, the price of the property for both the rental price and tax land increased along with the development of rehabilitation boosting the value of the area. Furthermore, gentrification is perceived as an injustice to the social class of society as it involves economic capitalism in urban areas boosting the value of land and property, capable of forcing the poor to move from their places of residence (Glass, 1964). This opinion is in accordance with the opinion of Medha (2017) explaining that gentrification is defined as a process of land-use transformation causing a change of middle-to-lower community settlement areas into middle-to-upper community settlement areas. In addition, gentrification is also related to a process of moving the middle community to a less viable location by undergoing improvement.

The effect of gentrification related to social aspect of the region is immense. As cited by Knox in Prayoga (2013), gentrification is described as the formation result of transformation and mixing of high-end community activities with the lower classes. Gentrification replaces the areas that have a low value, which will increase if most residents were those previously having a low economy level. When the area has not been concentrated, it is occupied by people with low economic and community groups who have the opportunity to exit the area concentrated with various causes such as economy-related matters. Areas that face development become attractors or magnets for immigrants, adding to competition, increasingly cornered the locals after the presence of immigrants (Prayoga, 2013). According to Prayoga (2013), groups with a newly arrived middle economic state will bring culture or habits from their origin. Along with the formation of development, the concentrated area is transformed into an area with a new characteristic socioculturally.

Furthermore, the phenomenon of gentrification is widely mentioned in relation to rehabilitation, revitalization, rejuvenation, and improvement of environmental quality in historic areas. The process of preserving historic areas, which is a process of physical improvement of environmental, social, and economic components of the private or public sector, leads to a gentrification phenomenon. According to (Arkaraprasertkul, 2018) gentrification that occurs in historic areas refers to a process where indigenous people improve buildings and environment toward the middle class by implementing the uniqueness of the building to maintain their economy. Elaborating further, 1 (Arkaraprasertkul, 2018) pointed out that the impact of gentrification in heritage areas is interpreted differently, conveying that there are two reasons for gentrification, including functional and symbolic reasons. The functional reason relates to a change in the function of the building, while the symbolic reason means that there is exotification of buildings and areas (symbolic) without any development affecting the area. Meanwhile, according to (Cesari & Dimova, 2019), gentrification is defined as a process that involves changes in the population of land use, thereby boosting a higher socioeconomic status for new users.

Research conducted by (Yuksel & Iclal, 2005) states that the development of cultural tourism with the case of a small town in Turkey as an alternative to large group tourism, where tourism becomes a "tool" to increase cultural tourism visits which at some point will become commodification of tourism (culture) and consumption (tourists). It is concluded by both authors that the opportunity to conserve through cultural tourism is only short-term, but it requires more effort, especially in the community approach for the long-term and sustainability.

2.2 Efforts of Preserve the Heritage

Preservation is the protection of objects or areas that have a history related to something that interferes with their existence and preservation. Referring to Law No. 11 of 2010 concerning cultural property, conservation becomes a dynamic effort in protecting, enhancing, and utilizing the existence of cultural property and its value. Preservation involves events in the past and present. The purpose of conservation as stated by prior study (Catanese & Snyder, 1989) is to provide protection to objects or areas with a history that has a special personality and to ensure the aesthetic and strategic diversity of the area.

The concept of preservation has been seen as an action framework, which means security, protection, maintenance, and care in a very small interpretation. A broader interpretation of conservation includes action management or life management in the form of natural areas, such as nature reserves or built-up areas such as urban areas, villages, or rural areas. The interpretation involves an effort to preserve historical relics, whether in the form of buildings or areas, which is an act of management of cultural resources, conducted through a decision-making process that is non-formal, or official political.

According to prior study (Catanese & Snyder, 1989), the protected areas continue to expand in relation to the growth and conservation of an area based on planning aspects, as well as the growing need for maintenance activities. During the process, preservation is not only in the form of maintenance, but it also relates to a sustainable new life addressing the needs of the population by engaging with the surrounding population and its new functions. Elaborating further, ultimate in Warsito (2017) affirms that studies which are theoretical and conceptual should be handled with comprehensive area planning and design, accompanied by marketing and implementation as well as targets related to the development/preparation of research areas implementing quality and distinctive character. Particularly, the implementation of conservation activities is adjusted to the level of needs and conditions of the conservation object and the surrounding environment. Efforts that should be actualized in relation with preservation activities of historic buildings or areas (ICOMOS, 1999), include:

2.2.1 Conservation

Conservation refers to the entire stage of preserving an object or place without compromising its existing historical value. According to (ICOMOS, 1999), the purposes of conservation are to improve, maintain, or exhibit historical records on objects as well as to preserve the state of future life for the object. Utilization in a building is important because all functions are required in order to make it effective.

2.2.2 Preservation

Preservation refers to an effort to defend objects or places in accordance with field conditions to make the building durable according to the original situation in the process of weathering the building. Preservation activities in buildings or historical areas are not intended to preserve and maintain buildings architecturally rather than to historical and cultural values.

2.2.3 Reconstruction

Reconstruction refers to a restoration effort to build and repair the same buildings and environments damaged by the time or disaster. Thus, it is not managed or must be moved by using the remaining elements, adding new building materials, and functioning the building properly by meeting the technical conditions.

2.2.4 Revitalization

Revitalization refers to an effort in obtaining the added value in terms of economic, social, and cultural value in utilizing historic buildings and areas, and in preventing the destruction of relics that have historical value due to the productivity of the region.

3 Methods

This research utilizes a qualitative approach with a case study approach. Through a case study research strategy referring to (Creswell, 2014), the steps are organized by preparing data management, interpreting information and giving codes, describing the case in detail according to its context, observing patterns or relationships between several categories, drawing conclusions from the case, and presenting them in a narrative. Case study research emphasizes the depth of analysis on a particular case that is more specific in understanding certain phenomena in a particular place and at a particular time (Rahardjo, 2017). This study implements thematic analysis method to provide the reports by emphasizing answers to research questions, resulting in reporting themes that are in accordance with research questions (Hancock & Algozzine, 2006). The analysis used in this study is content analysis of interviews that will form themes, thereby withdrawing conclusions regarding the relationship between gentrification and conservation efforts in the old city area of Semarang.

4 Result and Discussion

4.1 Results

This study was conducted in Semarang Old Town area (KKLS), Semarang City, and Central Java Province. Semarang Old Town area (KKLS) is acknowledged as a historic area that experienced conservation activities in 2015. In particular, Fig. 1 (Google Maps, 2020) shows the delegation in this study is located in the core area of Semarang Old Town, comprising an area of 31 ha. Semarang Old Town is included in the administrative area of Tanjung Mas Village, North Semarang District, Purwodinatan Village, and Central Semarang District.

The Old Town Area of Semarang comprises an area of 31 ha based on Semarang City Regulation Number 8 of 2003, concerning Building and Environmental Planning (RTBL) of the old town area. Buildings in the Old Town of Semarang are functioned as housing (2.04 ha), public park (0.48 ha) commercial area (3.11 ha), education and religion facility (0.09 ha), parking (0.87 ha), and factory, office, warehouse, and abandoned area (4.08 ha). Figure 2 illustrates the area of land use in Old Town of Semarang in 2020 (Puspitasari & Ramli, 2017).

Buildings in the Old Town of Semarang in 2020 were mostly empty or not functioning, although some are enabled for trade and services, influenced by the development of area and quality improvement due to conservation efforts in the Semarang Old Town area. Additionally, most buildings in the Old Town Area of Semarang serve as housing and offices. Figure 3 shows the function of the building in Semarang Old Town (Semarang City Planning Agency, 2019).

Concept of preserving the Old Town of Semarang is implemented as a reference for improving the state of buildings and environment and for developing the economy and social community. The present author believes that the concept of KKLS preservation should be adjusted to the existed conditions and problems. However, roles of government and stakeholders are deemed necessary in preserving and managing KKLS by also involving building owners, community, private, or government in order to support the acceleration of KKLS regulation.

KKLS space utilization consists of five land-use segmentation with different conservation and management concepts, based on Semarang City Regulation Number 8 of 2003 concerning Building and Environmental Planning (RTBL) of old town and a prior study (Puspitasari & Ramli, 2018). Figure 4 illustrates the five segmentations of land use in Old Town of Semarang. The segmentation of

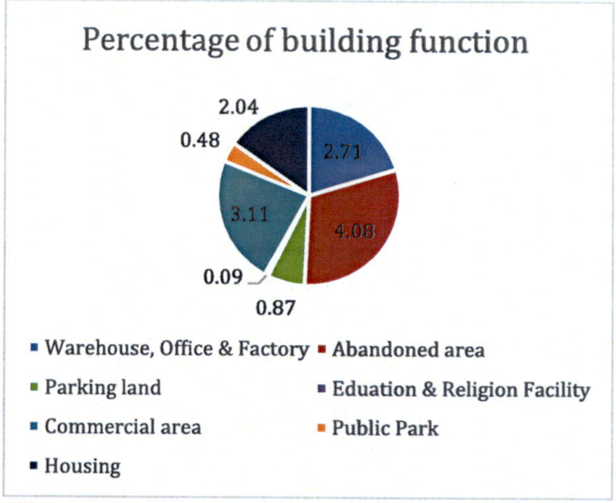

Fig. 2 Building function of Semarang Old Town area in 2020. *Source* Pupitasari and Ramli (2017) and Author (2020)

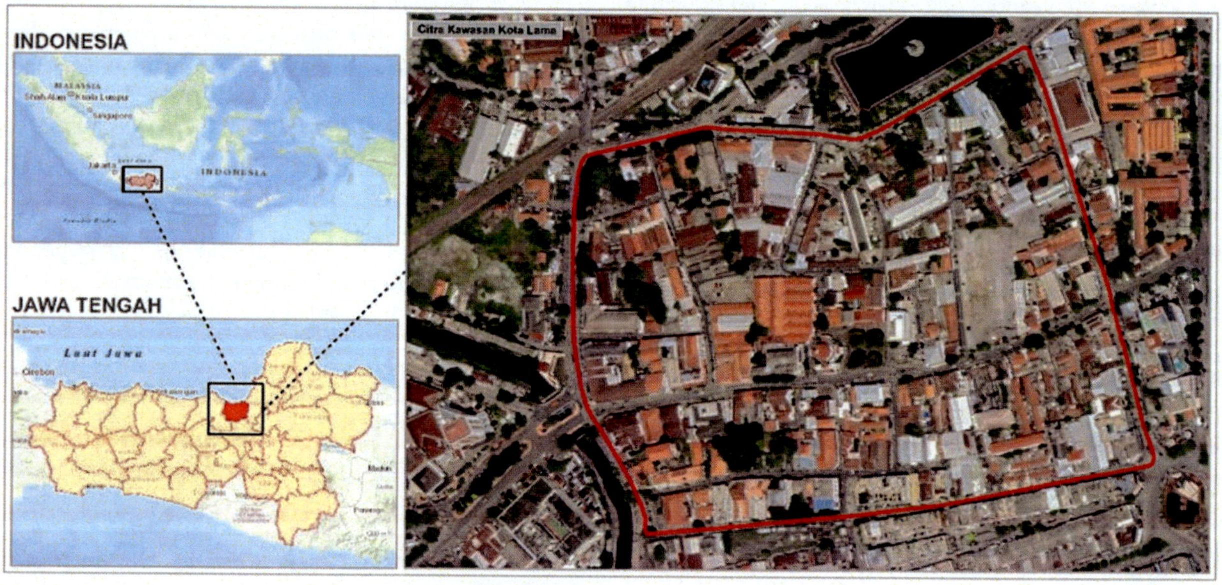

Fig. 1 Orientation of Semarang Old Town area. *Source* Google Map (2020)

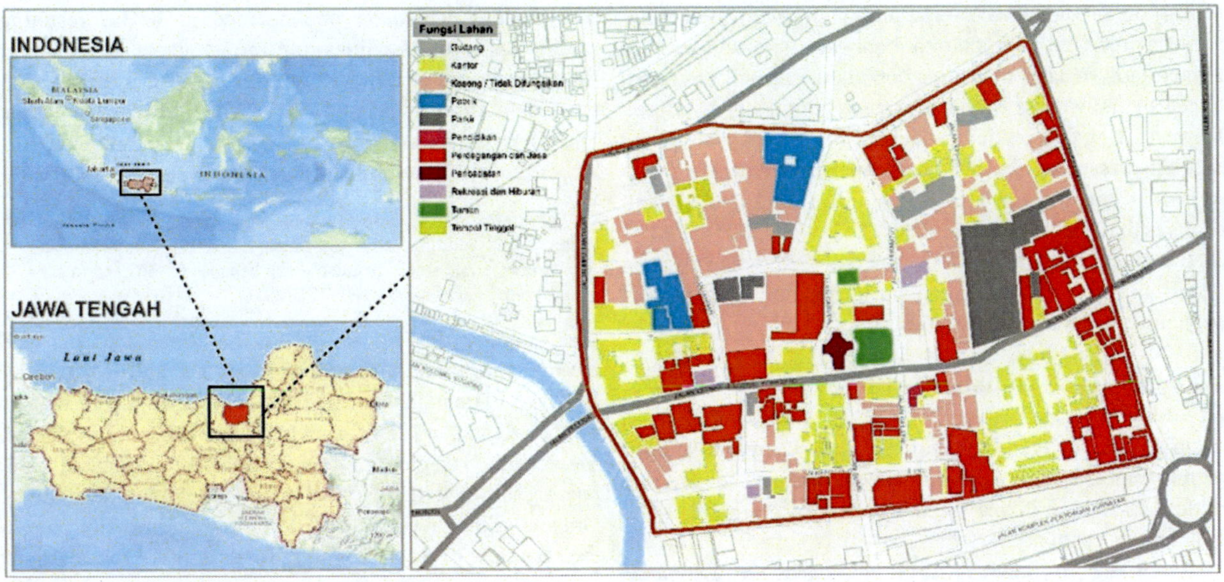

Fig. 3 Map of building functions of Semarang Old Town area. *Source* Semarang City Planning Agency (2019) and Author (2020)

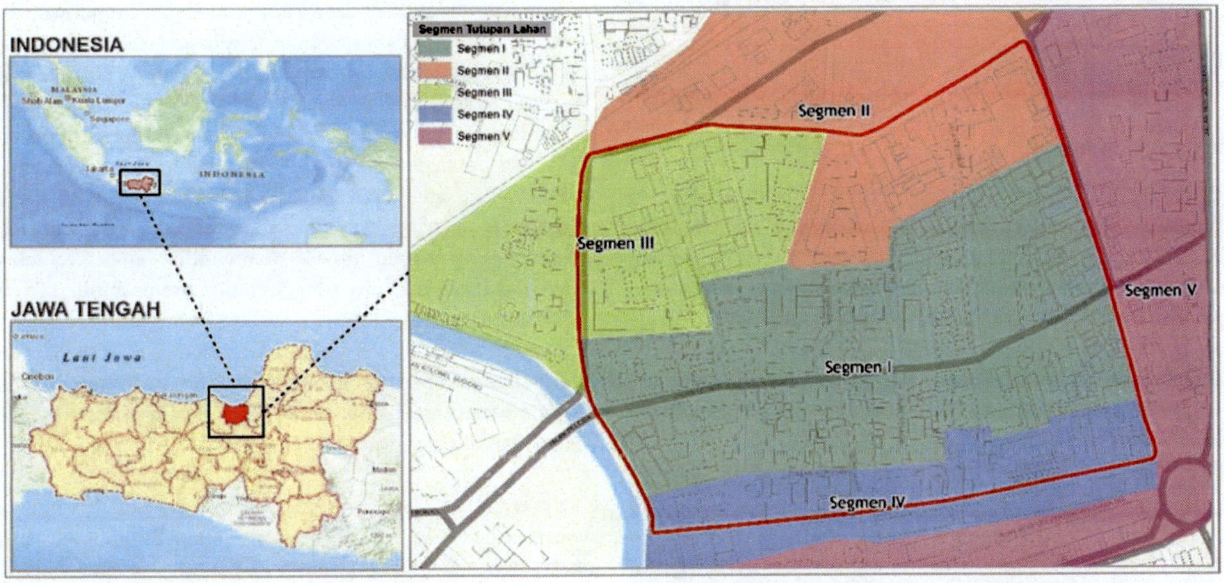

Fig. 4 Map of the land-use segment of Semarang Old Town area. *Source* Semarang City Planning Agency (2019)

Semarang Old Town based on urban design guideline that conduct by government in 2019 (Semarang City Planning Agency, 2019).

4.2 Discussion

The discussion of this study is referred to the theory of Bell et al. (2001) and Hardoyo (2013). According to Bell et al. (2001), adaptation is conducted by adjusting the limited response of stimulus consisting of three strategies in individual and environmental relationships, which are physical, social, and movement stimulus. Meanwhile, based on Hardoyo (2013), there are three adaptation strategies, which include physical, socioeconomic, and human resource adaptation strategies.

Humans do "*coping*" to adjust to their environment in order to sustain themselves in their environment. The results of an individual's relationship or interaction with their environment in an optimal boundary state will be in a homeostatic (balanced) state, a condition that will be maintained by individuals because it causes feelings of pleasure. However, if the perception occurs outside the optimal limits, then the pressure of the individual experience will do *coping* to his/her environment in order to adjust to environmental changes that occur around him/her (Bell et al., 2001).

Based on this study, it is apparent that gentrification occurs in the form of adaptation, in the form of physical, economic, and social adaptation.

4.2.1 Physical Adaptation

According to Arkaraprasertkul (2018), gentrification that occurs in historic areas involves a process where indigenous people improve more buildings and environment toward the middle class by utilizing the uniqueness of the building to maintain its economy.

According to previous study (Kennedy & Leonard, 2001), gentrification is characterized by changes in the land use and the emergence of new, more commercial activities that trigger new facilities or infrastructure. The arrangement of Semarang Old Town area (KKLS) resulted in people living in KKLS or people from outside, pursuing physical adaptation efforts to adjust to the gentrification phenomenon.

A. Conservation

Based on Old Town of Semarang Priority Heritage Planning Plan 2017, several strategies for structuring the Old City Priority Areas are conducted to improve the quality, quantity, and completeness of facilities and infrastructure within priority heritage areas, the development of wider parking areas, the addition of signage/information boards, and the development of heritage city galleries or museums.

The strategy has been implemented by Semarang City government, such as by improving the road network, street lights, and roadblocks. Meanwhile, several other facilities in the form of plant pots, hand washing places, and charger boxes are obtained through Corporate Social Responsibility (CSR) from stakeholders such as BNI, Gojek, BUMN, and other companies. This narration is explained by Anastasia Dwi Rahmi as follows:

> Preservation is performed by the government only by improving road infrastructure, lights, and roadblocks. As for the ornament (plant pot), such as the wash hands and charger box are from CSR companies such as BNI, BUMN, Gojek, etc. (ADR/221020)

The strategy has been implemented at KKLS resulting in changes in the facilities and infrastructure as a support for tourist activities at KKLS. This is in accordance with the vision of structuring KKLS, which reads "Semarang Old Town Area as a World Tourism Destination 2020". Accordingly, Dimas Hokka Pratama explained as follows:

> There are building improvements and changes in road pavements. In addition, after the preservation of clean water and wastewater, networks began to exist because of the need to support tourist activities. (DHP/301020)

The mobility network at KKLS has been laid out and arranged, in accordance with Semarang Old Town Priority Heritage Planning Plan 2017, planning that access will be limited to four-wheeled and two-wheeled motor vehicles. Meanwhile, vehicles on four wheels are prohibited from passing through the Core Area of Semarang Old Town. Gradually, the diversion of traffic flow that usually crosses the core area of Semarang Old Town to other roads around is on progress. The following details are the road section settings in KKLS, consisting of:

a. Free access which is limited, including Jalan Mpu Tantular, Jalan Cendrawasih, and Jalan Sendowo, in which access to these roads is allowed for two-wheelers, four-wheelers, tourism buses, and city transport and restricted to heavy vehicles.
b. Limited access includes Jalan Letjen Suprapto, Jalan Merak, Jalan Branjang, Jalan Kedasih, Jalan Perkutut, Jalan Nuri, and Jalan Gelatik, in which access to these roads is restricted to four-wheeled and two-wheeled vehicles at certain times. Meanwhile, public vehicles such as buses and city transportation are prohibited from passing this road.
c. Access which is very limited, including Jalan Garuda and Jalan Kepodang, in which access to these roads is not allowed across two-wheeled and four-wheeled vehicles, but allowed for cultural events at certain times.

However, Jalan Garuda is still crossed by heavy vehicles such as trucks. In addition, Jalan Letjen Suprapto is still passed by city transportation, damaging several paving roads.

After arranging activities at KKLS, the road network contained in KKLS has been repaired generating a change in the road pavement. However, the road pavement in KKLS is now in the form of paving that is easily damaged and not durable, and thus, more roads in KKLS have been damaged. There are changes in the old city of Semarang, which can be observed such as changes in road sections, some supporting facilities such as pedestrian paths, public toilets, and other facilities that are all prepared and changed due to efforts to attract visitors to the old city area. These changes can be seen further in Table 1.

B. Renovation

Gentrification at KKLS occurred as an impact of structuring activities in 2018 to the mid-2019. The arrangement of KKLS encourages the building owners to renovate or repair buildings that have been hit or damaged. However, if the landlord is unable or does not have the cost to renovate it, the landlord seeks investors to lease the building and renovate it. Semarang City Government through Semarang Old Town Area Management Agency (BPK2L) also suggests, if the building owner is unable to renovate the building, *corporate social responsibility* (CSR) will be engaged with a company as explained by Anastasia Dwi Rahmi:

> Landlords mostly renovate their own buildings. But there are also those who are looking for an investor to renovate it or looking for it to rent. It can also be done by means of engaging with CSR (Corporate Social Responsibility). (ADR/221020)
> Most of the buildings in KKLS are mainly owned by SOEs who prefer to lease their buildings, but there are also building owners who renovate their own buildings. (JS/261020)

Buildings in KKLS that have been renovated or repaired are primarily utilized as commercial activities to revive KKLS. In Fig. 5, several buildings were renovated and changed functions including building in Fig. 5a, which was initially a house into a café (Semar Café) Fig. 5b, then in Fig. 5b which was initially a building that functioned as a convenience store owned by Winkel Maatschappij which provided goods for the needs of residents/Dutch residents, established in 1885 (Yuliati et al., 2020), and currently transformed into Spiegel Bar and Bistro, and Fig. 5d was originally an Nederlandsche Indische Handelsbank (NIH) office building established in 1870 and engaged in foreign exchange business Yuliati et al. (2020) and currently (2020 to present) becomes the Mandiri Bank office in a well-maintained condition. Figure 5e and f shows the same function as a bank (Mandiri Bank) that own by the Indonesian Government; Fig. 5e shows before the renovation in 2019 and Fig. 5f after the renovation (2021).

C. Preservation

Gentrification, as an impact of structuring activities at KKLS, causes a shift in character and addition of land-use functions. Prior to the arrangement of KKLS, the building function at KKLS is in the form of offices, warehouses, and residences. Further, preservation activities are carried out, indicating efforts to defend objects or places according to original and old circumstances. Preservation activities in buildings or historical areas are not intended to preserve and maintain buildings architecturally, rather than to historical and cultural values. Thus, efforts are made to maintain the historical value of building by utilizing historic buildings for economic functions, open spaces, recreation, and entertainment. However, after arranging KKLS, the use of land in KKLS is in the form of cafés and restaurants (food and beverage), as expressed by Dimas Hokka Pratama and Albertus Kriswandhono:

> Gentrification at KKLS is not a land use change, rather than character shifts and the addition of land use functions. Because from the beginning, land use planning in KKLS has not changed and it has remained in the form of office areas, trade and services to date. It used to be a pure commercial area such as machinery, factory offices, and warehouses. Now it is shifting to commercial cafés (food and beverage). At first, it was pure trade areas, through government policies of becoming tourist attractions, gentrification emerged as a response to a mere shifting land use. (DHP/301020)

Based on the results of the interview, strengthened by comparing land use in KKLS spatially, it is obvious that land use in KKLS in 2017 was dominantly in the form of warehouses, offices, empty or not functioning, factories, education, trade and services, worship, recreation and entertainment, nest wallets, parks, and residence. The largest land use in KKLS has been in the form of empty or non-functioning buildings amounting to 3.58 ha. Meanwhile, the use of land in KKLS in 2020 was in the form of warehouses, offices, empty or not functioning, factories, parking, education, trade and services, worship, recreation and entertainment, parks, and residences.

However, the use of land as warehouses, offices, recreation, and entertainment as well as dwellings has been decreasing. The warehouse at KKLS in 2017 amounted to 1.21 ha, but by 2020, it decreased into 0.48 ha. Warehouses in KKLS were reduced because they had been moved outside KKLS, for example, Archive Warehouse of Semarang City Education Office which was moved outside KKLS because the road in front of the warehouse was used as a park. Meanwhile, land use for trade and services increased from 2.16 ha in 2017 to 3.11 ha in 2020.

Figure 6 shows some of the buildings that have been preserved; then, the buildings are used as a place to showcase the handicrafts of several economic actors in the city of Semarang (UMKM Gallery) (see Fig. 6a and b). Figure 6a shows situation building that was abandoned, and Fig. 6b after adaptive reuse for the outlet of several handicrafts of Semarang (UMKM Gallery). Figures 6c and d are Marabunta building; the first Marabunta building is a

Table 1 Changes in facilities and infrastructure in Old Town Area of Semarang

No.	Infrastructure	Condition of Facilities and Infrastructure		Description
		2018	2020	
1	Street			Arrangement of vehicle access in KKLS and the presence of traffic flow switching The road network contained in KKLS has been repaired generating a change in the road pavement. However, the road pavement in KKLS is now in the form of paving that is easily damaged and not durable, and thus, more roads in KKLS have been damaged Arrangement of movement networks in KKLS resulted in more regular accessibility in KKLS
2	Pedestrian Trail			Every road section in KKLS has a pedestrian path, equipped with a *yellow line* for disabled users and a barrier pole between pedestrian path and road Completeness of facilities on pedestrian pathway encouraging the utilization of pedestrian pathway by tourists
3	Parking area			Prior to the arrangement of KKLS, parking lot at KKLS was on-site parking that was parked on the side of the road. After arranging KKLS, there was an addition of parking lots to three pockets of parking lots located on Jalan Letjen Suprapto, Jalan Garuda, and Jalan Suari in front of the shop. However, there is still on-site parking area such as in front of empty buildings or on the edge of Jalan Suari, Jalan Garuda, and the side street of Taman Srigunting
4	Drainage			Drainage at KKLS has been repaired after KKLS arrangement. Before KKLS structuring activities, there was still open drainage, causing an unpleasant odor. However, repairs have recently been performed, and drainage contained in KKLS is in the form of closed drainage

(continued)

Table 1 (continued)

No.	Infrastructure	Condition of Facilities and Infrastructure		Description
		2018	2020	
5	Waste Bin			Every part and section of the road in KKLS has a waste bin, preventing the visitors from littering
6	Public toilets			Prior to KKLS arrangement, the public toilet next to Srigunting Park had a small room made of zinc array. To date, the public toilets at KKLS have been repaired and made of iron and steel, appearing better for the aesthetics of area
7	Other infrastructure (seating, charger box, drinking water faucet)	–		The addition of new facilities at KKLS is in the form of seating on every road section, information boards, charger box, and drinking water faucet. The addition of new facilities is accomplished to support tourism activities at KKLS

Source Author (2018 and 2020)

Fig. 5 Renovation of buildings. *Source* Author (2020) and Yuliati, et al. (2020)

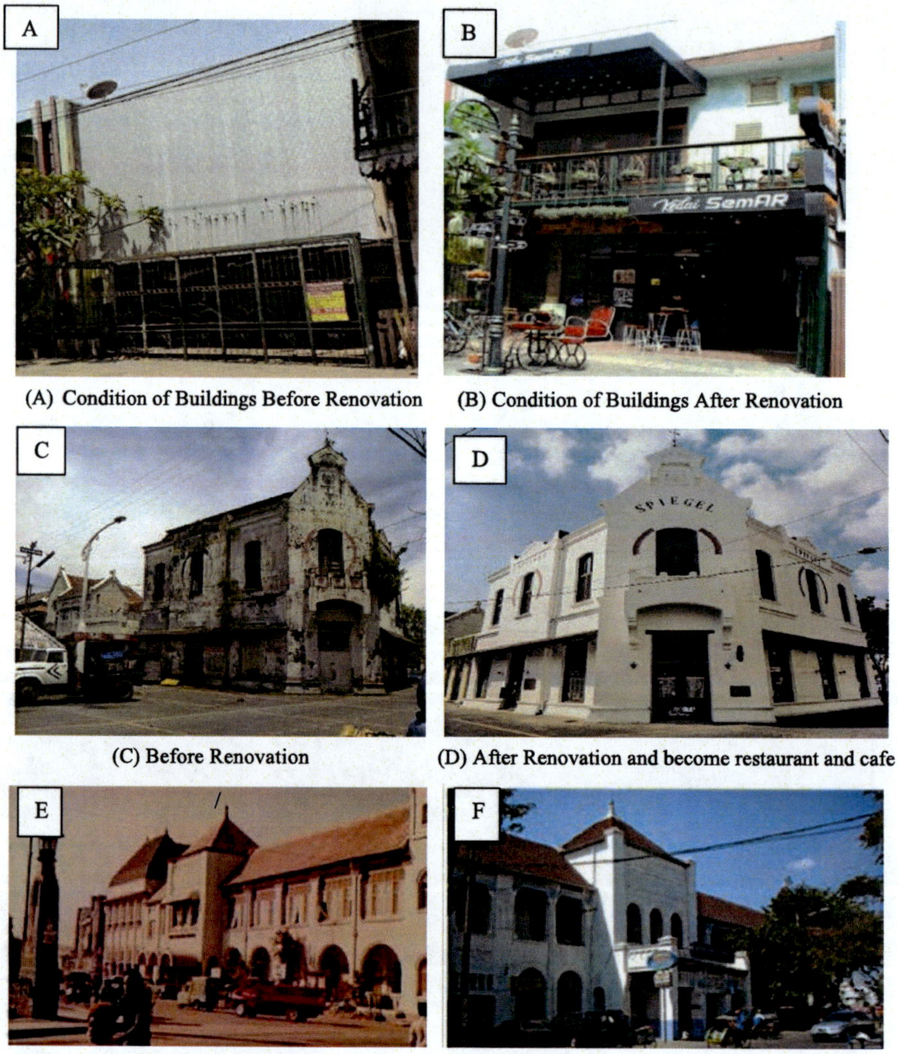

(A) Condition of Buildings Before Renovation
(B) Condition of Buildings After Renovation
(C) Before Renovation
(D) After Renovation and become restaurant and cafe
(E) Before Renovation
(F) After renovation and became Mandiri Bank

performance building. After being renovated, the function was changed to become a restaurant (Fig. 6d).

D. Revitalization

After the improvement of road infrastructure through conservation activities, revitalization was carried out as an effort to restore the value added in terms of economic, social, and cultural values in utilizing historic areas. This effort is indicated by utilizing the pedestrian path by tourists more than before the preservation activities were carried out.

In addition to the use of pedestrian paths, other public spaces are more utilized, such as in Srigunting Park. After KKLS arranging activities, Taman Srigunting is more visited by tourists both to sit back and take photos. Srigunting Park has since become the center of tourist activities and center of crowds in KKLS (see Fig. 7). The utilization of park, street even terrace of building give a live in Semarang Old Town Area. Meanwhile, at the time before the arrangement of KKLS activities, Taman Srigunting rarely had visitors and usually those who came to Taman Srigunting were only the old people to sit around.

> In front of the intersection of Srigunting Park becomes the center of the crowd so that the former is quiet because there are tourist activities. (DHP/301020)
> In the past, Srigunting Park was mostly visited by old people to sit around, but now there are already many endings. (P/N 281220)

After revitalization was carried out in several areas, such as Srigunting Park (Fig. 7a and b), in the picture, Fig. 7a is the

Fig. 6 Changes in building functions in KKLS. *Source* **a** Author (2018), **c** Author (2020), **b** and **d** Author (2022)

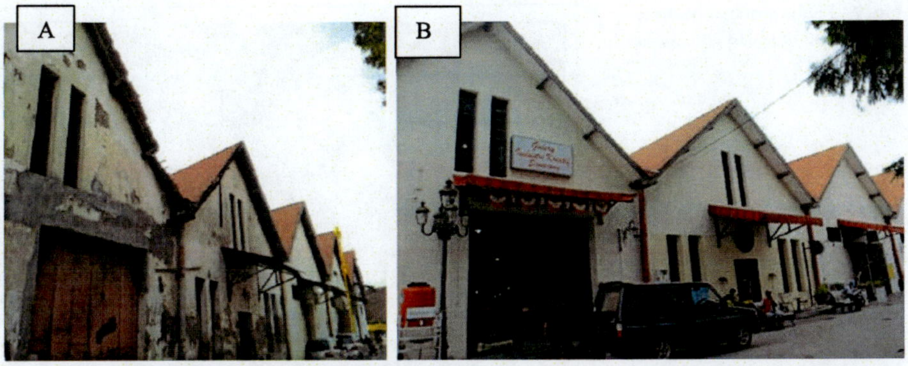

Fig A and B PPI warehouse became MSME (Semarang Creative Industry Gallery)

Fig C and D Marabunta Building

condition of the Srigunting Park before revitalization (2017) while Fig. 7B after revitalization in 2019, activities that have also developed in Srigunting Park and its surroundings such as fashion shows (Fig. 7c), painting (Fig. 7d), music (Fig. 7e), and used goods markets (*senthiling* market) (Fig. 7f).

Based on the aforementioned explanation, physical adaptations made by people who live inside and outside KKLS are in the form of building renovations, changes in building functions, changes in building ownership, and improvements in infrastructure and utilization of public space. This finding is consistent with (Kennedy & Leonard, 2001) theory of gentrification in historic areas, from which the land-use changes, thereby triggering more commercial activities, facilities, and infrastructure. However, based on the findings in the field, there is also renovation of buildings, changes in building ownership, and utilization of public space, found in KKLS.

4.2.2 Economic Adaptation

Prior to the arrangement of KKLS, the existing building in KKLS was functioned as a residence, warehouse, office, and others. Trade and services in the form of cafés and restaurants have not been significantly established. Investors have not been able to put their investments in KKLS because the condition of KKLS was deserted and slum. There were only seven business actors at the time before the arrangement of KKLS. Business or trade and services in the form of existing cafés or restaurants only include IBCs, Tekodeko Koffiehuis, Spiegel Bar & Bistro, Retro Café, Vercoffee, FiftySeven Café, and Black Meet White. However, there are also medium business actors currently residing in KKLS such as traveling food vendors and others.

During the arrangement of KKLS, several investors were also interested in investing in KKLS because investors are aware of the potential at KKLS. There are currently three investors who develop trading and service businesses at KKLS in the form of cafés or restaurants including Pringsewu, Hero Coffee, and Coffee Philosophy.

> I chose a place at KKLS, because it is researched that the prospects are good here and also some news that KKLS is expected to grow as well. (DA/201120)

After the arrangement of KKLS, investors in KKLS are increasing, indicated by the increasing trade and services in the form of cafés and restaurants at this stage. Examples of businesses or trade and services that stand after the arrangement of KKLS include MSME Gallery, Covare, Soto Kartosuro, Javara Culture, Promise of Soul, Old en Nieuw Ice Cream, Batik Jayakarta, DOL Kopi, Sukajaya, Koopman, Monggo Dhahar, Coffee Edge, Tahu Rasa, Semar, Arrow, Pak Kisno Fried Chicken, and Bobobox. Meanwhile, the number of small and medium-sized

Srigunting Park Conditions and other public spaces in Semarang Old Town

Fig. 7 Public space utilization. *Source* Author (2018 and 2021)

business actors has also been increasing in KKLS. Some people living inside and outside of KKLS started to open medium businesses because the community understands the potential of KKLS, thereby encouraging them to take advantage of this situation.

> ... to trigger the Area to come back to life and to involve the community, even if there is still a moderate community. For example, the community participates in selling and guarding the parking lot. It's the least we can do together. (AK/141220)

Medium business actors such as traveling food vendors, who are not settled, have existed since before the arrangement of KKLS, but at this time they are concentrated on Jalan Letjen Suprapto and around Taman Srigunting. Business actors who are settled only move places because the crowded center in KKLS is around Taman Srigunting. Meanwhile, business actors who are not on Jalan Letjen Suprapto remain less apparent.

Some buildings that changed function due to economic interests can be seen in the picture below (Fig. 8). Figure 8 shows a few buildings. For example, (Fig. 8a) originally was a warehouse owned by NV Hagemeyer & Co's Handel Mij (Yuliati et al., 2020), the condition was damaged, and after revitalization turned into a government bank (Bank Mandiri) (Fig. 8b) as well as in Fig. 8c, the building was originally the office of the exporters of agricultural products NV Trading Company "Adji Saka," (Yuliati, et al. (2020) (Fig. 8c) currently a Pringsewu restaurant (Fig. 8d). In Fig. 8e, this building was originally (1930–1950) a Dutch East Indies printing company (Van Dorp) (Yuliati et al., 2020) (Fig. 8e), then until 2016, it became an empty building, 2020, until now it became a 3D museum building (Fig. 8f). An empty building, 2020, until now has become a 3D museum building (Fig. 8F).

Fig. 8 Economic adaptation. *Source* Author (2021)

4.2.3 Social Adaptation

Prior to the arrangement of KKLS, there were no residents listed as *boro* residents (non-permanent residents). However, after the arrangement of KKLS, some people in KKLS remained, and some people moved out of KKLS. Some residents, who live in KKLS, move out of KKLS or there are also residents who are listed as *boro* residents (non-permanent residents). For example, it was in RT 3 RW 1 Tanjungmas Village and RT 4 RW 2 Purwodinatan Village that *boro* residents are those who are registered as residents in the KKLS area but residing outside the area or residents who live in the KKLS area but are not registered as residents in the area.

In addition, KKLS conditions remain deserted, dark, often flooded, and has high levels of criminality because street furniture such as streetlights is still inadequate. The image of KKLS area is still not attractive because there are homeless people who build houses from zinc or cardboard, such as on Jalan Kedasih and Jalan Cendrawasih I. In addition, there is an illegal house on Meliwis Street listed as residents of RT 1 RW 1 Tanjung Mas Village, which is included in the KKLS area, but their residence is illegal because it is established on a waterway.

On the other hand, public awareness of the surrounding environment remains low. Therefore, BPK2L arranges strategies to tackle social problems by inviting people living inside and outside of KKLS to hold activities at KKLS. Existing communities, such as the photography community, the drawing community, and so on, could revive activities at KKLS.

Fig. 9 Shifting condition after displacement (social adaptation). *Source* Author (2021)

(a) Illegal House in Meliwis Street (b) After the resident moves into Public Housing

> What is being done is to trigger a community movement and to invite people who want to participate so that it will determine the quality of an area. Because if people move, then the area will move. BPK2L makes a trigger so that people in KKLS and outside (community) are aware. All existing communities take their turns by doing activities at KKLS to trigger the area to come back to life and the community will be involved, even though there are still simple communities. (AK/141220)

However, in structuring KKLS, the community is not included in this process, causing a lot of protests from the community due to KKLS structuring activities. One of them is concerning with the time of work that is not over. Thus, the community has difficulty entering their residence at KKLS.

> Because the process takes a lot of time and the development is not very smooth. It was a very early promise of 3 months but in fact it lasts for 1.5 years. (JS/261020)

After the arrangement of KKLS activities, some people who live or work from the outside of KKLS start selling or working as park keepers. Thus, some stakeholders provide training to the community so that the community not only sells and maintains parking, but they can carry out other beneficial activities. The training was conducted by Monod Dephuis by inviting various sources in the form of tour guide training, make up training, bouquet training, and so forth.

After arranging KKLS, the image of the surrounding community in this area is good and well organized. Illegal houses on Meliwis Street have been moved to Kaligawe flats and Karangroto flats. The arrangement was also conducted by handling homeless people living in front of KKLS buildings (Fig. 9a) and the situation after the homeless moved out from the location (Fig. 9b).

Findings

Research shows that gentrification has different impacts and consequences for each business actor in the old city area of Semarang. Three types of adaptations exist physical adaptation (conservation, renovation, preservation, and revitalization), economic adaptation in the form of renting buildings or selling, and environmental adaptation. Meanwhile, owners and residents who cannot adapt to the location will either choose to move or leave the area by eviction. For more details, please see Fig. 10.

5 Conclusion

The author of this study concluded that the arrangement of urban areas resulted in the emergence of the phenomenon of gentrification due to a change in the character of the environment. In particular for Semarang Old Town, this gentrification requires adaptation from the people who live in the area or the communities around the area, which are in the forms of:

(a) Physical adaptation, in the form of conservation activities, renovation activities, preservation activities, and revitalization activities carried out on buildings and the environment of the Semarang Old Town area.
However, physical changes are often not in accordance with the original function and façade, so that the whole plan does not integrate with the area.
(b) Economic adaptation, in the form of investors by developing businesses in the Old Town Area of Semarang and changes in community activities because they take advantage of the existing tourism potential. In this adaptation, there is a process of displacement for residents/business actors who cannot afford to rent buildings, and thus, local residents consider that revitalization only benefits some parties, especially investors.
(c) Environmental adaptation, in the form of preserving the old town area to increase the value of the area,
(d) Social adaptation, in the form of moving out for some people and some remain living in the Old Town Area of Semarang, and the participation of people living inside

Finding

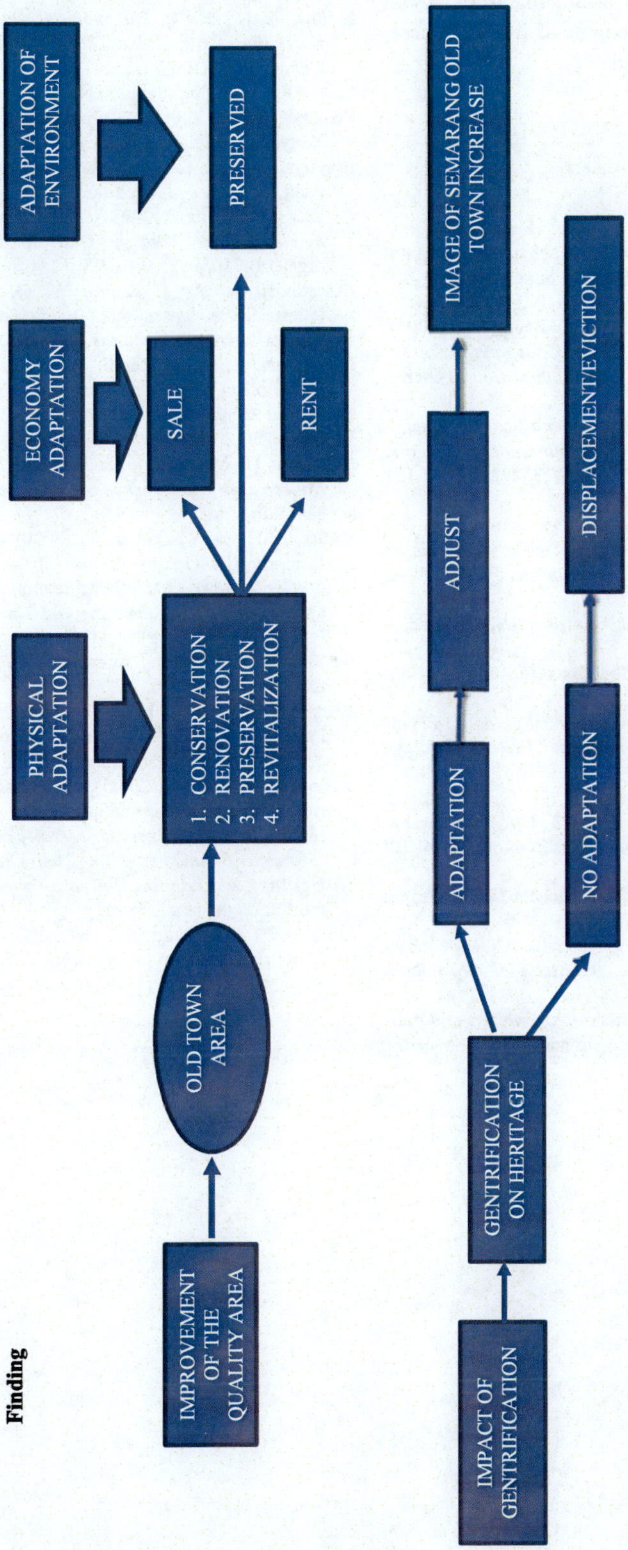

Fig. 10 Relation between heritage site–tourism–gentrification in *Kota Lama* Semarang (Semarang Old Town) Heritage Site. *Source* Author (2021)

and outside of the Old Town Area of Semarang to carry out activities, as well as the arrangement of social problems including the arrangement of illegal houses, homelessness, and illegal activities.

References

Arkaraprasertkul, N. (2018). Gentrifying heritage: How historic preservation drives gentrification in urban Shanghai. *International Journal of Heritage Studies, 25*(9), 882–896. https://doi.org/10.1080/13527258.2018.1460732

Bell, P. A., Greene, T. C., Fisher, J. D., & Baum, A. (2001). *Environmental psychology* (5th ed.). Harcourt College Publishers.

Catanese, A. J., & Snyder, J. C. (1989). *Perencanaan Kota*. Penerbit Erlangga.

Cesari, C. D., & Dimova, R. (2019). *Heritage, gentrification, participation: Remaking urban landscapes in the name of culture and historic preservation*. https://doi.org/10.1080/13527258.2018.1512515

Creswell, J. W. (2014). *Research design, qualitative, quantitative, and mixed methods Approaches atau Research Design: Pendekatan Metode Kualitatif, Kuantitatif, dan Campuran* (A. Fawaid & R. Kusmini P., Eds.). Pustaka Pelajar.

Glass, R. (1964). Aspects of change. *The Gentrification Debates: A Reader*, 19–30.

Google Maps. (2020). Google Maps directions for driving from Indonesia. Retrieved 2020.

Hancock, D. R., & Algozzine, B. (2006). *A practical guide for beginning researchers doing case study research*. Teachers College Press.

Harani, A. R., Werdiningsih, H., & Riskiyanto, R. (2016). Peningkatan Fungsi Guna Bangunan Di Kawasan Kota Lama Semarang. *E-Journal Undip, 17*(1), 22–26. https://doi.org/10.14710/mdl.17.1.2017.22-26

Hardoyo, S. R. (2013). *Aspek Sosial Banjir Genangan (Rob)*. Gadjah Mada University Press.

ICOMOS. (1999). The Burra Charter: The Australia ICOMOS charter for places of cultural significance. *International Council on Monuments and Sites*.

Kennedy, M., & Leonard, P. (2001). *Gentrification: Practice and politics*. The LISC Center for Home Ownership and LISC Knowledge Sharing Inititiative Online Resource Library. www.brookings.edu/urban

Medha, A. N. (2017). *Dukungan Modal Sosial Dalam Membangun Kolektifitas Masyarakat Untuk Beradaptasi Terhadap Fenomena Gentrifikasi Di Kawasan Medokan Semampir* [Tugas Akhir]. Institut Teknologi Sepuluh Nopember.

Prasetya, L. E. (2006). Konservasi Kawasan Panggung Krapyak Yogyakarta Melalui Pendekatan Urban Linkage. *Jurnal Arsitektur*.

Prayoga, I. N. T. (2013). Keberlangsungan Menetap Penduduk Asli pada Kawasan di Sekitar Kampus UNDIP Tembalang sebagai Permukiman Kota Semarang yang Tergentrifikasi. *Jurnal Pembangunan Wilayah Dan Kota, 9*(1), 1–10. https://doi.org/10.14710/pwk.v9i1.6499

Puspitasari, A. Y., & Ramli, W. O. S. K. (2017). Manajemen Pelestarian Kawasan Kota Lama Sebagai Bagian Penting Kota Pusaka Semarang. *Seminar Nasional SPACE, 3*, 328–341.

Puspitasari, A. Y., & Ramli, W. O. S. K. (2018). Masalah dalam Pengelolaan Kota Lama Semarang sebagai Nominasi Situs Warisan Dunia. *Jurnal Planologi, 15*(1), 96–114. https://doi.org/10.30659/jpsa.v15i1.2764

Rahardjo, H. M. (2017). *Studi Kasus Dalam Penelitian Kualitatif: Konsep Dan Prosedurnya oleh*. Universitas Islam Negeri Maulana Malik Ibrahim.

Paulla, S. D., & Wakhidah, K. (2020). Gentrification Impact to the Community's Quality of Life in Semarang Old Town. *E3S Web of Conferences, 202*. Gentrification Impact to the Community's Quality of Life in Semarang Old Town | E3S Web of Conferences (e3s-conferences.org).

Semarang City Planning Agency. (2019). Urban design guideline of Semarang Old Town.

Warsito, A. (2017). 19 Pengertian Arsitektur Menurut Para Ahli. https://ilmuseni.com/senirupa/arsitektur/pengertianarsitektur-menurut-para-ahli

Yuksel, D., & Iclal, D. (2005). Historical heritage-conservation-restoration in small towns and question of rural gentrification in Turkey. In *15th ICOMOS General Assembly and International Symposium: 'Monuments and Sites in Their Setting—Conserving Cultural Heritage in Changing Townscapes and Landscapes.'* http://www.mudurnu.gov.tr/tarih.html

Yuliati, D., Susilowari, E., & Suliyati, T. (2020). Riwayat Kota Lama dan Keunggulannya sebagai Warisan Dunia. *Publikasi Tim Departemen Sejarah Universitas Diponegoro*, 318–574.

Mural Arts in Historic Cairo: A Critical Review

Farah A. ElDin, Mona A. Abdelwahab, and Lobna Sherif

Abstract

In recent years, mural arts have secured a place within the urban space of metropolitan cities around the world, including Cairo. These murals, created by numerous artists, appear in different areas throughout the city. The literature shows that they have an impact on the social and physical aspects of urban space. This paper investigates the murals located within the boundaries of Historic Cairo, focusing on the higher concentration within Al-Khalifa neighborhood. The paper discusses the messages conveyed by them and their addressed audience: either the local community or a different audience all together. It critically analyzes eight murals through an urban design lens, using visual and architectural methods including mapping, annotated photographs, and architectural diagrams. In between the envisioned purpose of these murals and their realized outcome, the paper questions the murals' interaction with streetscape, community, and time. The first question discusses whether the streets act as the murals' canvas or background or are shaped and identified by them. Simultaneously, it questions whether these murals represent the community's identity or if the artists impose their artistic vision through them. Lastly, whether these murals are ephemeral and disappear over time, or leave traces that mark the urban space.

F. A. ElDin (✉) · M. A. Abdelwahab · L. Sherif
Arab Academy for Science, Technology and Maritime Transport, Cairo, Egypt
e-mail: farahalaaeldin@aast.edu

M. A. Abdelwahab
e-mail: mona.abdelwahab@aast.edu

L. Sherif
e-mail: lobna.sherif@aast.edu

Keywords

Murals · Urban space · Visual image · Historic Cairo

1 Introduction

In metropolitan cities, art has become obtainable within the public space, outside the confinement of museums and galleries (Kwon, 2001). This is due to the fact that over the last three decades, more cities have found that arts and the broader cultural setting could be useful tools for renewal and revitalization (Landry, 2020). Moreover, considering the city's culture and heritage, these non-physical factors are considered "soul of the city" (Abdelaziz et al., 2018). According to Lovell (2020), public art is considered a "kind of social practice that conditions the spatial memory of the city" (p. 56). These artistic interventions, of many forms, have permanently marked the urban space (Redaelli & Stevenson, 2022), most notably mural arts (McAuliffe, 2012). In contrast with its ephemeral characteristics, mural arts painted in a public space can impact the collective memories of common spaces (Türközü & Caliskan, 2021). How these art interventions affect the public's relation to the city or urban space vastly depends on how accessible these artworks, visually and physically (Phillips, 1988).

In Egypt, public art with its modern definition has roots that goes to the late nineteenth century. Usually, art in the city was mainly sculptural art that was established "by state-sanctioned artists in state-sanctioned public spaces" (Bseiso, 2017, para. 10). These public art pieces in Egypt took numerous shapes, including statues, wall arts, installations, and memorials. Wall arts emerged within the urban space, from mosaics to murals, and rose in popularity in major Egyptian cities after the events of 2011 as a form of pop art. Often mural arts are considered a subcategory of the larger umbrella of graffiti, yet the main difference between mural arts and graffiti is the aspect of legality. While graffiti is an illegal art piece that is

M. F. Shahidan et al. (eds.), *Urban Narratives: Exploring Identity, Heritage, and Sustainable Development in Cities*, Advances in Science, Technology & Innovation, https://doi.org/10.1007/978-3-031-48517-6_25

considered vandalism, street art is a permitted action that professional artists use to carry on overarching messages to the public and is often accepted and celebrated (Pogrmic & Dercan, 2021).

The aim of this paper is to add an understanding to the mural arts in the city in general, if and how they affect the streetscape, and how people perceive them. This paper investigates two mural projects conducted in the larger neighborhood of Al-Khalifa in the midst of old Cairo, each painted by a different artist. The paper includes a descriptive analysis of mural arts in Egypt, specifically old Cairo, and exploring the theoretical background that the physical and mural analysis is based on. Furthermore, the paper explores two sets of murals in two areas within Al-Khalifa neighborhood, using maps, photographs, semi-structured interviews with artists, and casual conversations with residents of both areas. To conclude, a discussion of the spatial, thematic, and temporal aspects is conducted.

2 Mural Arts in Egypt

Within the traditional Egyptian culture, drawing on walls is a form of folk art in villages, as well as through independent artists in the streets of major cities. Egyptians have always had a custom of decorating the houses of those preforming the Mecca pilgrimage as early as the sixteenth century (Kruk et al., 2015). These paintings are considered folk art and are believed to have pharaonic origins (Elsaeed & El Kabbani, 2006). In addition to those types of murals, commissioned murals were painted on walls of buildings in major cities, but they were few, sporadic, and did not receive attention from the public. The existence of murals in Egypt intensified after the 2011 events in Cairo, where protestors used painting on the street walls of Cairo and other main cities. These murals gained worldwide recognition and were covered by international news at the time, because of their visual qualities, artistic processes, and strong political messages. After the 2011 uprisings, street art and mural art became trendy and prevailed in major cities of Egypt (Badawi, 2021). On both international and local levels, there has been a growing interest in street art in general, and many artists and NGOs initiate events involving mural arts with different communities with the goal of developing the public spaces in the past years. These murals were mostly conducted as parts of art festivals, by street artists, students of fine art, art institutions and NGOs, government officials, and by the community itself. Most documented murals were painted in metropolitan cities like Cairo and Alexandria, while a smaller number of murals were painted in cities like Aswan, Kafr El Sheikh, and Fayoum. Figure 1 depicts a specific area within Historic Cairo proposed by the Urban Regeneration Project of Historic Cairo (URHC) conducted by the UNESCO in 2014, which has a high concentration of murals created over the past seven years by the community, NGOs, local and international street artists, among other actors.

The area, Al-Khalifa neighborhood, is located within the inner city of Cairo and has unique urban characters. The neighborhood is a vast mixed-use area, composed of a dense fabric of buildings, and has a rich mix of residential, retail, and light-industrial uses. Moreover, it holds significant monuments of Islamic Cairo, including Salah El-Din Citadel, Ibn Tulun Mosque, Al-Rifa'i Mosque, and the Mosque and Madrasa of al-Sultan Hasan. Most of which are a walking distance from Salah ElDin Square, or Maydan al-Qal'a, a significant square in the city of Cairo. In the streets surrounding those major monuments, murals were implemented at several scales, by local artists, corporations, and NGOs. Two sets of murals were chosen for analysis; the first is in Al-Arba'een alley next to Ahmed Kuhya Mosque, and the second is in front of Al-Rifa'i Mosque in Darb Al-Labanah street (Fig. 2).

3 Theoretical Framework

Images of urban spaces are viewed in terms of perception, where individuals experience the urban space as a whole rather than fragmented pieces (Carmona et al., 2010). To understand and analyze the urban image, two main categories of elements should be looked at visual and non-visual features. The visual features include the perceived physical setting, while the non-visual features are related to sociocultural and environmental factors. Both features communicate meaning, intended and/or perceived.

Streets are three-dimensional linear spaces surrounded by buildings from both sides and are either visually dynamic or static, enclosed or open, long or short, wide or narrow, straight or curved, and the formality or informality of architectural treatments. Any urban space is defined by the three surfaces: walls (the surrounding buildings), floors (the street path itself), and the sky (an imaginary ceiling). Thus, the character, sense of place, and visual qualities of an urban space are directly affected by the changes in those three physical surfaces (Carmona et al., 2010). The physical setting of an urban space, particularly the street, is defined by its geometry and scale, typology, functionality, architectural details, and its context or connection to surrounding streets and squares. The geometrical or Cartesian dimensions that define the urban image include the length, width, and height of the street or any object within it. The play of horizontal or vertical lines can affect the degree and nature of enclosure within the street; vertical emphasis on the walls would slow the eye and allow for a slower flow, while horizontal emphasis does the opposite. Moreover,

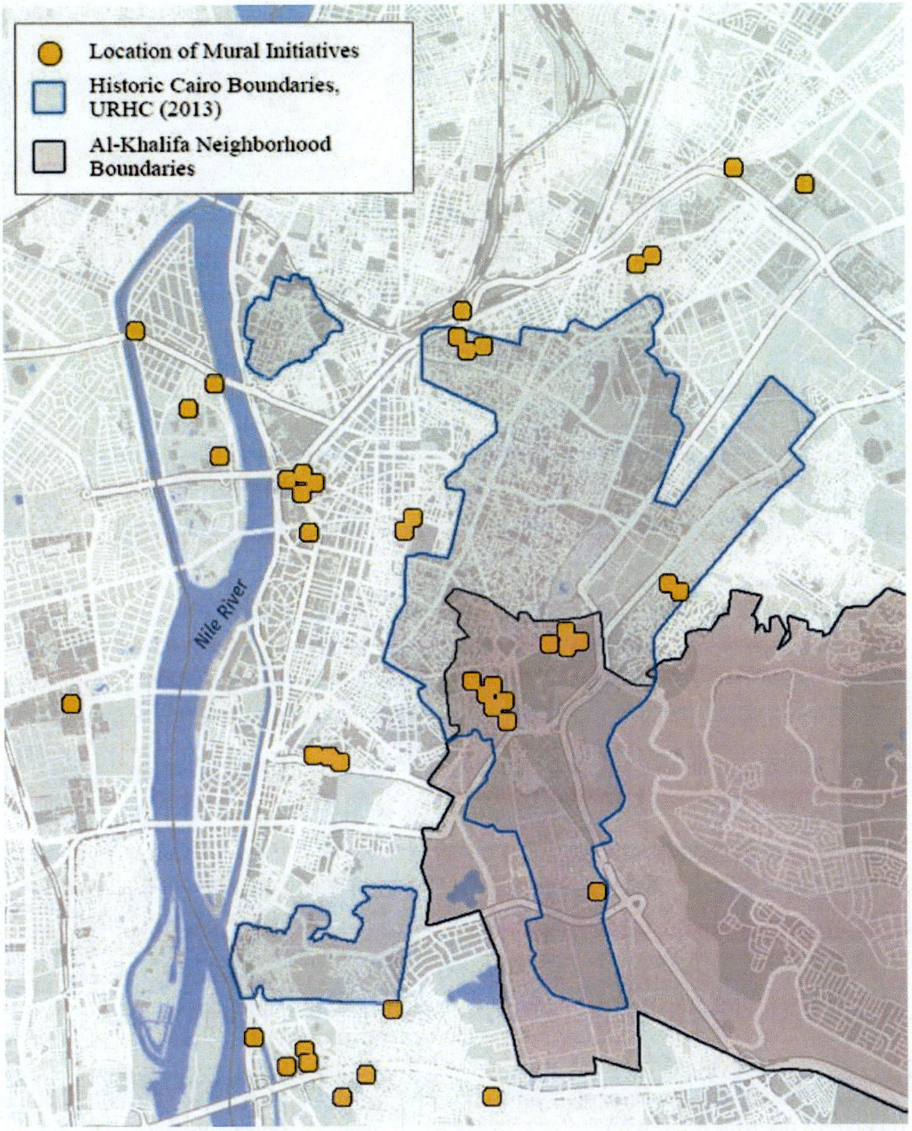

Fig. 1 Map of Historic Cairo adapted from UNESCO (2014), showing the murals' locations

the street volume or ratio between height and width define the street's spatial enclosure, while the street's width controls how much of the walls is seen. In wider streets, the facades can be seen as a whole, and their relationship with surrounding urban elements can be easily detected. While in narrower streets the vertical features become more dominant, projections more exaggerated, and eye-level details more significant. Additionally, typology of streets mainly depends on the modes of transportation used within them, from pedestrian-dominated streets to car-dominated streets and avenues. This distinction between pedestrian and car is most apparent when looking at traditional and modern cities, where the traditional city design is based on pedestrian and cart movement, influenced by topography, and became more organic over time, while the modern city design depends on the use of the vehicle (Carmona et al., 2010). This distinction is also apparent when we talk about functionality of streets, as pedestrian-dominated streets promote walkability and include more leisure and recreational activities. Also, the visual character of urban places derives from not just their spatial and dimensional qualities, but also the detailing of surfaces that define that urban space. Therefore, the analysis of the streetscape or any detailing element within it must include the analysis of the details of those three surfaces. These details include openings' location and size, architectural features, wall texture and material, and position. Also, the position and relational qualities of any urban element within its context conditions how it is perceived by people. The relationship between the here and there is what creates townscape effects, and each building

Fig. 2 Map of Al-Khalifa area showing the locations of murals and the main streets. *Source* Authors

or urban element is seen as a contributor to the larger picture (Cullen, 1961). Thus, any urban element within the streetscape (public artworks included) needs to be studied both within its context and in relation to its context's elements, as it is part of the larger image formed within the urban fabric. Similarly, the perception of street or mural art within the urban space is a multi-sensory experience, not only a visual one (Awad et al., 2017). Thus, the dimensions, colors, textures, details, scale, and choice of location relative to the pedestrian are all characteristics that need to be considered when analyzing mural arts within an urban context.

On the other hand, non-physical features represent an immense portion of how people perceive urban spaces. Cullen (1961) emphasized how the perception of the environment is connected to meaning: emotions, visuals, memories, and individual experiences. The constructed image could be a result of one's own experiences and therefore different people can construct different images with some similarities (Rapoport, 1977). Usually, the features of urban environments are formed through symbolic and spatial manifestations of larger social forces and therefore shape the observer's perception of space and subjective mental image (Pogrmic & Dercan, 2021). Mural arts, as a part of the features of urban environments, intrigue passers-by to stop and question their meaning. Symbolism behind that work of art largely affects how it is perceived by people and in turn how it could affect the public space.

To analyze how and if mural arts impact the urban space and the community, the visual and non-visual attributes of the urban space must be studied. For each set of murals, a qualitative method of analysis is used, divided into two parts: the mural as a part of the streetscape, and the mural as an artistic product. The first highlights the choice of location, dimensions, details, and colors and visibility of the murals in each area, while the second focuses on the symbolic and sensory codes of art present within these artworks and how they relate to the space they are located in.

4 Ahmed Kuhya Murals

The first set of murals is located within an area called Ahmed Kuhya, with a fifteenth century Mamluk mosque in its midst. It can be accessed through two main streets: Al-Saleeba, and Al-Ashraf. One hundred and sixty-one buildings are located within that area, with average building heights between 1 and 6 floors. The area continued to grow over traces of the historic city, with historic monuments in its midst. The street structure within the area is primarily pedestrian-dominated, as it follows the traditional city design. There is no clear distinction between the sidewalk and the street, and in some alleyways, there is no sidewalk at all, or it is taken over by the shops or by the building entrances. The secondary streets are pedestrian-dominated, due to their width that does not exceed 7–8 m from building to building, yet cars and motorcycles access them. Five murals in that area were painted in March of 2021, depicting local women wearing modest clothing and head coverings with children, along with elements from the traditional setting of the area, such as tricycles, pigeon houses, and traditional souqs, and with elements from historical Mamluk architecture (murals 1–2 and 4–5 in Fig. 3). They are drawn in black and white and were meant to be part of an augmented reality (AR) experience through a mobile application, where the application animated the mural through addition of colors, movement, and sound. The murals are 2.5–3 m high, painted on the ground floor of the buildings. Two of the murals were only 1.5 m wide, while the other three between 3.6 and 4 m wide.

The artist, Agnes Michalczyk, is a Polish expatriate who has been living in Cairo for more than 9 years. Working on multiple murals and artistic projects in Al-Khalifa neighborhood and specifically the area around Al-Ashraf Street, she has had several interactions with the locals. According to the artist, she had the artistic freedom to choose what can be painted on the walls and was primarily attempting to use the same themes she used in previous murals in that area, which include visuals from Bait al-Kretliya myths and scenes which she saw in the street itself. The choice of mural in that specific area, according to the artist, is rooted in her own experience in Cairo. She stated that it is essential for her that each mural is specific to the context and inspired by it and is for the people who will experience it in their everyday life.

Five murals are all located within the area surrounding Ahmed Kuhya Mosque (Fig. 3). The area is to the west of Al-Ashraf Street, the main and longest street in Al-Khalifa Neighborhood, and although most of these murals are located within secondary streets and alleys, mural 1 is visible from Al-Ashraf. The other four murals are all located in Al-Arba'een, a secondary street. The street form contains projections and recessions, increasing the blank walls. Mural 1 is seen at a distance, not very clear from the start of the street. As one moves closer into the alleyway, the mural becomes clearer, and its details are more pronounced. Although small and drawn in black and white, it is contrasting with the surroundings. The street, in which the mural is located, is narrow, 3–4 m, and the building heights vary between two and six floors. This allows for human-level details to be more pronounced and exaggerated for pedestrians. The mural is in front of the mosque's open space, which allows it to be more visible for pedestrians. Mural 2 is located at the intersection of two streets, Nadi alley and Al-Arba'een alley, and is drawn on a wall of a building projected from the street alignment. This mural is more visible from both alleys. In addition, the cornered projection upon which the mural is placed creates a pocket of space in front of the mural, which is directly adjacent to two workshops that utilize the surrounding open space in their production activities and display of products. Mural 3 is located further toward Al-Saleeba Street yet is also at a building's corner which adds to its visibility. Mural 4 is located on the side wall of a building in a small space that leads to the main entrance of Ahmed Kuhya Mosque, and one of the largest murals drawn in that area, on open space in front of it, with width around 7 m, the mural is easily visible from a distance. This open space is located at the entrance of the mosque and used by local food vendors in the early morning, and people gather around it to have breakfast. Moreover, mural 2 can be seen when standing in the open space in front of mural 4. Mural 5 is located at the intersection of Al-Arba'een and Darb Sengar alleys. Located on the corner of the building, it is visible from both alleys. It is also close to the secondary entrances of the mosque. The five murals are placed on the secondary or rear facades of the buildings, in a projection or an irregular part of the building itself. Such wall locations have vague qualities in being a part of the building but not fully identified with it. The murals are all located in secondary streets and are only visible to the residents or people going to the workshops located in those streets. Also, the area is pedestrian-dominated, increasing the visibility of murals for locals who usually walk to their destinations or to the nearest main street to use public transportation. Additionally, the murals' scale is small (between 2.5 and 3 m high), which adds to their visibility and the level of details seen by pedestrians. Moreover, all these murals are directly located next to a workshop or a local shop, and the spaces around them are appropriated by these shops or by other movable carts, as in mural 4's case.

Analyzing the themes presented in those murals can develop further understanding of an urban area. All the murals within this area were drawn by the same artist and as part of the same project and share common themes. According to the artist, the murals' main theme was place-based, which means they evoke a connection to the location and its history. This includes subcategories like local landmarks and scenes, local environments, local history, and local people. These categories are highlighted on four of the five murals (Fig. 4). In two out of the five murals, the artist represents local landmarks, including the famous twisted minaret of Ibn Tulun Mosque, and Mamluk Archways (see murals 2 and 4 in Fig. 4). Moreover, local environments are represented in the use of tuktuks (tricycles), pigeon towers that are common on the roofs of informal areas, and scenes depicting local markets and coffee shops. All of these are elements synonymous with the popular areas, which

Fig. 3 Detailed map showing a diagrammatic drawing of Ahmed Kuhya area and the murals' locations. *Source* Authors

Fig. 4 Highlighted thematic elements on the murals' photographs. *Source* Authors

Al-Khalifa is one of. Local oral history is also represented in most murals, as elements from myths of Al-Khalifa area are shown in the murals, such as the flying angels representing martyrs. Finally, the local people, especially the women, are featured in all murals. Local women are depicted in their everyday life: wearing head coverings, walking home carrying their groceries, selling their merchandise in the local market, sitting on a coffee shop, riding local transportation, or taking care of their children.

5 Darb Al-Labana Murals

In the area of Al-Labnah alley in front of Al-Rifa'i Mosque that is considered one of the most famous alleys in Islamic Cairo, as it has a special character and a unique location. It is located on a high slope, around 9 m high, from the main street and can be accessed through steps, making the area pedestrian-dominated. The street fabric of the area is traditional, with streets not exceeding 6 m wide. The area has many monuments, including Mosque of Qanibay Al-Rammah and Tekye of Taqi Al-Din, and many historical, residential, and cultural buildings. The three murals in the area were all painted by the same artist, in April of 2022, and he used a combination between Arabic calligraphy and graffiti writing. El Seed, a French-Tunisian artist, was visiting Cairo to look for a new space for his project, after he painted a large mural in Manshiyat Nasser in 2017, and he chose to apply the word "love" in calligraffiti style on some random walls. This endeavor was solely an artistic experiment by the artist, in collaboration with a cultural center located in the area and other local mural artists. The artist, however, took the approval of the building owners before starting to paint and engaged with the local community who were intrigued by the process as a whole.

The three murals are located in Darb Al-Labana Street (Fig. 5) in front of Al-Rifa'i mosque and can be accessed from the main street through flights of stairs. Mural 1 can be seen from the level of the main street (zero level), despite its small size and the difference in levels. It is painted on the corner of one of the buildings located on the main street and therefore can be seen when climbing the steps leading to Darb Al-Labana Street. The mural is around 2.5 m high and 1 m wide and is located between two openings. The outdoor area located in front of that painted corner is surrounded by a handrail and only accessible to pedestrians. Mural 2 is located at the crossing of two alleyways, next to the entrance of Tekye of Taqi Al-Din, a fifteenth century Mamluk monument that was renovated and reopened for public in 2018. The size of the mural is fairly small, with 1.5 m height and 1 m width, which does not allow it to stand out. However, the colors utilized in it, blue and black, contrast with the surrounding neutral built environment, allowing it to be seen. Mural 3, the smallest of the three, was an experiment of the artist to draw with an audience, and it measured approximately 60 cm high and wide. The choice of location was mainly based on the scene the artist wanted to create using his mural, located in front of an open space surrounded by buildings from three sides, one of which he chose to paint on. This open space was accessed through the street, and people were passing by constantly as the artist painted. According to him, the setting already dictated which colors to use and what type of lines will be drawn, using red as a contrasting color to the blue building in the background. This last mural was drawn without a permission from the building owners, as the building itself was depilated and scheduled for removal and the owners deserted it long ago.

The artist uses the same calligraffiti theme in all his murals, in any place around the world. The three murals

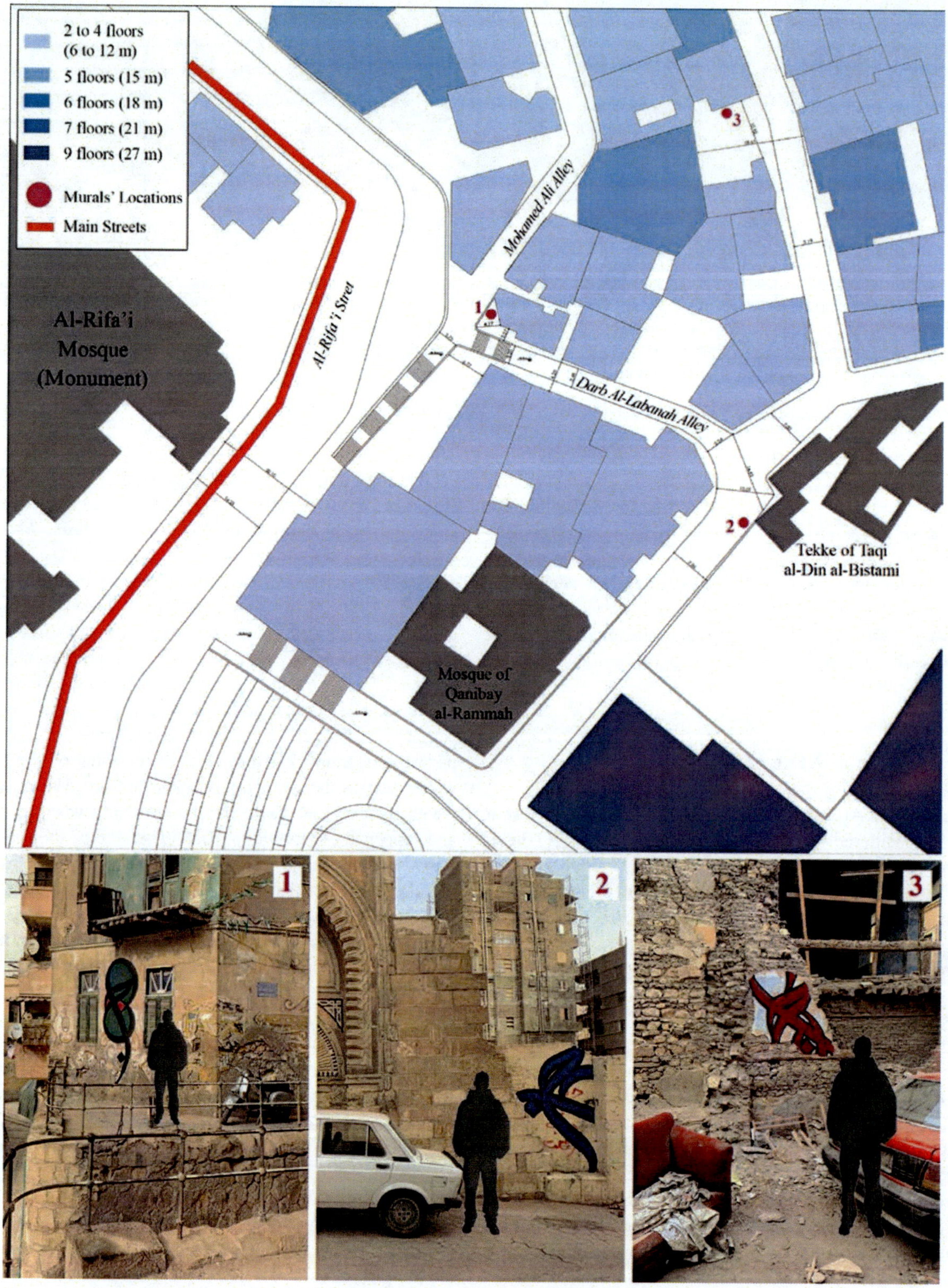

Fig. 5 Detailed map showing a diagrammatic drawing of Darb Al-Labana area and the murals' locations. *Source* Authors

resembled the word "love" in the Arabic language, which the artist painted in other countries before painting them in Egypt. The colors used in each mural varied, and according to the artist, the choice of colors and lines was based on the context in which the mural will be painted. Due to the murals being created in a particular calligraphic style, not all the residents could read them, which led to a range of responses. Some people were open-minded and willing to participate, while others were completely against it. One of the opposing residents said: "I do not understand these drawings, you come here as strangers and paint on our walls, and we do not understand what you are doing". The resident also stated that he finds these types of murals ambiguous and that he rejects the idea of painting on walls of his neighborhood.

6 Discussion

Public art could have a large effect on cities, and by engaging with the urban spaces, public art could optimize the image of the city as well as improve its living environment for the residents (Cheng & Worrall, 2021). As types of spatial art, murals need a level of visibility in order to reach the maximum number of audiences (Ferrell & Weide, 2010). Moreover, the perception of different types of audiences, whether residents or visitors, within the urban fabric is a question of placement and positioning that could contribute to the aesthetic qualities of the urban fabric (Cullen, 1961). For the two examples questioned in this paper, the positioning of murals within the streetscape has been extremely influential to their visibility. This visibility varies when looking at the pedestrians within the street versus the vehicles. Ahmed Kuhya murals are approachable to pedestrians as they are drawn at eye level and can easily get closer and look at the details. Although these murals are not visible from the main street and people had to be venture in the alleys to find them, this allowed for exploration of the secondary streets. This was not the case in actuality, as the murals were not easy to find from the main streets, and therefore did not encourage people to venture inwards. On the other hand, Darb Al-Labana murals were not all as physically accessible and did not contain details that require the pedestrian to get closer and explore. However, the main mural was visually accessible from the main street, which might lead pedestrians into the area. The size of the mural splayed a significant role in determining their level of visibility and in turn their impact on the street. In both areas, the murals are relatively small, no more than 3 m in height. This could be due to the available wall space to paint on, due to the building nature of the areas themselves. Both examples were mainly pedestrian oriented, located in the traditional street design in the area of old Cairo, and would be tremendously different when looking at murals drawn in areas that had contemporary street design.

On the other hand, Chuchuk (2021) argued that visual image of modern cities is "a direct projection of sociocultural processes" of its people, and they use them as a way of deciphering social ideological codes (p. 1). Januchta-Szostak (2010) also argued that mural arts, which rose from a need for public to freely express themselves, could be a material for building an image of a place. The creation of murals with the symbolic meaning within a certain environment can increase the people's sense of belonging to that place (Lang, 1987). The use of local and familiar elements situates art within its context, therefore creating a distinctive sense of place within the area (Salim, 2019). They can be considered a way to appropriate, represent, and restitute the urban space, enabling creative expressions among citizens (Pogrmic & Dercan, 2021). Usually painted in the forgotten spaces of the city, murals especially are considered a form of placemaking that "appropriates the underused wall surfaces" and when integrated with place identity helps redefine the place's nature (Dovey, 2016, p. 204) and carries ideological messages that grasp the pedestrian's attention within the urban space (Remesar, 2005). Both studied examples vary in the thematic aspect, although both artists are foreigners. The Ahmed Kuhya murals were more grounded in the context as the artist used scenes and visuals that were a part of the everyday life of the locals, this could be due to the artist living in Cairo for a long time and being inspired by it. According to Michalczyk, the artist, the impact of street art on the space and the people living in it is massive, the intention behind her work was to upgrade and attract more visitors to the area. She said during an interview: "Outsiders usually do not venture into the area; therefore, we created these murals to encourage people to explore it without so much as turning it into a sterilized museum space". Michalczyk has done several murals in that area over the years, and according to her, the feedback from the locals was generally positive, and the symbols in the murals were familiar to people and therefore resonated with them. Although the murals represent the artist's perception of the locals' environment, the locals were glad to be represented. From conversations with some youth living in the area, they were very excited to show the visitors the murals and how they can be seen through the mobile application. Furthermore, the symbols and cultural codes present in the murals signify a layer of representation and place-specificity, and the neighborhood is both a subject in and an object of public art. The second project, on the other hand, had varying reactions from the residents. Before drawing the first mural, the artist requested permission from the house owner, who allowed him to paint as her late husband painted hajj murals on the neighborhood walls before his passing, so she resonated with the idea of wall paintings. The artist did not request permission for either the second or third mural, as he drew them on public walls or on a depilated building, yet they evoked reactions from several residents, both positive and negative. The artist uses the urban space walls as a background for his specific style of painting without considering the contextual aspect. The art was not grounded in the area that has a rich and long history, and although he used Arabic in the mural, most residents could not read the word without the artist telling them what it meant.

Moreover, it is well known that mural arts are inherently ephemeral, due to being a part of the urban space, and being constantly subjected to wear and tear, and therefore, the temporal aspect must be investigated. The mural within the public space could be considered a snapshot, presenting a fraction of this space's history and how it was affected during one point in time. The artists themselves realize the temporality of their artwork: For example, the artist in Ahmed Kuhya acknowledged the temporality of the murals and stated that the beauty of working in an urban space is that it is constantly changing and evolving. Moreover, she detailed that art within public space brings forward power relations within the city, and by tracking what stays and what is removed you decipher those relations. On the other hand, the artist in Darb Al-Labana knows that the murals are not permanent and will wear down or be vandalized in a short time and thus makes sure to document them in photographs and videos to preserve them. This ephemeral quality of public art is not necessarily a negative aspect, as it keeps the urban space dynamic and constantly changing, constantly surprising to the passersby.

7 Conclusion

This paper focuses on two sets of mural projects conducted in Al-Khalifa neighborhood in Historic Cairo and aims to add an understanding to the mural arts in the city in general, if and how they affect the streetscape, how they are perceived by the community, and whether they are ephemeral or leave traces marking the urban space. Focusing on Egyptian cities, the paper looks into the murals conducted after the events of 2011, specifically in the area of Historic Cairo, where multiple murals were painted between the years of 2017 and 2022. The two sets of murals that are investigated in the paper were painted in two different areas by two foreigner artists, with varying intentions. The interactions the murals had with the streetscape and the community were varying between both areas, as well as the effect time had on them. The dimensions and positioning of murals in both areas greatly affected their visibility, making them either less or more accessible to a greater audience. Moreover, the messages behind the murals, the artists' intentions, and the symbolic codes present in them were important aspects guiding the community's acceptance or unacceptance of them. Lastly, the ephemeral qualities of mural art in general are inevitable and part of the ever-changing nature of the urban space, making an accurate analysis of their dynamics in time and space extremely difficult.

References

Abdelaziz, A., Abou ElEla, M., & Wanas, A. (2018). "Mural Arts" an Opportunity for Enhancing Cognitive Mapping with Respect to Cairo. *International Journal of Scientific & Engineering Research, 9*(2), 1378–1385.

Awad, S. H., Wagoner, B., & Glaveanu, V. (2017). The street art of resistance. In *Resistance in everyday life* (pp. 161–180). Springer.

Badawi, S. (2021). Public art from symbolism to randomness: The missed role of art in urban open spaces. In *Cities' vocabularies: The influences and formations* (pp. 259–267). Springer.

Bseiso, R. A. (2017). *Revolutionary art or 'revolutionizing art'?* (p. 23). Arab Media and Society.

Carmona, M., Tiesdell, S., Heath, T., & Oc, T. (2010). Public places–urban spaces: The dimensions of urban design.

Cheng, H., & Worrall, J. (2021). The influence of public art in developing chinese urban public space: Current trends and future directions. In *Cities' identity through architecture and arts* (pp. 157–165).

Chuchuk, S. (2021). Stylistic features and ideological codes of modern urban murals: The case of Ivano-Frankivsk city. *Paradigm of Knowledge, 6*(44).

Cullen, G. (1961). *The concise townscape*. Routledge.

Dovey, K. (2016). *Urban design thinking: A conceptual toolkit*. Bloomsbury.

Elsaeed, E., & El Kabbani, S. (2006). The Documentation of the Pilgrimage Arts in Upper Egypt–A comparative Study between Ancient and Islamic Egypt. EUT Edizioni Università di Trieste.

Ferrell, J., & Weide, R. D. (2010). Spot theory. *City, 14*(1–2), 48–62.

Januchta-Szostak, A. (2010). The role of public visual art in urban space recognition. In *Cognitive maps*. IntechOpen.

Kruk, R., Oort, F., Mols, L., & Buitelaar, M. (2015). Hajj murals in Dakhla Oasis (Egypt).

Kwon, M. (2001). *Public art and urban identities*. European Institute for Progressive Cultural Policies.

Landry, C. (2020). Arts, culture, and the city: An overview. *Built Environment, 46*(2), 10–21.

Lang, J. (1987). Creating architectural theory. The role of the behavioral sciences in environmental design.

Lovell, V. (2020). Artists and the public spaces of the city. *Built Environment, 46*(2), 54–68.

McAuliffe, C. (2012). Graffiti or street art? Negotiating the moral geographies of the creative city. *Journal of Urban Affairs, 34*(2), 189–206.

Phillips, P. (1988). *Out of order: The public art machine, Artforum, December*. Reprinted in: Miles, M., Borden, I. & Hall (pp. 96–102).

Pogrmic, Z., & Dercan, B. (2021). The role of Street art in urban space recognition. *Zbornik Radova Departmana Za Geografiju, Turizam i Hotelijerstvo, 50–2*, 122–131.

Rapoport, A. (1977). *Human aspects of urban form: Towards a man—Environment approach to urban form and design*. Elsevier.

Redaelli, E., & Stevenson, D. (2022). Arts in the city: Debates in the *Journal of Urban Affairs*. *Journal of Urban Affairs, 44*(4–5), 456–470.

Remesar, A. (2005). Public art. An ethical approach. In *Urban regeneration: A challenge for public art* (pp. 7–12). Publicacions de la Universitat de Barcelona.

Salim, Z. (2019). The contours of creativity: Public art, cultural landscapes, and urban space in Venice, California. *California Geographer, 58*.

Türközü, C. D., & Caliskan, O. (2021). Mural as public art in urban fabric: An attempt to link configurational approach to perceptual morphology. *Journal of Design for Resilience in Architecture and Planning, 2*, 147–170.

UNESCO World Heritage Centre—Management of World Heritage Sites in Egypt. (2014, September). Urban regeneration project for Historic Cairo. In https://www.urhcproject.org/Studies. Retrieved August 20, 2022, from https://www.urhcproject.org